Introduction to Aircraft Structural Analysis

Second Edition

Introduction to Aircraft Structural Analysis

Second Edition

T. H. G. Megson

AMSTERDAM • BOSTON • HEIDELBERG • LONDON
NEW YORK • OXFORD • PARIS • SAN DIEGO
SAN FRANCISCO • SINGAPORE • SYDNEY • TOKYO

Butterworth-Heinemann is an imprint of Elsevier

ELSEVIER

Butterworth-Heinemann is an imprint of Elsevier
The Boulevard, Langford Lane, Kidlington, Oxford, OX5 1 GB, UK
225 Wyman Street, Waltham, MA 02451, USA

Second edition 2014

Library of Congress Cataloging-in-Publication Data
Application submitted

British Library Cataloguing-in-Publication Data
A catalogue record for this book is available from the British Library.

ISBN: 978-0-08-098201-4

For information on all Butterworth–Heinemann publications, visit our website: www.elsevierdirect.com

Printed in the United States of America

Contents

Preface

During my experience of teaching aircraft structures, I have felt the need for a textbook written specifically for students of aeronautical engineering. Although there have been a number of excellent books written on the subject, they are now either out of date or too specialized in content to fulfill the requirements of an undergraduate textbook. With that in mind, I wrote *Aircraft Structures for Engineering Students*, the text on which this one is based. Users of that text have supplied many useful comments to the publisher, including comments that a briefer version of the book might be desirable, particularly for programs that do not have the time to cover all the material in the "big" book. That feedback, along with a survey done by the publisher, resulted in this book, *An Introduction to Aircraft Structural Analysis 2nd Edition*, designed to meet the needs of more time-constrained courses.

Much of the content of this book is similar to that of *Aircraft Structures for Engineering Students*, but the chapter on "Vibration of Structures" has been removed since this is most often covered in a separate standalone course. The topic of Aeroelasticity has also been removed, leaving detailed treatment to the graduate-level curriculum. The section on "Structural Loading and Discontinuities" remains in the big book but not this "intro" one. While these topics help develop a deeper understanding of load transfer and constraint effects in aircraft structures, they are often outside the scope of an undergraduate text. The reader interested in learning more on those topics should refer to the "big" book. In the interest of saving space, the appendix on "Design of a Rear Fuselage" is available for download from the book's companion Web site. Please visit *http://booksite.elsevier.com/9780080982014/* to view the downloadable content.

Supplementary materials, including solutions to end-of-chapter problems, are available for registered instructors who adopt this book as a course text. Please visit *www.textbooks.elsevier.com* for information and to register for access to these resources.

T.H.G. Megson

SUPPORTING MATERIAL ACCOMPANYING THIS BOOK

A full set of worked solutions for this book are available for teaching purposes.
Please visit *http://booksite.elsevier.com/9780080982014* and follow the registration instructions to access this material, which is intended for use by lecturers and tutors.

Fundamentals of structural analysis

Elasticity

Basic elasticity

We consider, in this chapter, the basic ideas and relationships of the theory of elasticity. The treatment is divided into three broad sections: stress, strain, and stress–strain relationships. The third section is deferred until the end of the chapter to emphasize the fact that the analysis of stress and strain, for example, the equations of equilibrium and compatibility, does not assume a particular stress–strain law. In other words, the relationships derived in Sections 1.1–1.14 are applicable to nonlinear as well as linearly elastic bodies.

1.1 STRESS

Consider the arbitrarily shaped, three-dimensional body shown in Fig. 1.1. The body is in equilibrium under the action of externally applied forces $P_1, P_2, \ldots$ and is assumed to constitute a continuous and deformable material, so that the forces are transmitted throughout its volume. It follows that, at any internal point O, there is a resultant force δP. The particle of material at O subjected to the force δP is in equilibrium, so that there must be an equal but opposite force δP (shown dotted in Fig. 1.1) acting on the particle at the same time. If we now divide the body by any plane nn containing O, then these two forces δP may be considered as being uniformly distributed over a small area δA of each face of the plane at the corresponding point O, as in Fig. 1.2. The *stress* at O is defined by the equation

$$\text{Stress} = \lim_{\delta A \to 0} \frac{\delta P}{\delta A} \tag{1.1}$$

The directions of the forces δP in Fig. 1.2 are such as to produce *tensile* stresses on the faces of the plane nn. It must be realized here that, while the direction of δP is absolute, the choice of plane is arbitrary, so that, although the direction of the stress at O is always in the direction of δP, its magnitude depends upon the actual plane chosen, since a different plane has a different inclination and therefore a different value for the area δA. This may be more easily understood by reference to the bar in simple tension in Fig. 1.3. On the cross-sectional plane mm, the uniform stress is given by P/A, *while on the inclined plane $m'm'$* the stress is of magnitude P/A'. In both cases, the stresses are parallel to the direction of P.

Generally, the direction of δP is not normal to the area δA, in which case, it is usual to resolve δP into two components: one, δP_n, normal to the plane and the other, δP_s, acting in the plane itself (see Fig. 1.2). Note that, in Fig. 1.2, the plane containing δP is perpendicular to δA. The stresses associated with these components are a *normal* or *direct stress* defined as

$$\sigma = \lim_{\delta A \to 0} \frac{\delta P_n}{\delta A} \tag{1.2}$$

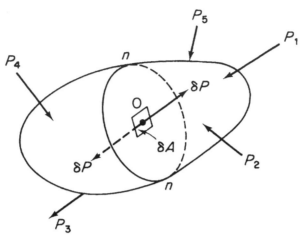

FIGURE 1.1 Internal Force at a Point in an Arbitrarily Shaped Body

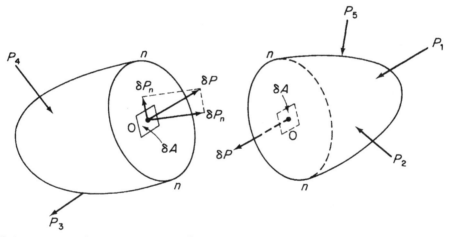

FIGURE 1.2 Internal Force Components at the Point O

and a *shear stress* defined as

$$\tau = \lim_{\delta A \to 0} \frac{\delta P_s}{\delta A} \tag{1.3}$$

The resultant stress is computed from its components by the normal rules of vector addition, i.e.:

$$\text{Resultant stress} = \sqrt{\sigma^2 + \tau^2}$$

Generally, however, as indicated previously, we are interested in the separate effects of σ and τ.

However, to be strictly accurate, stress is not a vector quantity for, in addition to magnitude and direction, we must specify the plane on which the stress acts. Stress is therefore a *tensor*, its complete description depending on the two vectors of force and surface of action.

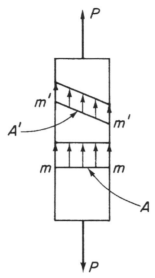

FIGURE 1.3 Values of Stress on Different Planes in a Uniform Bar

1.2 NOTATION FOR FORCES AND STRESSES

It is usually convenient to refer the state of stress at a point in a body to an orthogonal set of axes $Oxyz$. In this case we cut the body by planes parallel to the direction of the axes. The resultant force δP acting at the point O on one of these planes may then be resolved into a normal component and two in-plane components, as shown in Fig. 1.4, thereby producing one component of direct stress and two components of shear stress.

The direct stress component is specified by reference to the plane on which it acts, but the stress components require a specification of direction in addition to the plane. We therefore allocate a single subscript to direct stress to denote the plane on which it acts and two subscripts to shear stress, the first specifying the plane, the second direction. Therefore, in Fig. 1.4, the shear stress components are τ_{zx} and τ_{zy} acting on the z plane and in the x and y directions, respectively, while the direct stress component is σ_z.

We may now completely describe the state of stress at a point O in a body by specifying components of shear and direct stress on the faces of an element of side δx, δy, δz, formed at O by the cutting planes as indicated in Fig. 1.5.

The sides of the element are infinitesimally small, so that the stresses may be assumed to be uniformly distributed over the surface of each face. On each of the opposite faces there will be, to a first simplification, equal but opposite stresses.

We now define the directions of the stresses in Fig. 1.5 as positive, so that normal stresses directed away from their related surfaces are tensile and positive; opposite compressive stresses are negative. Shear stresses are positive when they act in the positive direction of the relevant axis in a plane on which the direct tensile stress is in the positive direction of the axis. If the tensile stress is in the

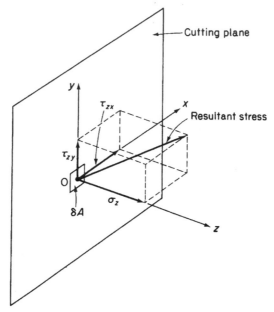

FIGURE 1.4 Components of Stress at a Point in a Body

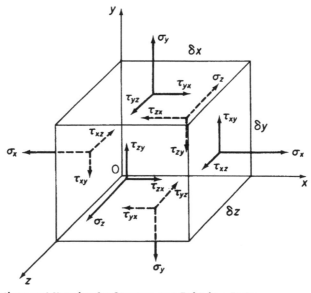

FIGURE 1.5 Sign Conventions and Notation for Stresses at a Point in a Body

opposite direction, then positive shear stresses are in directions opposite to the positive directions of the appropriate axes.

Two types of external force may act on a body to produce the internal stress system we have already discussed. Of these, *surface forces* such as $P_1, P_2, \ldots$, or hydrostatic pressure, are distributed over the surface area of the body. The surface force per unit area may be resolved into components parallel to our orthogonal system of axes, and these are generally given the symbols $\overline{X}, \overline{Y}$, and $\overline{Z}$. The second force system derives from gravitational and inertia effects, and the forces are known as *body forces*. These are distributed over the volume of the body and the components of body force per unit volume are designated X, Y, and Z.

1.3 EQUATIONS OF EQUILIBRIUM

Generally, except in cases of uniform stress, the direct and shear stresses on opposite faces of an element are not equal, as indicated in Fig. 1.5, but differ by small amounts. Therefore if, say, the direct stress acting on the z plane is σ_z, then the direct stress acting on the $z + \delta z$ plane is, from the first two terms of a Taylor's series expansion, $\sigma_z + (\partial \sigma_z / \partial z)\delta z$.

We now investigate the equilibrium of an element at some internal point in an elastic body where the stress system is obtained by the method just described.

In Fig. 1.6, the element is in equilibrium under forces corresponding to the stresses shown and the components of body forces (not shown). Surface forces acting on the boundary of the body, although contributing to the production of the internal stress system, do not directly feature in the equilibrium equations.

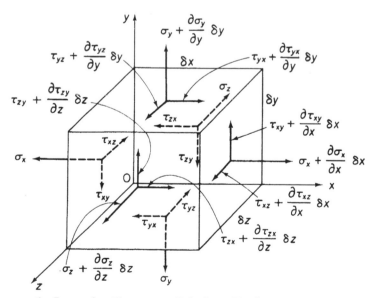

FIGURE 1.6 Stresses on the Faces of an Element at a Point in an Elastic Body

Taking moments about an axis through the center of the element parallel to the z axis,

$$\tau_{xy}\delta y\delta z\frac{\delta x}{2} + \left(\tau_{xy} + \frac{\partial\tau_{xy}}{\partial x}\delta x\right)\delta y\delta z\frac{\delta x}{2} - \tau_{yx}\delta x\delta z\frac{\delta y}{2}$$

$$- \left(\tau_{yx} + \frac{\partial\tau_{yx}}{\partial y}\delta y\right)\delta x\delta z\frac{\delta y}{2} = 0$$

which simplifies to

$$\tau_{xy}\delta y\delta z\delta x + \frac{\partial\tau_{xy}}{\partial x}\delta y\delta z\frac{(\delta x)^2}{2} - \tau_{yx}\delta x\delta z\delta y - \frac{\partial\tau_{yx}}{\partial y}\delta x\delta z\frac{(\delta y)^2}{2} = 0$$

dividing through by $\delta x\delta y\delta z$ and taking the limit as δx and δy approach zero.

Similarly,
$$\left.\begin{array}{c} \tau_{xy} = \tau_{yx} \\ \tau_{xz} = \tau_{zx} \\ \tau_{yz} = \tau_{zy} \end{array}\right\} \qquad (1.4)$$

We see, therefore, that a shear stress acting on a given plane (τ_{xy}, τ_{xz}, τ_{yz}) is always accompanied by an equal *complementary shear stress* (τ_{yx}, τ_{zx}, τ_{zy}) acting on a plane perpendicular to the given plane and in the opposite sense.

Now, considering the equilibrium of the element in the x direction,

$$\left(\sigma_x + \frac{\partial\sigma_x}{\partial x}\delta x\right)\delta y\,\delta z - \sigma_x\delta y\delta z + \left(\tau_{yx} + \frac{\partial\tau_{yx}}{\partial y}\delta y\right)\delta x\delta z$$

$$- \tau_{yx}\delta x\delta z + \left(\tau_{zx} + \frac{\partial\tau_{zx}}{\partial z}\delta z\right)\delta x\delta y$$

$$- \tau_{zx}\delta x\delta y + X\delta x\delta y\delta z = 0$$

which gives

$$\frac{\partial\sigma_x}{\partial x} + \frac{\partial\tau_{yx}}{\partial y} + \frac{\partial\tau_{zx}}{\partial z} + X = 0$$

Or, writing $\tau_{xy} = \tau_{yx}$ and $\tau_{xz} = \tau_{zx}$ from Eq. (1.4),

similarly,
$$\left.\begin{array}{c} \dfrac{\partial\sigma_x}{\partial x} + \dfrac{\partial\tau_{xy}}{\partial y} + \dfrac{\partial\tau_{xz}}{\partial z} + X = 0 \\[2mm] \dfrac{\partial\sigma_y}{\partial y} + \dfrac{\partial\tau_{yx}}{\partial x} + \dfrac{\partial\tau_{yz}}{\partial z} + Y = 0 \\[2mm] \dfrac{\partial\sigma_z}{\partial z} + \dfrac{\partial\tau_{zx}}{\partial x} + \dfrac{\partial\tau_{zy}}{\partial y} + Z = 0 \end{array}\right\} \qquad (1.5)$$

The *equations of equilibrium* must be satisfied at all interior points in a deformable body under a three-dimensional force system.

1.4 PLANE STRESS

Most aircraft structural components are fabricated from thin metal sheet, so that stresses across the thickness of the sheet are usually negligible. Assuming, say, that the z axis is in the direction of the thickness, then the three-dimensional case of Section 1.3 reduces to a two-dimensional case in which σ_z, τ_{xz}, and τ_{yz} are all zero. This condition is known as *plane stress*; the equilibrium equations then simplify to

$$\left.\begin{array}{l}\dfrac{\partial \sigma_x}{\partial x} + \dfrac{\partial \tau_{xy}}{\partial y} + X = 0 \\[2mm] \dfrac{\partial \sigma_y}{\partial y} + \dfrac{\partial \tau_{yx}}{\partial x} + Y = 0\end{array}\right\} \tag{1.6}$$

1.5 BOUNDARY CONDITIONS

The equations of equilibrium (1.5)—and also (1.6), for a two-dimensional system—satisfy the requirements of equilibrium at all internal points of the body. Equilibrium must also be satisfied at all positions on the boundary of the body, where the components of the surface force per unit area are $\overline{X}$, $\overline{Y}$, and $\overline{Z}$. The triangular element of Fig. 1.7 at the boundary of a two-dimensional body of unit thickness is then in equilibrium under the action of surface forces on the elemental length AB of the boundary and internal forces on internal faces AC and CB.

Summation of forces in the x direction gives

$$\overline{X}\delta s - \sigma_x \delta y - \tau_{yx}\delta x + X\frac{1}{2}\delta x \delta y = 0$$

which, by taking the limit as δx approaches zero, becomes

$$\overline{X} = \sigma_x \frac{dy}{ds} + \tau_{yx}\frac{dx}{ds}$$

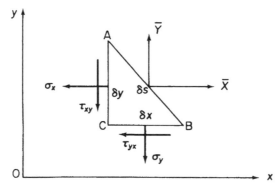

FIGURE 1.7 Stresses on the Faces of an Element at the Boundary of a Two-Dimensional Body

The derivatives dy/ds and dx/ds are the direction cosines l and m of the angles that a normal to AB makes with the x and y axes, respectively. It follows that

$$\overline{X} = \sigma_x l + \tau_{yx} m$$

and in a similar manner

$$\overline{Y} = \sigma_y m + \tau_{xy} l$$

A relatively simple extension of this analysis produces the boundary conditions for a three-dimensional body, namely,

$$\left.\begin{array}{l} \overline{X} = \sigma_x l + \tau_{yx} m + \tau_{zx} n \\ \overline{Y} = \sigma_y m + \tau_{xy} l + \tau_{zy} n \\ \overline{Z} = \sigma_z n + \tau_{yz} m + \tau_{xz} l \end{array}\right\} \tag{1.7}$$

where l, m, and n become the direction cosines of the angles that a normal curvature to the surface of the body makes with the x, y, and z axes, respectively.

1.6 DETERMINATION OF STRESSES ON INCLINED PLANES

The complex stress system of Fig. 1.6 is derived from a consideration of the actual loads applied to a body and is referred to a predetermined, though arbitrary, system of axes. The values of these stresses may not give a true picture of the severity of stress at that point, so that it is necessary to investigate the state of stress on other planes on which the direct and shear stresses may be greater.

We restrict the analysis to the two-dimensional system of plane stress defined in Section 1.4.

Figure 1.8(a) shows a complex stress system at a point in a body referred to axes Ox, Oy. All stresses are positive, as defined in Section 1.2. The shear stresses τ_{xy} and τ_{yx} were shown to be equal in Section 1.3. We now, therefore, designate them both τ_{xy}. The element of side δx, δy and of unit

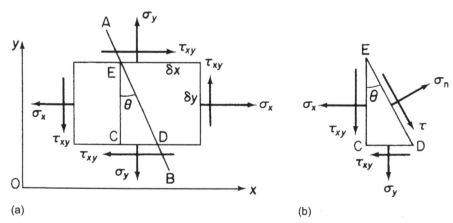

(a)   (b)

FIGURE 1.8 (a) Stresses on a Two-Dimensional Element; (b) Stresses on an Inclined Plane at the Point

thickness is small, so that stress distributions over the sides of the element may be assumed to be uniform. Body forces are ignored, since their contribution is a second-order term.

Suppose that we need to find the state of stress on a plane AB inclined at an angle θ to the vertical. The triangular element EDC formed by the plane and the vertical through E is in equilibrium under the action of the forces corresponding to the stresses shown in Fig. 1.8(b), where σ_n and τ are the direct and shear components of the resultant stress on AB. Then, resolving forces in a direction perpendicular to ED, we have

$$\sigma_n ED = \sigma_x EC \, \cos\theta + \sigma_y CD \, \sin\theta + \tau_{xy} EC \, \sin\theta + \tau_{xy} CD \, \cos\theta$$

Dividing through by ED and simplifying,

$$\sigma_n = \sigma_x \cos^2\theta + \sigma_y \sin^2\theta + \tau_{xy} \sin2\theta \tag{1.8}$$

Now, resolving forces parallel to ED,

$$\tau ED = \sigma_x EC \, \sin\theta - \sigma_y CD \, \cos\theta - \tau_{xy} EC \, \cos\theta + \tau_{xy} CD \sin\theta$$

Again, dividing through by ED and simplifying,

$$\tau = \frac{(\sigma_x - \sigma_y)}{2} \sin2\theta - \tau_{xy} \, \cos2\theta \tag{1.9}$$

Example 1.1

A cylindrical pressure vessel has an internal diameter of 2 m and is fabricated from plates 20 mm thick. If the pressure inside the vessel is 1.5 N/mm² and, in addition, the vessel is subjected to an axial tensile load of 2500 kN, calculate the direct and shear stresses on a plane inclined at an angle of 60° to the axis of the vessel. Calculate also the maximum shear stress.

The expressions for the longitudinal and circumferential stresses produced by the internal pressure may be found in any text on stress analysis[1] and are

$$\text{Longitudinal stress } (\sigma_x) = \frac{pd}{4t} = 1.5 \times 2 \times 10^3/4 \times 20 = 37.5 \text{ N/mm}^2$$

$$\text{Circumferential stress } (\sigma_y) = \frac{pd}{2t} = 1.5 \times 2 \times 10^3/2 \times 20 = 75 \text{ N/mm}^2$$

The direct stress due to the axial load will contribute to σ_x and is given by

$$\sigma_x \text{ (axial load)} = 2500 \times 10^3/\pi \times 2 \times 10^3 \times 20 = 19.9 \text{ N/mm}^2$$

A rectangular element in the wall of the pressure vessel is then subjected to the stress system shown in Fig. 1.9. Note that no shear stresses act on the x and y planes; in this case, σ_x and σ_y form a *biaxial* stress system.

The direct stress, σ_n, and shear stress, τ, on the plane AB, which makes an angle of 60° with the axis of the vessel, may be found from first principles by considering the equilibrium of the triangular element ABC or by direct substitution in Eqs. (1.8) and (1.9). Note that, in the latter case, $\theta = 30°$ and $\tau_{xy} = 0$. Then,

$$\sigma_n = 57.4 \cos^2 30° + 75 \sin^2 30° = 61.8 \text{ N/mm}^2$$
$$\tau = (57.4 - 75)[\sin(2 \times 30°)]/2 = -7.6 \text{ N/mm}^2$$

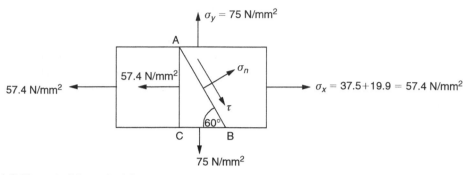

FIGURE 1.9 Element of Example 1.1

The negative sign for τ indicates that the shear stress is in the direction BA and not AB.

From Eq. (1.9), when $\tau_{xy} = 0$,

$$\tau = (\sigma_x - \sigma_y)(\sin 2\theta)/2 \qquad \text{(i)}$$

The maximum value of τ therefore occurs when $\sin 2\theta$ is a maximum, that is, when $\sin 2\theta = 1$ and $\theta = 45°$. Then, substituting the values of σ_x and σ_y in Eq. (i),

$$\tau_{max} = (57.4 - 75)/2 = -8.8 \, \text{N/mm}^2$$

Example 1.2

A cantilever beam of solid, circular cross-section supports a compressive load of 50 kN applied to its free end at a point 1.5 mm below a horizontal diameter in the vertical plane of symmetry together with a torque of 1200 Nm (Fig. 1.10). Calculate the direct and shear stresses on a plane inclined at 60° to the axis of the cantilever at a point on the lower edge of the vertical plane of symmetry. See Ex. 1.1.

The direct loading system is equivalent to an axial load of 50 kN together with a bending moment of $50 \times 10^3 \times 1.5 = 75,000$ Nmm in a vertical plane. Therefore, at any point on the lower edge of the vertical plane of symmetry,

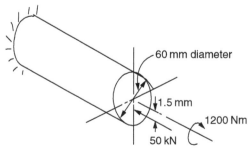

FIGURE 1.10 Cantilever Beam of Example 1.2.

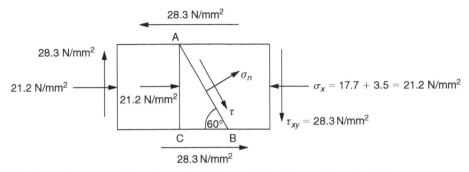

FIGURE 1.11 Stress System on a Two-Dimensional Element of the Beam of Example 1.2

there are compressive stresses due to the axial load and bending moment that act on planes perpendicular to the axis of the beam and are given, respectively, by Eqs. (1.2) and (16.9); that is,

$$\sigma_x \text{ (axial load)} = 50 \times 10^3 / \pi \times (60^2 / 4) = 17.7 \, \text{N/mm}^2$$
$$\sigma_x \text{ (bending moment)} = 75,000 \times 30 / \pi \times (60^4 / 64) = 3.5 \, \text{N/mm}^2$$

The shear stress, τ_{xy}, at the same point due to the torque is obtained from Eq. (iv) in Example 3.1; that is,

$$\tau_{xy} = 1200 \times 10^3 \times 30 / \pi \times (60^4 / 32) = 28.3 \, \text{N/mm}^2$$

The stress system acting on a two-dimensional rectangular element at the point is shown in Fig. 1.11. Note that, since the element is positioned at the bottom of the beam, the shear stress due to the torque is in the direction shown and is negative (see Fig. 1.8).

Again, σ_n and τ may be found from first principles or by direct substitution in Eqs. (1.8) and (1.9). Note that $\theta = 30°$, $\sigma_y = 0$, and $\tau_{xy} = -28.3 \, \text{N/mm}^2$, the negative sign arising from the fact that it is in the opposite direction to τ_{xy} in Fig. 1.8.

Then,

$$\sigma_n = -21.2 \cos^2 30° - 28.3 \sin 60° = -40.4 \, \text{N/mm}^2 \text{(compression)}$$
$$\tau = (-21.2/2) \sin 60° + 28.3 \cos 60° = 5.0 \, \text{N/mm}^2 \text{ (acting in the direction AB)}$$

Different answers are obtained if the plane AB is chosen on the opposite side of AC.

1.7 PRINCIPAL STRESSES

For given values of σ_x, σ_y, and τ_{xy}, in other words, given loading conditions, σ_n varies with the angle θ and attains a maximum or minimum value when $d\sigma_n/d\theta = 0$. From Eq. (1.8),

$$\frac{d\sigma_n}{d\theta} = -2\sigma_x \cos\theta \, \sin\theta + 2\sigma_y \, \sin\theta \, \cos\theta + 2\tau_{xy} \, \cos 2\theta = 0$$

Hence,

$$-(\sigma_x - \sigma_y) \sin 2\theta + 2\tau_{xy} \cos 2\theta = 0$$

or

$$\tan 2\theta = \frac{2\tau_{xy}}{\sigma_x - \sigma_y} \tag{1.10}$$

Two solutions, θ and $\theta + \pi/2$, are obtained from Eq. (1.10), so that there are two mutually perpendicular planes on which the direct stress is either a maximum or a minimum. Further, by comparison with Eqs. (1.9) and (1.10), it will be observed that these planes correspond to those on which there is no shear stress. The direct stresses on these planes are called *principal stresses* and the planes themselves, *principal planes*.

From Eq. (1.10),

$$\sin 2\theta = \frac{2\tau_{xy}}{\sqrt{(\sigma_x - \sigma_y)^2 + 4\tau_{xy}^2}} \qquad \cos 2\theta = \frac{\sigma_x - \sigma_y}{\sqrt{(\sigma_x - \sigma_y)^2 + 4\tau_{xy}^2}}$$

and

$$\sin 2(\theta + \pi/2) = \frac{-2\tau_{xy}}{\sqrt{(\sigma_x - \sigma_y)^2 + 4\tau_{xy}^2}} \qquad \cos 2(\theta + \pi/2) = \frac{-(\sigma_x - \sigma_y)}{\sqrt{(\sigma_x - \sigma_y)^2 + 4\tau_{xy}^2}}$$

Rewriting Eq. (1.8) as

$$\sigma_n = \frac{\sigma_x}{2}(1 + \cos 2\theta) + \frac{\sigma_y}{2}(1 - \cos 2\theta) + \tau_{xy} \sin 2\theta$$

and substituting for $\{\sin 2\theta, \cos 2\theta\}$ and $\{\sin 2(\theta + \pi/2), \cos 2(\theta + \pi/2)\}$ in turn gives

$$\sigma_{\mathrm{I}} = \frac{\sigma_x + \sigma_y}{2} + \frac{1}{2}\sqrt{(\sigma_x - \sigma_y)^2 + 4\tau_{xy}^2} \tag{1.11}$$

and

$$\sigma_{\mathrm{II}} = \frac{\sigma_x + \sigma_y}{2} - \frac{1}{2}\sqrt{(\sigma_x - \sigma_y)^2 + 4\tau_{xy}^2} \tag{1.12}$$

where σ_{I} is the *maximum* or *major principal stress* and σ_{II} is the *minimum* or *minor principal stress*. Note that σ_{I} is algebraically the greatest direct stress at the point while σ_{II} is algebraically the least. Therefore, when σ_{II} is negative, that is, compressive, it is possible for σ_{II} to be numerically greater than σ_{I}.

The maximum shear stress at this point in the body may be determined in an identical manner. From Eq. (1.9),

$$\frac{d\tau}{d\theta} = (\sigma_x - \sigma_y)\cos 2\theta + 2\tau_{xy} \sin 2\theta = 0$$

giving

$$\tan 2\theta = -\frac{(\sigma_x - \sigma_y)}{2\tau_{xy}} \tag{1.13}$$

It follows that

$$\sin2\theta = \frac{-(\sigma_x - \sigma_y)}{\sqrt{(\sigma_x - \sigma_y)^2 + 4\tau_{xy}^2}} \qquad \cos2\theta = \frac{2\tau_{xy}}{\sqrt{(\sigma_x - \sigma_y)^2 + 4\tau_{xy}^2}}$$

$$\sin2(\theta + \pi/2) = \frac{(\sigma_x - \sigma_y)}{\sqrt{(\sigma_x - \sigma_y)^2 + 4\tau_{xy}^2}} \qquad \cos2(\theta + \pi/2) = \frac{-2\tau_{xy}}{\sqrt{(\sigma_x - \sigma_y)^2 + 4\tau_{xy}^2}}$$

Substituting these values in Eq. (1.9) gives

$$\tau_{\text{max,min}} = \pm\frac{1}{2}\sqrt{(\sigma_x - \sigma_y)^2 + 4\tau_{xy}^2} \tag{1.14}$$

Here, as in the case of principal stresses, we take the maximum value as being the greater algebraic value.

Comparing Eq. (1.14) with Eqs. (1.11) and (1.12), we see that

$$\tau_{\text{max}} = \frac{\sigma_{\text{I}} - \sigma_{\text{II}}}{2} \tag{1.15}$$

Equations (1.14) and (1.15) give the maximum shear stress at the point in the body in *the plane of the given stresses*. For a three-dimensional body supporting a two-dimensional stress system, this is not necessarily the maximum shear stress at the point.

Since Eq. (1.13) is the negative reciprocal of Eq. (1.10), the angles 2θ given by these two equations differ by 90° or, alternatively, the planes of maximum shear stress are inclined at 45° to the principal planes.

1.8 MOHR'S CIRCLE OF STRESS

The state of stress at a point in a deformable body may be determined graphically by *Mohr's circle of stress*.

In Section 1.6, the direct and shear stresses on an inclined plane were shown to be given by

$$\sigma_n = \sigma_x \cos^2\theta + \sigma_y \sin^2\theta + \tau_{xy} \sin2\theta \tag{1.8}$$

and

$$\tau = \frac{(\sigma_x - \sigma_y)}{2} \sin2\theta - \tau_{xy} \cos2\theta \tag{1.9}$$

respectively. The positive directions of these stresses and the angle θ are defined in Fig. 1.12(a). Equation (1.8) may be rewritten in the form

$$\sigma_n = \frac{\sigma_x}{2}(1 + \cos2\theta) + \frac{\sigma_y}{2}(1 - \cos2\theta) + \tau_{xy} \sin2\theta$$

or

$$\sigma_n - \frac{1}{2}(\sigma_x + \sigma_y) = \frac{1}{2}(\sigma_x - \sigma_y)\cos2\theta + \tau_{xy} \sin2\theta$$

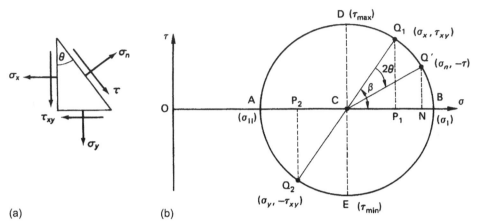

FIGURE 1.12 (a) Stresses on a Triangular Element; (b) Mohr's Circle of Stress for the Stress System Shown in (a)

Squaring and adding this equation to Eq. (1.9), we obtain

$$\left[\sigma_n - \frac{1}{2}(\sigma_x + \sigma_y)\right]^2 + \tau^2 = \left[\frac{1}{2}(\sigma_x - \sigma_y)\right]^2 + \tau_{xy}^2$$

which represents the equation of a circle of radius $\frac{1}{2}\sqrt{(\sigma_x - \sigma_y)^2 + 4\tau_{xy}^2}$ and having its center at the point $[(\sigma_x - \sigma_y)/2, 0]$.

The circle is constructed by locating the points $Q_1 (\sigma_x, \tau_{xy})$ and $Q_2 (\sigma_y, -\tau_{xy})$ referred to axes $O\sigma\tau$, as shown in Fig. 1.12(b). The center of the circle then lies at C, the intersection of Q_1Q_2 and the $O\sigma$ axis; clearly C is the point $[(\sigma_x - \sigma_y)/2, 0]$ and the radius of the circle is $\frac{1}{2}\sqrt{(\sigma_x - \sigma_y)^2 + 4\tau_{xy}^2}$, as required. CQ' is now set off at an angle 2θ (positive clockwise) to CQ_1, Q' is then the point $(\sigma_n, -\tau)$, as demonstrated next. From Fig. 1.12(b), we see that

$$ON = OC + CN$$

or, since $OC = (\sigma_x + \sigma_y)/2$, $CN = CQ' \cos(\beta - 2\theta)$, and $CQ' = CQ_1$, we have

$$\sigma_n = \frac{\sigma_x + \sigma_y}{2} + CQ_1(\cos\beta \cos2\theta + \sin\beta \sin2\theta)$$

But,

$$CQ_1 = \frac{CP_1}{\cos\beta} \quad \text{and} \quad CP_1 = \frac{(\sigma_x - \sigma_y)}{2}$$

Hence,

$$\sigma_n = \frac{\sigma_x + \sigma_y}{2} + \left(\frac{\sigma_x - \sigma_y}{2}\right)\cos2\theta + CP_1 \tan\beta \sin2\theta$$

which, on rearranging, becomes

$$\sigma_n = \sigma_x \cos^2\theta + \sigma_y \sin^2\theta + \tau_{xy} \sin2\theta$$

as in Eq. (1.8). Similarly, it may be shown that

$$Q'N = \tau_{xy} \cos 2\theta - \left(\frac{\sigma_x - \sigma_y}{2}\right) \sin 2\theta = -\tau$$

as in Eq. (1.9). Note that the construction of Fig. 1.12(b) corresponds to the stress system of Fig. 1.12 (a), so that any sign reversal must be allowed for. Also, the $O\sigma$ and $O\tau$ axes must be constructed to the same scale or the equation of the circle is not represented.

The maximum and minimum values of the direct stress, that is, the major and minor principal stresses σ_I and σ_{II}, occur when N (and Q') coincide with B and A, respectively. Thus,

$$\sigma_I = OC + \text{radius of circle}$$

$$= \frac{(\sigma_x + \sigma_y)}{2} + \sqrt{CP_1^2 + P_1 Q_1^2}$$

or

$$\sigma_I = \frac{(\sigma_x + \sigma_y)}{2} + \frac{1}{2}\sqrt{(\sigma_x - \sigma_y)^2 + 4\tau_{xy}^2}$$

and, in the same fashion,

$$\sigma_{II} = \frac{(\sigma_x + \sigma_y)}{2} - \frac{1}{2}\sqrt{(\sigma_x - \sigma_y)^2 + 4\tau_{xy}^2}$$

The principal planes are then given by $2\theta = \beta(\sigma_I)$ and $2\theta = \beta + \pi(\sigma_{II})$.

Also, the maximum and minimum values of shear stress occur when Q' coincides with D and E at the upper and lower extremities of the circle.

At these points, Q'N is equal to the radius of the circle, which is given by

$$CQ_1 = \sqrt{\frac{(\sigma_x - \sigma_y)^2}{4} + \tau_{xy}^2}$$

Hence, $\tau_{\text{max,min}} = \pm\frac{1}{2}\sqrt{(\sigma_x - \sigma_y)^2 + 4\tau_{xy}^2}$, as before. The planes of maximum and minimum shear stress are given by $2\theta = \beta + \pi/2$ and $2\theta = \beta + 3\pi/2$, these being inclined at 45° to the principal planes.

Example 1.3

Direct stresses of 160 N/mm^2 (tension) and 120 N/mm^2 (compression) are applied at a particular point in an elastic material on two mutually perpendicular planes. The principal stress in the material is limited to 200 N/mm^2 (tension). Calculate the allowable value of shear stress at the point on the given planes. Determine also the value of the other principal stress and the maximum value of shear stress at the point. Verify your answer using Mohr's circle. See Ex. 1.1.

The stress system at the point in the material may be represented as shown in Fig. 1.13 by considering the stresses to act uniformly over the sides of a triangular element ABC of unit thickness. Suppose that the direct stress on the principal plane AB is σ. For horizontal equilibrium of the element,

$$\sigma AB \cos\theta = \sigma_x BC + \tau_{xy} AC$$

which simplifies to

$$\tau_{xy} \tan\theta = \sigma - \sigma_x \qquad \text{(i)}$$

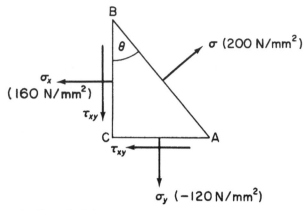

FIGURE 1.13 Stress System for Example 1.3

Considering vertical equilibrium gives

$$\sigma AB \sin\theta = \sigma_y AC + \tau_{xy} BC$$

or

$$\tau_{xy} \cot\theta = \sigma - \sigma_y \tag{ii}$$

Hence, from the product of Eqs. (i) and (ii),

$$\tau_{xy}^2 = (\sigma - \sigma_x)(\sigma - \sigma_y)$$

Now, substituting the values $\sigma_x = 160$ N/mm^2, $\sigma_y = -120$ N/mm^2, and $\sigma = \sigma_1 = 200$ N/mm^2, we have

$$\tau_{xy} = \pm 113 \text{ N/mm}^2$$

Replacing $\cot\theta$ in Eq. (ii) with $1/\tan\theta$ from Eq. (i) yields a quadratic equation in σ:

$$\sigma^2 - \sigma(\sigma_x - \sigma_y) + \sigma_x \sigma_y - \tau_{xy}^2 = 0 \tag{iii}$$

The numerical solutions of Eq. (iii) corresponding to the given values of σ_x, σ_y, and τ_{xy} are the principal stresses at the point, namely,

$$\sigma_1 = 200 \text{ N/mm}^2$$

given

$$\sigma_{II} = -160 \text{ N/mm}^2$$

Having obtained the principal stresses, we now use Eq. (1.15) to find the maximum shear stress, thus

$$\tau_{max} = \frac{200 + 160}{2} = 180 \text{ N/mm}^2$$

The solution is rapidly verified from Mohr's circle of stress (Fig. 1.14). From the arbitrary origin O, OP$_1$, and OP$_2$ are drawn to represent $\sigma_x = 160$ N/mm^2 and $\sigma_y = -120$ N/mm^2. The mid-point C of P$_1$P$_2$ is then located. Next, OB $= \sigma_1 = 200$ N/mm^2 is marked out and the radius of the circle is then CB. OA is the required principal stress. Perpendiculars P$_1$Q$_1$ and P$_2$Q$_2$ to the circumference of the circle are equal to $\pm\tau_{xy}$ (to scale), and the radius of the circle is the maximum shear stress.

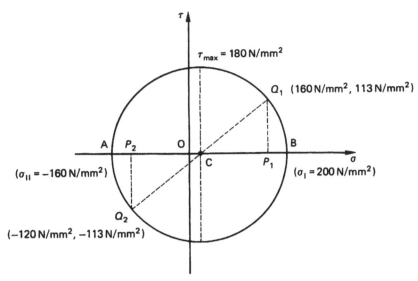

FIGURE 1.14 Solution of Example 1.3 Using Mohr's Circle of Stress

Example 1.3 MATLAB®

Repeat the derivations presented in Example 1.3 using the Symbolic Math Toolbox in MATLAB®. Do not recreate Mohr's circle. See Ex. 1.1.

Using the element shown in Fig. 1.13, derivations of the principal stresses and maximum shear stress are obtained through the following MATLAB file:

```
% Declare any needed symbolic variables
syms sig tau_xy sig_x sig_y theta AB BC AC

% Define known stress values
sig_x = sym(160);
sig_y = sym(-120);
sig_val = sym(200);

% Define relationships between AB, BC, and AC
BC = AB*cos(theta);
AC = AB*sin(theta);

% For horizonatal equalibrium of the element
eqI = sig*AB*cos(theta)-sig_x*BC-tau_xy*AC;

% For vertical equalibrium of the element
eqII = sig*AB*sin(theta)-sig_y*AC-tau_xy*BC;
```

```
% Solve eqI and eqII for tau_xy
tau_xyI = solve(eqI,tau_xy);
tau_xyII = solve(eqII,tau_xy);

% Take the square-root of tau_xyI times tau_xyII to get tau_xy
tau_xy_val = sqrt(tau_xyI*tau_xyII);

% Substitite the given value of sig into tau_xy
tau_xy_val = subs(tau_xy_val,sig,sig_val);

% Solve eqI for theta and substitute into eqII
eqI = simplify(eqI/cos(theta));
theta_I = solve(eqI,theta);
eqIII = subs(eqII,theta,theta_I);

% Substitute the value of tau_xy into eqIII and solve for the principle stresses (sig_p)
sig_p = solve(subs(eqIII,tau_xy,tau_xy_val),sig);
sig_I = max(double(sig_p));
sig_II = min(double(sig_p));

% Calculate the maximum shear stress using Eq. (1.15)
tau_max = (sig_I-sig_II)/2;

% Output tau_xy, the principle stresses, and tau_max to the Command Window
disp(['tau_xy = +/-' num2str(double(tau_xy_val)) 'N/mm^2'])
disp(['sig_I =' num2str(sig_I) 'N/mm^2'])
disp(['sig_II =' num2str(sig_II) 'N/mm^2'])
disp(['tau_max =' num2str(tau_max) 'N/mm^2'])
```

The Command Window outputs resulting from this MATLAB file are as follows:

```
tau_xy = +/- 113.1371 N/mm^2
sig_I = 200 N/mm^2
sig_II = -160 N/mm^2
tau_max = 180 N/mm^2
```

1.9 STRAIN

The external and internal forces described in the previous sections cause linear and angular displacements in a deformable body. These displacements are generally defined in terms of *strain*. *Longitudinal* or *direct strains* are associated with direct stresses σ and relate to changes in length, while *shear strains* define changes in angle produced by shear stresses. These strains are designated, with appropriate suffixes, by the symbols ε and γ, respectively, and have the same sign as the associated stresses.

Consider three mutually perpendicular line elements OA, OB, and OC at a point O in a deformable body. Their original or unstrained lengths are δx, δy, and δz, respectively. If, now, the body is subjected

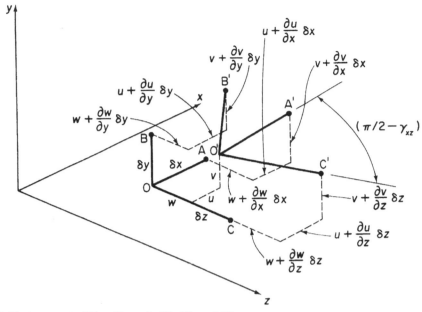

FIGURE 1.15 Displacement of Line Elements OA, OB, and OC

to forces that produce a complex system of direct and shear stresses at O, such as that in Fig. 1.6, then the line elements deform to the positions O'A', O'B', and O'C' shown in Fig. 1.15.

The coordinates of O in the unstrained body are (x, y, z) so that those of A, B, and C are $(x + \delta x, y, z)$, $(x, y + \delta y, z)$, and $(x, y, z + \delta z)$. The components of the displacement of O to O' parallel to the x, y, and z axes are u, v, and w. These symbols are used to designate these displacements throughout the book and are defined as positive in the positive directions of the axes. We again employ the first two terms of a Taylor's series expansion to determine the components of the displacements of A, B, and C. Thus, the displacement of A in a direction parallel to the x axis is $u + (\partial u/\partial x)\delta x$. The remaining components are found in an identical manner and are shown in Fig. 1.15.

We now define direct strain in more quantitative terms. If a line element of length L at a point in a body suffers a change in length ΔL, then the longitudinal strain at that point in the body in the direction of the line element is

$$\varepsilon = \lim_{L \to 0} \frac{\Delta L}{L}$$

The change in length of the element OA is (O'A' – OA), so that the direct strain at O in the x direction is obtained from the equation

$$\varepsilon_x = \frac{OA' - OA}{OA} = \frac{O'A' - \delta x}{\delta x} \tag{1.16}$$

Now,

$$(Q'A')^2 = \left(\delta x + u + \frac{\partial u}{\partial x}\delta x - u\right)^2 + \left(v + \frac{\partial v}{\partial x}\delta x - v\right)^2 + \left(w + \frac{\partial w}{\partial x}\delta x - w\right)^2$$

or

$$O'A' = \delta x \sqrt{\left(1 + \frac{\partial u}{\partial x}\right)^2 + \left(\frac{\partial v}{\partial x}\right)^2 + \left(\frac{\partial w}{\partial x}\right)^2}$$

which may be written, when second-order terms are neglected, as

$$O'A' = \delta x \left(1 + 2\frac{\partial u}{\partial x}\right)^{\frac{1}{2}}$$

Applying the binomial expansion to this expression, we have

$$O'A' = \delta x \left(1 + \frac{\partial u}{\partial x}\right) \tag{1.17}$$

in which squares and higher powers of $\partial u/\partial x$ are ignored. Substituting for $O'A'$ in Eq. (1.16), we have

It follows that

$$\left.\begin{aligned} \varepsilon_x &= \frac{\partial u}{\partial x} \\ \varepsilon_y &= \frac{\partial v}{\partial y} \\ \varepsilon_z &= \frac{\partial w}{\partial z} \end{aligned}\right\} \tag{1.18}$$

The shear strain at a point in a body is defined as the change in the angle between two mutually perpendicular lines at the point. Therefore, if the shear strain in the xz plane is γ_{xz}, then the angle between the displaced line elements $O'A'$ and $O'C'$ in Fig. 1.15 is $\pi/2 - \gamma_{xz}$ radians.

Now, $\cos A'O'C' = \cos(\pi/2 - \gamma_{xz}) = \sin\gamma_{xz}$ and as γ_{xz} is small, $\cos A'O'C' = \gamma_{xz}$. From the trigonometrical relationships for a triangle,

$$\cos A'O'C' = \frac{(O'A')^2 + (O'C')^2 - (A'C')^2}{2(O'A')(O'C')} \tag{1.19}$$

We showed in Eq. (1.17) that

$$O'A' = \delta x \left(1 + \frac{\partial u}{\partial x}\right)$$

Similarly,

$$(O'C') = \delta z \left(1 + \frac{\partial w}{\partial z}\right)$$

But, for small displacements, the derivatives of u, v, and w are small compared with l, so that, as we are concerned here with actual length rather than change in length, we may use the approximations

$$O'A' \approx \delta x, \quad O'C' \approx \delta z$$

Again, to a first approximation,

$$(A'C')^2 = \left(\delta z - \frac{\partial w}{\partial x}\delta x\right)^2 + \left(\delta x - \frac{\partial u}{\partial z}\delta z\right)^2$$

Substituting for $O'A'$, $O'C'$, and $A'C'$ in Eq. (1.19), we have

$$\cos A'O'C' = \frac{(\delta x^2) + (\delta z)^2 - [\delta z - (\partial w/\partial x)\delta x]^2 - [\delta x - (\partial u/\partial z)\delta z]^2}{2\delta x \delta z}$$

Expanding and neglecting fourth-order powers gives

$$\cos A'O'C' = \frac{2(\partial w/\partial x)\delta x \delta z + 2(\partial u/\partial z)\delta x \delta z}{2\delta x \delta z}$$

or,

Similarly,

$$\left.\begin{aligned} \gamma_{xz} &= \frac{\partial w}{\partial x} + \frac{\partial u}{\partial z} \\ \gamma_{xy} &= \frac{\partial v}{\partial x} + \frac{\partial u}{\partial y} \\ \gamma_{yz} &= \frac{\partial w}{\partial y} + \frac{\partial v}{\partial z} \end{aligned}\right\} \tag{1.20}$$

It must be emphasized that Eqs. (1.18) and (1.20) are derived on the assumption that the displacements involved are small. Normally, these linearized equations are adequate for most types of structural problem, but in cases where deflections are large, for example, types of suspension cable, the full, nonlinear, large deflection equations, given in many books on elasticity, must be employed.

1.10 COMPATIBILITY EQUATIONS

In Section 1.9, we expressed the six components of strain at a point in a deformable body in terms of the three components of displacement at that point, u, v, and w. We supposed that the body remains continuous during the deformation, so that no voids are formed. It follows that each component, u, v, and w, must be a continuous, single-valued function or, in quantitative terms,

$$u = f_1(x, y, z), \quad v = f_2(x, y, z), \quad w = f_3(x, y, z)$$

If voids are formed, then displacements in regions of the body separated by the voids are expressed as different functions of x, y, and z. The existence, therefore, of just three single-valued functions for displacement is an expression of the continuity or *compatibility* of displacement, which we presupposed.

Since the six strains are defined in terms of three displacement functions, they must bear some relationship to each other and cannot have arbitrary values. These relationships are found as follows. Differentiating γ_{xy} from Eq. (1.20) with respect to x and y gives

$$\frac{\partial^2 \gamma_{xy}}{\partial x\, \partial y} = \frac{\partial^2}{\partial x\, \partial y}\frac{\partial v}{\partial x} + \frac{\partial^2}{\partial x\, \partial y}\frac{\partial u}{\partial y}$$

or, since the functions of u and v are continuous,

$$\frac{\partial^2 \gamma_{xy}}{\partial x \, \partial y} = \frac{\partial^2}{\partial x^2} \frac{\partial v}{\partial y} + \frac{\partial^2}{\partial y^2} \frac{\partial u}{\partial x}$$

which may be written, using Eq. (1.18), as

$$\frac{\partial^2 \gamma_{xy}}{\partial x \, \partial y} = \frac{\partial^2 \varepsilon_y}{\partial x^2} + \frac{\partial^2 \varepsilon_x}{\partial y^2} \tag{1.21}$$

In a similar manner,

$$\frac{\partial^2 \gamma_{yz}}{\partial y \partial z} = \frac{\partial^2 \varepsilon_y}{\partial z^2} + \frac{\partial^2 \varepsilon_z}{\partial y^2} \tag{1.22}$$

$$\frac{\partial^2 \gamma_{xz}}{\partial x \partial z} = \frac{\partial^2 \varepsilon_z}{\partial x^2} + \frac{\partial^2 \varepsilon_x}{\partial z^2} \tag{1.23}$$

If we now differentiate γ_{xy} with respect to x and z and add the result to γ_{xz}, differentiated with respect to y and x, we obtain

$$\frac{\partial^2 \gamma_{xy}}{\partial x \partial z} + \frac{\partial^2 \gamma_{xz}}{\partial y \partial x} = \frac{\partial^2}{\partial x \partial z} \left(\frac{\partial u}{\partial y} + \frac{\partial v}{\partial x} \right) + \frac{\partial^2}{\partial y \partial x} \left(\frac{\partial w}{\partial x} + \frac{\partial u}{\partial z} \right)$$

or

$$\frac{\partial}{\partial x} \left(\frac{\partial \gamma_{xy}}{\partial z} + \frac{\partial \gamma_{xz}}{\partial y} \right) = \frac{\partial^2}{\partial z \partial y} \frac{\partial u}{\partial x} + \frac{\partial^2}{\partial x^2} \left(\frac{\partial v}{\partial z} + \frac{\partial w}{\partial y} \right) + \frac{\partial^2}{\partial y \partial z} + \frac{\partial u}{\partial x}$$

Substituting from Eqs. (1.18) and (1.21) and rearranging,

$$2 \frac{\partial^2 \varepsilon_x}{\partial y \partial z} = \frac{\partial}{\partial x} \left(-\frac{\partial \gamma_{yz}}{\partial x} + \frac{\partial \gamma_{xz}}{\partial y} + \frac{\partial \gamma_{xy}}{\partial z} \right) \tag{1.24}$$

Similarly,

$$2 \frac{\partial^2 \varepsilon_y}{\partial x \partial z} = \frac{\partial}{\partial y} \left(\frac{\partial \gamma_{yz}}{\partial x} - \frac{\partial \gamma_{xz}}{\partial y} + \frac{\partial \gamma_{xy}}{\partial z} \right) \tag{1.25}$$

and

$$2 \frac{\partial^2 \varepsilon_z}{\partial x \partial y} = \frac{\partial}{\partial z} \left(\frac{\partial \gamma_{yz}}{\partial x} + \frac{\partial \gamma_{xz}}{\partial y} - \frac{\partial \gamma_{xy}}{\partial z} \right) \tag{1.26}$$

Equations (1.21)–(1.26) are the six equations of *strain compatibility* which must be satisfied in the solution of three-dimensional problems in elasticity.

1.11 PLANE STRAIN

Although we derived the compatibility equations and the expressions for strain for the general three-dimensional state of strain, we shall be concerned mainly with the two-dimensional case described in Section 1.4. The corresponding state of strain, in which it is assumed that particles of the body suffer displacements in one plane only, is known as *plane strain*. We shall suppose that this plane is, as for plane stress, the xy plane. Then, ε_z, γ_{xz}, and γ_{yz} become zero and Eqs. (1.18) and (1.20) reduce to

$$\varepsilon_x = \frac{\partial u}{\partial x}, \quad \varepsilon_y = \frac{\partial v}{\partial y} \tag{1.27}$$

and

$$\gamma_{xy} = \frac{\partial v}{\partial x} + \frac{\partial u}{\partial y} \tag{1.28}$$

Further, by substituting $\varepsilon_z = \gamma_{xz} = \gamma_{yz} = 0$ in the six equations of compatibility and noting that ε_x, ε_y, and γ_{xy} are now purely functions of x and y, we are left with Eq. (1.21), namely,

$$\frac{\partial^2 \gamma_{xy}}{\partial x \partial y} = \frac{\partial^2 \varepsilon_y}{\partial x^2} + \frac{\partial^2 \varepsilon_x}{\partial y^2}$$

as the only equation of compatibility in the two-dimensional or plane strain case.

1.12 DETERMINATION OF STRAINS ON INCLINED PLANES

Having defined the strain at a point in a deformable body with reference to an arbitrary system of coordinate axes, we may calculate direct strains in any given direction and the change in the angle (shear strain) between any two originally perpendicular directions at that point. We shall consider the two-dimensional case of plane strain described in Section 1.11.

An element in a two-dimensional body subjected to the complex stress system of Fig. 1.16(a) distorts into the shape shown in Fig. 1.16(b). In particular, the triangular element ECD suffers distortion to the shape E′C′D′ with corresponding changes in the length FC and angle EFC. Suppose that the known direct and shear strains associated with the given stress system are ε_x, ε_y, and γ_{xy} (the actual

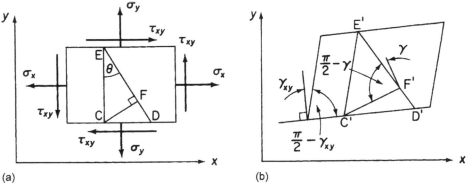

FIGURE 1.16 (a) Stress System on a Rectangular Element; (b) Distorted Shape of the Element Due to Stress System in (a)

relationships are investigated later) and we are required to find the direct strain ε_n in a direction normal to the plane ED and the shear strain γ produced by the shear stress acting on the plane ED.

To a first order of approximation,

$$\left.\begin{array}{l} C'D' = CD(1 + \varepsilon_x) \\ C'E' = CE(1 + \varepsilon_y) \\ E'D' = ED(1 + \varepsilon_{n+\pi/2}) \end{array}\right\} \tag{1.29}$$

where $\varepsilon_{n+\pi/2}$ is the direct strain in the direction ED. From the geometry of the triangle E'C'D' in which angle $E'C'D' = \pi/2 - \gamma_{xy}$,

$$(E'D')^2 = (C'D')^2 + (C'E')^2 - 2(C'D')(C'E')\cos(\pi/2 - \gamma_{xy})$$

or, substituting from Eqs. (1.29),

$$(ED)^2(1 + \varepsilon_{n+\pi/2})^2 = (CD)^2(1 + \varepsilon_x)^2 + (CE)^2(1 + \varepsilon_y)^2 \\ - 2(CD)(CE)(1 + \varepsilon_x)(1 + \varepsilon_y)\sin\gamma_{xy}$$

Noting that $(ED)^2 = (CD)^2 + (CE)^2$ and neglecting squares and higher powers of small quantities, this equation may be rewritten as

$$2(ED)^2\varepsilon_{n+\pi/2} = 2(CD)^2\varepsilon_x + 2(CE)^2\varepsilon_y - 2(CE)(CD)\gamma_{xy}$$

Dividing through by $2(ED)^2$ gives

$$\varepsilon_{n+\pi/2} = \varepsilon_x \sin^2\theta + \varepsilon_y \cos^2\theta - \cos\theta \sin\theta\gamma_{xy} \tag{1.30}$$

The strain ε_n in the direction normal to the plane ED is found by replacing the angle θ in Eq. (1.30) by $\theta - \pi/2$. Hence,

$$\varepsilon_n = \varepsilon_x \cos^2\theta + \varepsilon_y \sin^2\theta + \frac{\gamma_{xy}}{2}\sin 2\theta \tag{1.31}$$

Turning our attention to the triangle C'F'E', we have

$$(C'E')^2 = (C'F')^2 + (F'E')^2 - 2(C'F')(F'E')\cos(\pi/2 - \gamma) \tag{1.32}$$

in which

$$\begin{array}{l} C'E' = CE(1 + \varepsilon_y) \\ C'F' = CF(1 + \varepsilon_n) \\ F'E' = FE(1 + \varepsilon_{n+\pi/2}) \end{array}$$

Substituting for C'E', C'F', and F'E' in Eq. (1.32) and writing $\cos(\pi/2 - \gamma) = \sin\gamma$, we find

$$(CE)^2(1 + \varepsilon_y)^2 = (CF)^2(1 + \varepsilon_n)^2 + (FE)^2(1 + \varepsilon_{n+\pi/2})^2 \\ - 2(CF)(FE)(1 + \varepsilon_n)(1 + \varepsilon_{n+\pi/2})\sin\gamma \tag{1.33}$$

All the strains are assumed to be small, so that their squares and higher powers may be ignored. Further, $\sin\gamma \approx \gamma$ and Eq. (1.33) becomes

$$(CE)^2(1 + 2\varepsilon_y) = (CF)^2(1 + 2\varepsilon_n) + (FE)^2(1 + 2\varepsilon_{n+\pi/2}) - 2(CF)(FE)\gamma$$

From Fig. 1.16(a), $(CE)^2 = (CF)^2 + (FE)^2$ and the preceding equation simplifies to

$$2(CE)^2 \varepsilon_y = 2(CF)^2 \varepsilon_n + 2(FE)^2 \varepsilon_{n+\pi/2} - 2(CF)(FE)\gamma$$

Dividing through by $2(CE)^2$ and transposing,

$$\gamma = \frac{\varepsilon_n \ \sin^2\theta + \varepsilon_{n+\pi/2} \ \cos^2\theta - \varepsilon_y}{\sin\theta \cos\theta}$$

Substitution of ε_n and $\varepsilon_{n+\pi/2}$ from Eqs. (1.31) and (1.30) yields

$$\frac{\gamma}{2} = \frac{(\varepsilon_x - \varepsilon_y)}{2} \sin2\theta - \frac{\gamma_{xy}}{2} \cos2\theta \tag{1.34}$$

1.13 PRINCIPAL STRAINS

If we compare Eqs. (1.31) and (1.34) with Eqs. (1.8) and (1.9), we observe that they may be obtained from Eqs. (1.8) and (1.9) by replacing σ_n with ε_n, σ_x by ε_x, σ_y by ε_y, τ_{xy} by $\gamma_{xy}/2$, and τ by $\gamma/2$. Therefore, for each deduction made from Eqs. (1.8) and (1.9) concerning σ_n and τ, there is a corresponding deduction from Eqs. (1.31) and (1.34) regarding ε_n and $\gamma/2$.

Therefore, at a point in a deformable body, there are two mutually perpendicular planes on which the shear strain γ is zero and normal to which the direct strain is a maximum or minimum. These strains are the *principal strains* at that point and are given (from comparison with Eqs. (1.11) and (1.12)) by

$$\varepsilon_I = \frac{\varepsilon_x + \varepsilon_y}{2} + \frac{1}{2}\sqrt{(\varepsilon_x - \varepsilon_y)^2 + \gamma_{xy}^2} \tag{1.35}$$

and

$$\varepsilon_{II} = \frac{\varepsilon_x + \varepsilon_y}{2} - \frac{1}{2}\sqrt{(\varepsilon_x - \varepsilon_y)^2 + \gamma_{xy}^2} \tag{1.36}$$

If the shear strain is zero on these planes, it follows that the shear stress must also be zero; and we deduce, from Section 1.7, that the directions of the principal strains and principal stresses coincide. The related planes are then determined from Eq. (1.10) or from

$$\tan2\theta = \frac{\gamma_{xy}}{\varepsilon_x - \varepsilon_y} \tag{1.37}$$

In addition, the maximum shear strain at the point is

$$\left(\frac{\gamma}{2}\right)_{max} = \frac{1}{2}\sqrt{(\varepsilon_x - \varepsilon_y)^2 + \gamma_{xy}^2} \tag{1.38}$$

or

$$\left(\frac{\gamma}{2}\right)_{max} = \frac{\varepsilon_I - \varepsilon_{II}}{2} \tag{1.39}$$

(compare with Eqs. (1.14) and (1.15)).

1.14 MOHR'S CIRCLE OF STRAIN

We now apply the arguments of Section 1.13 to the Mohr's circle of stress described in Section 1.8. A circle of strain, analogous to that shown in Fig. 1.12(b), may be drawn when σ_x, σ_y, etc., are replaced by ε_x, ε_y, etc., as specified in Section 1.13. The horizontal extremities of the circle represent the principal strains, the radius of the circle, half the maximum shear strain, and so on.

1.15 STRESS–STRAIN RELATIONSHIPS

In the preceding sections, we developed, for a three-dimensional deformable body, three equations of equilibrium (Eqs. (1.5)) and six strain-displacement relationships (Eqs. (1.18) and (1.20)). From the latter, we eliminated displacements, thereby deriving six auxiliary equations relating strains. These compatibility equations are an expression of the continuity of displacement, which we have assumed as a prerequisite of the analysis. At this stage, therefore, we have obtained nine independent equations toward the solution of the three-dimensional stress problem. However, the number of unknowns totals 15, comprising six stresses, six strains, and three displacements. An additional six equations are therefore necessary to obtain a solution.

So far we have made no assumptions regarding the force–displacement or stress–strain relationship in the body. This will, in fact, provides us with the required six equations, but before these are derived, it is worthwhile considering some general aspects of the analysis.

The derivation of the equilibrium, strain–displacement, and compatibility equations does not involve any assumption as to the stress–strain behavior of the material of the body. It follows that these basic equations are applicable to any type of continuous, deformable body, no matter how complex its behavior under stress. In fact, we shall consider only the simple case of linearly elastic, *isotropic* materials, for which stress is directly proportional to strain and whose elastic properties are the same in all directions. A material possessing the same properties at all points is said to be *homogeneous*.

Particular cases arise where some of the stress components are known to be zero and the number of unknowns may then be no greater than the remaining equilibrium equations which have not identically vanished. The unknown stresses are then found from the conditions of equilibrium alone and the problem is said to be *statically determinate*. For example, the uniform stress in the member supporting a tensile load P in Fig. 1.3 is found by applying one equation of equilibrium and a boundary condition. This system is therefore statically determinate.

Statically indeterminate systems require the use of some, if not all, of the other equations involving strain–displacement and stress–strain relationships. However, whether the system be statically determinate or not, stress–strain relationships are necessary to determine deflections. The role of the six auxiliary compatibility equations will be discussed when actual elasticity problems are formulated in Chapter 2.

We now proceed to investigate the relationship of stress and strain in a three–dimensional, linearly elastic, isotropic body.

Experiments show that the application of a uniform direct stress, say σ_x, does not produce any shear distortion of the material and that the direct strain ε_x is given by the equation

$$\varepsilon_x = \frac{\sigma_x}{E} \tag{1.40}$$

where E is a constant known as the *modulus of elasticity* or *Young's modulus*. Equation (1.40) is an expression of *Hooke's law*. Further, ε_x is accompanied by lateral strains

$$\varepsilon_y = -\nu \frac{\sigma_x}{E}, \quad \varepsilon_z = -\nu \frac{\sigma_x}{E} \tag{1.41}$$

in which ν is a constant termed *Poisson's ratio*.

For a body subjected to direct stresses σ_x, σ_y, and σ_z, the direct strains are, from Eqs. (1.40) and (1.41) and the *principle of superposition* (see Chapter 5, Section 5.9),

$$\left. \begin{aligned} \varepsilon_x &= \frac{1}{E}[\sigma_x - \nu(\sigma_y + \sigma_z)] \\ \varepsilon_y &= \frac{1}{E}[\sigma_y - \nu(\sigma_x + \sigma_z)] \\ \varepsilon_z &= \frac{1}{E}[\sigma_z - \nu(\sigma_x + \sigma_y)] \end{aligned} \right\} \tag{1.42}$$

Equations (1.42) may be transposed to obtain expressions for each stress in terms of the strains. The procedure adopted may be any of the standard mathematical approaches and gives

$$\sigma_x = \frac{\nu E}{(1+\nu)(1-2\nu)} e + \frac{E}{(1+\nu)} \varepsilon_x \tag{1.43}$$

$$\sigma_y = \frac{\nu E}{(1+\nu)(1-2\nu)} e + \frac{E}{(1+\nu)} \varepsilon_y \tag{1.44}$$

$$\sigma_z = \frac{\nu E}{(1+\nu)(1-2\nu)} e + \frac{E}{(1+\nu)} \varepsilon_z \tag{1.45}$$

in which

$$e = \varepsilon_x + \varepsilon_y + \varepsilon_z$$

See Eq. (1.53).

For the case of plane stress in which $\sigma_z = 0$, Eqs. (1.43) and (1.44) reduce to

$$\sigma_x = \frac{E}{1-\nu^2}(\varepsilon_x + \nu\varepsilon_y) \tag{1.46}$$

$$\sigma_y = \frac{E}{1-\nu^2}(\varepsilon_y + \nu\varepsilon_x) \tag{1.47}$$

Suppose now that, at some arbitrary point in a material, there are principal strains ε_{I} and $\varepsilon_{\mathrm{II}}$ corresponding to principal stresses σ_{I} and σ_{II}. If these stresses (and strains) are in the direction of the coordinate axes x and y, respectively, then $\tau_{xy} = \gamma_{xy} = 0$ and, from Eq. (1.34), the shear strain on an arbitrary plane at the point inclined at an angle θ to the principal planes is

$$\gamma = (\varepsilon_{\mathrm{I}} - \varepsilon_{\mathrm{II}})\sin 2\theta \tag{1.48}$$

Using the relationships of Eqs. (1.42) and substituting in Eq. (1.48), we have

$$\gamma = \frac{1}{E}[(\sigma_I - \nu\sigma_{II}) - (\sigma_{II} - \nu\sigma_I)]\sin2\theta$$

or

$$\gamma = \frac{(1 + \nu)}{E}(\sigma_I - \sigma_{II})\sin2\theta \qquad (1.49)$$

Using Eq. (1.9) and noting that for this particular case $\tau_{xy} = 0$, $\sigma_x = \sigma_I$, and $\sigma_y = \sigma_{II}$,

$$2\tau = (\sigma_I - \sigma_{II})\sin2\theta$$

from which we may rewrite Eq. (1.49) in terms of τ as

$$\gamma = \frac{2(1 + \nu)}{E}\tau \qquad (1.50)$$

The term $E/2(1 + \nu)$ is a constant known as the *modulus of rigidity* G. Hence,

$$\gamma = \tau/G$$

and the shear strains γ_{xy}, γ_{xz}, and γ_{yz} are expressed in terms of their associated shear stresses as follows:

$$\gamma_{xy} = \frac{\tau_{xy}}{G}, \quad \gamma_{xz} = \frac{\tau_{xz}}{G}, \quad \gamma_{yz} = \frac{\tau_{yz}}{G} \qquad (1.51)$$

Equations (1.51), together with Eqs. (1.42), provide the additional six equations required to determine the 15 unknowns in a general three-dimensional problem in elasticity. They are, however, limited in use to a linearly elastic, isotropic body.

For the case of plane stress, they simplify to

$$\left.\begin{aligned} \varepsilon_x &= \frac{1}{E}(\sigma_x - \nu\sigma_y) \\ \varepsilon_y &= \frac{1}{E}(\sigma_y - \nu\sigma_x) \\ \varepsilon_z &= \frac{-\nu}{E}(\sigma_x - \sigma_y) \\ \gamma_{xy} &= \frac{\tau_{xy}}{G} \end{aligned}\right\} \qquad (1.52)$$

It may be seen from the third of Eqs. (1.52) that the conditions of plane stress and plane strain do not necessarily describe identical situations. See Ex. 1.1.

Changes in the linear dimensions of a strained body may lead to a change in volume. Suppose that a small element of a body has dimensions δx, δy, and δz. When subjected to a three-dimensional stress system, the element sustains a volumetric strain e (change in volume/unit volume) equal to

$$e = \frac{(1 + \varepsilon_x)\delta x(1 + \varepsilon_y)\delta y(1 + \varepsilon_z)\delta z - \delta x\delta y\delta z}{\delta x\delta y\delta z}$$

Neglecting products of small quantities in the expansion of the right-hand side of this equation yields

$$e = \varepsilon_x + \varepsilon_y + \varepsilon_z \tag{1.53}$$

Substituting for ε_x, ε_y, and ε_z from Eqs. (1.42), we find, for a linearly elastic, isotropic body,

$$e = \frac{1}{E}[\sigma_x + \sigma_y + \sigma_z - 2v(\sigma_x + \sigma_y + \sigma_z)]$$

or

$$e = \frac{(1-2v)}{E}(\sigma_x + \sigma_y + \sigma_z)$$

In the case of a uniform hydrostatic pressure, $\sigma_x = \sigma_y = \sigma_z = -p$ and

$$e = -\frac{3(1-2v)}{E}p \tag{1.54}$$

The constant $E/3(1-2v)$ is known as the *bulk modulus* or *modulus of volume expansion* and is often given the symbol K.

An examination of Eq. (1.54) shows that $v \leq 0.5$, since a body cannot increase in volume under pressure. Also, the lateral dimensions of a body subjected to uniaxial tension cannot increase, so that $v > 0$. Therefore, for an isotropic material $0 \leq v \leq 0.5$ and for most isotropic materials, v is in the range 0.25–0.33 below the elastic limit. Above the limit of proportionality, v increases and approaches 0.5.

Example 1.4

A rectangular element in a linearly elastic, isotropic material is subjected to tensile stresses of 83 and 65 N/mm^2 on mutually perpendicular planes. Determine the strain in the direction of each stress and in the direction perpendicular to both stresses. Find also the principal strains, the maximum shear stress, the maximum shear strain, and their directions at the point. Take $E = 200,000$ N/mm^2 and $v = 0.3$. See Ex. 1.1.

If we assume that $\sigma_x = 83$ N/mm^2 and $\sigma_y = 65$ N/mm^2, then from Eqs (1.52),

$$\varepsilon_x = \frac{1}{200,000}(83 - 0.3 \times 65) = 3.175 \times 10^{-4}$$

$$\varepsilon_y = \frac{1}{200,000}(65 - 0.3 \times 83) = 2.005 \times 10^{-4}$$

$$\varepsilon_z = \frac{-0.3}{200,000}(83 + 65) = -2.220 \times 10^{-4}$$

In this case, since there are no shear stresses on the given planes, σ_x and σ_y are principal stresses, so that ε_x and ε_y are the principal strains and are in the directions of σ_x and σ_y. It follows from Eq. (1.15) that the maximum shear stress (in the plane of the stresses) is

$$\tau_{max} = \frac{83 - 65}{2} = 9\,\text{N/mm}^2$$

acting on planes at 45° to the principal planes.

Further, using Eq. (1.50), the maximum shear strain is

$$\gamma_{max} = \frac{2 \times (1 + 0.3) \times 9}{200,000}$$

so that $\gamma_{max} = 1.17 \times 10^{-4}$ on the planes of maximum shear stress.

Example 1.5

At a particular point in a structural member, a two-dimensional stress system exists where $\sigma_x = 60$ N/mm^2, $\sigma_y = -40$ N/mm^2, and $\tau_{xy} = 50$ N/mm^2. If Young's modulus $E = 200,000$ N/mm^2 and Poisson's ratio $v = 0.3$, calculate the direct strain in the x and y directions and the shear strain at the point. Also calculate the principal strains at the point and their inclination to the plane on which σ_x acts; verify these answers using a graphical method. See Ex. 1.1.

From Eqs. (1.52),

$$\varepsilon_x = \frac{1}{200,000}(60 + 0.3 \times 40) = 360 \times 10^{-6}$$

$$\varepsilon_y = \frac{1}{200,000}(-40 - 0.3 \times 60) = -290 \times 10^{-6}$$

From Eq. (1.50), the shear modulus, G, is given by

$$G = \frac{E}{2(1+v)} = \frac{200,000}{2(1+0.3)} = 76\,923 \ \text{N/mm}^2$$

Hence, from Eqs. (1.52),

$$\gamma_{xy} = \frac{\tau_{xy}}{G} = \frac{50}{76,923} = 650 \times 10^{-6}$$

Now substituting in Eq. (1.35) for ε_x, ε_y, and γ_{xy},

$$\varepsilon_I = 10^{-6}\left[\frac{360 - 290}{2} + \frac{1}{2}\sqrt{(360 + 290)^2 + 650^2}\right]$$

which gives

$$\varepsilon_I = 495 \times 10^{-6}$$

Similarly, from Eq. (1.36),

$$\varepsilon_{II} = -425 \times 10^{-6}$$

From Eq. (1.37),

$$\tan 2\theta = \frac{650 \times 10^{-6}}{360 \times 10^{-6} + 290 \times 10^{-6}} = 1$$

Therefore,

$$2\theta = 45° \ \text{or} \ 225°$$

so that

$$\theta = 22.5° \ \text{or} \ 112.5°$$

The values of ε_I, ε_{II}, and θ are verified using Mohr's circle of strain (Fig. 1.17). Axes $O\varepsilon$ and $O\gamma$ are set up and the points $Q_1(360 \times 10^{-6}, \frac{1}{2} \times 650 \times 10^{-6})$ and $Q_2(-290 \times 10^{-6}, -\frac{1}{2} \times 650 \times 10^{-6})$

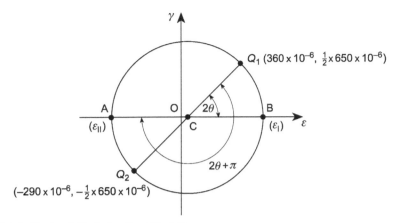

FIGURE 1.17 Mohr's Circle of Strain for Example 1.5

located. The center C of the circle is the intersection of Q_1Q_2 and the $O\varepsilon$ axis. The circle is then drawn with radius CQ_1 and the points $B(\varepsilon_I)$ and $A(\varepsilon_{II})$ located. Finally, angle $Q_1CB = 2\theta$ and angle $Q_1CA = 2\theta + \pi$.

1.15.1 Temperature effects

The stress–strain relationships of Eqs. (1.43)–(1.47) apply to a body or structural member at a constant uniform temperature. A temperature rise (or fall) generally results in an expansion (or contraction) of the body or structural member so that there is a change in size, that is, a strain.

Consider a bar of uniform section, of original length L_o, and suppose that it is subjected to a temperature change ΔT along its length; ΔT can be a rise ($+ve$) or fall ($-ve$). If the coefficient of linear expansion of the material of the bar is α, the final length of the bar is, from elementary physics,

$$L = L_o(1 + \alpha\Delta T)$$

so that the strain, ε, is given by

$$\varepsilon = \frac{L - L_o}{L_o} = \alpha\Delta T \tag{1.55}$$

Suppose now that a compressive axial force is applied to each end of the bar, such that the bar returns to its original length. The *mechanical* strain produced by the axial force is therefore just large enough to offset the *thermal* strain due to the temperature change, making the *total* strain zero. In general terms, the total strain, ε, is the sum of the mechanical and thermal strains. Therefore, from Eqs. (1.40) and (1.55),

$$\varepsilon = \frac{\sigma}{E} + \alpha\Delta T \tag{1.56}$$

In the case where the bar is returned to its original length or if the bar had not been allowed to expand at all, the total strain is zero and, from Eq. (1.56),

$$\sigma = -E\alpha\Delta T \tag{1.57}$$

Equations (1.42) may now be modified to include the contribution of thermal strain. Therefore, by comparison with Eq. (1.56),

$$
\left.
\begin{aligned}
\varepsilon_x &= \frac{1}{E}[\sigma_x - \nu(\sigma_y + \sigma_z)] + \alpha\Delta T \\[2mm]
\varepsilon_y &= \frac{1}{E}[\sigma_y - \nu(\sigma_x + \sigma_z)] + \alpha\Delta T \\[2mm]
\varepsilon_z &= \frac{1}{E}[\sigma_z - \nu(\sigma_x + \sigma_y)] + \alpha\Delta T
\end{aligned}
\right\} \tag{1.58}
$$

Equations (1.58) may be transposed in the same way as Eqs. (1.42) to give stress–strain relationships rather than strain–stress relationships; that is,

$$
\left.
\begin{aligned}
\sigma_x &= \frac{\nu E}{(1+\nu)(1-2\nu)}e + \frac{E}{(1+\nu)}\varepsilon_x - \frac{E}{(1-2\nu)}\alpha\Delta T \\[2mm]
\sigma_y &= \frac{\nu E}{(1+\nu)(1-2\nu)}e + \frac{E}{(1+\nu)}\varepsilon_y - \frac{E}{(1-2\nu)}\alpha\Delta T \\[2mm]
\sigma_z &= \frac{\nu E}{(1+\nu)(1-2\nu)}e + \frac{E}{(1+\nu)}\varepsilon_z - \frac{E}{(1-2\nu)}\alpha\Delta T
\end{aligned}
\right\} \tag{1.59}
$$

For the case of plane stress in which $\sigma_z = 0$, these equations reduce to

$$
\left.
\begin{aligned}
\sigma_x &= \frac{E}{(1-\nu^2)}(\varepsilon_x + \nu\varepsilon_y) - \frac{E}{(1-\nu)}\alpha\Delta T \\[2mm]
\sigma_y &= \frac{E}{(1-\nu^2)}(\varepsilon_y + \nu\varepsilon_x) - \frac{E}{(1-\nu)}\alpha\Delta T
\end{aligned}
\right\} \tag{1.60}
$$

Example 1.6

A composite bar of length L has a central core of copper loosely inserted in a sleeve of steel; the ends of the steel and copper are attached to each other by rigid plates. If the bar is subjected to a temperature rise ΔT, determine the stress in the steel and in the copper and the extension of the composite bar. The copper core has a Young's modulus E_c, a cross-sectional area A_c, and a coefficient of linear expansion α_c; the corresponding values for the steel are E_s, A_s, and α_s. See Ex. 1.1.

Assume that $\alpha_c > \alpha_s$. If the copper core and steel sleeve are allowed to expand freely, their final lengths would be different, since they have different values of the coefficient of linear expansion. However, since they are rigidly attached at their ends, one restrains the other and an axial stress is induced in each. Suppose that this stress is σ_x. Then, in Eqs. (1.58). $\sigma_x = \sigma_c$ or σ_s and $\sigma_y = \sigma_z = 0$; the total strain in the copper and steel is then, respectively,

$$
\varepsilon_c = \frac{\sigma_c}{E_c} + \alpha_c\Delta T \tag{i}
$$

$$
\varepsilon_s = \frac{\sigma_s}{E_s} + \alpha_s\Delta T \tag{ii}
$$

The total strain in the copper and steel is the same, since their ends are rigidly attached to each other. Therefore, from compatibility of displacement,

$$\frac{\sigma_c}{E_c} + \alpha_c \Delta T = \frac{\sigma_s}{E_s} + \alpha_s \Delta T \tag{iii}$$

No external axial load is applied to the bar, so that

$$\sigma_c A_c + \sigma_s A_s = 0$$

that is,

$$\sigma_s = -\frac{A_c}{A_s}\sigma_c \tag{iv}$$

Substituting for σ_s in Eq. (iii) gives

$$\sigma_c\left(\frac{1}{E_c} + \frac{A_c}{A_s E_s}\right) = \Delta T(\alpha_s - \alpha_c)$$

from which

$$\sigma_c = \frac{\Delta T(\alpha_s - \alpha_c)A_s E_s E_c}{A_s E_s + A_c E_c} \tag{v}$$

Also $\alpha_c > \alpha_s$, so that σ_c is negative and therefore compressive. Now substituting for σ_c in Eq. (iv),

$$\sigma_s = -\frac{\Delta T(\alpha_s - \alpha_c)A_c E_s E_c}{A_s E_s + A_c E_c} \tag{vi}$$

which is positive and therefore tensile, as would be expected by a physical appreciation of the situation.

Finally, the extension of the compound bar, δ, is found by substituting for σ_c in Eq. (i) or for σ_s in Eq. (ii). Then,

$$\delta = \Delta T L\left(\frac{\alpha_c A_c E_c + \alpha_s A_s E_s}{A_s E_s + A_c E_c}\right) \tag{vii}$$

∎

1.16 EXPERIMENTAL MEASUREMENT OF SURFACE STRAINS

Stresses at a point on the surface of a piece of material may be determined by measuring the strains at the point, usually by electrical resistance strain gauges arranged in the form of a rosette, as shown in Fig. 1.18. Suppose that ε_I and ε_{II} are the principal strains at the point, then if ε_a, ε_b, and ε_c are the measured strains in the directions θ, $(\theta + \alpha)$, $(\theta + \alpha + \beta)$ to ε_I, we have, from the general direct strain relationship of Eq. (1.31),

$$\varepsilon_a = \varepsilon_I \cos^2\theta + \varepsilon_{II} \sin^2\theta \tag{1.61}$$

since ε_x becomes ε_I, ε_y becomes ε_{II}, and γ_{xy} is zero, since the x and y directions have become principal directions. Rewriting Eq. (1.61), we have

$$\varepsilon_a = \varepsilon_I\left(\frac{1 + \cos2\theta}{2}\right) + \varepsilon_{II}\left(\frac{1 - \cos2\theta}{2}\right)$$

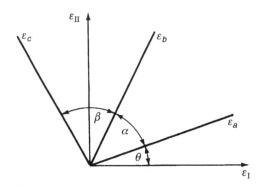

FIGURE 1.18 Strain Gauge Rosette

or

$$\varepsilon_a = \frac{1}{2}(\varepsilon_I + \varepsilon_{II}) + \frac{1}{2}(\varepsilon_I - \varepsilon_{II})\cos 2\theta \tag{1.62}$$

Similarly,

$$\varepsilon_b = \frac{1}{2}(\varepsilon_I + \varepsilon_{II}) + \frac{1}{2}(\varepsilon_I - \varepsilon_{II})\cos 2(\theta + \alpha) \tag{1.63}$$

and

$$\varepsilon_c = \frac{1}{2}(\varepsilon_I + \varepsilon_{II}) + \frac{1}{2}(\varepsilon_I - \varepsilon_{II})\cos 2(\theta + \alpha + \beta) \tag{1.64}$$

Therefore, if ε_a, ε_b, and ε_c are measured in given directions, that is, given angles α and β, then ε_I, ε_{II}, and θ are the only unknowns in Eqs. (1.62)–(1.64).

The principal stresses are now obtained by substitution of ε_I and ε_{II} in Eqs. (1.52).

Thus,

$$\varepsilon_I = \frac{1}{E}(\sigma_I - \nu\sigma_{II}) \tag{1.65}$$

and

$$\varepsilon_{II} = \frac{1}{E}(\sigma_{II} - \nu\sigma_I) \tag{1.66}$$

Solving Eqs. (1.65) and (1.66) gives

$$\sigma_I = \frac{E}{1 - \nu^2}(\varepsilon_I + \nu\varepsilon_{II}) \tag{1.67}$$

and

$$\sigma_{II} = \frac{E}{1 - \nu^2}(\varepsilon_{II} + \nu\varepsilon_I) \tag{1.68}$$

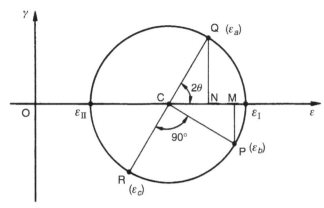

FIGURE 1.19 Experimental Values of Principal Strain Using Mohr's Circle

A typical rosette would have $\alpha = \beta = 45°$, in which case the principal strains are most conveniently found using the geometry of Mohr's circle of strain. Suppose that the arm a of the rosette is inclined at some unknown angle θ to the maximum principal strain, as in Fig. 1.18. Then, Mohr's circle of strain is as shown in Fig. 1.19; the shear strains γ_a, γ_b, and γ_c do not feature in the analysis and are therefore ignored. From Fig. 1.19,

$$\mathrm{OC} = \frac{1}{2}(\varepsilon_a + \varepsilon_c)$$

$$\mathrm{CN} = \varepsilon_a - \mathrm{OC} = \frac{1}{2}(\varepsilon_a - \varepsilon_c)$$

$$\mathrm{QN} = \mathrm{CM} = \varepsilon_b - \mathrm{OC} = \varepsilon_b - \frac{1}{2}(\varepsilon_a + \varepsilon_c)$$

The radius of the circle is CQ and

$$\mathrm{CQ} = \sqrt{\mathrm{CN}^2 + \mathrm{QN}^2}$$

Hence,

$$\mathrm{CQ} = \sqrt{\left[\tfrac{1}{2}(\varepsilon_a - \varepsilon_c)\right]^2 + \left[\varepsilon_b - \tfrac{1}{2}(\varepsilon_a + \varepsilon_c)\right]^2}$$

which simplifies to

$$\mathrm{CQ} = \frac{1}{\sqrt{2}}\sqrt{(\varepsilon_a - \varepsilon_b)^2 + (\varepsilon_c - \varepsilon_b)^2}$$

Therefore, ε_I, which is given by

$$\varepsilon_\mathrm{I} = \mathrm{OC} + \text{radius of circle}$$

is

$$\varepsilon_I = \frac{1}{2}(\varepsilon_a + \varepsilon_c) + \frac{1}{\sqrt{2}}\sqrt{(\varepsilon_a - \varepsilon_b)^2 + (\varepsilon_c - \varepsilon_b)^2} \tag{1.69}$$

Also,

$$\varepsilon_{II} = OC - \text{radius of circle}$$

that is,

$$\varepsilon_{II} = \frac{1}{2}(\varepsilon_a + \varepsilon_c) - \frac{1}{\sqrt{2}}\sqrt{(\varepsilon_a - \varepsilon_b)^2 + (\varepsilon_c - \varepsilon_b)^2} \tag{1.70}$$

Finally, the angle θ is given by

$$\tan 2\theta = \frac{QN}{CN} = \frac{\varepsilon_b - \frac{1}{2}(\varepsilon_a + \varepsilon_c)}{\frac{1}{2}(\varepsilon_a - \varepsilon_c)}$$

that is,

$$\tan 2\theta = \frac{2\varepsilon_b - \varepsilon_a - \varepsilon_c}{\varepsilon_a - \varepsilon_c} \tag{1.71}$$

A similar approach may be adopted for a 60° rosette.

Example 1.7
A bar of solid circular cross-section has a diameter of 50 mm and carries a torque, T, together with an axial tensile load, P. A rectangular strain gauge rosette attached to the surface of the bar gives the following strain readings: $\varepsilon_a = 1000 \times 10^{-6}$, $\varepsilon_b = -200 \times 10^{-6}$, and $\varepsilon_c = -300 \times 10^{-6}$, where the gauges a and c are in line with, and perpendicular to, the axis of the bar, respectively. If Young's modulus, E, for the bar is 70,000 N/mm² and Poisson's ratio, v, is 0.3; calculate the values of T and P. See Ex. 1.1.

Substituting the values of ε_a, ε_b, and ε_c in Eq. (1.69),

$$\varepsilon_I = \frac{10^{-6}}{2}(1000 - 300) + \frac{10^{-6}}{\sqrt{2}}\sqrt{(1000 + 200)^2 + (-200 + 300)^2}$$

which gives

$$\varepsilon_I = 1202 \times 10^{-6}$$

Similarly, from Eq. (1.70),

$$\varepsilon_{II} = -502 \times 10^{-6}$$

Now, substituting for ε_I and ε_{II} in Eq. (1.67),

$$\sigma_I = 70,000 \times 10^{-6}[1202 - 0.3(502)]/[1 - (0.3)^2] = 80.9 \text{ N/mm}^2$$

Similarly, from Eq. (1.68),

$$\sigma_{II} = -10.9 \text{ N/mm}^2$$

Since $\sigma_y = 0$, Eqs. (1.11) and (1.12) reduce to

$$\sigma_{\mathrm{I}} = \frac{\sigma_x}{2} + \frac{1}{2}\sqrt{\sigma_x^2 + 4\tau_{xy}^2} \qquad (\mathrm{i})$$

and

$$\sigma_{\mathrm{II}} = \frac{\sigma_x}{2} - \frac{1}{2}\sqrt{\sigma_x^2 + 4\tau_{xy}^2} \qquad (\mathrm{ii})$$

respectively. Adding Eqs. (i) and (ii), we obtain

$$\sigma_{\mathrm{I}} + \sigma_{\mathrm{II}} = \sigma_x$$

Therefore,

$$\sigma_x = 80.9 - 10.9 = 70\,\mathrm{N/mm^2}$$

For an axial load P,

$$\sigma_x = 70\ \mathrm{N/mm^2} = \frac{P}{A} = \frac{P}{\pi \times 50^2/4}$$

from which

$$P = 137.4\,\mathrm{kN}$$

Substituting for σ_x in either of Eq. (i) or (ii) gives

$$\tau_{xy} = 29.7\,\mathrm{N/mm^2}$$

From the theory of the torsion of circular section bars (see Eq. (iv) in Example 3.1),

$$\tau_{xy} = 29.7\ \mathrm{N/mm^2} = \frac{Tr}{J} = \frac{T \times 25}{\pi \times 50^4/32}$$

from which

$$T = 0.7\,\mathrm{kNm}$$

Note that P could have been found directly in this particular case from the axial strain. Thus, from the first of Eqs. (1.52),

$$\sigma_x = E\varepsilon_a = 70\,000 \times 1000 \times 10^{-6} = 70\,\mathrm{N/mm^2}$$

as before.

∎

Example 1.7 MATLAB

Repeat the derivations presented in Example 1.7 using the Symbolic Math Toolbox in MATLAB. To obtain the same results as Example 1.7, set the number of significant digits used in calculations to four. See Ex. 1.1.

Calculations of T and P are obtained through the following MATLAB file:

```
% Declare any needed symbolic variables
syms e_a e_b e_c E v sig_y sig_x A J r T tau_xy
```

```
% Set the significant digits
digits(4);

% Define known variable values
e_a = sym(1000*10^(-6));
e_b = sym(-200*10^(-6));
e_c = sym(-300*10^(-6));
E = sym(70000);
v = sym(0.3);
sig_y = sym(0);
A = sym(pi*50^2/4,'d');
J = sym(pi*50^4/32,'d');
r = sym(25);

% Evaluate Eqs (1.69) and (1.70)
e_I = vpa((e_a + e_c)/2) + vpa(sqrt(((e_a-e_b)^2 + (e_c - e_c)^2)))/vpa(sqrt(2));
e_II = vpa((e_a + e_c)/2) - vpa(sqrt(((e_a-e_b)^2 + (e_c - e_c)^2)))/vpa(sqrt(2));

% Substitute e_I and e_II into Eqs (1.67) and (1.68)
sig_I = vpa(E*(e_I+v*e_II)/(1-v^2));
sig_II = vpa(E*(e_II+v*e_I)/(1-v^2));

% Evaluate Eqs (1.11) and (1.12)
eqI = vpa(-sig_I + (sig_x+sig_y)/2 + sqrt((sig_x-sig_y)^2 + 4*tau_xy^2)/2);
eqII = vpa(-sig_II + (sig_x+sig_y)/2 - sqrt((sig_x-sig_y)^2 + 4*tau_xy^2)/2);

% Add eqI and eqII and solve for the value of sig_x
sig_x_val = vpa(solve(eqI+eqII,sig_x));

% Calculate the axial load (P in kN)
P = vpa(sig_x_val*A/1000);
P = round(double(P)*10)/10;

% Substitute sig_x back into eqI and solve for tau_xy
tau_xy_val = sym(max(double(solve(subs(eqI,sig_x,sig_x_val),tau_xy))));

% Calculate the applied torsion (T in kN-m) using Eq. (iv) in Example 3.1
T = vpa(tau_xy_val*J/r/1000/1000);
T = round(double(T)*10)/10;

% Output values for P and T to the Command Window
disp(['P =' num2str(P) 'kN'])
disp(['T =' num2str(T) 'kN m'])
```

The Command Window outputs resulting from this MATLAB file are as follows:

```
P = 137.4 kN
T = 0.7 kN m
```

References

[1] Megson THG. Structural and stress analysis. 2nd ed. Burlington, MA: Elsevier; 2005.

Additional Reading

Timoshenko S, Goodier JN. Theory of elasticity. 2nd ed. New York: McGraw-Hill; 1951.
Wang CT. Applied elasticity. New York: McGraw-Hill; 1953.

PROBLEMS

P.1.1. A structural member supports loads that produce, at a particular point, a direct tensile stress of 80 N/mm^2 and a shear stress of 45 N/mm^2 on the same plane. Calculate the values and directions of the principal stresses at the point and also the maximum shear stress, stating on which planes this acts.

Answer: $\sigma_I = 100.2 \, \text{N/mm}^2$, $\theta = 24°11'$
$\sigma_{II} = -20.2 \, \text{N/mm}^2$, $\theta = 114°11'$
$\tau_{max} = 60.2 \, \text{N/mm}^2$ at 45° to principal planes

P.1.2. At a point in an elastic material, there are two mutually perpendicular planes, one of which carries a direct tensile stress of 50 N/mm^2 and a shear stress of 40 N/mm^2, while the other plane is subjected to a direct compressive stress of 35 N/mm^2 and a complementary shear stress of 40 N/mm^2. Determine the principal stresses at the point, the position of the planes on which they act, and the position of the planes on which there is no normal stress.

Answer: $\sigma_I = 65.9 \, \text{N/mm}^2$, $\theta = 21°38'$
$\sigma_{II} = -50.9 \, \text{N/mm}^2$, $\theta = 111°38'$

No normal stress on planes at 70°21′ and –27°5′ to vertical.

P.1.3. The following are varying combinations of stresses acting at a point and referred to axes x and y in an elastic material. Using Mohr's circle of stress determine the principal stresses at the point and their directions for each combination.

	$\sigma_x(\text{N/mm}^2)$	$\sigma_y(\text{N/mm}^2)$	$\tau_{xy}(\text{N/mm}^2)$
(i)	+54	+30	+5
(ii)	+30	+54	−5
(iii)	−60	−36	+5
(iv)	+30	−50	+30

Answer: (i) $\sigma_I = +55 \, \text{N/mm}^2$ $\sigma_{II} = +29 \, \text{N/mm}^2$ σ_I at 11.5° to x axis.
(ii) $\sigma_I = +55 \, \text{N/mm}^2$ $\sigma_{II} = +29 \, \text{N/mm}^2$ σ_{II} at 11.5° to x axis.
(iii) $\sigma_I = -34.5 \, \text{N/mm}^2$ $\sigma_{II} = -61 \, \text{N/mm}^2$ σ_I at 79.5° to x axis.
(iv) $\sigma_I = +40 \, \text{N/mm}^2$ $\sigma_{II} = -60 \, \text{N/mm}^2$ σ_I at 18.5° to x axis.

P.1.3 MATLAB Repeat P.1.3 by creating a script in MATLAB in place of constructing Mohr's circle. In addition to calculating the principal stresses, also calculate the maximum shear stress at the point for each combination. Do not repeat the calculation of the principal stress directions.

Answer: (i) $\tau_{max} = 13 \text{ N/mm}^2$, (ii) $\tau_{max} = 13 \text{ N/mm}^2$
(ii) $\tau_{max} = 13 \text{ N/mm}^2$, (iv) $\tau_{max} = 50 \text{ N/mm}^2$

σ_I and σ_{II} for each combination are as shown in P.1.3

P.1.4. The state of stress at a point is caused by three separate actions, each of which produces a pure, unidirectional tension of 10 N/mm² individually but in three different directions, as shown in Fig. P.1.4. By transforming the individual stresses to a common set of axes (x, y), determine the principal stresses at the point and their directions.

Answer: $\sigma_I = \sigma_{II} = 15 \text{ N/mm}^2$. All directions are principal directions.

P.1.5. A shear stress τ_{xy} acts in a two-dimensional field in which the maximum allowable shear stress is denoted by τ_{max} and the major principal stress by σ_I. Derive, using the geometry of Mohr's circle of stress, expressions for the maximum values of direct stress which may be applied to the x and y planes in terms of these three parameters.

Answer: $\sigma_x = \sigma_I - \tau_{max} + \sqrt{\tau_{max}^2 - \tau_{xy}^2}$
$\sigma_y = \sigma_I - \tau_{max} - \sqrt{\tau_{max}^2 - \tau_{xy}^2}$

P.1.6. A solid shaft of circular cross-section supports a torque of 50 kNm and a bending moment of 25 kNm. If the diameter of the shaft is 150 mm, calculate the values of the principal stresses and their directions at a point on the surface of the shaft.

Answer: $\sigma_I = 121.4 \text{ N/mm}^2$, $\theta = 31°43'$
$\sigma_{II} = -46.4 \text{ N/mm}^2$, $\theta = 121°43'$

P.1.7. An element of an elastic body is subjected to a three-dimensional stress system σ_x, σ_y, and σ_z. Show that, if the direct strains in the directions x, y, and z are ε_x, ε_y, and ε_z, then

$$\sigma_x = \lambda e + 2G\varepsilon_x, \quad \sigma_y = \lambda e + 2G\varepsilon_y, \quad \sigma_z = \lambda e + 2G\varepsilon_z$$

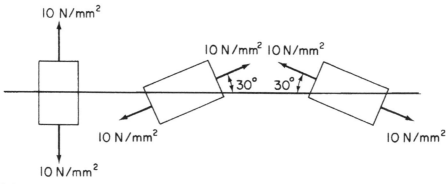

FIGURE P.1.4

where

$$\lambda = \frac{\nu E}{(1+\nu)(1-2\nu)} \quad \text{and} \quad e = \varepsilon_x + \varepsilon_y + \varepsilon_z$$

the volumetric strain.

P.1.8. Show that the compatibility equation for the case of plane strain, namely,

$$\frac{\partial^2 \gamma_{xy}}{\partial x\, \partial y} = \frac{\partial^2 \varepsilon_y}{\partial x^2} + \frac{\partial^2 \varepsilon_x}{\partial y^2}$$

may be expressed in terms of direct stresses σ_x and σ_y in the form

$$\left(\frac{\partial^2}{\partial x^2} + \frac{\partial^2}{\partial y^2} \right)(\sigma_x + \sigma_y) = 0$$

P.1.9. A bar of mild steel has a diameter of 75 mm and is placed inside a hollow aluminum cylinder of internal diameter 75 mm and external diameter 100 mm; both bar and cylinder are the same length. The resulting composite bar is subjected to an axial compressive load of 1000 kN. If the bar and cylinder contract by the same amount, calculate the stress in each. The temperature of the compressed composite bar is then reduced by 150°C but no change in length is permitted. Calculate the final stress in the bar and in the cylinder if E (steel) = 200,000 N/mm^2, E (aluminum) = 80,000 N/mm^2, α (steel) = 0.000012/°C and α (aluminum) = 0.000005/°C.

Answer: Due to load: σ (steel) = 172.6 N/mm^2 (compression)
   σ (aluminum) = 69.1 N/mm^2 (compression).
 Final stress: σ (steel) = 187.4 N/mm^2 (tension)
   σ (aluminum) = 9.1 N/mm^2 (compression).

P.1.9 MATLAB Repeat P.1.9 by creating a script in MATLAB using the Symbolic Math Toolbox for an axial tension load of 1000 kN.

Answer: Due to load: σ (steel) = 172.6 N/mm^2 (tension)
   σ (aluminum) = 69.1 N/mm^2 (tension)
 Final stress: σ (steel) = 532.6 N/mm^2 (tension)
   σ (aluminum) = 129.1 N/mm^2 (tension)

P.1.10. In Fig. P.1.10, the direct strains in the directions a, b, c are –0.002, –0.002, and +0.002, respectively. If I and II denote principal directions, find ε_{I}, $\varepsilon_{\mathrm{II}}$, and θ.

Answer: $\varepsilon_{\mathrm{I}} = +0.00283$, $\varepsilon_{\mathrm{II}} = -0.00283$, $\theta = -22.5°$ or $+67.5°$

P.1.11. The simply supported rectangular beam shown in Fig. P.1.11 is subjected to two symmetrically placed transverse loads each of magnitude Q. A rectangular strain gauge rosette located at a point P on the centroidal axis on one vertical face of the beam gives strain readings as follows: $\varepsilon_a = -222 \times 10^{-6}$, $\varepsilon_b = -213 \times 10^{-6}$, and $\varepsilon_c = +45 \times 10^{-6}$. The longitudinal stress σ_x at the point P due to an external

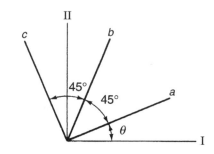

FIGURE P.1.10

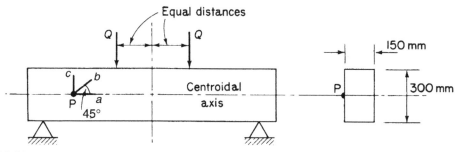

FIGURE P.1.11

compressive force is 7 N/mm^2. Calculate the shear stress τ at the point P in the vertical plane and hence the transverse load Q:

$$(Q = 2bd\tau/3, \quad \text{where } b = \text{breadth}, \ d = \text{depth of beam})$$
$$E = 31,000 \text{ N/mm}^2 \quad \nu = 0.2$$

Answer: $\tau_{xy} = 3.16 \text{ N/mm}^2, \quad Q = 94.8 \text{ kN}$

P.1.11. MATLAB Repeat P.1.11 by creating a script in MATLAB using the Symbolic Math Toolbox for the following cross-section dimensions and longitudinal stress combinations:

	b	d	σ_x
(i)	150 mm	300 mm	7 N/mm^2
(ii)	100 mm	250 mm	11 N/mm^2
(iii)	270 mm	270 mm	9 N/mm^2

Answer: (i) $\tau_{xy} = 3.16 \text{ N/mm}^2$ $Q = 94.8 \text{ kN}$
(ii) $\tau_{xy} = 2.5 \text{ N/mm}^2$ $Q = 41.7 \text{ kN}$
(iii) $\tau_{xy} = 1.41 \text{ N/mm}^2$ $Q = 68.5 \text{ kN}$

Two-dimensional problems in elasticity

Theoretically, we are now in a position to solve any three-dimensional problem in elasticity having derived three equilibrium conditions, Eqs. (1.5); six strain–displacement equations, Eqs. (1.18) and (1.20); and six stress–strain relationships, Eqs. (1.42) and (1.46). These equations are sufficient, when supplemented by appropriate boundary conditions, to obtain unique solutions for the six stress, six strain, and three displacement functions. It is found, however, that exact solutions are obtainable only for some simple problems. For bodies of arbitrary shape and loading, approximate solutions may be found by numerical methods (e.g., finite differences) or by the Rayleigh–Ritz method based on energy principles (Chapter 7).

Two approaches are possible in the solution of elasticity problems. We may solve initially for either the three unknown displacements or the six unknown stresses. In the former method, the equilibrium equations are written in terms of strain by expressing the six stresses as functions of strain (see Problem P.1.7). The strain–displacement relationships are then used to form three equations involving the three displacements u, v, and w. The boundary conditions for this method of solution must be specified as displacements. Determination of u, v, and w enables the six strains to be computed from Eqs. (1.18) and (1.20); the six unknown stresses follow from the equations expressing stress as functions of strain. It should be noted here that no use has been made of the compatibility equations. The fact that u, v, and w are determined directly ensures that they are single-valued functions, thereby satisfying the requirement of compatibility.

In most structural problems the object is usually to find the distribution of stress in an elastic body produced by an external loading system. It is therefore more convenient in this case to determine the six stresses before calculating any required strains or displacements. This is accomplished by using Eqs. (1.42) and (1.46) to rewrite the six equations of compatibility in terms of stress. The resulting equations, in turn, are simplified by making use of the stress relationships developed in the equations of equilibrium. The solution of these equations automatically satisfies the conditions of compatibility and equilibrium throughout the body.

2.1 TWO-DIMENSIONAL PROBLEMS

For the reasons discussed in Chapter 1, we shall confine our actual analysis to the two-dimensional cases of plane stress and plane strain. The appropriate equilibrium conditions for plane stress are given by Eqs. (1.6); that is,

$$\frac{\partial \sigma_x}{\partial x} + \frac{\partial \tau_{xy}}{\partial y} + X = 0$$

$$\frac{\partial \sigma_y}{\partial y} + \frac{\partial \tau_{yx}}{\partial x} + Y = 0$$

and the required stress–strain relationships obtained from Eqs. (1.47):

$$\varepsilon_x = \frac{1}{E}(\sigma_x - \nu\sigma_y)$$

$$\varepsilon_y = \frac{1}{E}(\sigma_y - \nu\sigma_x)$$

$$\gamma_{xy} = \frac{2(1 + \nu)}{E}\tau_{xy}$$

We find that, although ε_z exists, Eqs. (1.22)–(1.26) are identically satisfied, leaving Eq. (1.21) as the required compatibility condition. Substitution in Eq. (1.21) of the preceding strains gives

$$2(1 + \nu)\frac{\partial^2 \tau_{xy}}{\partial x\,\partial y} = \frac{\partial^2}{\partial x^2}(\sigma_y - \nu\sigma_x) + \frac{\partial^2}{\partial y^2}(\sigma_x - \nu\sigma_y) \tag{2.1}$$

From Eqs. (1.6),

$$\frac{\partial^2 \tau_{xy}}{\partial y\,\partial x} = -\frac{\partial^2 \sigma_x}{\partial x^2} - \frac{\partial X}{\partial x} \tag{2.2}$$

and

$$\frac{\partial^2 \tau_{xy}}{\partial x\,\partial y} = -\frac{\partial^2 \sigma_y}{\partial y^2} - \frac{\partial Y}{\partial y} \quad (\tau_{yx} = \tau_{xy}) \tag{2.3}$$

Adding Eqs. (2.2) and (2.3), then substituting in Eq. (2.1) for $2\partial^2\tau_{xy}/\partial_x\partial_y$, we have

$$-(1 + \nu)\left(\frac{\partial X}{\partial x} + \frac{\partial Y}{\partial y}\right) = \frac{\partial^2 \sigma_x}{\partial x^2} + \frac{\partial^2 \sigma_y}{\partial y^2} + \frac{\partial^2 \sigma_y}{\partial x^2} + \frac{\partial^2 \sigma_x}{\partial y^2}$$

or

$$\left(\frac{\partial^2}{\partial x^2} + \frac{\partial^2}{\partial y^2}\right)(\sigma_x + \sigma_y) = -(1 + \nu)\left(\frac{\partial X}{\partial x} + \frac{\partial Y}{\partial y}\right) \tag{2.4}$$

The alternative two-dimensional problem of plane strain may also be formulated in the same manner. We saw in Section 1.11 that the six equations of compatibility reduce to the single equation (1.21) for the plane strain condition. Further, from the third of Eqs. (1.42),

$$\sigma_z = \nu(\sigma_x + \sigma_y) \qquad \text{(since } \varepsilon_z = 0 \text{ for plane strain)}$$

so that

$$\varepsilon_x = \frac{1}{E}\left[(1 - \nu^2)\sigma_x - \nu(1 + \nu)\sigma_y\right]$$

and

$$\varepsilon_y = \frac{1}{E}\left[(1 - v^2)\sigma_y - v(1 + v)\sigma_x\right]$$

Also,

$$\gamma_{xy} = \frac{2(1 + v)}{E}\tau_{xy}$$

Substituting as before in Eq. (1.21) and simplifying by use of the equations of equilibrium, we have the compatibility equation for plane strain:

$$\left(\frac{\partial^2}{\partial x^2} + \frac{\partial^2}{\partial y^2}\right)(\sigma_x + \sigma_y) = -\frac{1}{1 - v}\left(\frac{\partial X}{\partial x} + \frac{\partial Y}{\partial y}\right) \tag{2.5}$$

The two equations of equilibrium together with the boundary conditions, from Eqs. (1.7), and one of the compatibility equations, (2.4) or (2.5), are generally sufficient for the determination of the stress distribution in a two-dimensional problem.

2.2 STRESS FUNCTIONS

The solution of problems in elasticity presents difficulties, but the procedure may be simplified by the introduction of a *stress function*. For a particular two-dimensional case, the stresses are related to a single function of x and y such that substitution for the stresses in terms of this function automatically satisfies the equations of equilibrium no matter what form the function may take. However, a large proportion of the infinite number of functions that fulfill this condition are eliminated by the requirement that the form of the stress function must also satisfy the two-dimensional equations of compatibility, (2.4) and (2.5), plus the appropriate boundary conditions.

For simplicity, let us consider the two-dimensional case for which the body forces are zero. The problem is now to determine a stress–stress function relationship that satisfies the equilibrium conditions of

$$\left.\begin{array}{l} \dfrac{\partial \sigma_x}{\partial x} + \dfrac{\partial \tau_{xy}}{\partial y} = 0 \\[3mm] \dfrac{\partial \sigma_y}{\partial y} + \dfrac{\partial \tau_{yx}}{\partial x} = 0 \end{array}\right\} \tag{2.6}$$

and a form for the stress function giving stresses which satisfy the compatibility equation

$$\left(\frac{\partial^2}{\partial x^2} + \frac{\partial^2}{\partial y^2}\right)(\sigma_x + \sigma_y) = 0 \tag{2.7}$$

The English mathematician Airy proposed a stress function ϕ defined by the equations

$$\sigma_x = \frac{\partial^2 \phi}{\partial y^2}, \quad \sigma_y = \frac{\partial^2 \phi}{\partial x^2}, \quad \tau_{xy} = -\frac{\partial^2 \phi}{\partial x\, \partial y} \tag{2.8}$$

Clearly, substitution of Eqs. (2.8) into Eq. (2.6) verifies that the equations of equilibrium are satisfied by this particular stress–stress function relationship. Further substitution into Eq. (2.7) restricts the possible forms of the stress function to those satisfying the *biharmonic equation*:

$$\frac{\partial^4 \phi}{\partial x^4} + 2 \frac{\partial^4 \phi}{\partial x^2 \partial y^2} + \frac{\partial^4 \phi}{\partial y^4} = 0 \tag{2.9}$$

The final form of the stress function is then determined by the boundary conditions relating to the actual problem. Therefore, a two-dimensional problem in elasticity with zero body forces reduces to the determination of a function ϕ of x and y which satisfies Eq. (2.9) at all points in the body and Eqs. (1.7) reduced to two dimensions at all points on the boundary of the body.

2.3 INVERSE AND SEMI-INVERSE METHODS

The task of finding a stress function satisfying the preceding conditions is extremely difficult in the majority of elasticity problems, although some important classical solutions have been obtained in this way. An alternative approach, known as the *inverse method*, is to specify a form of the function ϕ satisfying Eq. (2.9), assume an arbitrary boundary, and then determine the loading conditions that fit the assumed stress function and chosen boundary. Obvious solutions arise in which ϕ is expressed as a polynomial. Timoshenko and Goodier[1] consider a variety of polynomials for ϕ and determine the associated loading conditions for a variety of rectangular sheets. Some of these cases are quoted here.

Example 2.1

Consider the stress function

$$\phi = Ax^2 + Bxy + Cy^2$$

where A, B, and C are constants. Equation (2.9) is identically satisfied, since each term becomes zero on substituting for ϕ. The stresses follow from

$$\sigma_x = \frac{\partial^2 \phi}{\partial y^2} = 2C$$

$$\sigma_y = \frac{\partial^2 \phi}{\partial x^2} = 2A$$

$$\tau_{xy} = -\frac{\partial^2 \phi}{\partial x\, \partial y} = -B$$

To produce these stresses at any point in a rectangular sheet, we require loading conditions providing the boundary stresses shown in Fig. 2.1.

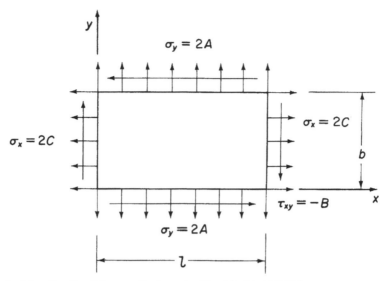

FIGURE 2.1 Required Loading Conditions on Rectangular Sheet in Example 2.1

Example 2.2

A more complex polynomial for the stress function is

$$\phi = \frac{Ax^3}{6} + \frac{Bx^2y}{2} + \frac{Cxy^2}{2} + \frac{Dy^3}{6}$$

As before,

$$\frac{\partial^4 \phi}{\partial x^4} = \frac{\partial^4 \phi}{\partial x^2 \partial y^2} = \frac{\partial^4 \phi}{\partial y^4} = 0$$

so that the compatibility equation (2.9) is identically satisfied. The stresses are given by

$$\sigma_x = \frac{\partial^2 \phi}{\partial y^2} = Cx + Dy$$

$$\sigma_y = \frac{\partial^2 \phi}{\partial x^2} = Ax + By$$

$$\tau_{xy} = -\frac{\partial^2 \phi}{\partial x \, \partial y} = -Bx - Cy$$

We may choose any number of values of the coefficients A, B, C, and D to produce a variety of loading conditions on a rectangular plate. For example, if we assume $A = B = C = 0$, then $\sigma_x = Dy$, $\sigma_y = 0$, and $\tau_{xy} = 0$, so that, for axes referred to an origin at the mid-point of a vertical side of the plate, we obtain the state of pure bending shown in Fig. 2.2(a). Alternatively, Fig. 2.2(b) shows the loading conditions corresponding to $A = C = D = 0$ in which $\sigma_x = 0$, $\sigma_y = By$, and $\tau_{xy} = -Bx$.

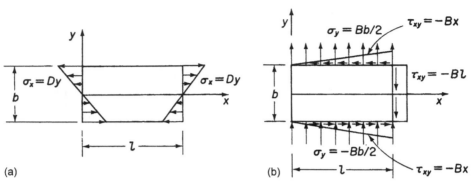

FIGURE 2.2 **(a) Required Loading Conditions on Rectangular Sheet in Example 2.2 for $A = B = C = 0$; (b) as in (a) but $A = C = D = 0$**

By assuming polynomials of the second or third degree for the stress function, we ensure that the compatibility equation is identically satisfied whatever the values of the coefficients. For polynomials of higher degrees, compatibility is satisfied only if the coefficients are related in a certain way. For example, for a stress function in the form of a polynomial of the fourth degree,

$$\phi = \frac{Ax^4}{12} + \frac{Bx^3y}{6} + \frac{Cx^2y^2}{2} + \frac{Dxy^3}{6} + \frac{Ey^4}{12}$$

and

$$\frac{\partial^4\phi}{\partial x^4} = 2A, \quad 2\frac{\partial^4\phi}{\partial x^2\,\partial y^2} = 4C, \quad \frac{\partial^4\phi}{\partial y^4} = 2E$$

Substituting these values in Eq. (2.9), we have

$$E = -(2C + A)$$

The stress components are then

$$\sigma_x = \frac{\partial^2\phi}{\partial y^2} = Cx^2 + Dxy - (2C + A)y^2$$

$$\sigma_y = \frac{\partial^2\phi}{\partial x^2} = Ax^2 + Bxy + Cy^2$$

$$\tau_{xy} = -\frac{\partial^2\phi}{\partial y} = -\frac{Bx^2}{2} - 2Cxy - \frac{Dy^2}{2}$$

The coefficients A, B, C, and D are arbitrary and may be chosen to produce various loading conditions, as in the previous examples.

Example 2.3

A cantilever of length L and depth $2h$ is in a state of plane stress. The cantilever is of unit thickness is rigidly supported at the end $x = L$, and loaded as shown in Fig. 2.3. Show that the stress function

$$\phi = Ax^2 + Bx^2 y + Cy^3 + D(5x^2 y^3 - y^5)$$

is valid for the beam and evaluate the constants A, B, C, and D. See Ex. 1.1.

The stress function must satisfy Eq. (2.9). From the expression for ϕ,

$$\frac{\partial \phi}{\partial x} = 2Ax + 2Bxy + 10Dxy^3$$

$$\frac{\partial^2 \phi}{\partial x^2} = 2A + 2By + 10Dy^3 = \sigma_y \tag{i}$$

Also,

$$\frac{\partial \phi}{\partial y} = Bx^2 + 3Cy^2 + 15Dx^2 y^2 - 5Dy^4$$

$$\frac{\partial^2 \phi}{\partial y^2} = 6Cy + 30Dx^2 y - 20Dy^3 = \sigma_x \tag{ii}$$

and

$$\frac{\partial^2 \phi}{\partial x \partial y} = 2Bx + 30Dxy^2 = -\tau_{xy} \tag{iii}$$

Further,

$$\frac{\partial^4 \phi}{\partial x^4} = 0, \quad \frac{\partial^4 \phi}{\partial y^4} = -120Dy, \quad \frac{\partial^4 \phi}{\partial x^2 \partial y^2} = 60Dy$$

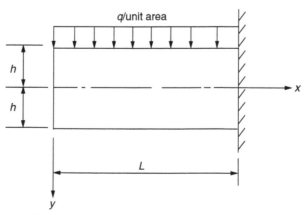

FIGURE 2.3 Beam of Example 2.3

Substituting in Eq. (2.9) gives

$$\frac{\partial^4 \phi}{\partial x^4} + 2\frac{\partial^4 \phi}{\partial x^2 \partial y^2} + \frac{\partial^4 \phi}{\partial y^4} = 2 \times 60Dy - 120Dy = 0$$

The biharmonic equation is therefore satisfied and the stress function is valid.

From Fig. 2.3, $\sigma_y = 0$ at $y = h$, so that, from Eq. (i),

$$2A + 2BH + 10Dh^3 = 0 \tag{iv}$$

Also, from Fig. 2.3, $v_y = -q$ at $y = -h$, so that, from Eq. (i),

$$2A - 2BH - 10Dh^3 = -q \tag{v}$$

Again, from Fig. 2.3, $\tau_{xy} = 0$ at $y = \pm h$, giving, from Eq. (iii),

$$2Bx + 30Dxh^2 = 0$$

so that

$$2B + 30Dh^2 = 0 \tag{vi}$$

At $x = 0$, no resultant moment is applied to the beam; that is,

$$M_{x=0} = \int_{-h}^{h} \sigma_x y \, \mathrm{d}y = \int_{-h}^{h} (6Cy^2 - 20Dy^4) \, \mathrm{d}y = 0$$

so that

$$M_{x=0} = \left[2Cy^3 - 4Dy^5 \right]_{-h}^{h} = 0$$

or

$$C - 2Dh^2 = 0 \tag{vii}$$

Subtracting Eq. (v) from (iv),

$$4Bh + 20Dh^3 = q$$

or

$$B + 5Dh^2 = \frac{q}{4h} \tag{viii}$$

From Eq. (vi),

$$B + 15Dh^2 = 0 \tag{ix}$$

so that, subtracting Eq. (viii) from Eq. (ix),

$$D = -\frac{q}{40h^3}$$

Then,

$$B = \frac{3q}{8h}, \quad A = -\frac{q}{4}, \quad C = -\frac{q}{20h}$$

and

$$\phi = \frac{q}{40h^3}\left[-10h^3x^2 + 15h^2x^2y - 2h^2y^3 - (5x^2y^3 - y^5)\right]$$

The obvious disadvantage of the inverse method is that we are determining problems to fit assumed solutions, whereas in structural analysis the reverse is the case. However, in some problems, the shape of the body and the applied loading allow simplifying assumptions to be made, thereby enabling a solution to be obtained. St. Venant suggested a *semi-inverse method* for the solution of this type of problem, in which assumptions are made as to stress or displacement components. These assumptions may be based on experimental evidence or intuition. St. Venant first applied the method to the torsion of solid sections (Chapter 3) and to the problem of a beam supporting shear loads (Section 2.6).

Example 2.3 MATLAB

Repeat Example 2.3 using the Symbolic Math Toolbox in MATLAB. See Ex. 1.1.

Using Fig. 2.3, derivations of the constants A, B, C, and D, along with validation of the stress function ϕ are obtained through the following MATLAB file:

```
% Declare any needed symbolic variables
syms A B C D x y L h q sig_y sig_x tau_xy

% Define the given stress function
phi = A*x^2 + B*x^2*y + C*y^3 + D*(5*x^2*y^3 - y^5);

% Check Eq. (2.9)
check = diff(phi,x,4) + 2*diff(diff(phi,x,2),y,2) + diff(phi,y,4);
if double(check) == 0
   disp('The stress function is valid')
   disp(' ')

% Calculate expressions for sig_y, sig_x, and tau_xy
eqI = diff(phi,x,2) - sig_y;
eqII = diff(phi,y,2) - sig_x;
eqIII = diff(diff(phi,x),y) + tau_xy;

% From Fig. 2.3, sig_y=0 at y=h so that from eqI
eqIV = subs(subs(eqI,y,h),sig_y,0);

% From Fig. 2.3, sig_y=-q at y=-h so that from eqI
eqV = subs(subs(eqI,y,-h),sig_y,-q);

% From Fig. 2.3, tau_xy=0 at y=+/-h so that from eqIII
eqVI = subs(subs(eqIII,y,h),tau_xy,0);

% Calculate the expression of the applied moment
M = int(solve(eqII,sig_x)*y, y, -h, h);
```

```
% From Fig. 2.3, M=0 at x=0
eqVII = subs(M,x,0);

% Solve eqIV, eqV, eqVI, and eqVII for A, B, C, and D
D_expr = solve(subs(eqIV-eqV,B,solve(eqVI,B)), D);
C_expr = solve(subs(eqVII,D,D_expr), C);
B_expr = solve(subs(eqVI,D,D_expr), B);
A_expr = solve(subs(subs(eqIV,D,D_expr),B,B_expr), A);

% Substitute the expressions for A, B, C, and D into phi
phi = subs(subs(subs(subs(phi,A,A_expr),B,B_expr),C,C_expr),D,D_expr);

% Output the expression for phi to the Command Window
phi = factor(phi)
else
 disp('The stress function does not satisfy the biharmonic equation (Eq. (2.9))')
 disp(' ')
end
```

The Command Window outputs resulting from this MATLAB file are as follows.

The stress function is valid.

$$phi = -(q*(10*h^3*x^2 - 15*h^2*x^2*y + 2*h^2*y^3 + 5*x^2*y^3 - y^5))/(40*h^3)$$

2.4 ST. VENANT'S PRINCIPLE

In the examples of Section 2.3, we have seen that a particular stress function form may be applicable to a variety of problems. Different problems are deduced from a given stress function by specifying, in the first instance, the shape of the body then assigning a variety of values to the coefficients. The resulting stress functions give stresses which satisfy the equations of equilibrium and compatibility *at all points within and on the boundary of the body*. It follows that the applied loads must be distributed around the boundary of the body in the same manner as the internal stresses at the boundary. In the case of pure bending, for example (Fig. 2.2(a)), the applied bending moment must be produced by tensile and compressive forces on the ends of the plate, their magnitudes being dependent on their distance from the neutral axis. If this condition is invalidated by the application of loads in an arbitrary fashion or by preventing the free distortion of any section of the body, then the solution of the problem is no longer exact. As this is the case in practically every structural problem, it would appear that the usefulness of the theory is strictly limited. To surmount this obstacle, we turn to the important *principle of St. Venant*, which may be summarized as stating

> *that while statically equivalent systems of forces acting on a body produce substantially different local effects, the stresses at sections distant from the surface of loading are essentially the same.*

Therefore, at a section AA close to the end of a beam supporting two point loads *P*, the stress distribution varies as shown in Fig. 2.4, while at the section BB, a distance usually taken to be greater than the dimension of the surface to which the load is applied, the stress distribution is uniform.

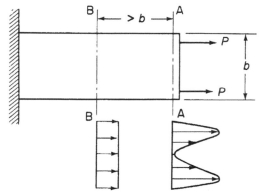

FIGURE 2.4 Stress Distributions Illustrating St. Venant's Principle

We may therefore apply the theory to sections of bodies away from points of applied loading or constraint. The determination of stresses in these regions requires, for some problems, separate calculation.

2.5 DISPLACEMENTS

Having found the components of stress, Eqs. (1.47) (for the case of plane stress) are used to determine the components of strain. The displacements follow from Eqs. (1.27) and (1.28). The integration of Eqs. (1.27) yields solutions of the form

$$u = \varepsilon_x x + a - by \tag{2.10}$$

$$v = \varepsilon_y y + c + bx \tag{2.11}$$

in which a, b, and c are constants representing movement of the body as a whole or *rigid body* displacements. Of these, a and c represent pure translatory motions of the body, while b is a small angular rotation of the body in the xy plane. If we assume that b is positive in a counterclockwise sense, then in Fig. 2.5, the displacement v' due to the rotation is given by

$$v' = P'Q' - PQ$$
$$= OP \, \sin(\theta + b) - OP \sin\theta$$

which, since b is a small angle, reduces to

$$v' = bx$$

Similarly,

$$u' = -by, \text{ as stated}$$

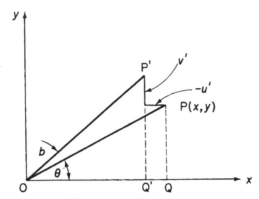

FIGURE 2.5 Displacements Produced by Rigid Body Rotation

2.6 BENDING OF AN END-LOADED CANTILEVER

In his semi-inverse solution of this problem, St. Venant based his choice of stress function on the reasonable assumptions that the direct stress is directly proportional to the bending moment (and therefore distance from the free end) and height above the neutral axis. The portion of the stress function giving shear stress follows from the equilibrium condition relating σ_x and τ_{xy}. The appropriate stress function for the cantilever beam shown in Fig. 2.6 is then

$$\phi = Axy + \frac{Bxy^3}{6} \qquad (2.12)$$

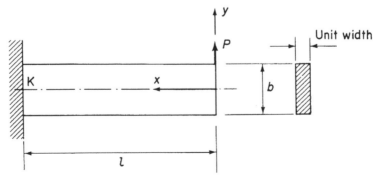

FIGURE 2.6 Bending of an End-Loaded Cantilever

where A and B are unknown constants. Hence,

$$
\left.\begin{array}{l}
\sigma_x = \dfrac{\partial^2 \phi}{\partial y^2} = Bxy \\[2mm]
\sigma_y = \dfrac{\partial^2 \phi}{\partial x^2} = 0 \\[2mm]
\tau_{xy} = -\dfrac{\partial^2 \phi}{\partial x\,\partial y} = -A - \dfrac{By^2}{2}
\end{array}\right\}
\tag{2.13}
$$

Substitution for ϕ in the biharmonic equation shows that the form of the stress function satisfies compatibility for all values of the constants A and B. The actual values of A and B are chosen to satisfy the boundary condition, namely, $\tau_{xy} = 0$ along the upper and lower edges of the beam, and the resultant shear load over the free end is equal to P.

From the first of these,

$$
\tau_{xy} = -A - \frac{By^2}{2} = 0 \text{ at } y = \pm\frac{b}{2}
$$

giving

$$
A = -\frac{Bb^2}{8}
$$

From the second,

$$
-\int_{-b/2}^{b/2} \tau_{xy}\,dy = P \ (\text{see sign convention for } \tau_{xy})
$$

or

$$
-\int_{-b/2}^{b/2} \left(\frac{Bb^2}{8} - \frac{By^2}{2}\right) dy = P
$$

from which

$$
B = \frac{12P}{b^3}
$$

The stresses follow from Eqs. (2.13):

$$
\left.\begin{array}{l}
\sigma_x = -\dfrac{12Pxy}{b^3} = -\dfrac{Px}{I}y \\[2mm]
\sigma_y = 0 \\[2mm]
\tau_{xy} = -\dfrac{12P}{8b^3}(b^2 - 4y^2) = -\dfrac{P}{8I}(b^2 - 4y^2)
\end{array}\right\}
\tag{2.14}
$$

where $I = b^3/12$, the second moment of area of the beam cross-section.

We note from the discussion of Section 2.4 that Eqs. (2.14) represent an exact solution subject to the following conditions:

1. The shear force P is distributed over the free end in the same manner as the shear stress τ_{xy} given by Eqs. (2.14);
2. The distribution of shear and direct stresses at the built-in end is the same as those given by Eqs. (2.14);
3. All sections of the beam, including the built-in end, are free to distort.

In practical cases, none of these conditions is satisfied, but by virtue of St. Venant's principle, we may assume that the solution is exact for regions of the beam away from the built-in end and the applied load. For many solid sections, the inaccuracies in these regions are small. However, for thin-walled structures, with which we are primarily concerned, significant changes occur and we consider the effects of structural and loading discontinuities on this type of structure in Chapters 26 and 27.

We now proceed to determine the displacements corresponding to the stress system of Eqs. (2.14). Applying the strain–displacement and stress–strain relationships, Eqs. (1.27), (1.28), and (1.47), we have

$$\varepsilon_x = \frac{\partial u}{\partial x} = \frac{\sigma_x}{E} = -\frac{Pxy}{EI} \tag{2.15}$$

$$\varepsilon_y = \frac{\partial v}{\partial y} = -\frac{\nu\sigma_x}{E} = \frac{\nu Pxy}{EI} \tag{2.16}$$

$$\gamma_{xy} = \frac{\partial u}{\partial y} + \frac{\partial v}{\partial x} = \frac{\tau_{xy}}{G} = -\frac{P}{8IG}(b^2 - 4y^2) \tag{2.17}$$

Integrating Eqs. (2.15) and (2.16) and noting that ε_x and ε_y are partial derivatives of the displacements, we find

$$u = -\frac{Px^2y}{2EI} + f_1(y), \quad v = \frac{\nu Pxy^2}{2EI} + f_2x \tag{2.18}$$

where $f_1(y)$ and $f_2(x)$ are unknown functions of x and y. Substituting these values of u and v in Eq. (2.17),

$$-\frac{Px^2}{2EI} + \frac{\partial f_1(y)}{\partial y} + \frac{\nu Py^2}{2EI} + \frac{\partial f_2(x)}{\partial x} = -\frac{P}{8IG}(b^2 - 4y^2)$$

Separating the terms containing x and y in this equation and writing

$$F_1(x) = -\frac{Px^2}{2EI} + \frac{\partial f_2(x)}{\partial x} \quad F_2(y) = \frac{\nu Py^2}{2EI} - \frac{Py^2}{2IG} + \frac{\partial f_1(y)}{\partial y}$$

we have

$$F_1(x) + F_2(y) = -\frac{Pb^2}{8IG}$$

The term on the right-hand side of this equation is a constant, which means that $F_1(x)$ and $F_2(y)$ must be constants; otherwise, a variation in either x or y destroys the equality. Denoting $F_1(x)$ by C and $F_2(y)$ by D gives

$$C + D = -\frac{Pb^2}{8IG} \tag{2.19}$$

and

$$\frac{\partial f_2(x)}{\partial x} = \frac{Px^2}{2EI} + C \qquad \frac{\partial f_1(y)}{\partial y} = \frac{Py^2}{2IG} - \frac{vPy^2}{2EI} + D$$

so that

$$f_2(x) = \frac{Px^3}{6EI} + Cx + F$$

and

$$f_1(y) = \frac{Py^3}{6IG} - \frac{vPy^3}{6EI} + Dy + H$$

Therefore, from Eqs. (2.18),

$$u = -\frac{Px^2y}{2EI} - \frac{vPy^3}{6EI} + \frac{Py^3}{6IG} + Dy + H \tag{2.20}$$

$$v = \frac{vPxy^2}{2EI} + \frac{Px^3}{6EI} + Cx + F \tag{2.21}$$

The constants C, D, F, and H are now determined from Eq. (2.19) and the displacement boundary conditions imposed by the support system. Assuming that the support prevents movement of the point K in the beam cross-section at the built-in end, $u = v = 0$ at $x = l$, $y = 0$, and from Eqs. (2.20) and (2.21),

$$H = 0, \quad F = -\frac{Pl^3}{6EI} - Cl$$

If we now assume that the slope of the neutral plane is zero at the built-in end, then $\partial v/\partial x = 0$ at $x = l$, $y = 0$ and, from Eq. (2.21),

$$C = -\frac{Pl^2}{2EI}$$

It follows immediately that

$$F = \frac{Pl^3}{2EI}$$

and, from Eq. (2.19),

$$D = \frac{Pl^2}{2EI} - \frac{Pb^2}{8IG}$$

Substitution for the constants C, D, F, and H in Eqs. (2.20) and (2.21) now produces the equations for the components of displacement at any point in the beam:

$$u = -\frac{Px^2y}{2EI} - \frac{vPy^3}{6EI} + \frac{Py^3}{6IG} + \left(\frac{Pl^2}{2EI} - \frac{Pb^2}{8IG}\right)y \tag{2.22}$$

$$v = \frac{vPxy^2}{2EI} + \frac{Px^3}{6EI} - \frac{Pl^2x}{2EI} + \frac{Pl^3}{3EI} \tag{2.23}$$

The deflection curve for the neutral plane is

$$(v)_{y=0} = \frac{Px^3}{6EI} - \frac{Pl^2x}{2EI} + \frac{Pl^3}{3EI} \tag{2.24}$$

from which the tip deflection ($x = 0$) is $Pl^3/3EI$. This value is predicted by simple beam theory (Chapter 15) and does not include the contribution to deflection of the shear strain. This was eliminated when we assumed that the slope of the neutral plane at the built-in end was zero. A more detailed examination of this effect is instructive. The shear strain at any point in the beam is given by Eq. (2.17):

$$\gamma_{xy} = -\frac{P}{8IG}(b^2 - 4y^2)$$

and is obviously independent of x. Therefore, at all points on the neutral plane, the shear strain is constant and equal to

$$\gamma_{xy} = -\frac{Pb^2}{8IG}$$

which amounts to a rotation of the neutral plane, as shown in Fig. 2.7. The deflection of the neutral plane due to this shear strain at any section of the beam is therefore equal to

$$\frac{Pb^2}{8IG}(l - x)$$

and Eq. (2.24) may be rewritten to include the effect of shear as

$$(v)_{y=0} = \frac{Px^3}{6EI} - \frac{Pl^2x}{2EI} + \frac{Pl^3}{3EI} + \frac{Pb^2}{8IG}(l - x) \tag{2.25}$$

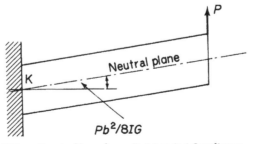

FIGURE 2.7 Rotation of Neutral Plane Due to Shear in an End-Loaded Cantilever

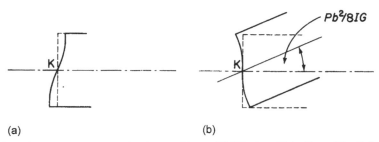

FIGURE 2.8 (a) Distortion of the Cross-section Due to Shear; (b) Effect on Distortion of the Rotation Due to Shear

Let us now examine the distorted shape of the beam section, which the analysis assumes is free to take place. At the built-in end, when $x = l$, the displacement of any point is, from Eq. (2.22),

$$u = \frac{vPy^3}{6EI} + \frac{Py^3}{6IG} - \frac{Pb^2y}{8IG} \tag{2.26}$$

The cross-section would, therefore, if allowed, take the shape of the shallow reversed S shown in Fig. 2.8(a). We have not included in Eq. (2.26) the previously discussed effect of rotation of the neutral plane caused by shear. However, this merely rotates the beam section as indicated in Fig. 2.8(b).

The distortion of the cross-section is produced by the variation of shear stress over the depth of the beam. Therefore, the basic assumption of simple beam theory that plane sections remain plane is not valid when shear loads are present, although for long, slender beams, bending stresses are much greater than shear stresses and the effect may be ignored.

We observe from Fig. 2.8 that an additional direct stress system is imposed on the beam at the support, where the section is constrained to remain plane. For most engineering structures, this effect is small but, as mentioned previously, may be significant in thin-walled sections.

Reference
[1] Timoshenko S, Goodier JN. Theory of elasticity. 2nd ed. New York: McGraw-Hill; 1951.

PROBLEMS

P.2.1 A metal plate has rectangular axes Ox, Oy marked on its surface. The point O and the direction of Ox are fixed in space and the plate is subjected to the following uniform stresses:

Compressive, $3p$, parallel to Ox,

Tensile, $2p$, parallel to Oy,

Shearing, $4p$, in planes parallel to Ox and Oy, in a sense tending to decrease the angle xOy.

Determine the direction in which a certain point on the plate is displaced; the coordinates of the point are (2, 3) before straining. Poisson's ratio is 0.25.

Answer: 19.73° to Ox.

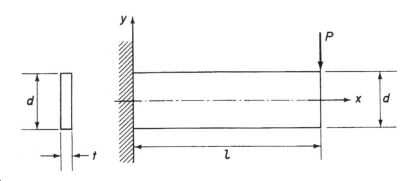

FIGURE P.2.2

P.2.2 What do you understand by an Airy stress function in two dimensions? A beam of length l, with a thin rectangular cross-section, is built in at the end $x = 0$ and loaded at the tip by a vertical force P (Fig. P.2.2). Show that the stress distribution, as calculated by simple beam theory, can be represented by the expression

$$\phi = Ay^3 + By^3x + Cyx$$

as an Airy stress function and determine the coefficients A, B, and C.

   Answer: $A = 2Pl/td^3$, $B = -2P/td^3$, $C = 3P/2td$

P.2.3 The cantilever beam shown in Fig. P.2.3 is in a state of plane strain and is rigidly supported at $x = L$. Examine the following stress function in relation to this problem:

$$\phi = \frac{w}{20h^3} \left(15h^2x^2y - 5x^2y^3 - 2h^2y^3 + y^5\right)$$

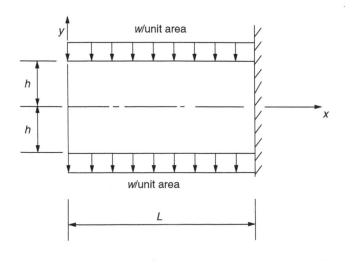

FIGURE P.2.3

Show that the stresses acting on the boundaries satisfy the conditions except for a distributed direct stress at the free end of the beam, which exerts no resultant force or bending moment.

Answer: The stress function satisfies the biharmonic equation:
- At $y = h$, $\sigma_y = w$ and $\tau_{xy} = 0$, boundary conditions satisfied,
- At $y = -h$, $\sigma_y = -w$ and $\tau_{xy} = 0$, boundary conditions satisfied.

Direct stress at free end of beam is not zero, there is no resultant force or bending moment at the free end.

P.2.4 A thin rectangular plate of unit thickness (Fig. P.2.4) is loaded along the edge $y = +d$ by a linearly varying distributed load of intensity $w = px$ with corresponding equilibrating shears along the vertical edges at $x = 0$ and l. As a solution to the stress analysis problem, an Airy stress function ϕ is proposed, where

$$\phi = \frac{p}{120d^3}\left[5(x^3 - l^2x)(y + d)^2(y - 2d) - 3yx(y^2 - d^2)^2\right]$$

Show that ϕ satisfies the internal compatibility conditions and obtain the distribution of stresses within the plate. Determine also the extent to which the static boundary conditions are satisfied.

Answer: $\sigma_x = \dfrac{px}{20d^3}\left[5y(x^2 - l^2) - 10y^3 + 6d^2y\right]$

$\sigma_y = \dfrac{px}{4d^3}\left(y^3 - 3yd^2 - 2d^3\right)$

$\tau_{xy} = \dfrac{-p}{40d^3}\left[5(3x^2 - l^2)(y^2 - d^2) - 5y^4 + 6y^2d^2 - d^4\right]$

The boundary stress function values of τ_{xy} do not agree with the assumed constant equilibrating shears at $x = 0$ and l.

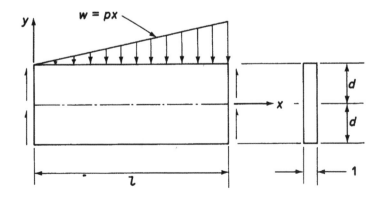

FIGURE P.2.4

P.2.4 MATLAB Use the Symbolic Math Toolbox in MATLAB to show the Airy stress function proposed in Problem P.2.4 satisfies the internal compatibility conditions. Also repeat the derivation of the distribution of stresses within the plate, assuming that $y = -d, -d/2, 0, d/2,$ and d. Assume that for all cases $x = l$.

Answer:

(i) $\sigma_y = 0,$ $\qquad\qquad\qquad \sigma_x = lp/5,$
$\quad\tau_{xy} = 0$

(ii) $\sigma_y = -5lp/32,$ $\qquad \sigma_x = -7lp/80,$
$\quad\tau_{xy} = -p((9d^4)/16 - (45d^2l^2)/2)/(120d^3)$

(iii) $\sigma_y = -lp/2,$ $\qquad\qquad \sigma_x = 0,$
$\quad\tau_{xy} = p(3d^4 + 30d^2l^2)/(120d^3)$

(iv) $\sigma_y = -27lp/32,$ $\qquad \sigma_x = 7lp/80,$
$\quad\tau_{xy} = -p((9d^4)/16 - (45d^2l^2)/2)/(120d^3)$

(v) $\sigma_y = -lp,$ $\qquad\qquad\qquad \sigma_x = -lp/5,$
$\quad\tau_{xy} = 0$

P.2.5 The cantilever beam shown in Fig. P.2.5 is rigidly fixed at $x = L$ and carries loading such that the Airy stress function relating to the problem is

$$\phi = \frac{w}{40bc^3}\left(-10c^3x^2 - 15c^2x^2y + 2c^2y^3 + 5x^2y^3 - y^5\right)$$

Find the loading pattern corresponding to the function and check its validity with respect to the boundary conditions.

Answer: The stress function satisfies the biharmonic equation. The beam is a cantilever under a uniformly distributed load of intensity w/unit area with a self-equilibrating stress application given by $\sigma_x = w(12c^3y - 20y^3)/40bc^3$ at $x = 0$. There is zero shear stress at $y = \pm c$ and $x = 0$. At $y = +c, \sigma_y = -w/b$; and at $y = -c, \sigma_y = 0$.

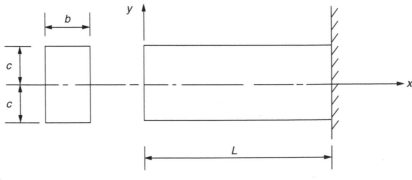

FIGURE P.2.5

P.2.6 A two-dimensional isotropic sheet, having a Young's modulus E and linear coefficient of expansion α, is heated nonuniformly, the temperature being $T\,(x,\,y)$. Show that the Airy stress function ϕ satisfies the differential equation

$$\nabla^2(\nabla^2\phi + E\alpha T) = 0$$

where

$$\nabla^2 = \frac{\partial^2}{\partial x^2} + \frac{\partial^2}{\partial y^2}$$

is the Laplace operator.

P.2.7 Investigate the state of plane stress described by the following Airy stress function:

$$\phi = \frac{3Qxy}{4a} - \frac{Qxy^3}{4a^3}$$

over the square region $x = -a$ to $x = +a$, $y = -a$ to $y = +a$. Calculate the stress resultants per unit thickness over each boundary of the region.

Answer: The stress function satisfies the biharmonic equation. Also,

when $x = a$,

$$\sigma_x = \frac{-3Qy}{2a^2}$$

when $x = -a$,

$$\sigma_x = \frac{3Qy}{2a^2}$$

and

$$\tau_{xy} = \frac{-3Q}{4a}\left(1 - \frac{y^2}{a^2}\right)$$

Torsion of solid sections

The elasticity solution of the torsion problem for bars of arbitrary but uniform cross-section is accomplished by the semi-inverse method (Section 2.3), in which assumptions are made regarding either stress or displacement components. The former method owes its derivation to Prandtl, the latter to St. Venant. Both methods are presented in this chapter, together with the useful membrane analogy introduced by Prandtl.

3.1 PRANDTL STRESS FUNCTION SOLUTION

Consider the straight bar of uniform cross-section shown in Fig. 3.1. It is subjected to equal but opposite torques T at each end, both of which are assumed to be free from restraint so that warping displacements w, that is, displacements of cross-sections normal to and out of their original planes, are unrestrained. Further, we make the reasonable assumptions that, since no direct loads are applied to the bar,

$$\sigma_x = \sigma_y = \sigma_z = 0$$

and that the torque is resisted solely by shear stresses in the plane of the cross-section, giving

$$\tau_{xy} = 0$$

To verify these assumptions, we must show that the remaining stresses satisfy the conditions of equilibrium and compatibility at all points throughout the bar and, in addition, fulfill the equilibrium boundary conditions at all points on the surface of the bar.

If we ignore body forces, the equations of equilibrium (1.5), reduce, as a result of our assumptions, to

$$\frac{\partial \tau_{xz}}{\partial z} = 0, \quad \frac{\partial \tau_{yz}}{\partial z} = 0, \quad \frac{\partial \tau_{zx}}{\partial x} + \frac{\partial \tau_{yz}}{\partial y} = 0 \tag{3.1}$$

The first two equations of Eqs. (3.1) show that the shear stresses τ_{xz} and τ_{yz} are functions of x and y only. They are therefore constant at all points along the length of the bar which have the same x and y coordinates. At this stage, we turn to the stress function to simplify the process of solution. Prandtl introduced a stress function ϕ defined by

$$\frac{\partial \phi}{\partial x} = -\tau_{zy}, \quad \frac{\partial \phi}{\partial y} = \tau_{zx} \tag{3.2}$$

which identically satisfies the third of the equilibrium equations (3.1) whatever form ϕ may take. We therefore have to find the possible forms of ϕ which satisfy the compatibility equations and the

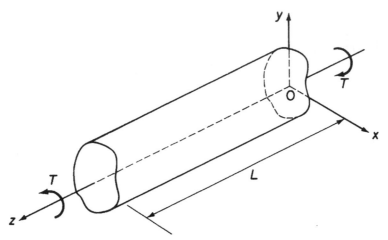

FIGURE 3.1 Torsion of a Bar of Uniform, Arbitrary Cross-section

boundary conditions, the latter being, in fact, the requirement that distinguishes one torsion problem from another.

From the assumed state of stress in the bar, we deduce that

$$\varepsilon_x = \varepsilon_y = \varepsilon_z = \gamma_{xy} = 0$$

(see Eqs. (1.42) and (1.46)). Further, since τ_{xz} and τ_{yz} and hence γ_{xz} and γ_{yz} are functions of x and y only, the compatibility equations (1.21)–(1.23) are identically satisfied, as is Eq. (1.26). The remaining compatibility equations, (1.24) and (1.25), are then reduced to

$$\frac{\partial}{\partial x}\left(-\frac{\partial \gamma_{yz}}{\partial x} + \frac{\partial \gamma_{xz}}{\partial y}\right) = 0$$

$$\frac{\partial}{\partial y}\left(\frac{\partial \gamma_{yz}}{\partial x} - \frac{\partial \gamma_{xz}}{\partial y}\right) = 0$$

Substituting initially for γ_{yz} and γ_{xz} from Eqs. (1.46) and for $\tau_{zy}\ (= \tau_{yz})$ and $\tau_{zx}\ (= \tau_{xz})$ from Eqs. (3.2) gives

$$\frac{\partial}{\partial x}\left(\frac{\partial^2 \phi}{\partial x^2} + \frac{\partial^2 \phi}{\partial y^2}\right) = 0$$

$$-\frac{\partial}{\partial y}\left(\frac{\partial^2 \phi}{\partial x^2} + \frac{\partial^2 \phi}{\partial y^2}\right) = 0$$

or

$$\frac{\partial}{\partial x}\nabla^2 \phi = 0, \quad -\frac{\partial}{\partial y}\nabla^2 \phi = 0 \tag{3.3}$$

where ∇^2 is the two-dimensional Laplacian operator

$$\left(\frac{\partial^2}{\partial x^2} + \frac{\partial^2}{\partial y^2}\right)$$

The parameter $\nabla^2 \phi$ is therefore constant at any section of the bar, so that the function ϕ must satisfy the equation

$$\frac{\partial^2 \phi}{\partial x^2} + \frac{\partial^2 \phi}{\partial y^2} = \text{constant} = F(\text{say}) \qquad (3.4)$$

at all points within the bar.

Finally, we must ensure that ϕ fulfills the boundary conditions specified by Eqs. (1.7). On the cylindrical surface of the bar, there are no externally applied forces, so that $\overline{X} = \overline{Y} = \overline{Z} = 0$. The direction cosine n is also zero, and therefore the first two equations of Eqs. (1.7) are identically satisfied, leaving the third equation as the boundary condition; that is,

$$\tau_{yz}m + \tau_{xz}l = 0 \qquad (3.5)$$

The direction cosines l and m of the normal N to any point on the surface of the bar are, by reference to Fig. 3.2,

$$l = \frac{dy}{ds}, \quad m = -\frac{dx}{ds} \qquad (3.6)$$

Substituting Eqs. (3.2) and (3.6) into (3.5), we have

$$\frac{\partial \phi}{\partial x}\frac{dx}{ds} + \frac{\partial \phi}{\partial y}\frac{dy}{ds} = 0$$

or

$$\frac{\partial \phi}{\partial s} = 0$$

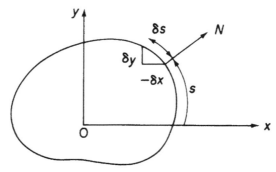

FIGURE 3.2 Formation of the Direction Cosines l and m of the Normal to the Surface of the Bar

Therefore, ϕ is constant on the surface of the bar and, since the actual value of this constant does not affect the stresses of Eq. (3.2), we may conveniently take the constant to be zero. Hence, on the cylindrical surface of the bar, we have the boundary condition

$$\phi = 0 \tag{3.7}$$

On the ends of the bar, the direction cosines of the normal to the surface have the values $l = 0$, $m = 0$, and $n = 1$. The related boundary conditions, from Eqs. (1.7), are then

$$\overline{X} = \tau_{zx}$$
$$\overline{Y} = \tau_{zy}$$
$$\overline{Z} = 0$$

We now observe that the forces on each end of the bar are shear forces which are distributed over the ends of the bar in the same manner as the shear stresses are distributed over the cross-section. The resultant shear force in the positive direction of the x axis, which we shall call S_x, is

$$S_x = \iint \overline{X}\, dx\, dy = \iint \tau_{zx}\, dx\, dy$$

or, using the relationship of Eqs. (3.2),

$$S_x = \iint \frac{\partial \phi}{\partial y}\, dx\, dy = \int dx \int \frac{\partial \phi}{\partial y}\, dy = 0$$

as $\phi = 0$ at the boundary. In a similar manner, S_y, the resultant shear force in the y direction, is

$$S_y = -\int dy \int \frac{\partial \phi}{\partial y}\, dx = 0$$

It follows that there is no resultant shear force on the ends of the bar and the forces represent a torque of magnitude, referring to Fig. 3.3,

$$T = \iint (\tau_{zy}\, x - \tau_{zx}\, y)\, dx\, dy$$

in which we take the sign of T as being positive in the counterclockwise sense.

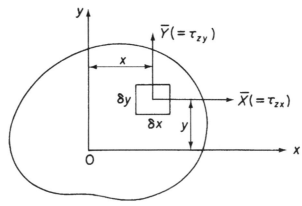

FIGURE 3.3 Derivation of Torque on Cross-section of Bar

Rewriting this equation in terms of the stress function ϕ,

$$T = -\iint \frac{\partial \phi}{\partial x} x \, dx \, dy - \iint \frac{\partial \phi}{\partial y} y \, dx \, dy$$

Integrating each term on the right-hand side of this equation by parts and noting again that $\phi = 0$ at all points on the boundary, we have

$$T = 2 \iint \phi \, dx \, dy \tag{3.8}$$

We are therefore in a position to obtain an exact solution to a torsion problem if a stress function $\phi(x, y)$ can be found which satisfies Eq. (3.4) at all points within the bar and vanishes on the surface of the bar, and providing that the external torques are distributed over the ends of the bar in an identical manner to the distribution of internal stress over the cross-section. Although the last proviso is generally impracticable, we know from St. Venant's principle that only stresses in the end regions are affected; therefore, the solution is applicable to sections at distances from the ends usually taken to be greater than the largest cross-sectional dimension. We have now satisfied all the conditions of the problem without the use of stresses other than τ_{zy} and τ_{xz}, demonstrating that our original assumptions were justified.

Usually, in addition to the stress distribution in the bar, we need to know the angle of twist and the warping displacement of the cross-section. First, however, we shall investigate the mode of displacement of the cross-section. We have seen that, as a result of our assumed values of stress,

$$\varepsilon_x = \varepsilon_y = \varepsilon_z = \gamma_{xy} = 0$$

It follows, from Eqs. (1.18) and the second of Eqs. (1.20), that

$$\frac{\partial u}{\partial x} = \frac{\partial v}{\partial y} = \frac{\partial w}{\partial z} = \frac{\partial v}{\partial x} + \frac{\partial u}{\partial y} = 0$$

which result leads to the conclusions that each cross-section rotates as a rigid body in its own plane about a center of rotation or twist and that, although cross-sections suffer warping displacements normal to their planes, the values of this displacement at points having the same coordinates along the length of the bar are equal. Each longitudinal fiber of the bar therefore remains unstrained, as we have in fact assumed.

Let us suppose that a cross-section of the bar rotates through a small angle θ about its center of twist assumed coincident with the origin of the axes Oxy (see Fig. 3.4). Some point $P(r, \alpha)$ will be displaced to $P'(r, \alpha + \theta)$, the components of its displacement being

$$u = -r\theta \sin\alpha, \quad v = r\theta \cos\alpha$$

or

$$u = -\theta y, \quad v = \theta x \tag{3.9}$$

Referring to Eqs. (1.20) and (1.46),

$$\gamma_{zx} = \frac{\partial u}{\partial z} + \frac{\partial w}{\partial x} = \frac{\tau_{zx}}{G}, \quad \gamma_{zy} = \frac{\partial w}{\partial y} + \frac{\partial v}{\partial z} = \frac{\tau_{zy}}{G}$$

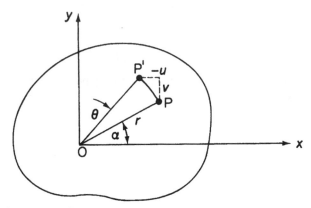

FIGURE 3.4 Rigid Body Displacement in the Cross-section of the Bar

Rearranging and substituting for u and v from Eqs. (3.9),

$$\frac{\partial w}{\partial x} = \frac{\tau_{zx}}{G} + \frac{d\theta}{dz}y, \quad \frac{\partial w}{\partial y} = \frac{\tau_{zy}}{G} - \frac{d\theta}{dz}x \tag{3.10}$$

For a particular torsion problem, Eqs. (3.10) enable the warping displacement w of the originally plane cross-section to be determined. Note that, since each cross-section rotates as a rigid body, θ is a function of z only.

Differentiating the first of Eqs. (3.10) with respect to y, the second with respect to x, and subtracting, we have

$$0 = \frac{1}{G}\left(\frac{\partial \tau_{zx}}{\partial y} - \frac{\partial \tau_{zy}}{\partial x}\right) + 2\frac{d\theta}{dz}$$

Expressing τ_{zx} and τ_{zy} in terms of ϕ gives

$$\frac{\partial^2 \phi}{\partial x^2} + \frac{\partial^2 \phi}{\partial y^2} = -2G\frac{d\theta}{dz}$$

or, from Eq. (3.4),

$$-2G\frac{d\theta}{dz} = \nabla^2\phi = F(\text{constant}) \tag{3.11}$$

It is convenient to introduce a *torsion constant J*, defined by the general torsion equation

$$T = GJ\frac{d\theta}{dz} \tag{3.12}$$

The product GJ is known as the *torsional rigidity* of the bar and may be written, from Eqs. (3.8) and (3.11),

$$GJ = -\frac{4G}{\nabla^2\phi}\iint \phi \, dx \, dy \tag{3.13}$$

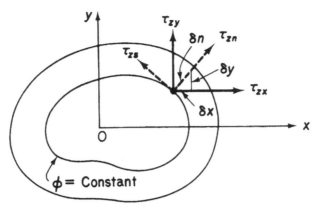

FIGURE 3.5 Lines of Shear Stress

Consider now the line of constant ϕ in Fig. 3.5. If s is the distance measured along this line from some arbitrary point, then

$$\frac{\partial \phi}{\partial s} = 0 = \frac{\partial \phi}{\partial y}\frac{dy}{ds} + \frac{\partial \phi}{\partial x}\frac{dx}{ds}$$

Using Eqs. (3.2) and (3.6), we may rewrite this equation as

$$\frac{\partial \phi}{\partial s} = \tau_{zx}l + \tau_{zy}m = 0 \tag{3.14}$$

From Fig. 3.5, the normal and tangential components of shear stress are

$$\tau_{zn} = \tau_{zx}l + \tau_{zy}m \quad \tau_{zs} = \tau_{zy}l - \tau_{zx}m \tag{3.15}$$

Comparing the first of Eqs. (3.15) with Eq. (3.14), we see that the normal shear stress is zero, so that the resultant shear stress at any point is tangential to a line of constant ϕ. These are known as *lines of shear stress* or *shear lines*.

Substituting ϕ in the second of Eqs. (3.15), we have

$$\tau_{zs} = -\frac{\partial \phi}{\partial x}l - \frac{\partial \phi}{\partial y}m$$

which may be written, from Fig. 3.5, as

$$\tau_{zx} = \frac{\partial \phi}{\partial x}\frac{dx}{dn} - \frac{\partial \phi}{\partial y}\frac{dy}{dn} = -\frac{\partial \phi}{\partial n} \tag{3.16}$$

where, in this case, the direction cosines l and m are defined in terms of an elemental normal of length δn.

We have therefore shown that the resultant shear stress at any point is tangential to the line of shear stress through the point and has a value equal to minus the derivative of ϕ in a direction normal to the line.

Example 3.1

Determine the rate of twist and the stress distribution in a circular section bar of radius R which is subjected to equal and opposite torques T at each of its free ends. See Ex. 1.1.

If we assume an origin of axes at the center of the bar, the equation of its surface is given by

$$x^2 + y^2 = R^2$$

If we now choose a stress function of the form

$$\phi = C\left(x^2 + y^2 - R^2\right) \tag{i}$$

the boundary condition $\phi = 0$ is satisfied at every point on the boundary of the bar and the constant C may be chosen to fulfill the remaining requirement of compatibility. Therefore, from Eqs. (3.11) and (i),

$$4C = -2G\frac{d\theta}{dz}$$

so that

$$C = -\frac{G}{2}\frac{d\theta}{dz}$$

and

$$\phi = -G\frac{d\theta}{dz}\left(x^2 + y^2 - R^2\right)/2 \tag{ii}$$

Substituting for ϕ in Eq. (3.8),

$$T = -G\frac{d\theta}{dz}\left(\iint x^2 \, dx \, dy + \iint y^2 \, dx \, dy - R^2 \iint dx \, dy\right)$$

Both the first and second integrals in this equation have the value $\pi R^4/4$, while the third integral is equal to πR^2, the area of cross-section of the bar. Then,

$$T = -G\frac{d\theta}{dz}\left(\frac{\pi R^4}{4} + \frac{\pi R^4}{4} - \pi R^4\right)$$

which gives

$$T = \frac{\pi R^4}{2}G\frac{d\theta}{dz}$$

that is,

$$T = GJ\frac{d\theta}{dz} \tag{iii}$$

in which $J = \pi R^4/2 = \pi D^4/32$ (D is the diameter), the *polar second moment of area* of the bar's cross-section.

Substituting for $G(d\theta/dz)$ in Eq. (ii) from (iii),

$$\phi = -\frac{T}{2J}\left(x^2 + y^2 - R^2\right)$$

and from Eqs. (3.2),

$$\tau_{zy} = -\frac{\partial \phi}{\partial x} = \frac{Tx}{J}, \quad \tau_{zx} = \frac{\partial \phi}{\partial y} = -\frac{Ty}{J}$$

The resultant shear stress at any point on the surface of the bar is then given by

$$\tau = \sqrt{\tau_{zy}^2 + \tau_{zx}^2}$$

that is,

$$\tau = \frac{T}{J}\sqrt{x^2 + y^2}$$

so that.

$$\tau = \frac{TR}{J} \tag{iv}$$

This argument may be applied to any annulus of radius r within the cross-section of the bar, so that the stress distribution is given by

$$\tau = \frac{Tr}{J}$$

and therefore increases linearly from zero at the center of the bar to a maximum TR/J at the surface.

Example 3.2

A uniform bar has the elliptical cross-section shown in Fig. 3.6(a) and is subjected to equal and opposite torques T at each of its free ends. Derive expressions for the rate of twist in the bar, the shear stress distribution, and the warping displacement of its cross-section. See Ex. 1.1.

The semi-major and semi-minor axes are a and b, respectively, so that the equation of its boundary is

$$\frac{x^2}{a^2} + \frac{y^2}{b^2} = 1$$

If we choose a stress function of the form

$$\phi = C\left(\frac{x^2}{a^2} + \frac{y^2}{b^2} - 1\right) \tag{i}$$

then the boundary condition $\phi = 0$ is satisfied at every point on the boundary and the constant C may be chosen to fulfill the remaining requirement of compatibility. Thus, from Eqs. (3.11) and (i),

$$2C\left(\frac{1}{a^2} + \frac{1}{b^2}\right) = -2G\frac{d\theta}{dz}$$

or

$$C = -G\frac{d\theta}{dz}\frac{a^2b^2}{(a^2 + b^2)} \tag{ii}$$

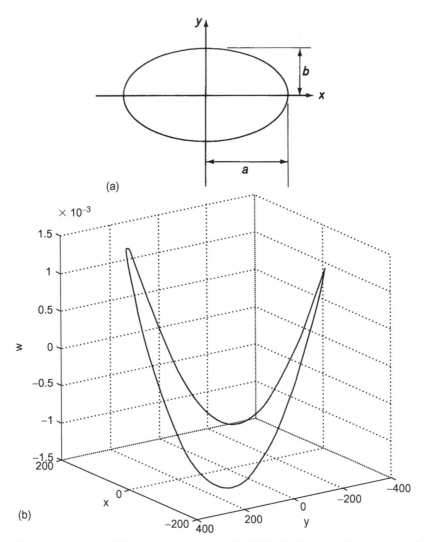

(a)

(b)

FIGURE 3.6 (a) Torsion of a Bar of Elliptical Cross-section; (b) (MATLAB) Warping Displacement Distribution

giving

$$\phi = -G\frac{d\theta}{dz}\frac{a^2b^2}{(a^2+b^2)}\left(\frac{x^2}{a^2}+\frac{y^2}{b^2}-1\right) \tag{iii}$$

Substituting this expression for ϕ in Eq. (3.8) establishes the relationship between the torque T and the rate of twist:

$$T = -2G\frac{d\theta}{dz}\frac{a^2b^2}{(a^2+b^2)}\left(\frac{1}{a^2}\iint x^2\,dx\,dy+\frac{1}{b^2}\iint y^2\,dx\,dy-\iint dx\,dy\right)$$

The first and second integrals in this equation are the second moments of area $I_{yy} = \pi a^3 b/4$ and $I_{xx} = \pi a b^3/4$, while the third integral is the area of the cross-section $A = \pi ab$. Replacing the integrals by these values gives

$$T = G\frac{d\theta}{dz}\frac{\pi a^3 b^3}{(a^2 + b^2)} \tag{iv}$$

from which (see Eq. (3.12))

$$J = \frac{\pi a^3 b^3}{(a^2 + b^2)} \tag{v}$$

The shear stress distribution is obtained in terms of the torque by substituting for the product $G(d\theta/dz)$ in Eq. (iii) from (iv) and then differentiating as indicated by the relationships of Eqs. (3.2). Thus,

$$\tau_{zx} = -\frac{2Ty}{\pi ab^3}, \quad \tau_{zy} = \frac{2Tx}{\pi a^3 b} \tag{vi}$$

So far we have solved for the stress distribution, Eqs. (vi), and the rate of twist, Eq. (iv). It remains to determine the warping distribution w over the cross-section. For this, we return to Eqs. (3.10), which become, on substituting from the preceding for τ_{zx}, τ_{zy}, and $d\theta/dz$,

$$\frac{\partial w}{\partial x} = -\frac{2Ty}{\pi ab^3 G} + \frac{T}{G}\frac{(a^2 + b^2)}{\pi a^3 b^3}y, \quad \frac{\partial w}{\partial y} = \frac{2Tx}{\pi a^3 b^3 G} - \frac{T}{G}\frac{(a^2 + b^2)}{\pi a^3 b^3}x$$

or

$$\frac{\partial w}{\partial x} = \frac{T}{\pi a^3 b^3 G}(b^2 - a^2)y, \quad \frac{\partial w}{\partial y} = \frac{T}{\pi a^3 b^3 G}(b^2 - a^2)x \tag{vii}$$

Integrating both of Eqs. (vii),

$$w = \frac{T(b^2 - a^2)}{\pi a^3 b^3 G}yx + f_1(y) \quad w\frac{T(b^2 - a^2)}{\pi a^3 b^3 G}xy + f_2(x)$$

The warping displacement given by each of these equations must have the same value at identical points (x, y). It follows that $f_1(y) = f_2(x) = 0$. Hence,

$$w = \frac{T(b^2 - a^2)}{\pi a^3 b^3 G}xy \tag{viii}$$

Lines of constant w therefore describe hyperbolas with the major and minor axes of the elliptical cross-section as asymptotes. Further, for a positive (counterclockwise) torque, the warping is negative in the first and third quadrants ($a > b$) and positive in the second and fourth.

Example 3.2 MATLAB

Calculate the warping displacement (*w*) along the boundary of the cross-section illustrated in Fig. 3.6(a) using the equation for *w* derived in Example 3.2. Plot the resulting values of *w* using the plot3 function in MATLAB. Assume the following variable values: a = 200 mm; b = 300 mm; G = 25,000 N/mm^2; T = 15 kNm. See Ex. 1.1.

Calculations of *w* along the cross-section boundary and the code required to create the needed plot are obtained through the following MATLAB file:

```
% Declare any needed variables
a = 200;
b = 300;
G = 25000;
T = 15*10^6;

% Define (x,y) values for 1001 points along the cross-section boundary
steps = 1000;
x = zeros(steps+1,1);
y = zeros(steps+1,1);
w = zeros(steps+1,1);
for i=1:1:steps/2
  x(i) = a - (4*a/steps)*(i-1);
  y(i) = b*sqrt(1-(x(i)/a)^2);
end
for i=1:1:steps/2+1
  x(steps/2+i) = -a + (4*a/steps)*(i-1);
  y(steps/2+i) = -b*sqrt(1-(x(i)/a)^2);
end

% Calculate the warping displacement (w) at each point (x,y)
for i=1:1:steps+1
  w(i) = T*(b^2-a^2)*x(i)*y(i)/(pi*a^3*b^3*G);
end

% Plot the values of w
figure(1)
  plot3(x,y,w)
  grid on
  axis square
  xlabel('x')
  ylabel('y')
  zlabel('w')
  view(-125,18)
```

The plot resulting from this MATLAB file is illustrated in Fig. 3.6(b) (MATLAB).

3.2 ST. VENANT WARPING FUNCTION SOLUTION

In formulating his stress function solution, Prandtl made assumptions concerning the stress distribution in the bar. The alternative approach presented by St. Venant involves assumptions as to the mode of displacement of the bar; namely, that cross-sections of a bar subjected to torsion maintain their original unloaded shape, although they may suffer warping displacements normal to their plane. The first of these assumptions leads to the conclusion that cross-sections rotate as rigid bodies about a center of rotation or twist. This fact was also found to derive from the stress function approach of Section 3.1, so that, referring to Fig. 3.4 and Eq. (3.9), the components of displacement in the x and y directions of a point P in the cross-section are

$$u = -\theta y, \; v = \theta x$$

It is also reasonable to assume that the warping displacement w is proportional to the rate of twist and is therefore constant along the length of the bar. Hence, we may define w by the equation

$$w = \frac{d\theta}{dz}\psi(x, y) \tag{3.17}$$

where $\psi(x, y)$ is the *warping function*.

The assumed form of the displacements u, v, and w must satisfy the equilibrium and force boundary conditions of the bar. We note here that it is unnecessary to investigate compatibility, as we are concerned with displacement forms that are single-valued functions and therefore automatically satisfy the compatibility requirement.

The components of strain corresponding to the assumed displacements are obtained from Eqs. (1.18) and (1.20) and are

$$\left.\begin{aligned}
\varepsilon_x = \varepsilon_y = \varepsilon_z = \gamma_{xy} = 0 \\
\gamma_{zx} = \frac{\partial w}{\partial x} + \frac{\partial u}{\partial z} = \frac{d\theta}{dz}\left(\frac{\partial \psi}{\partial x} - y\right) \\
\gamma_{zy} = \frac{\partial w}{\partial y} + \frac{\partial v}{\partial z} = \frac{d\theta}{dz}\left(\frac{\partial \psi}{\partial x} - x\right)
\end{aligned}\right\} \tag{3.18}$$

The corresponding components of stress are, from Eqs. (1.42) and (1.46),

$$\left.\begin{aligned}
\sigma_x = \sigma_y = \sigma_z = \tau_{xy} = 0 \\
\tau_{zx} = G\frac{d\theta}{dz}\left(\frac{\partial \psi}{\partial x} - y\right) \\
\tau_{zy} = G\frac{d\theta}{dz}\left(\frac{\partial \psi}{\partial y} + x\right)
\end{aligned}\right\} \tag{3.19}$$

Ignoring body forces, we see that these equations identically satisfy the first two of the equilibrium equations (1.5) and also that the third is fulfilled if the warping function satisfies the equation

$$\frac{\partial^2 \psi}{\partial x^2} + \frac{\partial^2 \psi}{\partial y^2} = \nabla^2 \psi = 0 \tag{3.20}$$

The direction cosine n is zero on the cylindrical surface of the bar and so the first two of the boundary conditions (Eqs. (1.7)) are identically satisfied by the stresses of Eqs. (3.19). The third equation simplifies to

$$\left(\frac{\partial \psi}{\partial y} + x\right) m + \left(\frac{\partial \psi}{\partial x} - y\right) l = 0 \tag{3.21}$$

It may be shown, but not as easily as in the stress function solution, that the shear stresses defined in terms of the warping function in Eqs. (3.19) produce zero resultant shear force over each end of the bar.[1] The torque is found in a similar manner to that in Section 3.1, where, by reference to Fig. 3.3, we have

$$T = \iint (\tau_{zy}x - \tau_{zx}y)\,dx\,dy$$

or

$$T = G\frac{d\theta}{dz} \iint \left[\left(\frac{\partial \psi}{\partial y} + x\right)x - \left(\frac{\partial \psi}{\partial x} - y\right)y\right] dx\,dy \tag{3.22}$$

By comparison with Eq. (3.12), the torsion constant J is now, in terms of ψ,

$$J = \iint \left[\left(\frac{\partial \psi}{\partial y} + x\right)x - \left(\frac{\partial \psi}{\partial x} - y\right)y\right] dx\,dy \tag{3.23}$$

The warping function solution to the torsion problem reduces to the determination of the warping function τ, which satisfies Eqs. (3.20) and (3.21). The torsion constant and the rate of twist follow from Eqs. (3.23) and (3.22); the stresses and strains from Eqs. (3.19) and (3.18); and finally, the warping distribution from Eq. (3.17).

3.3 THE MEMBRANE ANALOGY

Prandtl suggested an extremely useful analogy relating the torsion of an arbitrarily shaped bar to the deflected shape of a membrane. The latter is a thin sheet of material which relies for its resistance to transverse loads on internal in-plane or membrane forces.

Suppose that a membrane has the same external shape as the cross-section of a torsion bar (Fig. 3.7(a)). It supports a transverse uniform pressure q and is restrained along its edges by a uniform tensile force N/unit length, as shown in Figs. 3.7(a) and (b). It is assumed that the transverse displacements of the membrane are small, so that N remains unchanged as the membrane deflects. Consider the equilibrium of an element $\delta x \delta y$ of the membrane. Referring to Fig. 3.8 and summing forces in the z direction, we have

$$-N\delta y = \frac{\partial w}{\partial x} - N\delta y \left(-\frac{\partial w}{\partial x} - \frac{\partial^2 w}{\partial x^2}\delta x\right) - N\delta x \frac{\partial w}{\partial y} - N\delta x \left(-\frac{\partial w}{\partial y} - \frac{\partial^2 w}{\partial y^2}\delta x\right) + q\delta x\delta y = 0$$

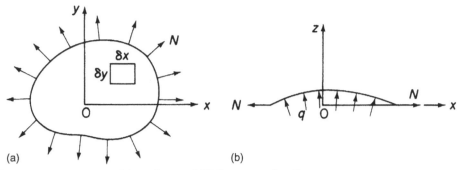

FIGURE 3.7 Membrane Analogy: (a) In-plane and (b) Transverse Loading

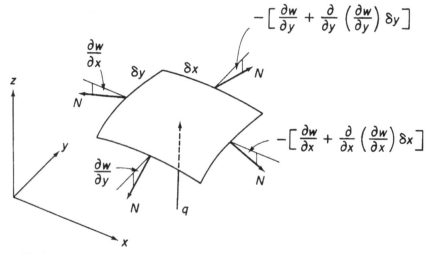

FIGURE 3.8 Equilibrium of Element of Membrane

or

$$\frac{\partial^2 w}{\partial x^2} + \frac{\partial^2 w}{\partial y^2} = \nabla^2 w = -\frac{q}{N} \tag{3.24}$$

Equation (3.24) must be satisfied at all points within the boundary of the membrane. Furthermore, at all points on the boundary,

$$w = 0 \tag{3.25}$$

and we see that, by comparing Eqs. (3.24) and (3.25) with Eqs. (3.11) and (3.7), w is analogous to ϕ when q is constant. Thus, if the membrane has the same external shape as the cross-section of the bar, then

$$w(x, y) = \phi(x, y)$$

and

$$\frac{q}{N} = -F = 2G\frac{d\theta}{dz}$$

The analogy now being established, we may make several useful deductions relating the deflected form of the membrane to the state of stress in the bar.

Contour lines or lines of constant w correspond to lines of constant ϕ or lines of shear stress in the bar. The resultant shear stress at any point is tangential to the membrane contour line and equal in value to the negative of the membrane slope, $\partial w/\partial n$, at that point, the direction n being normal to the contour line (see Eq. (3.16)). The volume between the membrane and the xy plane is

$$\text{Vol} = \iint w \, dx \, dy$$

and we see that, by comparison with Eq. (3.8),

$$T = 2 \, \text{Vol}$$

The analogy therefore provides an extremely useful method of analyzing torsion bars possessing irregular cross-sections for which stress function forms are not known. Hetényi[2] describes experimental techniques for this approach. In addition to the strictly experimental use of the analogy, it is also helpful in the visual appreciation of a particular torsion problem. The contour lines often indicate a form for the stress function, enabling a solution to be obtained by the method of Section 3.1. Stress concentrations are made apparent by the closeness of contour lines where the slope of the membrane is large. These are in evidence at sharp internal corners, cutouts, discontinuities, and the like.

3.4 TORSION OF A NARROW RECTANGULAR STRIP

In Chapter 17, we investigate the torsion of thin-walled open-section beams; the development of the theory being based on the analysis of a narrow rectangular strip subjected to torque. We now conveniently apply the membrane analogy to the torsion of such a strip, shown in Fig. 3.9. The corresponding membrane surface has the same cross-sectional shape at all points along its length except for small regions near its ends, where it flattens out. If we ignore these regions and assume that the shape of the membrane is independent of y, then Eq. (3.11) simplifies to

$$\frac{d^2\phi}{dx^2} = -2G\frac{d\theta}{dz}$$

Integrating twice,

$$\phi = -G\frac{d\theta}{dx}x^2 + Bx + C$$

Substituting the boundary conditions $\phi = 0$ at $x = \pm t/2$, we have

$$\phi = -G\frac{d\theta}{dz}\left[x^2 - \left(\frac{t}{2}\right)^2\right] \tag{3.26}$$

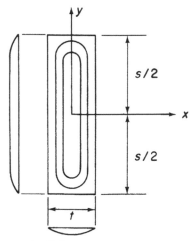

FIGURE 3.9 Torsion of a Narrow Rectangular Strip

Although ϕ does not disappear along the short edges of the strip and therefore does not give an exact solution, the actual volume of the membrane differs only slightly from the assumed volume, so that the corresponding torque and shear stresses are reasonably accurate. Also, the maximum shear stress occurs along the long sides of the strip, where the contours are closely spaced, indicating, in any case, that conditions in the end region of the strip are relatively unimportant.

The stress distribution is obtained by substituting Eq. (3.26) in Eqs. (3.2), then

$$\tau_{zy} = 2Gx\frac{d\theta}{dz} \quad \tau_{zx} = 0 \tag{3.27}$$

the shear stress varying linearly across the thickness and attaining a maximum

$$\tau_{zy,\text{max}} = \pm Gt\frac{d\theta}{dz} \tag{3.28}$$

at the outside of the long edges, as predicted. The torsion constant J follows from the substitution of Eq. (3.26) into (3.13), giving

$$J = \frac{st^3}{3} \tag{3.29}$$

and

$$\tau_{zy,\text{max}} = \frac{3T}{st^3}$$

These equations represent exact solutions when the assumed shape of the deflected membrane is the actual shape. This condition arises only when the ratio s/t approaches infinity; however, for ratios in excess of 10, the error is on the order of only 6 percent. Obviously, of the approximate nature of the solution increases as s/t decreases. Therefore, to retain the usefulness of the analysis, a factor μ is included in the torsion constant; that is,

$$J = \frac{\mu st^3}{3}$$

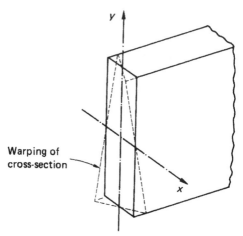

FIGURE 3.10 Warping of a Thin Rectangular Strip

Values of μ for different types of section are found experimentally and quoted in various references.[3,4] We observe that, as s/t approaches infinity, μ approaches unity.

The cross-section of the narrow rectangular strip of Fig. 3.9 does not remain plane after loading but suffers warping displacements normal to its plane; this warping may be determined using either of Eqs. (3.10). From the first of these equations,

$$\frac{\partial w}{\partial x} = y\frac{d\theta}{dz} \tag{3.30}$$

since $\tau_{zx} = 0$ (see Eqs. (3.27)). Integrating Eq. (3.30), we obtain

$$w = xy\frac{d\theta}{dz} + \text{constant} \tag{3.31}$$

Since the cross-section is doubly symmetrical, $w = 0$ at $x = y = 0$, so that the constant in Eq. (3.31) is zero. Therefore,

$$w = xy\frac{d\theta}{dz} \tag{3.32}$$

and the warping distribution at any cross-section is as shown in Fig. 3.10.

We should not close this chapter without mentioning alternative methods of solution of the torsion problem. These in fact provide approximate solutions for the wide range of problems for which exact solutions are not known. Examples of this approach are the numerical finite difference method and the Rayleigh–Ritz method based on energy principles.[5]

References

[1] Wang CT. Applied elasticity. New York: McGraw-Hill; 1953.
[2] Hetényi M. Handbook of experimental stress analysis. New York: John Wiley and Sons; 1950.
[3] Roark RJ. Formulas for stress and strain. 4th ed. New York: McGraw-Hill; 1965.

[4] Handbook of aeronautics, no. 1, Structural principles and data. 4th ed. Published under the authority of the Royal Aeronautical Society. London: New Era Publishing; 1952.

[5] Timoshenko S, Goodier JN. Theory of elasticity. 2nd ed. New York: McGraw-Hill; 1951.

PROBLEMS

P.3.1 Show that the stress function $\phi = k(r^2 - a^2)$ is applicable to the solution of a solid circular section bar of radius a. Determine the stress distribution in the bar in terms of the applied torque, the rate of twist, and the warping of the cross-section. Is it possible to use this stress function in the solution for a circular bar of hollow section?

Answer: $\tau = Tr/I_p$,   Where $I_p = \pi a^4/2$,
$d\theta/dz = 2T/G\pi a^4$, $w = 0$ everywhere.

P.3.2 Deduce a suitable warping function for the circular section bar of P.3.1 and derive the expressions for stress distribution and rate of twist.

Answer: $\psi = 0$, $\tau_{zx} = -\dfrac{Ty}{I_p}$, $\tau_{zy} = \dfrac{Tx}{I_p}$, $\tau_{zs} = \dfrac{Tr}{I_p}$, $\dfrac{d\theta}{dz} = \dfrac{T}{GI_P}$

P.3.3 Show that the warping function $\psi = kxy$, in which k is an unknown constant, may be used to solve the torsion problem for the elliptical section of Example 3.2.

P.3.4 Show that the stress function

$$\phi = -G\frac{d\theta}{dz}\left[\frac{1}{2}(x^2 + y^2) - \frac{1}{2a}(x^3 - 3xy^2) - \frac{2}{27}a^2\right]$$

is the correct solution for a bar having a cross-section in the form of the equilateral triangle shown in Fig. P.3.4. Determine the shear stress distribution, the rate of twist, and the warping of the cross-section. Find the position and magnitude of the maximum shear stress.

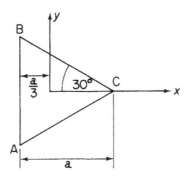

FIGURE P.3.4

Answer: $\tau_{zy} = G\dfrac{d\theta}{dz}\left(x - \dfrac{3x^2}{2a} + \dfrac{3y^2}{2a}\right)$

$\tau_{zx} = -G\dfrac{d\theta}{dz}\left(y + \dfrac{3xy}{a}\right)$

τ_{max} (at center of each side) $= -\dfrac{a}{2}G\dfrac{d\theta}{dz}$

$\dfrac{d\theta}{dz} = \dfrac{15\sqrt{3}T}{Ga^4}$

$w = \dfrac{1}{2a}\dfrac{d\theta}{dz}\left(y^3 - 3x^2y\right)$

P.3.5 Determine the maximum shear stress and the rate of twist in terms of the applied torque T for the section made up of narrow rectangular strips shown in Fig. P.3.5.

Answer: $\tau_{max} = 3T/(2a + b)t^2, \quad d\theta/dz = 3T/G(2a + b)t^3$

P.3.5 MATLAB Use the Symbolic Math Toolbox in MATLAB to repeat Problem P.3.5, assuming that the vertical rectangular strip of length b in Fig. P.3.5 has a thickness of $2t$.

Answer: $\tau_{max} = 3T/[2t^2(a + 4b)], \quad d\theta/dz = 3T[2Gt^3(a + 4b)]$

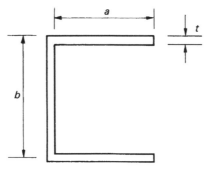

FIGURE P.3.5

Virtual work, energy, and matrix methods

Virtual work and energy methods

Many structural problems are statically determinate; that is, the support reactions and internal force systems may be found using simple statics, where the number of unknowns is equal to the number of equations of equilibrium available. In cases where the number of unknowns exceeds the possible number of equations of equilibrium, for example, a propped cantilever beam, other methods of analysis are required.

The methods fall into two categories and are based on two important concepts; the first, which is presented in this chapter, is *the principle of virtual work*. This is the most fundamental and powerful tool available for the analysis of statically indeterminate structures and has the advantage of being able to deal with conditions other than those in the elastic range. The second, based on *strain energy*, can provide approximate solutions of complex problems for which exact solutions do not exist and is discussed in Chapter 5. In some cases, the two methods are equivalent, since, although the governing equations differ, the equations themselves are identical.

In modern structural analysis, computer-based techniques are widely used; these include the flexibility and stiffness methods (see Chapter 6). However, the formulation of, say, stiffness matrices for the elements of a complex structure is based on one of the previous approaches, so that a knowledge and understanding of their application is advantageous.

4.1 WORK

Before we consider the principle of virtual work in detail, it is important to clarify exactly what is meant by *work*. The basic definition of *work* in elementary mechanics is that "work is done when a force moves its point of application." However, we require a more exact definition since we are concerned with work done by both forces and moments and with the work done by a force when the body on which it acts is given a displacement that is not coincident with the line of action of the force.

Consider the force, F, acting on a particle, A, in Fig. 4.1(a). If the particle is given a displacement, Δ, by some external agency so that it moves to A′ in a direction at an angle α to the line of action of F, the work, W_F, done by F is given by

$$W_F = F\,(\Delta\cos\alpha) \tag{4.1}$$

or

$$W_F = (F\cos\alpha)\Delta \tag{4.2}$$

We see therefore that the work done by the force, F, as the particle moves from A to A′ may be regarded as either the product of F and the component of Δ in the direction of F (Eq. (4.1)) or as the product of the component of F in the direction of Δ and Δ (Eq. (4.2)).

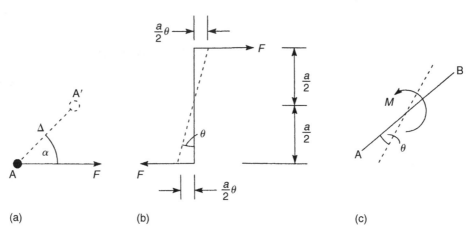

(a)   (b)   (c)

FIGURE 4.1 Work Done by a Force and a Moment

Now, consider the couple (pure moment) in Fig. 4.1(b) and suppose that the couple is given a small rotation of θ radians. The work done by each force F is then $F(a/2)\theta$, so that the total work done, W_C, by the couple is

$$W_C = F\frac{a}{2}\theta + F\frac{a}{2}\theta = Fa\theta$$

It follows that the work done, W_M, by the pure moment, M, acting on the bar AB in Fig. 4.1(c) as it is given a small rotation, θ, is

$$W_M = M\theta \tag{4.3}$$

Note that in this equation the force, F, and moment, M, are in position before the displacements take place and are not the cause of them. Also, in Fig. 4.1(a), the component of Δ parallel to the direction of F is in the same direction as F; if it had been in the opposite direction, the work done would have been negative. The same argument applies to the work done by the moment, M, where we see in Fig. 4.1(c) that the rotation, θ, is in the same sense as M. Note also that if the displacement, Δ, had been perpendicular to the force, F, no work would have been done by F.

Finally, it should be remembered that work is a scalar quantity, since it is not associated with direction (in Fig. 4.1(a), the force F does work if the particle is moved in any direction). Therefore, the work done by a series of forces is the algebraic sum of the work done by each force.

4.2 PRINCIPLE OF VIRTUAL WORK

The establishment of the principle is carried out in stages. First, we consider a particle, then a rigid body, and finally a deformable body, which is the practical application we require when analyzing structures.

4.2.1 Principle of virtual work for a particle

In Fig. 4.2, a particle, A, is acted upon by a number of concurrent forces, $F_1, F_2, \ldots, F_k, \ldots, F_r$; the resultant of these forces is R. Suppose that the particle is given a small arbitrary displacement, Δ_v, to A' in some specified direction; Δ_v is an imaginary or *virtual* displacement and is sufficiently small so that the directions of F_1, F_2, etc., are unchanged. Let θ_R be the angle that the resultant, R, of the forces makes with the direction of Δ_v and $\theta_1, \theta_2, \ldots, \theta_k, \ldots, \theta_r$ the angles that $F_1, F_2, \ldots, F_k, \ldots, F_r$ makes with the direction of Δ_v, respectively. Then, from either of Eqs. (4.1) or (4.2). the total virtual work, W_F, done by the forces F as the particle moves through the virtual displacement, Δ_v, is given by

$$W_F = F_1 \Delta_v \cos\theta_1 + F_2 \Delta_v \cos\theta_2 + \cdots + F_k \Delta_v \cos\theta_k + \cdots + F_r \Delta_v \cos\theta_r$$

Therefore,

$$W_F = \sum_{k=1}^{r} F_k \Delta_v \cos\theta_k$$

or, since Δ_v is a fixed, although imaginary displacement,

$$W_F = \Delta_v \sum_{k=1}^{r} F_k \cos\theta_k \tag{4.4}$$

In Eq. (4.4), $\sum_{k=1}^{r} F_k \cos\theta_k$ is the sum of all the components of the forces, F, in the direction of Δ_v and therefore must be equal to the component of the resultant, R, of the forces, F, in the direction of Δ_v; that is,

$$W_F = \Delta_v \sum_{k=1}^{r} F_k \cos\theta_k = \Delta_v R \cos\theta_R \tag{4.5}$$

If the particle, A, is in equilibrium under the action of the forces, $F_1, F_2, \ldots, F_k, \ldots, F_r$, the resultant, R, of the forces is zero. It follows from Eq. (4.5) that the virtual work done by the forces, F, during the virtual displacement, Δ_v, is zero.

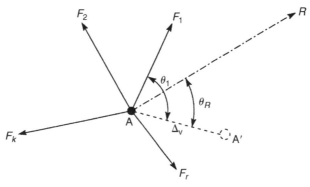

FIGURE 4.2 Virtual Work for a System of Forces Acting on a Particle

We can therefore state the *principle of virtual work* for a particle as follows:

> *If a particle is in equilibrium under the action of a number of forces, the total work done by the forces for a small arbitrary displacement of the particle is zero.*

It is possible for the total work done by the forces to be zero even though the particle is not in equilibrium, if the virtual displacement is taken to be in a direction perpendicular to their resultant, R. We cannot, therefore, state the converse of this principle unless we specify that the total work done must be zero for *any* arbitrary displacement. Thus,

> *A particle is in equilibrium under the action of a system of forces if the total work done by the forces is zero for any virtual displacement of the particle.*

Note that, in this, Δ_v is a purely imaginary displacement and is not related in any way to the possible displacement of the particle under the action of the forces, F. The virtual displacement Δ_v has been introduced purely as a device for setting up the work–equilibrium relationship of Eq. (4.5). The forces, F, therefore remain unchanged in magnitude and direction during this imaginary displacement; this would not be the case if the displacement were real.

4.2.2 Principle of virtual work for a rigid body

Consider the rigid body shown in Fig. 4.3, which is acted upon by a system of external forces, F_1, F_2, $\ldots$, F_k, $\ldots$, F_r. These external forces induce internal forces in the body, which may be regarded as comprising an infinite number of particles; on adjacent particles, such as A_1 and A_2, these internal forces are equal and opposite, in other words, self-equilibrating. Suppose now that the rigid body is given a small, imaginary, that is, virtual, displacement, Δ_v (or a rotation or a combination of both) in some specified direction. The external and internal forces then do virtual work and the total virtual work done, W_t, is the sum of the virtual work, W_e, done by the external forces and the virtual work, W_i, done by the internal forces. Thus,

$$W_t = W_e + W_i \tag{4.6}$$

Since the body is rigid, all the particles in the body move through the same displacement, Δ_v, so that the virtual work done on all the particles is numerically the same. However, for a pair of adjacent particles,

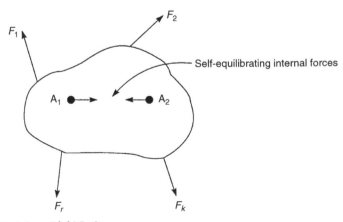

FIGURE 4.3 Virtual Work for a Rigid Body

such as A_1 and A_2 in Fig. 4.3, the self-equilibrating forces are in opposite directions, which means that the work done on A_1 is opposite in sign to the work done on A_2. Therefore, the sum of the virtual work done on A_1 and A_2 is zero. The argument can be extended to the infinite number of pairs of particles in the body, from which we conclude that the internal virtual work produced by a virtual displacement in a rigid body is zero. Equation (4.6) then reduces to

$$W_t = W_e \tag{4.7}$$

Since the body is rigid and the internal virtual work is therefore zero, we may regard the body as a large particle. It follows that if the body is in equilibrium under the action of a set of forces, $F_1, F_2, \ldots, F_k, \ldots, F_r$, the total virtual work done by the external forces during an arbitrary virtual displacement of the body is zero.

Example 4.1

Calculate the support reactions in the cantilever beam AB shown in Fig. 4.4(a).

The concentrated load, W, induces a vertical reaction, R_A, and also one of moment, M_A, at A.

Suppose that the beam is given a small imaginary, that is virtual, rotation, $\theta_{v,A}$, at A, as shown in Fig. 4.4(b). Since we are concerned here with only external forces, we may regard the beam as a rigid body, so that the beam remains straight and B is displaced to B′. The vertical displacement of B, $\Delta_{v,B}$, is then given by

$$\Delta_{v,B} = \theta_{v,A}L$$

or

$$\theta_{v,A} = \Delta_{v,B}/ \tag{i}$$

The total virtual work, W_t, done by all the forces acting on the beam is given by

$$W_t = W\Delta_{v,B} - M_A\theta_{v,A} \tag{ii}$$

Note that the contribution of M_A to the total virtual work done is negative, since the assumed direction of M_A is in the opposite sense to the virtual displacement, $\theta_{v,A}$. Note also that there is no linear movement of the beam at A so that R_A does no work. Substituting in Eq. (ii) for $\theta_{v,A}$ from Eq. (i), we have

$$W_t = W\Delta_{v,B} - M_A\Delta_{v,B}/L \tag{iii}$$

Since the beam is in equilibrium, $W_t = 0$, from the principle of virtual work. Therefore,

$$0 = W\Delta_{v,B} - M_A\Delta_{v,B}/L$$

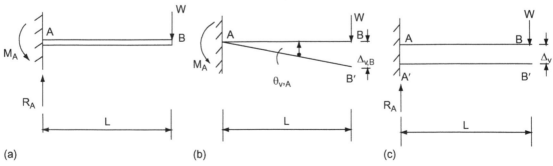

FIGURE 4.4 Use of the Principle of Virtual Work to Calculate Support Reactions

so that

$$M_A = WL$$

which is the result which would have been obtained from considering the moment equilibrium of the beam about A.

Suppose now that the complete beam is given a virtual displacement, Δ_v, as shown in Fig. 4.4(c). There is no rotation of the beam, so that M_A does no work. The total virtual work done is then given by

$$W_t = W\Delta_v - R_A\Delta_v \tag{iv}$$

The contribution of R_A is negative, since its line of action is in the direction opposite to Δ_v. The beam is in equilibrium, so that $W_t = 0$. Therefore, from Eq. (iv),

$$R_A = W$$

which is the result we would have obtained by resolving forces vertically.

Example 4.2

Calculate the support reactions in the cantilever beam shown in Fig. 4.5(a). See Ex. 1.1.

In this case, we obtain a solution by simultaneously giving the beam a virtual displacement, $\Delta_{v,A}$, at A and a virtual rotation, $\theta_{v,A}$, at A. The total deflection at B is then $\Delta_{v,A} + \theta_{v,A}L$ and at a distance x from A is $\Delta_{v,A} + \theta_{v,A}x$.

Since the beam carries a uniformly distributed load, we find the virtual work done by the load by first considering an elemental length, δx, of the load a distance x from A. The load on the element is $w\delta x$ and the virtual work done by this elemental load is given by

$$\delta W = w\delta x(\Delta_{v,A} + \theta_{v,A}x)$$

The total virtual work done on the beam is given by

$$W_t = \int_0^L w(\Delta_{v,A} + \theta_{v,A}x)dx - M_A\theta_{v,A} - R_A\Delta_{v,A}$$

which simplifies to

$$W_t = (wL - R_A)\Delta_{v,A} + [(wL^2/2) - M_A]\theta_{v,A} = 0 \tag{i}$$

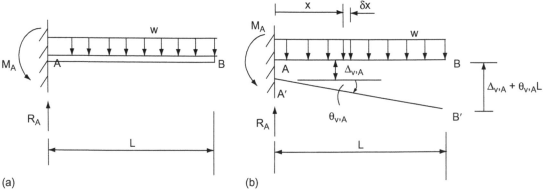

(a)   (b)

FIGURE 4.5 Calculation of Support Reactions Using the Principle of Virtual Work

since the beam is in equilibrium. Equation (i) is valid for all values of $\Delta_{v,A}$ and $\theta_{v,A}$, so that

$$wL - R_A = 0 \text{ and } (wL^2/2) - M_A = 0$$

Therefore,

$$R_A = wL \text{ and } M_A = wL^2/2$$

the results which would have been obtained by resolving forces vertically and taking moments about A. ■

Example 4.3
Calculate the reactions at the built-in end of the cantilever beam shown in Fig. 4.6(a). See Ex. 1.1.

In this example the load, W, produces reactions of vertical force, moment, and torque at the built-in end. The vertical force and moment are the same as in Example 4.1. To determine the torque reaction we impose a small, virtual displacement, $\Delta_{v,C}$, vertically downwards at C. This causes the beam AB to rotate as a rigid body through an angle, $\theta_{v,AB}$, which is given by

$$\theta_{v,AB} = \Delta_{v,C}/a \tag{i}$$

Alternatively, we could have imposed a small virtual rotation, $\theta_{v,AB}$, on the beam, which would have resulted in a virtual vertical displacement of C equal to $a\theta_{v,AB}$; clearly the two approaches produce identical results.

The total virtual work done on the beam is then given by

$$W_t = W\Delta_{v,C} - T_A\theta_{v,AB} = 0 \tag{ii}$$

since the beam is in equilibrium. Substituting for $\theta_{v,AB}$, in Eq. (ii) from Eq. (i), we have

$$T_A = Wa$$

which is the result which would have been obtained by considering the statical equilibrium of the beam. ■

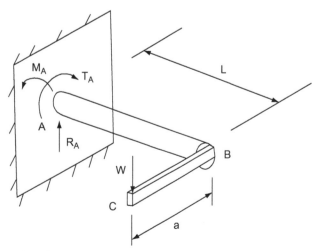

FIGURE 4.6 Beam of Example 4.3

Example 4.4

Calculate the support reactions in the simply supported beam shown in Fig. 4.7. See Ex. 1.1.

Only a vertical load is applied to the beam, so that only vertical reactions, R_A and R_C, are produced.

Suppose that the beam at C is given a small imaginary, that is, a virtual, displacement, $\Delta_{v,C}$, in the direction of R_C as shown in Fig. 4.7(b). Since we are concerned here solely with the *external* forces acting on the beam, we may

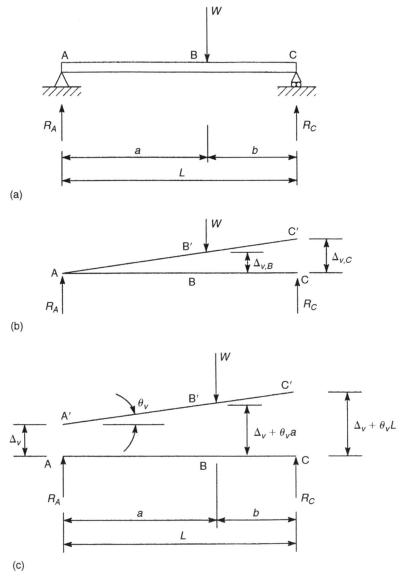

(a)

(b)

(c)

FIGURE 4.7 Use of the Principle of Virtual Work to Calculate Support Reactions

regard the beam as a rigid body. The beam therefore rotates about A so that C moves to C′ and B moves to B′. From similar triangles. we see that

$$\Delta_{v,\mathrm{B}} = \frac{a}{a+b}\Delta_{v,\mathrm{C}} = \frac{a}{L}\Delta_{v,\mathrm{C}} \tag{i}$$

The total virtual work, W_t, done by all the forces acting on the beam is given by

$$W_t = R_\mathrm{C}\Delta_{v,\mathrm{C}} - W\Delta_{v,\mathrm{B}} \tag{ii}$$

Note that the work done by the load, W, is negative, since $\Delta_{v,\mathrm{B}}$ is in the opposite direction to its line of action. Note also that the support reaction, R_A, does no work, since the beam rotates only about A. Now substituting for $\Delta_{v,\mathrm{B}}$ in Eq. (ii) from Eq. (i), we have

$$W_t = R_\mathrm{C}\Delta_{v,\mathrm{C}} - W\frac{a}{L}\Delta_{v,\mathrm{C}} \tag{iii}$$

Since the beam is in equilibrium, W_t is zero from the principle of virtual work. Hence, from Eq. (iii),

$$R_\mathrm{C}\Delta_{v,\mathrm{C}} - W\frac{a}{L}\Delta_{v,\mathrm{C}} = 0$$

which gives

$$R_\mathrm{C} = W\frac{a}{L}$$

which is the result which would have been obtained from a consideration of the moment equilibrium of the beam about A. The determination of R_A follows in a similar manner. Suppose now that, instead of the single displacement $\Delta_{v,\mathrm{C}}$, the complete beam is given a vertical virtual displacement, Δ_v, together with a virtual rotation, θ_v, about A, as shown in Fig. 4.7(c). The total virtual work, W_t, done by the forces acting on the beam is now given by

$$W_t = R_\mathrm{A}\Delta_v - W\left(\Delta_v + a\theta_v\right) + R_\mathrm{C}\left(\Delta_v + L\theta_v\right) = 0 \tag{iv}$$

since the beam is in equilibrium. Rearranging Eq. (iv),

$$(R_\mathrm{A} + R_\mathrm{C} - W)\Delta_v + (R_\mathrm{C}L - Wa)\theta_v = 0 \tag{v}$$

Equation (v) is valid for all values of Δ_v and θ_v so that

$$R_\mathrm{A} + R_\mathrm{C} - W = 0, \ R_\mathrm{C}L - Wa = 0$$

which are the equations of equilibrium we would have obtained by resolving forces vertically and taking moments about A.

It is not being suggested here that the application of the principles of statics should be abandoned in favor of the principle of virtual work. The purpose of Examples 4.1–4.4 is to illustrate the application of a virtual displacement and the manner in which the principle is used.

4.2.3 Virtual work in a deformable body

In structural analysis, we are not generally concerned with forces acting on a rigid body. Structures and structural members deform under load, which means that, if we assign a virtual displacement to a particular point in a structure, not all points in the structure suffer the same virtual displacement, as would be the case if the structure were rigid. This means that the virtual work produced by the internal forces is not zero, as it is in the rigid body case, since the virtual work produced by the self-equilibrating forces on adjacent particles does not cancel out. The total virtual work produced by applying a virtual displacement to a deformable body acted upon by a system of external forces is therefore given by Eq. (4.6).

If the body is in equilibrium under the action of the external force system, then every particle in the body is also in equilibrium. Therefore, from the principle of virtual work, the virtual work done by the forces acting on the particle is zero, irrespective of whether the forces are external or internal. It follows that, since the virtual work is zero for all particles in the body, it is zero for the complete body and Eq. (4.6) becomes

$$W_e + W_i = 0 \tag{4.8}$$

Note that, in this argument, only the conditions of equilibrium and the concept of work are employed. Equation (4.8) therefore does not require the deformable body to be linearly elastic (i.e., it need not obey Hooke's law), so that the principle of virtual work may be applied to any body or structure that is rigid, elastic, or plastic. The principle does require that displacements, whether real or imaginary, must be small, so that we may assume that external and internal forces are unchanged in magnitude and direction during the displacements. In addition, the virtual displacements must be compatible with the geometry of the structure and the constraints that are applied, such as those at a support. The exception is the situation we have in Examples 4.1–4.4, where we apply a virtual displacement at a support. This approach is valid, since we include the work done by the support reactions in the total virtual work equation.

4.2.4 **Work done by internal force systems**

The calculation of the work done by an external force is straightforward in that it is the product of the force and the displacement of its point of application in its own line of action (Eqs. (4.1), (4.2), or (4.3)), whereas the calculation of the work done by an internal force system during a displacement is much more complicated. Generally, no matter how complex a loading system is, it may be simplified to a combination of up to four load types: axial load, shear force, bending moment, and torsion; these in turn produce corresponding internal force systems. We now consider the work done by these internal force systems during arbitrary virtual displacements.

Axial force

Consider the elemental length, δx, of a structural member as shown in Fig. 4.8 and suppose that it is subjected to a positive internal force system comprising a normal force (i.e., axial force), N; a shear force, S; a bending moment, M; and a torque, T, produced by some external loading system acting on the structure of which the member is part. The stress distributions corresponding to these internal forces are related to an axis system whose origin coincides with the centroid of area of the cross-section. We are, in fact, using these stress distributions in the derivation of expressions for internal virtual work in linearly elastic structures, so that it is logical to assume the same origin of axes here; we also assume that the y axis is an axis of symmetry. Initially, we consider the normal force, N.

The direct stress, σ, at any point in the cross-section of the member is given by $\sigma = N/A$. Therefore the normal force on the element δA at the point (z, y) is

$$\delta N = \sigma \delta A = \frac{N}{A} \delta A$$

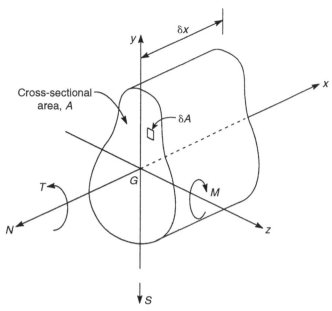

FIGURE 4.8 Virtual Work Due to Internal Force System

Suppose now that the structure is given an arbitrary virtual displacement, which produces a virtual axial strain, ε_v, in the element. The internal virtual work, $\delta w_{i,N}$, done by the axial force on the elemental length of the member is given by

$$\delta w_{i,N} = \int_A \frac{N}{A} \; dA\varepsilon_v\delta x$$

which, since $\int_A dA = A$, reduces to

$$\delta w_{i,N} = N\varepsilon_v\delta x \tag{4.9}$$

In other words, the virtual work done by N is the product of N and the virtual axial displacement of the element of the member. For a member of length L, the virtual work, $w_{i,N}$, done during the arbitrary virtual strain is then

$$w_{i,N} = \int_L N\varepsilon_v \; dx \tag{4.10}$$

For a structure comprising a number of members, the total internal virtual work, $W_{i,N}$, done by axial force is the sum of the virtual work of each of the members. Therefore,

$$w_{i,N} = \sum \int_L N\varepsilon_v \; dx \tag{4.11}$$

Note that, in the derivation of Eq. (4.11), we make no assumption regarding the material properties of the structure, so that the relationship holds for non-elastic as well as elastic materials. However, for a linearly elastic material, that is, one that obeys Hooke's law, we can express the virtual strain in terms of an equivalent virtual normal force:

$$\varepsilon_v = \frac{\sigma_v}{E} = \frac{N_v}{EA}$$

Therefore, if we designate the *actual* normal force in a member by N_A, Eq. (4.11) may be expressed in the form

$$w_{i,N} = \sum \int_L \frac{N_A N_v}{EA} \tag{4.12}$$

Shear force

The shear force, S, acting on the member section in Fig. 4.8 produces a distribution of vertical shear stress that depends upon the geometry of the cross-section. However, since the element, δA, is infinitesimally small, we may regard the shear stress, τ, as constant over the element. The shear force, δS, on the element is then

$$\delta S = \tau \delta A \tag{4.13}$$

Suppose that the structure is given an arbitrary virtual displacement which produces a virtual shear strain, γ_v, at the element. This shear strain represents the angular rotation in a vertical plane of the element $\delta A \times \delta x$ relative to the longitudinal centroidal axis of the member. The vertical displacement at the section being considered is therefore $\gamma_v \delta x$. The internal virtual work, $\delta w_{i,S}$, done by the shear force, S, on the elemental length of the member is given by

$$\delta w_{i,S} = \int_A \tau \, dA \gamma_v \delta x$$

A uniform shear stress through the cross-section of a beam may be assumed if we allow for the actual variation by including a form factor, β.[1] The expression for the internal virtual work in the member may then be written

$$\delta w_{i,S} = \int_A \beta \left(\frac{S}{A} \right) dA \gamma_v \delta x$$

or

$$\delta w_{i,S} = \beta S \gamma_v \delta x \tag{4.14}$$

Hence, the virtual work done by the shear force during the arbitrary virtual strain in a member of length L is

$$w_{i,S} = \beta \int_L S \gamma_v \, dx \tag{4.15}$$

For a linearly elastic member, as in the case of axial force, we may express the virtual shear strain, γ_v, in terms of an equivalent virtual shear force, S_v:

$$\gamma_v = \frac{\tau_v}{G} = \frac{S_v}{GA}$$

so that, from Eq. (4.15),

$$w_{i,S} = \beta \int_L \frac{S_A S_v}{GA}\ dx \tag{4.16}$$

For a structure comprising a number of linearly elastic members, the total internal work, $W_{i,S}$, done by the shear forces is

$$W_{i,S} = \sum \beta \int_L \frac{S_A S_v}{GA}\ dx \tag{4.17}$$

Bending moment

The bending moment, M, acting on the member section in Fig. 4.8 produces a distribution of direct stress, σ, through the depth of the member cross-section. The normal force on the element, δA, corresponding to this stress is therefore $\sigma \delta A$. Again, we suppose that the structure is given a small arbitrary virtual displacement which produces a virtual direct strain, ε_v, in the element $\delta A \times \delta x$. Therefore, the virtual work done by the normal force acting on the element δA is $\sigma \delta A \varepsilon_v \delta x$. Hence, integrating over the complete cross-section of the member, we obtain the internal virtual work, $\delta w_{i,M}$, done by the bending moment, M, on the elemental length of member:

$$\delta w_{i,M} = \int_A \sigma\ dA\varepsilon_v \delta x \tag{4.18}$$

The virtual strain, ε_v, in the element $\delta A \times \delta x$ is, from Eq. (16.2), given by

$$\varepsilon_v = \frac{y}{R_v}$$

where R_v is the radius of curvature of the member produced by the virtual displacement. Thus, substituting for ε_v in Eq. (4.18), we obtain

$$\delta w_{i,M} = \int_A \sigma \frac{y}{R_v}\ dA\delta x$$

or, since $\sigma y \delta A$ is the moment of the normal force on the element, δA, about the z axis,

$$\delta w_{i,M} = \frac{M}{R_v}\delta x$$

Therefore, for a member of length L, the internal virtual work done by an actual bending moment, M_A, is given by

$$w_{i,M} = \int_L \frac{M}{R_v}\ dx \tag{4.19}$$

In the derivation of Eq. (4.19), no specific stress–strain relationship has been assumed, so that it is applicable to a non-linear system. For the particular case of a linearly elastic system, the virtual curvature $1/R_v$ may be expressed in terms of an equivalent virtual bending moment, M_v, using the relationship of Eq. (16.8):

$$\frac{1}{R_v} = \frac{M_v}{EI}$$

Substituting for $1/R_v$ in Eq. (4.19), we have

$$w_{im} = \int_L \frac{M_A M_v}{EI} \, dx \qquad (4.20)$$

so that, for a structure comprising a number of members, the total internal virtual work, $W_{i,m}$, produced by bending is

$$W_{im} = \sum \int_L \frac{M_A M_v}{EI} \, dx \qquad (4.21)$$

Torsion

The internal virtual work, $w_{i,T}$, due to torsion in the particular case of a linearly elastic circular section bar may be found in a similar manner and is given by

$$w_{i,T} = \int_L \frac{T_A T_v}{GI_o} \, dx \qquad (4.22)$$

in which I_o is the polar second moment of area of the cross-section of the bar (see Example 3.1). For beams of a non-circular cross-section, I_o is replaced by a torsion constant, J, which, for many practical beam sections, is determined empirically.

Hinges

In some cases, it is convenient to impose a virtual rotation, θ_v, at some point in a structural member where, say, the actual bending moment is M_A. The internal virtual work done by M_A is then $M_A \theta_v$ (see Eq. (4.3)); physically this situation is equivalent to inserting a hinge at the point.

Sign of internal virtual work

So far, we have derived expressions for internal work without considering whether it is positive or negative in relation to external virtual work.

Suppose that the structural member, AB, in Fig. 4.9(a) is, say, a member of a truss and that it is in equilibrium under the action of two externally applied axial tensile loads, P; clearly the internal axial, that is normal, force at any section of the member is P. Suppose now that the member is given a virtual extension, δ_v, such that B moves to B′. Then, the virtual work done by the applied load, P, is positive, since the displacement, δ_v, is in the same direction as its line of action. However, the virtual work done by the internal force, $N\,(=P)$, is negative, since the displacement of B is in the opposite direction to its line of action; in other words, work is done *on* the member. Thus, from Eq. (4.8), we see that, in this case,

$$W_e = W_i \qquad (4.23)$$

Equation (4.23) applies if the virtual displacement is a contraction and not an extension, in which case, the signs of the external and internal virtual work in Eq. (4.8) are reversed. Clearly, this applies equally if P is a compressive load. The arguments may be extended to structural members subjected to shear, bending, and torsional loads, so that Eq. (4.23) is generally applicable.

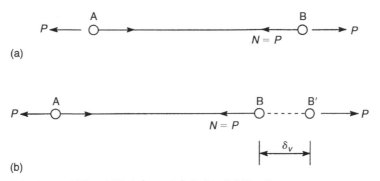

FIGURE 4.9 Sign of the Internal Virtual Work in an Axially Loaded Member

4.2.5 Virtual work due to external force systems

So far in our discussion, we have only considered the virtual work produced by externally applied concentrated loads. For completeness, we must also consider the virtual work produced by moments, torques, and distributed loads.

In Fig. 4.10, a structural member carries a distributed load, $w(x)$, and, at a particular point, a concentrated load, W; a moment, M; and a torque, T; together with an axial force, P. Suppose that, at the point, a virtual displacement is imposed having translational components, $\Delta_{v,y}$ and $\Delta_{v,x}$, parallel to the y and x axes, respectively, and rotational components, θ_v and ϕ_v, in the yx and zy planes, respectively.

If we consider a small element, δx, of the member at the point, the distributed load may be regarded as constant over the length δx and acting, in effect, as a concentrated load $w(x)\delta x$. The virtual work, w_e, done by the complete external force system is therefore given by

$$w_e = W\Delta_{v,y} + P\Delta_{v,x} + M\theta_v + T\phi_v + \int_L w(x)\Delta_{v,y}\ dx$$

For a structure comprising a number of load positions, the total external virtual work done is then

$$W_e = \sum \left[W\Delta_{v,y} + P\Delta_{v,x} + M\theta_v + T\phi_v + \int_L w(x)\Delta_{v,y}\ dx \right] \tag{4.24}$$

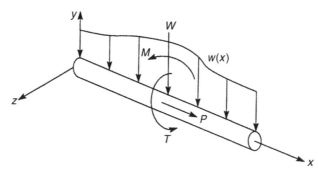

FIGURE 4.10 Virtual Work Due to Externally Applied Loads

In Eq. (4.24), a complete set of external loads need not be applied at every loading point so, in fact, the summation is for the appropriate number of loads. Further, the virtual displacements in the equation are related to forces and moments applied in a vertical plane. We could, of course, have forces and moments and components of the virtual displacement in a horizontal plane, in which case Eq. (4.24) is extended to include their contribution.

The internal virtual work equivalent of Eq. (4.24) for a linear system is, from Eqs. (4.12), (4.17), (4.21), and (4.22),

$$W_i = \sum \left[\int_L \frac{N_A N_v}{EA} \, dx + \beta \int_L \frac{S_A s_v}{GA} \, dx + \int_L \frac{M_A M_v}{EI} \, dx + \int_L \frac{T_A T_v}{GJ} \, dx + M_A \theta_v \right] \tag{4.25}$$

in which the last term on the right-hand side is the virtual work produced by an actual internal moment at a hinge (see previous text). Note that the summation in Eq. (4.25) is taken over all the *members* of the structure.

4.2.6 Use of virtual force systems

So far, in all the structural systems we have considered, virtual work is produced by actual forces moving through imposed virtual displacements. However, the actual forces are not related to the virtual displacements in any way, since, as we have seen, the magnitudes and directions of the actual forces are unchanged by the virtual displacements so long as the displacements are small. Thus, the principle of virtual work applies for *any* set of forces in equilibrium and *any* set of displacements. Equally, therefore, we could specify that the forces are a set of virtual forces *in equilibrium* and that the displacements are actual displacements. Therefore, instead of relating actual external and internal force systems through virtual displacements, we can relate actual external and internal displacements through virtual forces.

If we apply a virtual force system to a deformable body it induces an internal virtual force system that moves through the actual displacements; internal virtual work therefore is produced. In this case, for example, Eq. (4.10) becomes

$$w_{i,N} = \int_L N_v \varepsilon_A \, dx$$

in which N_v is the internal virtual normal force and ε_A is the actual strain. Then, for a linear system, in which the actual internal normal force is N_A, $\varepsilon_A = N_A/EA$, so that, for a structure comprising a number of members, the total internal virtual work due to a virtual normal force is

$$W_{i,N} = \sum \int_L \frac{N_v N_A}{EA} \, dx$$

which is identical to Eq. (4.12). Equations (4.17), (4.21), and (4.22) may be shown to apply to virtual force systems in a similar manner.

4.3 APPLICATIONS OF THE PRINCIPLE OF VIRTUAL WORK

We have now seen that the principle of virtual work may be used in the form of either imposed virtual displacements or imposed virtual forces. Generally, the former approach, as we saw in Example 4.4, is used to determine forces, while the latter is used to obtain displacements.

For statically determinate structures, the use of virtual displacements to determine force systems is a relatively trivial use of the principle, although problems of this type provide a useful illustration of the method. The real power of this approach lies in its application to the solution of statically indeterminate structures. However, the use of virtual forces is particularly useful in determining actual displacements of structures. We shall illustrate both approaches by examples.

Example 4.5

Determine the bending moment at the point B in the simply supported beam ABC shown in Fig. 4.11(a). See Ex. 1.1.

We determined the support reactions for this particular beam in Example 4.4. In this example, however, we are interested in the actual internal moment, M_B, at the point of application of the load. We must therefore impose a virtual displacement that relates the internal moment at B to the applied load and excludes other unknown external forces, such as the support reactions, and unknown internal force systems, such as the bending moment distribution along the length of the beam. Therefore, if we imagine that the beam is hinged at B and that the lengths AB and BC are rigid, a virtual displacement, $\Delta_{v,B}$, at B results in the displaced shape shown in Fig. 4.11(b).

Note that the support reactions at A and C do no work and that the internal moments in AB and BC do no work because AB and BC are rigid links. From Fig. 4.11(b),

$$\Delta_{v,B} = \alpha\beta = b\alpha \qquad (i)$$

Hence,

$$\alpha = \frac{a}{b}\beta$$

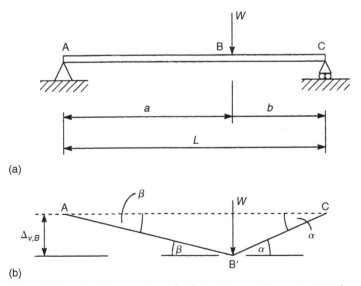

(a)

(b)

FIGURE 4.11 Determination of Bending Moment at a Point in the Beam of Example 4.5 Using Virtual Work

and the angle of rotation of BC relative to AB is then

$$\theta_B = \beta + \alpha = \beta\left(1 + \frac{a}{b}\right) = \frac{L}{b}\beta \tag{ii}$$

Now, equating the external virtual work done by W to the internal virtual work done by M_B (see Eq. (4.23)), we have

$$W\Delta_{v,B} = M_B\theta_B \tag{iii}$$

Substituting in Eq. (iii) for $\Delta_{v,B}$ from Eq. (i) and for θ_B from Eq. (ii), we have

$$Wa\beta = M_B\frac{L}{b}\beta$$

which gives

$$M_B = \frac{Wab}{L}$$

which is the result we would have obtained by calculating the moment of R_C ($= Wa/L$ from Example 4.4) about B.
◼

Example 4.6

Determine the force in the member AB of the truss shown in Fig. 4.12(a). See Ex. 1.1.

We are required to calculate the force in the member AB, so that again we need to relate this internal force to the externally applied loads without involving the internal forces in the remaining members of the truss. We therefore impose a virtual extension, $\Delta_{v,B}$, at B in the member AB, such that B moves to B'. If we assume that the remaining members are rigid, the forces in them do no work. Further, the triangle BCD rotates as a rigid body about D to B'C'D, as shown in Fig. 4.12(b). The horizontal displacement of C, Δ_C, is then given by

$$\Delta_C = 4\alpha$$

while

$$\Delta_{v,B} = 3\alpha$$

Hence,

$$\Delta_C = \frac{4\Delta_{v,B}}{3} \tag{i}$$

Equating the external virtual work done by the 30 kN load to the internal virtual work done by the force, F_{BA}, in the member, AB, we have (see Eq. (4.23) and Fig. 4.9)

$$30\Delta_C = F_{BA}\Delta_{v,B} \tag{ii}$$

Substituting for Δ_C from Eq. (i) in Eq. (ii),

$$30 \times \frac{4}{3}\Delta_{v,B} = F_{BA}\Delta_{v,B}$$

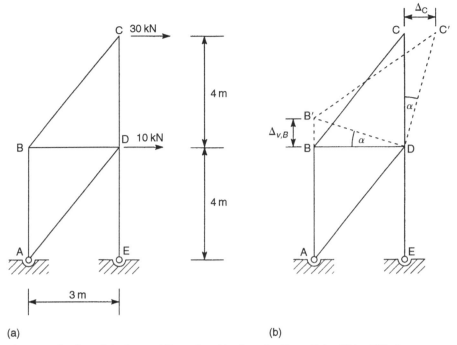

FIGURE 4.12 Determination of the Internal Force in a Member of a Truss Using Virtual Work

from which

$$F_{BA} = +40 \text{ kN}$$

In the preceding, we are, in effect, assigning a positive (that is, F_{BA} is tensile) sign to F_{BA} by imposing a virtual extension on the member AB.

The actual sign of F_{BA} is then governed by the sign of the external virtual work. Thus, if the 30 kN load were in the opposite direction to Δ_C, the external work done would have been negative, so that F_{BA} would be negative and therefore compressive. This situation can be verified by inspection. Alternatively, for the loading shown in Fig. 4.12(a), a contraction in AB implies that F_{BA} is compressive. In this case, DC would have rotated in a counter-clockwise sense, Δ_C would have been in the opposite direction to the 30 kN load so that the external virtual work done would be negative, resulting in a negative value for the compressive force F_{BA}; F_{BA} would therefore be ten-sile, as before. Note also that the 10 kN load at D does no work, since D remains undisplaced.

We now consider problems involving the use of virtual forces. Generally, we require the displacement of a particular point in a structure, so that, if we apply a virtual force to the structure at the point and in the direction of the required displacement, the external virtual work done will be the product of the virtual force and the actual displacement, which may then be equated to the internal virtual work produced by the internal virtual force system moving through actual displacements. Since the choice of the virtual force is arbitrary, we may give it any con-venient value; the simplest type of virtual force is therefore a unit load and the method then becomes the *unit load method* (see also Section 5.5).

Example 4.7

Determine the vertical deflection of the free end of the cantilever beam shown in Fig. 4.13(a). See Ex. 1.1.

Let us suppose that the actual deflection of the cantilever at B produced by the uniformly distributed load is v_B and that a vertically downward virtual unit load is applied at B before the actual deflection takes place. The external virtual work done by the unit load is, from Fig. 4.13(b), $1 v_B$. The deflection, v_B, is assumed to be caused by bending only; that is, we ignore any deflections due to shear. The internal virtual work is given by Eq. (4.21), which, since only one member is involved, becomes

$$W_{i,M} = \int_0^L \frac{M_A M_v}{EI} \, dx \tag{i}$$

The virtual moments, M_v, are produced by a unit load, so we replace M_v by M_1. Then,

$$W_{i,M} = \int_0^L \frac{M_A M_1}{EI} \, dx \tag{ii}$$

At any section of the beam a distance x from the built-in end,

$$M_A = -\frac{w}{2}(L - x)^2, \quad M_1 = -1(L - x)$$

Substituting for M_A and M_1 in Eq. (ii) and equating the external virtual work done by the unit load to the internal virtual work, we have

$$1 v_B = \int_0^L \frac{w}{2EI}(L - x)^3 \, dx$$

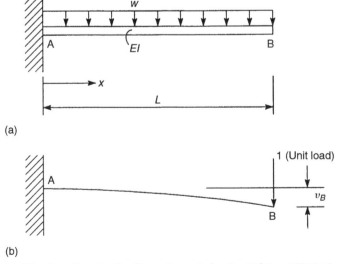

(a)

(b)

FIGURE 4.13 Deflection of the Free End of a Cantilever Beam Using the Unit Load Method

which gives

$$1\,v_B = -\frac{w}{2EI}\left[\frac{1}{4}(L-x)^4\right]_0^L$$

so that

$$v_B = \frac{wL^4}{8EI}$$

Note that v_B is in fact negative, but the positive sign here indicates that it is in the same direction as the unit load.

Example 4.7 MATLAB

Repeat Example 4.7 using the Symbolic Math Toolbox in MATLAB. See Ex. 1.1.

Calculation of the vertical deflection of the free end of the cantilever shown in Fig. 4.13(a) is obtained through the following MATLAB file:

```
% Declare any needed variables
syms M_A M_V W_e W_i E I w x L v_B

% Define the moments due to the applied load (M_A) and the unit virtual load (M_V)
M_A = -0.5*w*(L-x)^2;
M_V = -1*(L-x);

% Define equations for the external (W_e) and internal (W_i) virtual work
W_e = 1*v_B;
W_i = int(M_A*M_V/(E*I),x,0,L); % From Eq. (4.21)

% Equate W_e and W_i, and solve for the free end vertical displacement (v_B)
v_B = solve(W_e-W_i,v_B);

% Output v_B to the Command Window
disp(['v_B =' char(v_B)])
```

The Command Window output resulting from this MATLAB file is as follows:

```
v_B = (L^4*w)/(8*E*I)
```

Example 4.8

Determine the rotation, that is, the slope, of the beam ABC shown in Fig. 4.14(a) at A. See Ex. 1.1.

The actual rotation of the beam at A produced by the actual concentrated load, W, is θ_A. Let us suppose that a virtual unit moment is applied at A before the actual rotation takes place, as shown in Fig. 4.14(b). The virtual unit moment induces virtual support reactions of $R_{v,A}$ ($= 1/L$) acting downward and $R_{v,C}$ ($= 1/L$) acting upward. The actual internal bending moments are

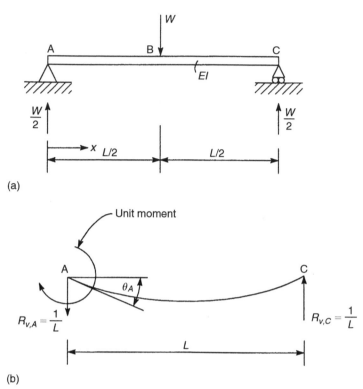

(a)

(b)

FIGURE 4.14 Determination of the Rotation of a Simply Supported Beam at a Support Using the Unit Load Method

$$M_A = +\frac{W}{2}x, \quad 0 \le x \le L/2$$

$$M_A = +\frac{W}{2}(L - x), \quad L/2 \le x \le L$$

The internal virtual bending moment is

$$M_v = 1 - \frac{1}{L}x, \quad 0 \le x \le L$$

The external virtual work done is $1\theta_A$ (the virtual support reactions do no work, as there is no vertical displacement of the beam at the supports), and the internal virtual work done is given by Eq. (4.21). Hence,

$$1\theta_A = \frac{1}{EI}\left[\int_0^{L/2} \frac{W}{2}x\left(1 - \frac{x}{L}\right) \, dx + \int_{L/2}^L \frac{W}{2}(L - x)\left(1 - \frac{x}{L}\right) \, dx\right] \tag{i}$$

Simplifying Eq. (i), we have

$$\theta_A = \frac{W}{2EIL}\left[\int_0^{L/2} (Lx - x^2) \, dx + \int_{L/2}^L (L - x)^2 \, dx\right] \tag{ii}$$

Hence,

$$\theta_A = \frac{W}{2EIL}\left\{\left[L\frac{x^2}{2} - \frac{x^3}{3}\right]_0^{L/2} - \frac{1}{3}\left[(L-x)^3\right]_{L/2}^L\right\}$$

from which

$$\theta_A = \frac{WL^2}{16EI}$$

Example 4.8 MATLAB

Repeat Example 4.8 using the Symbolic Math Toolbox in MATLAB. See Ex. 1.1.

Calculation of the slope of the beam shown in Fig. 4.14(a) at A is obtained through the following MATLAB file:

```
% Declare any needed variables
syms M_A M_V W x L theta_A E I

% Define the moments due to the applied load (M_A) and the unit virtual load (M_V)
M_A = [W/2*x; % 0 >= x >= L/2
W/2*(L-x)]; % L/2 >= x >= L
M_V = 1 - x/L; % 0 >= x >= L

% Define equations for the external (W_e) and internal (W_i) virtual work
W_e = 1*theta_A;
W_i = (int(M_A(1)*M_V,x,0,L/2) + int(M_A(2)*M_V,x,L/2,L))/(E*I); % From Eq. (4.21)

% Equate W_e and W_i, and solve for the slope of the beam (theta_A)
theta_A = solve(W_e-W_i,theta_A);

% Output theta_A to the Command Window
disp(['theta_A =' char(theta_A)])
```

The Command Window output resulting from this MATLAB file is as follows:

```
theta_A=(L^2*W)/(16*E*I)
```

Example 4.9

Calculate the vertical deflection of the joint B and the horizontal movement of the support D in the truss shown in Fig. 4.15(a). The cross-sectional area of each member is 1800 mm^2 and Young's modulus, E, for the material of the members is 200,000 N/mm^2. See Ex. 1.1.

The virtual force systems, that is, unit loads, required to determine the vertical deflection of B and the horizontal deflection of D are shown in Figs. 4.15(b) and (c), respectively. Therefore, if the actual vertical deflection at

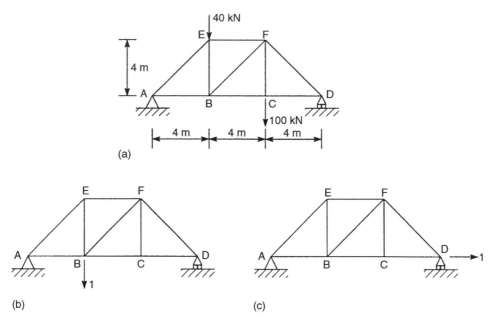

FIGURE 4.15 Deflection of a Truss Using the Unit Load Method

B is $\delta_{B,v}$ and the horizontal deflection at D is $\delta_{D,h}$, the external virtual work done by the unit loads is $1\delta_{B,v}$ and $1\delta_{D,h}$, respectively. The internal actual and virtual force systems constitute axial forces in all the members.

These axial forces are constant along the length of each member, so that, for a truss comprising n members, Eq. (4.12) reduces to

$$W_{i,N} = \sum_{j=1}^{n} \frac{F_{A,j} F_{v,j} L_j}{E_j A_j} \tag{i}$$

in which $F_{A,j}$ and $F_{v,j}$ are the actual and virtual forces in the jth member, which has a length L_j, an area of cross-section A_j, and a Young's modulus E_j.

Since the forces $F_{v,j}$ are due to a unit load, we write Eq. (i) in the form

$$W_{i,N} = \sum_{j=1}^{n} \frac{F_{A,j} F_{1,j} L_j}{E_j A_j} \tag{ii}$$

Also, in this particular example, the area of cross-section, A, and Young's modulus, E, are the same for all members, so that it is sufficient to calculate $\sum_{j=1}^{n} F_{A,j} F_{1,j} L_j$ then divide by EA to obtain $W_{i,N}$.

The forces in the members, whether actual or virtual, may be calculated by the method of joints.[1] Note that the support reactions corresponding to the three sets of applied loads (one actual and two virtual) must be calculated before the internal force systems can be determined. However, in Fig. 4.15(c), it is clear from inspection that $F_{1,AB} = F_{1,BC} = F_{1,CD} = +1$, while the forces in all other members are zero. The calculations are presented in Table 4.1; note that positive signs indicate tension and negative signs compression.

Table 4.1 Example 4.9

Member	L (m)	F_A (kN)	$F_{1,B}$	$F_{1,D}$	$F_A F_{1,B} L$ (kN m)	$F_A F_{1,D} L$ (kN m)
AE	5.7	−84.9	−0.94	0	+451.4	0
AB	4.0	+60.0	+0.67	+1.0	+160.8	+240.0
EF	4.0	−60.0	−0.67	0	+160.8	0
EB	4.0	+20.0	+0.67	0	+53.6	0
BF	5.7	−28.3	+0.47	0	−75.2	0
BC	4.0	+80.0	+0.33	+1.0	+105.6	+320.0
CD	4.0	+80.0	+0.33	+1.0	+105.6	+320.0
CF	4.0	+100.0	0	0	0	0
DF	5.7	−113.1	−0.47	0	+301.0	0
					$\Sigma = +1263.6$	$\Sigma = +880.0$

Equating internal and external virtual work done (Eq. (4.23)), we have

$$1\delta_{B,v} = \frac{1263.6 \times 10^6}{200,000 \times 1,800}$$

from which

$$\delta_{B,v} = 3.51 \text{ mm}$$

and

$$1\delta_{D,h} = \frac{880 \times 10^6}{200,000 \times 1,800}$$

which gives

$$\delta_{D,h} = 2.44 \text{ mm}$$

Both deflections are positive, which indicates that the deflections are in the directions of the applied unit loads. Note that, in the preceding, it is unnecessary to specify units for the unit load, since the unit load appears, in effect, on both sides of the virtual work equation (the internal F_1 forces are directly proportional to the unit load). ■

Example 4.10

Determine the components of the deflection of the point C in the frame shown in Fig. 4.16(a); consider the effect of bending only. See Ex. 1.1.

The horizontal and vertical components of the deflection of C may be found by applying unit loads in turn at C, as shown in Figs. 4.16(b) and (c). The internal work done by this virtual force system, that is, the unit loads acting through real displacements, is given by Eq. (4.20), in which, for AB,

$$M_A = WL - (wx^2/2), \quad M_v = 1L(\text{horiz.}), \quad M_v = -1x \text{ (vert.)}$$

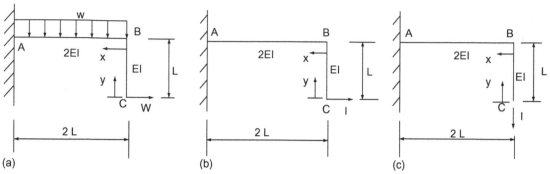

FIGURE 4.16 Frame of Example 4.10

for BC,

$$M_A = Wy, \quad M_v = 1y \text{ (horiz.)}, \quad M_v = 0 \text{ (vert.)}$$

Considering the horizontal deflection first, the total internal virtual work is

$$W_i = \int_0^L (Wy^2/EI) \; dy + \int_0^{2L} [WL - (wx^2/2)](L/2EI) \; dx$$

that is,

$$W_i = (W/EI)[y^3/3]_0^L + (L/2EI)[WLx - (wx^3/6)]_0^{2L}$$

which gives

$$W_i = 2L^3(2W - wL)/3EI \tag{i}$$

The virtual external work done by the unit load is $1\delta_{C,H}$, where $\delta_{C,H}$ is the horizontal component of the deflection of C. Equating this to the internal virtual work given by Eq. (i) gives

$$\delta_{C,H} = 2L^3(2W - wL)/3EI \tag{ii}$$

Now, considering the vertical component of deflection,

$$W_i = (1/2EI)\int_0^{2L} [WL - (wx^2/2)](-x)dx$$

Note that, for BC, $M_v = 0$. Integrating this expression and substituting the limits gives

$$W_i = L^3(-W + wL)/EI \tag{iii}$$

The external virtual work done is $1\delta_{C,V}$, where $\delta_{C,V}$ is the vertical component of the deflection of C. Equating the internal and external virtual work gives

$$\delta_{C,V} = L^3(-W + wL)/EI \tag{iv}$$

Note that the components of deflection can be either positive or negative, depending on the relative magnitudes of W and w. A positive value indicates a deflection in the direction of the applied unit load, a negative one indicates a deflection in the opposite direction to the applied unit load.

Example 4.11

A cantilever beam AB takes the form of a quadrant of a circle of radius, R, and is positioned on a horizontal plane. If the beam carries a vertically downward load, W, at its free end and its bending and torsional stiffnesses are EI and GJ, respectively, calculate the vertical component of the deflection at its free end. See Ex. 1.1.

A plan view of the beam is shown in Fig. 4.17. To determine the vertical displacement of B, we apply a virtual unit load at B vertically downward (i.e., into the plane of the paper).

At a section of the beam where the radius at the section makes an angle, α, with the radius through B,

$$M_A = Wp = WR\sin\alpha, \quad M_v = 1R\sin\alpha,$$
$$T_A = W(R - R\cos\alpha), \quad T_v = 1(R - R\cos\alpha)$$

The total internal virtual work done is given by the summation of Eqs. (4.20) and (4.22), that is,

$$W_i = (1/EI)\int_0^{\pi/2} WR^2\sin^2\alpha R d\alpha + (1/GJ)\int_0^{\pi/2} WR^2(1 - \cos\alpha)^2 R\,d\alpha \qquad \text{(i)}$$

Integrating and substituting the limits in Eq. (i) gives

$$W_i = WR^3\{(\pi/4EI) + (1/GJ)[(3\pi/4) - 2]\} \qquad \text{(ii)}$$

The external virtual work done by the unit load is $1\delta_B$, so that equating with Eq. (ii), we obtain

$$\delta_B = WR^3\{(\pi/4EI) + (1/GJ)[(3\pi/4) - 2]\}$$

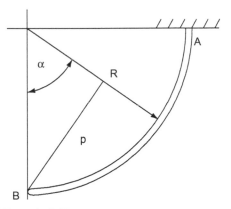

FIGURE 4.17 Cantilever Beam of Example 4.11

Reference

[1] Megson THG. Structural and stress analysis. 2nd ed. Oxford: Elsevier; 2005.

PROBLEMS

P.4.1 Use the principle of virtual work to determine the support reactions in the beam ABCD shown in Fig. P.4.1.

Answer: $R_A = 1.25W$, $R_D = 1.75W$

P.4.2 Find the support reactions in the beam ABC shown in Fig. P.4.2 using the principle of virtual work.

Answer: $R_A = (W + 2wL)/4$, $R_c = (3W + 2wL)/4$

P.4.2 MATLAB Use the Symbolic Math Toolbox in MATLAB to repeat Problem P.4.2, assuming that the vertical force W in Fig. P.4.2 is located $3L/5$ from A.

Answer: $R_A = 2W/5 + wL/2$, $R_C = 3W/5 + wL/2$

P.4.3 Determine the reactions at the built-in end of the cantilever beam ABC shown in Fig. P.4.3 using the principle of virtual work.

Answer: $R_A = 3W$, $M_A = 2.5WL$

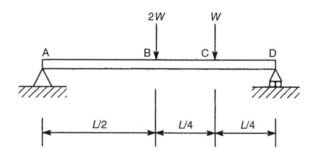

FIGURE P.4.1

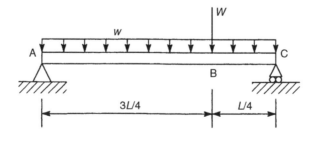

FIGURE P.4.2

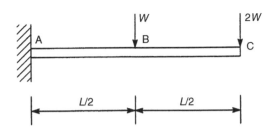

FIGURE P.4.3

P.4.4 Find the bending moment at the three-quarter-span point in the beam shown in Fig. P.4.4. Use the principle of virtual work.

Answer: $3wL^2/32$

P.4.5 Calculate the forces in the members FG, GD, and CD of the truss shown in Fig. P.4.5 using the principle of virtual work. All horizontal and vertical members are 1 m long.

Answer: $FG = +20$ kN, $GD = +28.3$ kN, $CD = -20$ kN

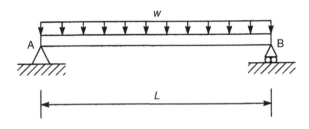

FIGURE P.4.4

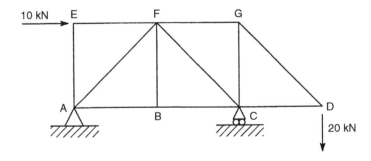

FIGURE P.4.5

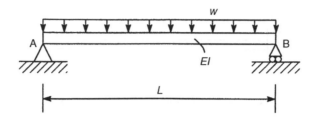

FIGURE P.4.6

P.4.6 Use the principle of virtual work to calculate the vertical displacements at the quarter- and mid-span points in the beam shown in Fig. P.4.6.

   Answer: $19wL^4/2048EI$, $5wL^4/384EI$ (both downward)

P.4.6 MATLAB Use the Symbolic Math Toolbox in MATLAB to repeat Problem P.4.6. Calculate the vertical displacements at increments of $L/8$ along the beam.

   Answer:

Distance from A	Vertical Displacement
0	0
$L/8$	$497L^4w/98,304EI$
$L/4$	$19L^4w/2,048EI$
$3L/8$	$395L^4w/32,768EI$
$L/2$	$5L^4w/384EI$
$5L/8$	$395L^4w/32,768EI$
$3L/5$	$19L^4w/2,048EI$
$7L/8$	$497L^4w/9,8304EI$
L	0

P.4.7 The frame shown in Fig. P.4.7 consists of a cranked beam simply supported at A and F and reinforced by a tie bar pinned to the beam at B and E. The beam carries a uniformly distributed load of intensity, w, over the outer parts AB and EF. Considering the effects of bending and axial load only determine the axial force in the tie bar and the bending moments at B and C. The bending stiffness of the beam is EI and its cross-sectional area is $3A$, while the corresponding values for the tie bar are EI and A.

   Answer: Force in tie bar is $9wL^3/8[L^2 + (4I/A)]$, $M(\text{at B}) = wL^2/2$ (counterclockwise),
   $M(\text{at C}) = (TL - wL^2)/2$ (clockwise).

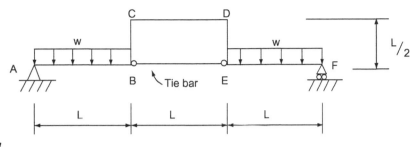

FIGURE P.4.7

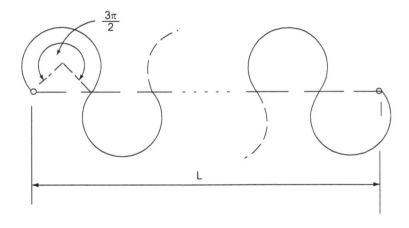

FIGURE P.4.8

P.4.8 The flat tension spring shown in Fig. P.4.8 consists of a length of wire of circular cross-section having a diameter, d, and Young's modulus, E. The spring consists of n open loops each of which subtends an angle of $3\pi/2$ radians at its center; the length between the ends of the spring is L. Considering bending and axial strains only calculate the stiffness of the spring.

Answer: $(\sqrt{2})\pi E d^2/L\{[48L^2(\pi+1)/n^2 d^2]+(3\pi-2)\}$

P.4.9 The circular fuselage frame shown in Fig. P.4.9 has a cut-out at the bottom and is loaded as shown. The member AB is pinned to the frame at A and B. If the second moment of area of the cross-section of the frame is 416,000 mm^4 and it has a Young's modulus of 267,000 N/mm^2 while the area of cross-section of the member AB is 130 mm^2 and its Young's modulus is 44,800 N/mm^2, calculate the axial force in the member AB.

Answer: 15.8 kN

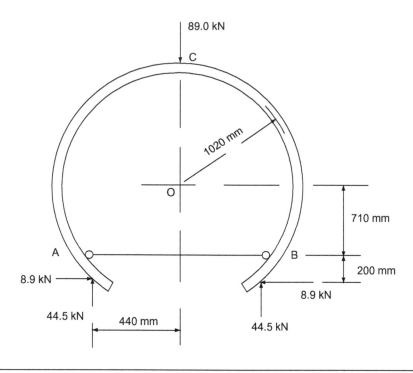

FIGURE P.4.9

Energy methods

In Chapter 2, we saw that the elasticity method of structural analysis embodies the determination of stresses and displacements by employing equations of equilibrium and compatibility in conjunction with the relevant force–displacement or stress–strain relationships. In addition, in Chapter 4, we investigated the use of virtual work in calculating forces, reactions, and displacements in structural systems. A powerful alternative but equally fundamental approach is the use of energy methods. These, while providing exact solutions for many structural problems, find their greatest use in the rapid approximate solution of problems for which exact solutions do not exist. Also, many structures which are statically indeterminate, that is, they cannot be analyzed by the application of the equations of statical equilibrium alone, may be conveniently analyzed using an energy approach. Further, energy methods provide comparatively simple solutions for deflection problems not readily solved by more elementary means.

Generally, as we shall see, modern analysis[1] uses the methods of *total complementary energy* and *total potential energy*. Either method may be employed to solve a particular problem, although as a general rule, deflections are more easily found using complementary energy and forces by potential energy.

Although energy methods are applicable to a wide range of structural problems and may even be used as indirect methods of forming equations of equilibrium or compatibility,[1,2] we shall be concerned in this chapter with the solution of deflection problems and the analysis of statically indeterminate structures. We also include some methods restricted to the solution of linear systems, that is, the *unit load method*, the *principle of superposition*, and the *reciprocal theorem*.

5.1 STRAIN ENERGY AND COMPLEMENTARY ENERGY

Figure 5.1(a) shows a structural member subjected to a steadily increasing load P. As the member extends, the load P does work, and from the law of conservation of energy, this work is stored in the member as *strain energy*. A typical load–deflection curve for a member possessing nonlinear elastic characteristics is shown in Fig. 5.1(b). The strain energy U produced by a load P and corresponding extension y is then

$$U = \int_0^y P \, dy \tag{5.1}$$

and is clearly represented by the area OBD under the load–deflection curve. Engesser (1889) called the area OBA above the curve the *complementary energy C*, and from Fig. 5.1(b),

$$C = \int_0^P y \, dP \tag{5.2}$$

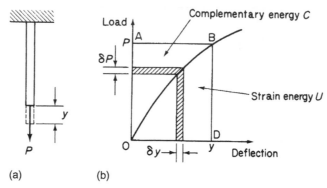

FIGURE 5.1 (a) Strain Energy of a Member Subjected to Simple Tension; (b) Load–Deflection Curve for a Nonlinearly Elastic Member

Complementary energy, as opposed to strain energy, has no physical meaning, being purely a convenient mathematical quantity. However, it is possible to show that complementary energy obeys the law of conservation of energy in the type of situation usually arising in engineering structures, so that its use as an energy method is valid.

Differentiation of Eqs. (5.1) and (5.2) with respect to y and P, respectively, gives

$$\frac{dU}{dy} = P, \quad \frac{dC}{dP} = y$$

Bearing these relationships in mind, we can now consider the interchangeability of strain and complementary energy. Suppose that the curve of Fig. 5.1(b) is represented by the function

$$P = by^n$$

where the coefficient b and exponent n are constants. Then,

$$U = \int_0^y P \, dy = \frac{1}{n} \int_0^P \left(\frac{P}{b} \right)^{1/n} dP$$

$$C = \int_0^P y \, dP = n \int_0^y by^n \, dy$$

Hence,

$$\frac{dU}{dy} = P, \quad \frac{dU}{dP} = \frac{1}{n} \left(\frac{P}{b} \right)^{1/n} = \frac{1}{n} y \tag{5.3}$$

$$\frac{dC}{dP} = y, \quad \frac{dC}{dy} = bny^n = nP \tag{5.4}$$

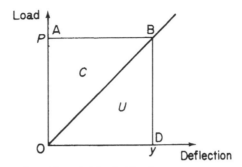

FIGURE 5.2 Load–Deflection Curve for a Linearly Elastic Member

When $n = 1$,

$$\left.\begin{array}{l} \dfrac{dU}{dy} = \dfrac{dC}{dy} = P \\[2mm] \dfrac{dU}{dP} = \dfrac{dC}{dP} = y \end{array}\right\} \qquad (5.5)$$

and the strain and complementary energies are completely interchangeable. Such a condition is found in a linearly elastic member; its related load–deflection curve being that shown in Fig. 5.2. Clearly, area OBD (U) is equal to area OBA (C).

We see that the latter of Eqs. (5.5) is in the form of what is commonly known as Castigliano's first theorem, in which the differential of the strain energy U of a structure with respect to a load is equated to the deflection of the load. To be mathematically correct, however, the differential of the complementary energy C is what should be equated to deflection (compare Eqs. (5.3) and (5.4)).

5.2 PRINCIPLE OF THE STATIONARY VALUE OF THE TOTAL COMPLEMENTARY ENERGY

Consider an elastic system in equilibrium supporting forces $P_1, P_2, \ldots, P_n$, which produce real corresponding displacements $\Delta_1, \Delta_2, \ldots, \Delta_n$. If we impose virtual forces $\delta P_1, \delta P_2, \ldots, \delta P_n$ on the system acting through the real displacements, then the total virtual work done by the system is (see Chapter 4)

$$-\int_{\text{vol}} y \, dP + \sum_{r=1}^{n} \Delta_r \delta P_r$$

The first term in this expression is the negative virtual work done by the particles in the elastic body, while the second term represents the virtual work of the externally applied virtual forces. From the principle of virtual work,

$$-\int_{\text{vol}} y \, dP + \sum_{r=1}^{n} \Delta_r \delta P_r = 0 \qquad (5.6)$$

Comparing Eq. (5.6) with Eq. (5.2), we see that each term represents an increment in complementary energy; the first, of the internal forces, the second, of the external loads. Equation (5.6) may therefore be rewritten

$$\delta(C_i + C_e) = 0 \qquad (5.7)$$

where

$$C_i = \int_{\text{vol}} \int_0^P y \, dP \quad \text{and} \quad C_e = -\sum_{r=1}^n \Delta_r P_r \qquad (5.8)$$

We now call the quantity $(C_i + C_e)$ the *total complementary energy* C of the system.

The displacements specified in Eq. (5.6) are real displacements of a continuous elastic body; they therefore obey the condition of compatibility of displacement, so that Eqs. (5.6) and (5.7) are equations of geometrical compatibility. The *principle of the stationary value of the total complementary energy* may then be stated as follows:

> *For an elastic body in equilibrium under the action of applied forces, the true internal forces (or stresses) and reactions are those for which the total complementary energy has a stationary value.*

In other words, the true internal forces (or stresses) and reactions are those which satisfy the condition of compatibility of displacement. This property of the total complementary energy of an elastic system is particularly useful in the solution of statically indeterminate structures, in which an infinite number of stress distributions and reactive forces may be found to satisfy the requirements of equilibrium.

5.3 APPLICATION TO DEFLECTION PROBLEMS

Generally, deflection problems are most readily solved by the complementary energy approach, although for linearly elastic systems, there is no difference between the methods of complementary and potential energy, since, as we have seen, complementary and strain energy then become completely interchangeable. We illustrate the method by reference to the deflections of frames and beams, which may or may not possess linear elasticity.

Let us suppose that we must find the deflection Δ_2 of the load P_2 in the simple pin-jointed framework consisting, say, of k members and supporting loads $P_1, P_2, \ldots, P_n$, as shown in Fig. 5.3. From Eqs. (5.8), the total complementary energy of the framework is given by

$$C = \sum_{i=1}^k \int_0^{F_i} \lambda_i \, dF_i - \sum_{r=1}^n \Delta_r P_r \qquad (5.9)$$

where λ_i is the extension of the ith member, F_i is the force in the ith member, and Δ_r is the corresponding displacement of the rth load P_r. From the principle of the stationary value of the total complementary energy,

$$\frac{\partial C}{\partial P_2} = \sum_{i=1}^k \lambda_i \frac{\partial F_i}{\partial P_2} - \Delta_2 = 0 \qquad (5.10)$$

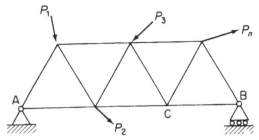

FIGURE 5.3 Determination of the Deflection of a Point on a Framework by the Method of Complementary Energy

from which

$$\Delta_2 = \sum_{i=1}^{k} \lambda_i \frac{\partial F_i}{\partial P_2} \tag{5.11}$$

Equation (5.10) is seen to be identical to the principle of virtual forces in which virtual forces δF and δP act through real displacements λ and Δ. Clearly, the partial derivatives with respect to P_2 of the constant loads $P_1, P_2, \ldots, P_n$ vanish, leaving the required deflection Δ_2 as the unknown. At this stage, before Δ_2 can be evaluated, the load–displacement characteristics of the members must be known. For linear elasticity,

$$\lambda_i = \frac{F_i L_i}{A_i E_i}$$

where L_i, A_i, and E_i are the length, cross-sectional area, and modulus of elasticity of the ith member. On the other hand, if the load–displacement relationship is of a nonlinear form, say,

$$F_i = b(\lambda_i)^c$$

in which b and c are known, then Eq. (5.11) becomes

$$\Delta_2 = \sum_{i=1}^{k} \left(\frac{F_i}{b}\right)^{1/c} \frac{\partial F_i}{\partial P_2}$$

The computation of Δ_2 is best accomplished in tabular form, but before the procedure is illustrated by an example, some aspects of the solution merit discussion.

We note that the support reactions do not appear in Eq. (5.9). This convenient absence derives from the fact that the displacements $\Delta_1, \Delta_2, \ldots, \Delta_n$ are the real displacements of the frame and fulfill the conditions of geometrical compatibility and boundary restraint. The complementary energy of the reaction at A and the vertical reaction at B is therefore zero, since both their corresponding displacements are zero. If we examine Eq. (5.11), we note that λ_i is the extension of the ith member of the framework due to the applied loads $P_1, P_2, \ldots, P_n$. Therefore, the loads F_i in the substitution for λ_i in Eq. (5.11) are those corresponding to the loads $P_1, P_2, \ldots, P_n$. The term $\partial F_i/\partial P_2$ in Eq. (5.11) represents the rate of change of F_i with P_2 and is calculated by applying the load P_2 to the *unloaded* frame and determining the corresponding member loads in terms of P_2. This procedure indicates a method for obtaining the displacement of either a point on the frame in a direction not coincident with the line of action of a load or, in fact, a point such as C that carries no load at all. We place at the point and in the required direction a *fictitious* or *dummy* load, say P_f, the original loads being removed. The loads in the members due to P_f are then calculated and $\partial F/\partial P_f$ obtained for each member. Substitution in Eq. (5.11) produces the required deflection.

It must be pointed out that it is not absolutely necessary to remove the actual loads during the application of P_f. The force in each member is then calculated in terms of the actual loading and P_f. F_i follows by substituting $P_f = 0$ and $\partial F_i/\partial P_f$ is found by differentiation with respect to P_f. Obviously, the two approaches yield the same expressions for F_i and $\partial F_i/\partial P_f$, although the latter is arithmetically clumsier.

Example 5.1

Calculate the vertical deflection of the point B and the horizontal movement of D in the pin-jointed framework shown in Fig. 5.4(a). All members of the framework are linearly elastic and have cross-sectional areas of 1800 mm². The value of E for the material of the members is 200,000 N/mm². See Ex. 1.1.

The members of the framework are linearly elastic, so that Eq. (5.11) may be written

$$\Delta = \sum_{i=1}^{k} \frac{F_i L_i}{A_i E_i} \frac{\partial F_i}{\partial P} \tag{i}$$

or, since each member has the same cross-sectional area and modulus of elasticity,

$$\Delta = \frac{1}{AE} \sum_{i=1}^{k} F_i L_i \frac{\partial F_i}{\partial P} \tag{ii}$$

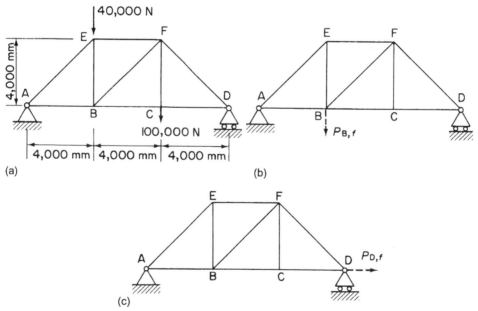

FIGURE 5.4 (a) Actual Loading of Framework; (b) Determination of Vertical Deflection of B; (c) Determination of Horizontal Deflection of D

Table 5.1 Example 5.1

① Member	② L (mm)	③ F (N)	④ $F_{B,f}$ (N)	⑤ $\partial F_{B,f}/\partial P_{B,f}$	⑥ $F_{D,f}$ (N)	⑦ $\partial F_{D,f}/\partial P_{D,f}$	⑧ $\times 10^6$ $FL\partial F_{B,f}/\partial P_{B,f}$	⑨ $\times 10^6$ $FL\partial F_{D,f}/\partial P_{D,f}$
AE	$4{,}000\sqrt{2}$	$-60{,}000\sqrt{2}$	$-2\sqrt{2}P_{B,f}/3$	$-2\sqrt{2}/3$	0	0	$320\sqrt{2}$	0
EF	$4{,}000$	$-60{,}000$	$-2P_{B,f}/3$	$-2/3$	0	0	160	0
FD	$4{,}000\sqrt{2}$	$-80{,}000\sqrt{2}$	$-\sqrt{2}P_{B,f}/3$	$-\sqrt{2}/3$	0	0	$640\sqrt{2}/3$	0
DC	$4{,}000$	$80{,}000$	$P_{B,f}/3$	$1/3$	$P_{D,f}$	1	$320/3$	320
CB	$4{,}000$	$80{,}000$	$P_{B,f}/3$	$1/3$	$P_{D,f}$	1	$320/3$	320
BA	$4{,}000$	$60{,}000$	$2P_{B,f}/3$	$2/3$	$P_{D,f}$	1	$480/3$	240
EB	$4{,}000$	$20{,}000$	$2P_{B,f}/3$	$2/3$	0	0	$160/3$	0
FB	$4{,}000\sqrt{2}$	$-20{,}000\sqrt{2}$	$\sqrt{2}P_{B,f}/3$	$\sqrt{2}/3$	0	0	$-160\sqrt{2}/3$	0
FC	$4{,}000$	$100{,}000$	0	0	0	0	0	0
							$\sum = 1268$	$\sum = 880$

The solution is completed in Table 5.1, in which F are the member forces due to the actual loading of Fig. 5.4(a), $F_{B,f}$ are the member forces due to the fictitious load $P_{B,f}$ in Fig. 5.4(b), and $F_{D,f}$ are the forces in the members produced by the fictitious load $P_{D,f}$ in Fig. 5.4(c). We take tensile forces as positive and compressive forces as negative.

The vertical deflection of B is

$$\Delta_{B,v} = \frac{1,268 \times 10^6}{1,800 \times 200,000} = 3.52 \text{ mm}$$

and the horizontal movement of D is

$$\Delta_{D,h} = \frac{880 \times 10^6}{1,800 \times 200,000} = 2.44 \text{ mm}$$

which agree with the virtual work solution (Example 4.9).

The positive values of $\Delta_{B,v}$ and $\Delta_{D,h}$ indicate that the deflections are in the directions of $P_{B,f}$ and $P_{D,f}$.

The analysis of beam deflection problems by complementary energy is similar to that of pin-jointed frameworks, except that we assume initially that displacements are caused primarily by bending action. Shear force effects are discussed later in the chapter.

Example 5.2

Determine the deflection of the free end of the tip-loaded cantilever beam shown in Fig. 5.5; the bending stiffness of the beam is EI. See Ex. 1.1.

The total complementary energy C of the system is given by

$$C = \int_L \int_0^M d\theta \, dM - P\Delta_v \tag{i}$$

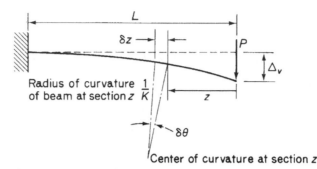

FIGURE 5.5 Beam Deflection by the Method of Complementary Energy

in which $\int_0^M d\theta\, dM$ is the complementary energy of an element δz of the beam. This element subtends an angle $\delta\theta$ at its center of curvature due to the application of the bending moment M. From the principle of the stationary value of the total complementary energy,

$$\frac{\partial C}{\partial P} = \int_L d\theta \frac{dM}{dP} - \Delta_v = 0$$

or

$$\Delta_v = \int_L d\theta \frac{dM}{dP} \tag{ii}$$

Equation (ii) is applicable to either a nonlinear or linearly elastic beam. To proceed further, therefore, we require the load–displacement (M–θ) and bending moment–load (M–P) relationships. It is immaterial for the purposes of this illustrative problem whether the system is linear or nonlinear, since the mechanics of the solution are the same in either case. We choose therefore a linear M–θ relationship, as this is the case in the majority of the problems we consider. Hence, from Fig. 5.5,

$$\delta\theta = K\delta z$$

or

$$d\theta = \frac{M}{EI} dz \quad \left(\frac{1}{K} = \frac{EI}{M} \text{ from simple beam theory}\right)$$

where the product *modulus of elasticity* × *second moment of area of the beam cross-section* is known as the *bending* or *flexural rigidity* of the beam. Also,

$$M = Pz$$

so that

$$\frac{dM}{dP} = z$$

Substitution for $d\theta$, M, and dM/dP in Eq. (ii) gives

$$\Delta_v = \int_0^L \frac{Pz^2}{EI} dz$$

or

$$\Delta_v = \frac{PL^3}{3EI}$$

The fictitious load method of the framework example may be employed in the solution of beam deflection problems where we require deflections at positions on the beam other than concentrated load points. ■

Example 5.3

Determine the deflection of the tip of the cantilever beam shown in Fig. 5.6; the bending stiffness of the beam is *EI*. See Ex. 1.1.

First, we apply a fictitious load P_f at the point where the deflection is required. The total complementary energy of the system is then

$$C = \int_L \int_0^M d\theta \, dM - \Delta_T P_f - \int_0^L \Delta w \, dz$$

where the symbols take their previous meanings and Δ is the vertical deflection of any point on the beam. Then,

$$\frac{\partial C}{\partial P_f} = \int_0^L d\theta \frac{\partial M}{\partial P_f} - \Delta_T = 0 \qquad\qquad \text{(i)}$$

As before,

$$d\theta = \frac{M}{EI} dz$$

but

$$M = P_f z + \frac{wz^2}{2} \quad (P_f = 0)$$

Hence,

$$\frac{\partial M}{\partial P_f} = z$$

Substituting in Eq. (i) for $d\theta$, M, and $\partial M/\partial P_f$ and remembering that $P_f = 0$, we have

$$\Delta_T = \int_0^L \frac{wz^3}{2EI} dz$$

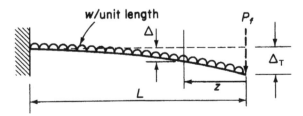

FIGURE 5.6 Deflection of a Uniformly Loaded Cantilever by the Method of Complementary Energy

giving

$$\Delta_T = \frac{wL^4}{8EI}$$

Note that here, unlike the method for the solution of the pin-jointed framework, the fictitious load is applied to the loaded beam. However, no arithmetical advantage is gained by the former approach, although the result obviously is the same, since M would equal $wz^2/2$ and $\partial M/\partial P_f$ would have the value z.

Example 5.4

Calculate the vertical displacements of the quarter- and mid-span points B and C of the simply supported beam of length L and flexural rigidity EI loaded, as shown in Fig. 5.7. See Ex. 1.1.

The total complementary energy C of the system including the fictitious loads $P_{B,f}$ and $P_{C,f}$ is

$$C = \int_L \int_0^M d\theta \, dM - P_{B,f}\Delta_B - P_{C,f}\Delta_C - \int_0^L \Delta w \, dz \tag{i}$$

Hence,

$$\frac{\partial C}{\partial P_{B,f}} = \int_L d\theta \frac{\partial M}{\partial P_{B,f}} - \Delta_B = 0 \tag{ii}$$

and

$$\frac{\partial C}{\partial P_{C,f}} = \int_L d\theta \frac{\partial M}{\partial P_{C,f}} - \Delta_C = 0 \tag{iii}$$

Assuming a linearly elastic beam, Eqs. (ii) and (iii) become

$$\Delta_B = \frac{1}{EI} \int_0^L M \frac{\partial M}{\partial P_{B,f}} \, dz \tag{iv}$$

$$\Delta_C = \frac{1}{EI} \int_0^L M \frac{\partial M}{\partial P_{C,f}} \, dz \tag{v}$$

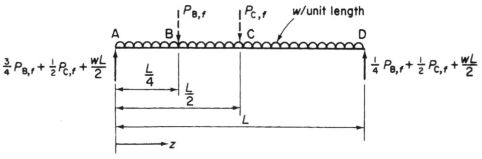

FIGURE 5.7 Deflection of a Simply Supported Beam by the Method of Complementary Energy

From A to B,

$$M = \left(\frac{3}{4} P_{B,f} + \frac{1}{2} P_{C,f} + \frac{wL}{2}\right) z - \frac{wz^2}{2}$$

so that

$$\frac{\partial M}{\partial P_{B,f}} = \frac{3}{4} z, \quad \frac{\partial M}{\partial P_{C,f}} = \frac{1}{2} z$$

From B to C,

$$M = \left(\frac{3}{4} P_{B,f} + \frac{1}{2} P_{C,f} + \frac{wL}{2}\right) z - \frac{wz^2}{2} - P_{B,f}\left(z - \frac{L}{4}\right)$$

giving

$$\frac{\partial M}{\partial P_{B,f}} = \frac{1}{4}(L - z), \quad \frac{\partial M}{\partial P_{C,f}} = \frac{1}{2} z$$

From C to D,

$$M = \left(\frac{1}{4} P_{B,f} + \frac{1}{2} P_{C,f} + \frac{wL}{2}\right)(L - z) - \frac{w}{2}(L - z)^2$$

so that

$$\frac{\partial M}{\partial P_{B,f}} = \frac{1}{4}(L - z), \quad \frac{\partial M}{\partial P_{C,f}} = \frac{1}{2}(L - z)$$

Substituting these values in Eqs. (iv) and (v) and remembering that $P_{B,f} = P_{C,f} = 0$, we have, from Eq. (iv)

$$\Delta_B = \frac{1}{EI} \left\{ \int_0^{L/4} \left(\frac{wLz}{2} - \frac{wz^2}{2}\right) \frac{3}{4} z\, dz + \int_{L/4}^{L/2} \left(\frac{wLz}{2} - \frac{wz^2}{2}\right) \frac{1}{4}(L - z) dz + \int_{L/2}^{L} \left(\frac{wLz}{2} - \frac{wz^2}{2}\right) \frac{1}{4}(L - z) dz \right\}$$

from which

$$\Delta_B = \frac{57wL^4}{6,144EI}$$

Similarly,

$$\Delta_C = \frac{5wL^4}{384EI}$$

Example 5.5

Use the principle of the stationary value of the total complementary energy of a system to calculate the horizontal displacement of the point C in the frame of Example 4.10. See Ex. 1.1.

Referring to Fig. 4.16, we apply a horizontal fictitious load, $P_{C,f}$, at C. The total complementary energy of the system, including the fictitious load, is given by

$$C = \int_L \int_0^M d\theta\, dM - P_{C,f}\Delta_{C,h} - \int_0^{2L} \Delta w\, dx - W\Delta_{C,h} \tag{i}$$

where $\Delta_{C,h}$ is the actual horizontal displacement of C. Then,

$$\partial C/\partial P_{C,f} = \int_L d\theta(\partial M/\partial P_{C,f}) - \Delta_{C,h} = 0 \qquad \text{(ii)}$$

Therefore, from Eq. (ii),

$$\Delta_{C,h} = (1/2EI)\int_0^{2L} M(\partial M/\partial P_{C,f})dx + (1/EI)\int_0^L M(\partial M/\partial P_{C,f})dy \qquad \text{(iii)}$$

In AB,

$$M = (W + P_{C,f})L - (wx^2/2)$$

so that

$$\partial M/\partial P_{C,f} = L$$

In BC,

$$M = (W + P_{C,f})y$$

and

$$\partial M/\partial P_{C,f} = y$$

Substituting these expressions in Eq. (iii) and remembering that $P_{C,f} = 0$, we have

$$\Delta_{C,h} = (1/2EI)\int_0^{2L} [WL - (wx^2/2)]L\, dx + (1/EI)\int_0^L Wy^2\, dy \qquad \text{(iv)}$$

Integrating Eq. (iv) and substituting the limits gives

$$\Delta_{C,h} = (2L^3/3EI)(2W - wL)$$

which is the solution produced in Example 4.10.

Example 5.6

Using the principle of the stationary value of the total complementary energy of a system, calculate the vertical displacement of the point B in the cantilever beam of Ex. 4.11. See Ex. 1.1.

In this example, we need to consider the torsional contribution to the total complementary energy in addition to that due to bending. By comparison with Eq. (i) of Example 5.2, we see that the internal complementary energy of a bar subjected to torsion is given by

$$C_{i,T} = \int_T \int_0^T d\gamma\, dT \qquad \text{(i)}$$

where γ is the angle of twist at any section of the bar.

Suppose that an imaginary load, $P_{B,f}$, is applied vertically at B (i.e., into the plane of the paper) in the cantilever beam of Example 4.11. The total complementary energy of the beam is then given by

$$C = \int_L \int_0^M d\theta\, dM + \int_L \int_0^T d\gamma\, dT - P_{B,f}\Delta_B \qquad \text{(ii)}$$

where Δ_B is the vertical deflection of B. Then,

$$\partial C/\partial P_{B,f} = \int_L d\theta\left(dM/dP_{B,f}\right) + \int_L d\gamma\left(dT/dP_{B,f}\right) - \Delta_B \tag{iii}$$

The first term on the right-hand side of Eq. (iii) is replaced as in Example 5.2. The second term is replaced in a similar manner using Eq. (3.12). Then,

$$\Delta_B = (1/EI)\int_0^L M\left(dM/dP_{B,f}\right) dz + (1/GJ)\int_0^L T\left(dT/dP_{B,f}\right) dz \tag{iv}$$

Referring to Fig. 4.17,

$$M = (W + P_{B,f})p = (W + P_{B,f})R\sin\alpha$$

Then,

$$\left(dM/dP_{B,f}\right) = R\sin\alpha$$

Also ,

$$T = (W + P_{B,f})(R - R\cos\alpha)$$

And

$$\left(dT/dP_{B,f}\right) = R(1 - \cos\alpha)$$

Substituting these expressions in Eq. (iv) and remembering that $P_{B,f} = 0$ gives

$$\Delta_B = (1/EI)\int_0^{\pi/2} WR^2\sin^2\alpha R\, d\alpha + (1/GJ)\int_0^{\pi/2} WR^2(1 - \cos\alpha)^2 R\, d\alpha \tag{v}$$

The right-hand side of Eq. (v) is identical to the expression for the internal work done in Example 4.11. Therefore,

$$\Delta_B = WR^3\{(\pi/4EI) + (1/GJ)[(3\pi/4) - 2]\}$$

as before.

The fictitious load method of determining deflections may be streamlined for linearly elastic systems and is then termed the *unit load method*; this we discuss later in the chapter.

■■■

5.4 APPLICATION TO THE SOLUTION OF STATICALLY INDETERMINATE SYSTEMS

In a statically determinate structure, the internal forces are determined uniquely by simple statical equilibrium considerations. This is not the case for a statically indeterminate system, in which, as already noted, an infinite number of internal force or stress distributions may be found to satisfy the conditions of equilibrium. The true force system is, as demonstrated in Section 5.2, the one satisfying the conditions of compatibility of displacement of the elastic structure or, alternatively, that for which the total complementary energy has a stationary value. We apply the principle to a variety of statically indeterminate structures, beginning with the relatively simple singly redundant pin-jointed frame of Example 5.7.

Example 5.7

Determine the forces in the members of the pin-jointed framework shown in Fig. 5.8. Each member has the same value of the product AE. See Ex. 1.1.

The first step is to choose the redundant member. In this example, no advantage is gained by the choice of any particular member, although in some cases careful selection can result in a decrease in the amount of arithmetical labor. Taking BD as the redundant member, we assume that it sustains a tensile force R due to the external loading. The total complementary energy of the framework is, with the notation of Eq. (5.9),

$$C = \sum_{i=1}^{k} \int_0^{F_i} \lambda_i \, dF_i - P\Delta$$

Hence,

$$\frac{\partial C}{\partial R} = \sum_{i=1}^{k} \lambda_i \frac{\partial F_i}{\partial R} = 0 \tag{i}$$

or, assuming linear elasticity,

$$\frac{1}{AE} \sum_{i=1}^{k} F_i L_i \frac{\partial F_i}{\partial R} = 0 \tag{ii}$$

The solution is now completed in Table 5.2, where, as in Table 5.1, positive signs indicate tension.

Hence, from Eq. (ii),

$$4.83RL + 2.707PL = 0$$

or

$$R = -0.56P$$

Substitution for R in column ③ of Table 5.2 gives the force in each member. Having determined the forces in the members, the deflection of any point on the framework may be found by the method described in Section 5.3.

Unlike the statically determinate type, statically indeterminate frameworks may be subjected to self-straining. Therefore, internal forces are present before external loads are applied. Such a situation may be caused by a local

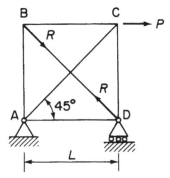

FIGURE 5.8 Analysis of a Statically Indeterminate Framework by the Method of Complementary Energy

Table 5.2 Example 5.7

① Member	② Length	③ F	④ $\partial F/\partial R$	⑤ $FL\partial F/\partial R$
AB	L	$-R/\sqrt{2}$	$-1/\sqrt{2}$	$RL/2$
BC	L	$-R/\sqrt{2}$	$-1/\sqrt{2}$	$RL/2$
CD	L	$-(P+R/\sqrt{2})$	$-1/\sqrt{2}$	$L(P+R/\sqrt{2})/\sqrt{2})$
DA	L	$-R/\sqrt{2}$	$-1/\sqrt{2}$	$RL/2$
AC	$\sqrt{2}L$	$\sqrt{2}P+R$	1	$L(2P+\sqrt{2}R)$
BD	$\sqrt{2}L$	R	1	$\sqrt{2}RL$
				$\sum = 4.83RL + 2.707PL$

temperature change or by an initial lack of fit of a member. Suppose that the member BD of the framework of Fig. 5.8 is short by a known amount Δ_R when the framework is assembled but is forced to fit. The load R in BD has suffered a displacement Δ_R in addition to that caused by the change in length of BD produced by the load P. The total complementary energy is then

$$C = \sum_{i=1}^{k} \int_0^{F_i} \lambda_i \, dF_i - P\Delta - R\Delta_R$$

and

$$\frac{\partial C}{\partial R} = \sum_{i=1}^{k} \lambda_i \frac{\partial F_i}{\partial R} - \Delta_R = 0$$

or

$$\Delta_R = \frac{1}{AE} \sum_{i=1}^{k} F_i L_i \frac{\partial F_i}{\partial R} \tag{iii}$$

Obviously, the summation term in Eq. (iii) has the same value as in the previous case, so that

$$R = -0.56P + \frac{AE}{4.83L}\Delta_R$$

Hence, the forces in the members are due to both applied loads and an initial lack of fit.

Some care should be given to the sign of the lack of fit Δ_R. We note here that the member BD is short by an amount Δ_R, so that the assumption of a positive sign for Δ_R is compatible with the tensile force R. If BD were initially too long, then the total complementary energy of the system would be written

$$C = \sum_{i=1}^{k} \int_0^{F_i} \lambda_i \, dF_i - P\Delta - R(-\Delta_R)$$

giving

$$-\Delta_R = \frac{1}{AE} \sum_{i=1}^{k} F_i L_i \frac{\partial F_i}{\partial R}$$

Example 5.8

Calculate the loads in the members of the singly redundant pin-jointed framework shown in Fig. 5.9. The members AC and BD are 30 mm^2 in cross-section, and all other members are 20 mm^2 in cross-section. The members AD, BC, and DC are each 800 mm long. $E = 200{,}000$ N/mm^2. See Ex. 1.1.

From the geometry of the framework $A\hat{B}D = C\hat{B}D = 30°$; therefore, $BD = AC = 800\sqrt{3}$ mm. Choosing CD as the redundant member and proceeding from Eq. (ii) of Example 5.7, we have

$$\frac{1}{E}\sum_{i=1}^{k}\frac{F_i L_i}{A_i}\frac{\partial F_i}{\partial R} = 0 \tag{i}$$

From Table 5.3, we have

$$\sum_{i=1}^{k}\frac{F_i L_i}{A_i}\frac{\partial F_i}{\partial R} = -268 + 129.2R = 0$$

Hence, $R = 2.1$ N and the forces in the members are tabulated in column ⑦ of Table 5.3.

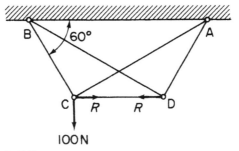

FIGURE 5.9 Framework of Example 5.8

Table 5.3 Example 5.8 (tension positive)						
① **Member**	② L **(mm)**	③ A **(mm^2)**	④ F **(N)**	⑤ $\partial F/\partial R$	⑥ $(FL/A)\partial F/\partial R$	⑦ **Force (N)**
AC	$800\sqrt{3}$	30	$50 - \sqrt{3}\,R/2$	$-\sqrt{3}\,/2$	$-2000 + 20\sqrt{3}\,R$	48.2
CB	800	20	$86.6 + R/2$	1/2	$1732 + 10R$	87.6
BD	$800\sqrt{3}$	30	$-\sqrt{3}\,R/2$	$-\sqrt{3}\,/2$	$20\sqrt{3}\,R$	−1.8
CD	800	20	R	1	$40R$	2.1
AD	800	20	$R/2$	1/2	$10R$	1.0
					$\sum = -268 + 129.2R$	

Example 5.8 MATLAB

Repeat Example 5.8 using MATLAB and columns 1–4 of Table 5.3. See Ex. 1.1.

Values for R and the member forces, rounded to the first decimal place, are obtained through the following MATLAB file:

```
% Declare any needed variables
syms F L A E R
E = 200000;

% Define L, A, and F using columns 2-4 of Table 5.3
L = [800*sqrt(3);      % L_AC
     800;              % L_CB
     800*sqrt(3);      % L_BD
     800;              % L_CD
     800];             % L_AD
A = sym([30;           % A_AC
         20;           % A_CB
         30;           % A_BD
         20;           % A_CD
         20]);         % A_AD
F = [50-sqrt(3)*R/2;   % F_AC
     86.6+R/2;         % F_CB
     -sqrt(3)*R/2;     % F_BD
     R;                % F_CD
     R/2];             % F_AD

% Use Eq. (5.16) to solve for R
eqI = F.*L.*diff(F,R)./(A.*E);    % Eq. (5.16)
R_val = solve(sum(eqI),R);

% Substitute R_val into F to calculate the member forces
Force = subs(F,R,R_val);

% Output R and the member forces, rounded to the first decimal place, to the Command
Window
R = round(double(R_val)*10)/10;
Force = round(double(Force)*10)/10;

disp(['R =' num2str(R) 'N'])
disp(['Force = ['num2str(Force(1))', 'num2str(Force(2))', 'num2str(Force(3))',
'num2str(Force(4))', 'num2str(Force(5))'] N'])
```

The Command Window outputs resulting from this MATLAB file are as follows:

```
R = 2.1 N
Force = [48.2, 87.6, -1.8, 2.1, 1] N
```

Example 5.9

A plane, pin-jointed framework consists of six bars forming a rectangle ABCD 4000 mm × 3000 mm with two diagonals, as shown in Fig. 5.10. The cross-sectional area of each bar is 200 mm^2 and the frame is unstressed when the temperature of each member is the same. Due to local conditions, the temperature of one of the 3000 mm members is raised by 30°C. Calculate the resulting forces in all the members if the coefficient of linear expansion α of the bars is 7×10^{-6}/°C and $E = 200,000$ N/mm^2. See Ex. 1.1.

Suppose that BC is the heated member, then the increase in length of $BC = 3000 \times 30 \times 7 \times 10^{-6} = 0.63$ mm. Therefore, from Eq. (iii) of Example 5.7,

$$-0.63 = \frac{1}{200 \times 200,000} \sum_{i=1}^{k} F_i L_i \frac{\partial F_i}{\partial R} \tag{i}$$

Substitution from the summation of column ⑤ in Table 5.4 into Eq. (i) gives

$$R = \frac{-0.63 \times 200 \times 200,000}{48,000} = -525 \text{ N}$$

Column ⑥ of Table 5.4 is now completed for the force in each member.

So far, our analysis has been limited to singly redundant frameworks, although the same procedure may be adopted to solve a multi-redundant framework of, say, m redundancies. Therefore, instead of a single equation of the type (i) in Example 5.7, we would have m simultaneous equations

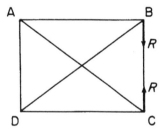

FIGURE 5.10 Framework of Example 5.9

Table 5.4 Example 5.9 (tension positive)

① Member	② L (mm)	③ F(N)	④ F/∂R	⑤ FL∂F/∂R	⑥ Force (N)
AB	4,000	4R/3	4/3	64,000R/9	−700
BC	3,000	R	1	3,000R	−525
CD	4,000	4R/3	4/3	64,000R/9	−700
DA	3,000	R	1	3,000R	−525
AC	5,000	−5R/3	−5/3	125,000R/9	875
DB	5,000	−5R/3	−5/3	125,000R/9	875
				$\sum = 48,000R$	

$$\frac{\partial C}{\partial R_j} = \sum_{i=1}^{k} \lambda_i \frac{\partial F_i}{\partial R_j} = 0 \quad (j = 1, 2, \ldots, m)$$

from which the m unknowns $R_1, R_2, \ldots, R_m$ are obtained. The forces F in the members follow, being expressed initially in terms of the applied loads and $R_1, R_2, \ldots, R_m$.

Other types of statically indeterminate structure are solved by the application of total complementary energy with equal facility. The propped cantilever of Fig. 5.11 is an example of a singly redundant beam structure for which total complementary energy readily yields a solution.

The total complementary energy of the system is, with the notation of Eq. (i) of Example 5.2,

$$C = \int_L \int_0^M d\theta \, dM - P\Delta_C - R_B\Delta_B$$

where Δ_C and Δ_B are the deflections at C and B, respectively. Usually, in problems of this type, Δ_B is either zero for a rigid support or a known amount (sometimes in terms of R_B) for a sinking support. Hence, for a stationary value of C,

$$\frac{\partial C}{\partial R_B} = \int_L d\theta \frac{\partial M}{\partial R_B} - \Delta_B = 0$$

from which equation R_B may be found; R_B being contained in the expression for the bending moment M.

Obviously, the same procedure is applicable to a beam having a multiredundant support system, for example, a continuous beam supporting a series of loads $P_1, P_2, \ldots, P_n$. The total complementary energy of such a beam is given by

$$C = \int_L \int_0^M d\theta \, dM - \sum_{j=1}^{m} R_j\Delta_j - \sum_{r=1}^{n} P_r\Delta_r$$

where R_j and Δ_j are the reaction and known deflection (at least in terms of R_j) of the jth support point in a total of m supports. The stationary value of C gives

$$\frac{\partial C}{\partial R_j} = \int_L d\theta \frac{\partial M}{\partial R_j} - \Delta_j = 0 \quad (j = 1, 2, \ldots, m)$$

producing m simultaneous equations for the m unknown reactions.

The intention here is not to suggest that continuous beams are best or most readily solved by the energy method; the moment distribution method produces a more rapid solution, especially for beams in which the degree of redundancy is large. Instead the purpose is to demonstrate the versatility and power of energy methods in their ready solution of a wide range of structural problems. A complete investigation of this versatility is impossible here due to restrictions of space; in fact, whole books have been devoted to this topic. We therefore limit our

FIGURE 5.11 Analysis of a Propped Cantilever by the Method of Complementary Energy

analysis to problems peculiar to the field of aircraft structures, with which we are primarily concerned. The remaining portion of this section is therefore concerned with the solution of frames and rings possessing varying degrees of redundancy.

The frameworks we considered in the earlier part of this section and in Section 5.3 comprised members capable of resisting direct forces only. Of a more general type are composite frameworks, in which some or all of the members resist bending and shear loads in addition to direct loads. It is usual, however, except for the thin-walled structures in Part B of this book, to ignore deflections produced by shear forces. We, therefore, only consider bending and direct force contributions to the internal complementary energy of such structures. The method of analysis is illustrated in Ex. 5.10.

Example 5.9 MATLAB

Repeat Example 5.9 using MATLAB and columns 1–3 of Table 5.4.

Values for R and the member forces are obtained through the following MATLAB file:

```
% Declare any needed variables
syms delta_R alpha F L A E R T
A = 200;
alpha = 7*10^(-6);
E = 200000;
T = 30;

% Define L and F using columns 2-3 of Table 5.4
L = [4000;          % L_AB
     3000;          % L_BC
     4000;          % L_CD
     3000;          % L_DA
     5000;          % L_AC
     5000];         % L_DB
F = [4*R/3;         % F_AB
      R;            % F_BC
     4*R/3;         % F_CD
      R;            % F_DA
     -5*R/3;        % F_AC
     -5*R/3];       % F_DB

% Use Eq. (5.17) to solve for R assuming BC is the heated member
delta_R = L(2)* T*alpha;
eqI = F.*L.*diff(F,R)./(A*E);   % Eq. (5.17)
R_val = solve(sum(eqI)+delta_R,R);

% Substitute R_val into F to calculate the member forces
Force = subs(F,R,R_val);
```

```
% Output R and the member forces to the Command Window
R = round(double(R_val));
Force = round(double(Force));

disp(['R =' num2str(R) 'N'])
disp(['Force = ['num2str(Force(1))', 'num2str(Force(2))', 'num2str(Force(3))',
'num2str(Force(4))', 'num2str(Force(5))', 'num2str(Force(6))'] N'])
```

The Command Window output resulting from this MATLAB file is as follows:

```
R = -525 N
Force = [-700, -525, -700, -525, 875, 875] N
```

Example 5.10

The simply supported beam ABC shown in Fig. 5.12 is stiffened by an arrangement of pin-jointed bars capable of sustaining axial loads only. If the cross-sectional area of the beam is A_B and that of the bars is A, calculate the forces in the members of the framework, assuming that displacements are caused by bending and direct force action only. See Ex. 1.1.

We observe that, if the beam were capable of supporting only direct loads, then the structure would be a relatively simple statically determinate pin-jointed framework. Since the beam resists bending moments (we are ignoring shear effects), the system is statically indeterminate with a single redundancy, the bending moment at any section of the beam. The total complementary energy of the framework is given, with the notation previously developed, by

$$C = \int_{ABC} \int_0^M d\theta \, dM + \sum_{i=1}^{k} \lambda_j \, dF_i - P\Delta \tag{i}$$

If we suppose that the tensile load in the member ED is R, then, for C to have a stationary value,

$$\frac{\partial C}{\partial R} = \int_{ABC} d\theta \frac{\partial M}{\partial R} + \sum_{i=1}^{k} \lambda_j \frac{\partial F_i}{\partial R} = 0 \tag{ii}$$

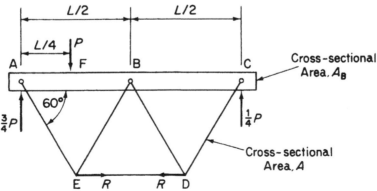

FIGURE 5.12 Analysis of a Trussed Beam by the Method of Complementary Energy

At this point, we assume the appropriate load–displacement relationships; again we take the system to be linear so that Eq. (ii) becomes

$$\int_0^L \frac{M}{EI} \frac{\partial M}{\partial R} \, dz + \sum_{i=1}^k \frac{F_i L_i}{A_i E} \frac{\partial F_i}{\partial R} = 0 \qquad \text{(iii)}$$

The two terms in Eq. (iii) may be evaluated separately, bearing in mind that only the beam ABC contributes to the first term, while the complete structure contributes to the second. Evaluating the summation term by a tabular process, we have Table 5.5.

Summation of column ⑥ in Table 5.5 gives

$$\sum_{i=1}^k \frac{F_i L_i}{A_i E} \frac{\partial F_i}{\partial R} = \frac{RL}{4E} \left(\frac{1}{A_B} + \frac{10}{A} \right) \qquad \text{(iv)}$$

The bending moment at any section of the beam between A and F is

$$M = \frac{3}{4} Pz - \frac{\sqrt{3}}{2} Rz; \quad \text{hence,} \quad \frac{\partial M}{\partial R} = -\frac{\sqrt{3}}{2} z$$

between F and B is

$$M = \frac{P}{4}(L - z) - \frac{\sqrt{3}}{2} Rz; \quad \text{hence,} \quad \frac{\partial M}{\partial R} = -\frac{\sqrt{3}}{2} z$$

and between B and C is

$$M = \frac{P}{4}(L - z) - \frac{\sqrt{3}}{2} R(L - z); \quad \text{hence,} \quad \frac{\partial M}{\partial R} = -\frac{\sqrt{3}}{2}(L - z)$$

Therefore,

$$\int_0^L \frac{M}{EI} \frac{\partial M}{\partial R} \, dz = \frac{1}{EI} \left\{ \int_{L/4}^{L/2} -\left(\frac{3}{4} Pz - \frac{\sqrt{3}}{2} Rz \right) \frac{\sqrt{3}}{2} z \, dz \right.$$

$$+ \int_{L/4}^{L/2} \left[\frac{P}{4}(L - z) - \frac{\sqrt{3}}{2} Rz \right] \left(-\frac{\sqrt{3}}{2} z \right) dz$$

$$\left. + \int_{L/2}^{L} -\left[\frac{P}{4}(L - z) - \frac{\sqrt{3}}{2} R(L - z) \right] \frac{\sqrt{3}}{2}(L - z) dz \right\}$$

Table 5.5 Example 5.10 (tension positive)

① Member	② Length	③ Area	④ F	⑤ ∂F/∂R	⑥ (F/A)∂F/∂R
AB	$L/2$	A_B	$-R/2$	$-1/2$	$R/4A_B$
BC	$L/2$	A_B	$-R/2$	$-1/2$	$R/4A_B$
CD	$L/2$	A	R	1	R/A
DE	$L/2$	A	R	1	R/A
BD	$L/2$	A	$-R$	-1	R/A
EB	$L/2$	A	$-R$	-1	R/A
AE	$L/2$	A	R	1	R/A

giving

$$\int_0^L \frac{M}{EI} \frac{\partial M}{\partial R} \, dz = \frac{-11\sqrt{3}PL^3}{768EI} + \frac{RL^3}{16EI} \qquad (v)$$

Substituting from Eqs. (iv) and (v) into Eq. (iii)

$$-\frac{11\sqrt{3}PL^3}{768EI} + \frac{RL^3}{16EI} + \frac{RL}{4E}\left(\frac{A + 10A_B}{A_B A}\right) = 0$$

from which

$$R = \frac{11\sqrt{3}PL^2 A_B A}{48[L^2 A_B A + 4I(A + 10A_B)]}$$

hence, the forces in each member of the framework. The deflection Δ of the load P or any point on the framework may be obtained by the method of Section 5.3. For example, the stationary value of the total complementary energy of Eq. (i) gives Δ; that is,

$$\frac{\partial C}{\partial P} = \int_{ABC} d\theta \frac{\partial M}{\partial R} + \sum_{i=1}^k \lambda_i \frac{\partial F_i}{\partial P} - \Delta = 0$$

Although braced beams are still found in modern light aircraft in the form of braced wing structures, a much more common structural component is the ring frame. The role of this particular component is discussed in detail in Chapter 11; it is therefore sufficient for the moment to say that ring frames form the basic shape of semi-monocoque fuselages reacting to shear loads from the fuselage skins, point loads from wing spar attachments, and distributed loads from floor beams. Usually, a ring is two dimensional, supporting loads applied in its own plane. Our analysis is limited to the two-dimensional case.

Example 5.11

Determine the bending moment distribution in the two-dimensional ring shown in Fig. 5.13; the bending stiffness of the ring is EI. See Ex. 1.1.

A two-dimensional ring has redundancies of direct load, bending moment, and shear at any section, as shown in Fig. 5.13. However, in some special cases of loading, the number of redundancies may be reduced. For example, on a plane of symmetry, the shear loads and sometimes the normal or direct loads are zero, while on a plane of anti-symmetry the direct loads and bending moments are zero. Let us consider the simple case of the doubly symmetrical ring shown in Fig. 5.14(a). At a section in the vertical plane of symmetry, the internal shear and direct loads vanish, leaving one redundancy, the bending moment M_A (Fig. 5.14(b)). Note that, in the horizontal plane of symmetry, the internal shears are zero but the direct loads have a value $P/2$. The total complementary energy of the system (again ignoring shear strains) is

$$C = \int_{ring} \int_0^M d\theta \, dM - 2\left(\frac{P}{2}\Delta\right)$$

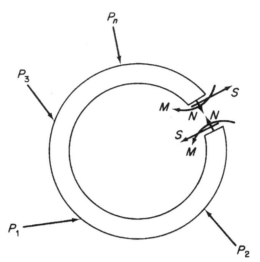

FIGURE 5.13 Internal Force System in a Two-Dimensional Ring

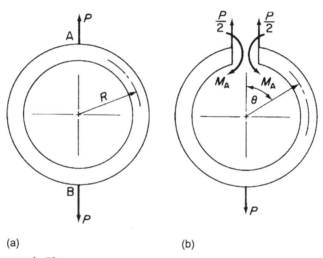

(a)   (b)

FIGURE 5.14 Doubly Symmetric Ring

taking the bending moment as positive when it increases the curvature of the ring. In this expression for C, Δ is the displacement of the top, A, of the ring relative to the bottom, B. Assigning a stationary value to C, we have

$$\frac{\partial C}{\partial M_A} = \int_{ring} d\theta \frac{\partial M}{\partial M_A} = 0$$

or assuming linear elasticity and considering, from symmetry, half the ring,

$$\int_0^{\pi R} \frac{M}{EI} \frac{\partial M}{\partial M_A} ds = 0$$

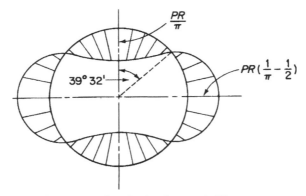

FIGURE 5.15 Distribution of Bending Moment in a Doubly Symmetric Ring

Therefore, since

$$M = M_A - \frac{P}{2}R\sin\theta, \quad \frac{\partial M}{\partial M_A} = 1$$

and we have

$$\int_0^\pi \left(M_A - \frac{P}{2}R\sin\theta \right) R\,d\theta = 0$$

or

$$\left[M_A\theta + \frac{P}{2}R\cos\theta \right]_0^\pi = 0$$

from which

$$M_A = \frac{PR}{\pi}$$

The bending moment distribution is then

$$M = PR\left(\frac{1}{\pi} - \frac{\sin\theta}{2} \right)$$

and is shown diagrammatically in Fig. 5.15.

We shall now consider a more representative aircraft structural problem.

Example 5.12

The circular fuselage frame shown in Fig. 5.16(a) supports a load, P, which is reacted to by a shear flow, q (i.e., a shear force per unit length, see Chapter 16), distributed around the circumference of the frame from the fuselage skin. If the bending stiffness of the frame is EI, calculate the distribution of bending moment round the frame. See Ex. 1.1.

The value and direction of this shear flow are quoted here but are derived from theory established in Section 17.3. From our previous remarks on the effect of symmetry, we observe that there is no shear force at

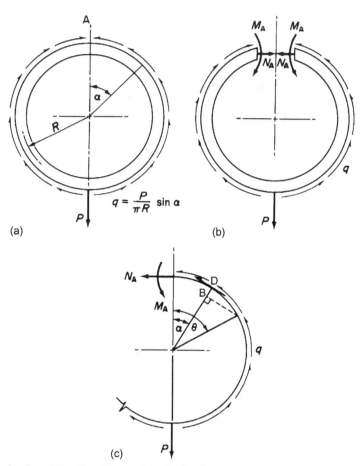

(a)

(b)

(c)

FIGURE 5.16 Determination of Bending Moment Distribution in a Shear and Direct Loaded Ring

the section A on the vertical plane of symmetry. The unknowns are therefore the bending moment M_A and normal force N_A. We proceed, as in the previous example, by writing down the total complementary energy C of the system. Then, neglecting shear strains,

$$C = \int_{ring} \int_0^M d\theta \, dM - P\Delta \tag{i}$$

in which Δ is the deflection of the point of application of P relative to the top of the frame. Note that M_A and N_A do not contribute to the complement of the potential energy of the system, since, by symmetry, the rotation and horizontal displacements at A are zero. From the principle of the stationary value of the total complementary energy,

$$\frac{\partial C}{\partial M_A} = \int_{ring} d\theta \frac{\partial M}{\partial M_A} = 0 \tag{ii}$$

and

$$\frac{\partial C}{\partial N_A} = \int_{ring} d\theta \, \frac{\partial M}{\partial N_A} = 0 \tag{iii}$$

The bending moment at a radial section inclined at an angle θ to the vertical diameter is, from Fig. 5.16(c),

$$M = M_A + N_A R(1 - \cos\theta) + \int_0^\theta q B D R \, d\alpha$$

or

$$M = M_A + N_A R(1 - \cos\theta) + \int_0^\theta \frac{P}{\pi R} \sin \alpha [R - R\cos(\theta - \alpha)] R \, d\alpha$$

which gives

$$M = M_A + N_A R(1 - \cos\theta) + \frac{PR}{\pi}(1 - \cos\theta - \frac{1}{2}\theta \sin \theta) \tag{iv}$$

Hence,

$$\frac{\partial M}{\partial M_A} = 1, \quad \frac{\partial M}{\partial N_A} = R(1 - \cos\theta) \tag{v}$$

Assuming that the fuselage frame is linearly elastic, we have, from Eqs. (ii) and (iii),

$$2 \int_0^\pi \frac{M}{EI} \frac{\partial M}{\partial M_A} R \, d\theta = 2 \int_0^\pi \frac{M}{EI} \frac{\partial M}{\partial N_A} R \, d\theta = 0 \tag{vi}$$

Substituting from Eqs. (iv) and (v) into Eq. (vi) gives two simultaneous equations:

$$-\frac{PR}{2\pi} = M_A + N_A R \tag{vii}$$

$$-\frac{7PR}{8\pi} = M_A + \frac{3}{2} N_A R \tag{viii}$$

These equations may be written in matrix form as follows:

$$\frac{PR}{\pi} \begin{Bmatrix} -1/2 \\ -7/8 \end{Bmatrix} = \begin{bmatrix} 1 & R \\ 1 & 3R/2 \end{bmatrix} \begin{Bmatrix} M_A \\ N_A \end{Bmatrix} \tag{ix}$$

so that

$$\begin{Bmatrix} M_A \\ N_A \end{Bmatrix} = \frac{PR}{\pi} \begin{bmatrix} 1 & R \\ 1 & 3R/2 \end{bmatrix}^{-1} \begin{Bmatrix} -1/2 \\ -7/8 \end{Bmatrix}$$

or

$$\begin{Bmatrix} M_A \\ N_A \end{Bmatrix} = \frac{PR}{\pi} \begin{bmatrix} 3 & -2 \\ -2/R & 2/R \end{bmatrix} \begin{Bmatrix} -1/2 \\ -7/8 \end{Bmatrix}$$

which gives

$$M_A = \frac{PR}{4\pi}, \quad N_A = \frac{-3P}{4\pi}$$

The bending moment distribution follows from Eq. (iv) and is

$$M = \frac{PR}{4\pi} \left(1 - \frac{1}{2} \cos\theta - \theta \sin\theta \right) \tag{x}$$

The solution of Eq. (ix) involves the inversion of the matrix

$$\begin{bmatrix} 1 & R \\ 1 & 3R/2 \end{bmatrix}$$

which may be carried out using any of the standard methods detailed in texts on matrix analysis. In this example, Eqs. (vii) and (viii) are clearly most easily solved directly; however, the matrix approach illustrates the technique and serves as a useful introduction to the more detailed discussion in Chapter 6.

Example 5.13

A two-cell fuselage has circular frames with a rigidly attached straight member across the middle. The bending stiffness of the lower half of the frame is $2EI$, while that of the upper half and also the straight member is EI.

Calculate the distribution of the bending moment in each part of the frame for the loading system shown in Fig. 5.17(a). Illustrate your answer by means of a sketch and show clearly the bending moment carried by each part of the frame at the junction with the straight member. Deformations due only to bending strains need be taken into account. See Ex. 1.1.

The loading is antisymmetrical, so that there are no bending moments or normal forces on the plane of antisymmetry; there remain three shear loads S_A, S_D, and S_C, as shown in Fig. 5.17(b). The total complementary energy of the half-frame is then (neglecting shear strains)

$$C = \int_{\text{half-frame}} \int_0^M d\theta \, dM - M_0 \alpha_B - \frac{M_0}{r} \Delta_B \tag{i}$$

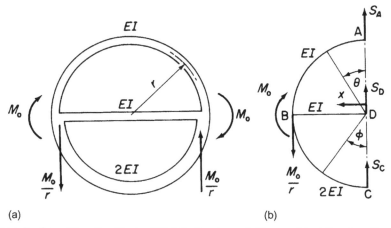

(a)   (b)

FIGURE 5.17 Determination of Bending Moment Distribution in an Antisymmetrical Fuselage Frame

where α_B and Δ_B are the rotation and deflection of the frame at B caused by the applied moment M_0 and concentrated load M_0/r, respectively. From antisymmetry, there is no deflection at A, D, or C so that S_A, S_C, and S_D make no contribution to the total complementary energy. In addition, overall equilibrium of the half-frame gives

$$S_A + S_D + S_C = \frac{M_0}{r} \tag{ii}$$

Assigning stationary values to the total complementary energy and considering the half-frame only, we have

$$\frac{\partial C}{\partial S_A} = \int_{\text{half-frame}} d\theta \frac{\partial M}{\partial S_A} = 0$$

and

$$\frac{\partial C}{\partial S_D} = \int_{\text{half-frame}} d\theta \frac{\partial M}{\partial S_D} = 0$$

or, assuming linear elasticity,

$$\int_{\text{half-frame}} \frac{M}{EI} \frac{\partial M}{\partial S_A} \, ds = \int_{\text{half-frame}} \frac{M}{EI} \frac{\partial M}{\partial S_D} \, ds = 0 \tag{iii}$$

In AB,

$$M = -S_A r \sin\theta \quad \text{and} \quad \frac{\partial M}{\partial S_A} = -r\sin\theta, \quad \frac{\partial M}{\partial S_D} = 0$$

In DB,

$$M = S_D x \quad \text{and} \quad \frac{\partial M}{\partial S_A} = 0, \quad \frac{\partial M}{\partial S_D} = x$$

In CB,

$$M = S_C r \sin\phi = \left(\frac{M_0}{r} - S_A - S_D \right) r \sin\phi$$

Thus,

$$\frac{\partial M}{\partial S_A} = -r\sin\phi \quad \text{and} \quad \frac{\partial M}{\partial S_{D'}} = -r\sin\phi$$

Substituting these expressions in Eq. (iii) and integrating, we have

$$3.365 S_A + S_C = M_0/r \tag{iv}$$

$$S_A + 2.178 S_C = M_0/r \tag{v}$$

which, with Eq. (ii), enable S_A, S_D, and S_C to be found. In matrix form, these equations are written

$$\begin{Bmatrix} M_0/r \\ M_0/r \\ M_0/r \end{Bmatrix} = \begin{bmatrix} 1 & 1 & 1 \\ 3.356 & 0 & 1 \\ 1 & 0 & 2.178 \end{bmatrix} \begin{Bmatrix} S_A \\ S_D \\ S_C \end{Bmatrix} \tag{vi}$$

from which we obtain

$$\begin{Bmatrix} S_A \\ S_D \\ S_C \end{Bmatrix} = \begin{bmatrix} 0 & 0.345 & -0.159 \\ 1 & -0.187 & -0.373 \\ 0 & -0.159 & 0.532 \end{bmatrix} \begin{Bmatrix} M_0/r \\ M_0/r \\ M_0/r \end{Bmatrix} \tag{vii}$$

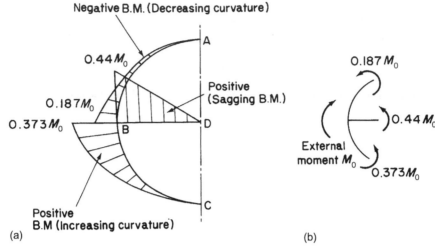

FIGURE 5.18 Distribution of Bending Moment in the Frame of Example 5.13

which give

$$S_A = 0.187 \, M_0/r, \quad S_D = 0.44 \, M_0/r, \quad S_C = 0.373 \, M_0/r$$

Again, the square matrix of Eq. (vi) has been inverted to produce Eq. (vii).

The bending moment distribution with directions of bending moment is shown in Fig. 5.18.

So far in this chapter we considered the application of the principle of the stationary value of the total complementary energy of elastic systems in the analysis of various types of structure. Although the majority of the examples used to illustrate the method are of linearly elastic systems, it was pointed out that generally they may be used with equal facility for the solution of nonlinear systems.

In fact, the question of whether a structure possesses linear or non-linear characteristics arises only after the initial step of writing down expressions for the total potential or complementary energies. However, a great number of structures are linearly elastic and possess unique properties which enable solutions, in some cases, to be more easily obtained. The remainder of this chapter is devoted to these methods.

5.5 UNIT LOAD METHOD

In Section 5.3, we discussed the *dummy* or *fictitious* load method of obtaining deflections of structures. For a linearly elastic structure, the method may be streamlined as follows.

Consider the framework of Fig. 5.3, in which we require, say, to find the vertical deflection of the point C. Following the procedure of Section 5.3, we would place a vertical dummy load P_f at C and write down the total complementary energy of the framework (see Eq. (5.9)):

$$C = \sum_{I=1}^{k} \int_{0}^{F_i} \lambda_i \, dF_i - \sum_{r=1}^{n} \Delta_r P_r$$

For a stationary value of C,

$$\frac{\partial C}{\partial P_f} = \sum_{i=1}^{k} \lambda_i \frac{\partial F_i}{\partial P_f} - \Delta_C = 0 \tag{5.12}$$

from which

$$\Delta_C = \sum_{i=1}^{k} \lambda_i \frac{\partial F_i}{\partial P_f} \tag{5.13}$$

as before. If, instead of the arbitrary dummy load P_f, we had placed a unit load at C, then the load in the ith linearly elastic member would be

$$F_i = \frac{\partial F_i}{\partial P_f} 1$$

Therefore, the term $\partial F_i / \partial P_f$ in Eq. (5.13) is equal to the load in the ith member due to a unit load at C, and Eq. (5.13) may be written

$$\Delta_C = \sum_{i=1}^{k} \frac{F_{i,0} F_{i,1} L_i}{A_i E_i} \tag{5.14}$$

where $F_{i,0}$ is the force in the ith member due to the actual loading and $F_{i,1}$ is the force in the ith member due to a unit load placed at the position and in the direction of the required deflection. Thus, in Example 5.1, columns ④ and ⑥ in Table 5.1 would be eliminated, leaving column ⑤ as $F_{B,1}$ and column ⑦ as $F_{D,1}$. Obviously column ③ is F_0.

Similar expressions for deflection due to bending and torsion of linear structures follow from the well-known relationships between bending and rotation and torsion and rotation. Hence, for a member of length L and flexural and torsional rigidities EI and GJ, respectively,

$$\Delta_{B.M} = \int_L \frac{M_0 M_1}{EI} dz, \quad \Delta_T = \int_L \frac{T_0 T_1}{GJ} dz \tag{5.15}$$

where M_0 is the bending moment at any section produced by the actual loading and M_1 is the bending moment at any section due to a unit load applied at the position and in the direction of the required deflection; similarly for torsion.

Generally, shear deflections of slender beams are ignored but may be calculated when required for particular cases. Of greater interest in aircraft structures is the calculation of the deflections produced by the large shear stresses experienced by thin-walled sections. This problem is discussed in Chapter 19.

Example 5.14

A steel rod of uniform circular cross-section is bent as shown in Fig. 5.19, AB and BC being horizontal and CD vertical. The arms AB, BC, and CD are of equal length. The rod is encastré at A and the other end D is free. A uniformly distributed load covers the length BC. Find the components of the displacement of the free end D in terms of EI and GJ. See Ex. 1.1.

Since the cross-sectional area A and modulus of elasticity E are not given, we assume that displacements due to axial distortion are to be ignored. We place, in turn, unit loads in the assumed positive directions of the axes xyz.

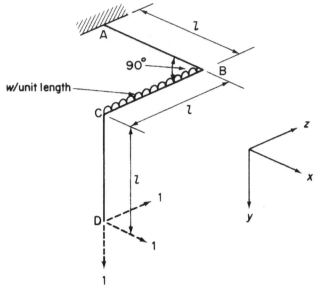

FIGURE 5.19 Deflection of a Bent Rod

First, consider the displacement in the direction parallel to the x axis. From Eqs. (5.15),

$$\Delta_x = \int_L \frac{M_0 M_1}{EI}\, ds + \int_L \frac{T_0 T_1}{GJ}\, ds$$

Employing a tabular procedure,

Plane	M_0			M_1			T_0			T_1		
	xy	xz	yz	xy	xz	yz	xy	xz	yz	xy	xz	yz
CD	0	0	0	y	0	0	0	0	0	0	0	0
CB	0	0	$-wz^2/2$	0	z	0	0	0	0	l	0	0
BA	$-wlx$	0	0	l	l	0	0	0	$wl^2/2$	0	0	0

Hence,

$$\Delta_x = \int_0^l -\frac{wl^2 x}{EI}\, dx$$

or

$$\Delta_x = -\frac{wl^4}{2EI}$$

Similarly,

$$\Delta_y = wl^4 \left(\frac{11}{24EI} + \frac{1}{2GJ} \right)$$

$$\Delta_z = wl^4 \left(\frac{1}{6EI} + \frac{1}{2GJ} \right)$$

5.6 FLEXIBILITY METHOD

An alternative approach to the solution of statically indeterminate beams and frames is to release the structure, that is, remove redundant members or supports, until the structure becomes statically determinate. The displacement of some point in the released structure is then determined by, say, the unit load method. The actual loads on the structure are removed and unknown forces applied to the points where the structure has been released; the displacement at the point produced by these unknown forces must, from compatibility, be the same as that in the released structure. The unknown forces are then obtained; this approach is known as *the flexibility method*.

Example 5.15

Determine the forces in the members of the truss shown in Fig. 5.20(a); the cross-sectional area A and Young's modulus E are the same for all members. See Ex. 1.1.

The truss in Fig. 5.20(a) is clearly externally statically determinate but has a degree of internal statical indeterminacy equal to 1. We therefore release the truss so that it becomes statically determinate by "cutting" one of the members, say BD, as shown in Fig. 5.20(b). Due to the actual loads (P in this case) the cut ends of the member BD separate or come together, depending on whether the force in the member (before it was cut) is tensile or compressive; we shall assume that it was tensile.

We are assuming that the truss is linearly elastic, so that the relative displacement of the cut ends of the member BD (in effect, the movement of B and D away from or toward each other along the diagonal BD) may be found using, say, the unit load method. Thus, we determine the forces $F_{a,j}$ in the members produced by the actual loads. We then apply equal and opposite unit loads to the cut ends of the member BD, as shown in Fig 5.20(c), and calculate the forces, $F_{1,j}$ in the members. The displacement of B relative to D, Δ_{BD} (see Eq. (ii) in Ex. 4.9), is given by

$$\Delta_{BD} = \sum_{j=1}^{n} \frac{F_{a,j} F_{1,j} L_j}{AE}$$

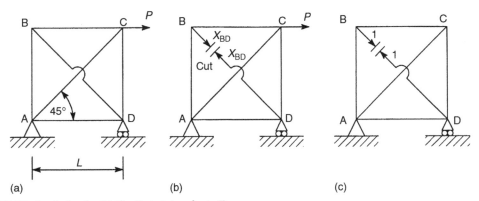

(a)   (b)   (c)

FIGURE 5.20 Analysis of a Statically Indeterminate Truss

The forces, $F_{a,j}$, are the forces in the members of the released truss due to the actual loads and are not, therefore, the actual forces in the members of the complete truss. We therefore redesignate the forces in the members of the released truss as $F_{0,j}$. The expression for Δ_{BD} becomes

$$\Delta_{BD} = \sum_{j=1}^{n} \frac{F_{0j}F_{1j}L_j}{AE} \tag{i}$$

In the actual structure, this displacement is prevented by the force, X_{BD}, in the redundant member BD. If, therefore, we calculate the displacement, a_{BD}, in the direction of BD produced by a unit value of X_{BD}, the displacement due to X_{BD} is $X_{BD}a_{BD}$. Clearly, from compatibility,

$$\Delta_{BD} + X_{BD}a_{BD} = 0 \tag{ii}$$

from which X_{BD} is found, a_{BD} is a *flexibility coefficient*. Having determined X_{BD}, the actual forces in the members of the complete truss may be calculated by, say, the method of joints or the method of sections.

In Eq. (ii), a_{BD} is the displacement of the released truss in the direction of BD produced by a unit load. Therefore, in using the unit load method to calculate this displacement, the actual member forces ($F_{1,j}$) and the member forces produced by the unit load ($F_{1,j}$) are the same. Therefore, from Eq. (i),

$$a_{BD} = \sum_{j=1}^{n} \frac{F_{1,j}^2 L_j}{AE} \tag{iii}$$

The solution is completed in Table 5.6.

From Table 5.6,

$$\Delta_{BD} = \frac{2.71\,PL}{AE}, \quad a_{BD} = \frac{4.82L}{AE}$$

Substituting these values in Eq. (i), we have

$$\frac{2.71PL}{AE} + X_{BD}\frac{4.82L}{AE} = 0$$

from which

$$X_{BD} = -0.56P$$

(i.e., compression). The actual forces, $F_{a,j}$, in the members of the complete truss of Fig. 5.20(a) are now calculated using the method of joints and are listed in the final column of Table 5.6.

Table 5.6 Example 5.15

Member	L_j (m)	$F_{0,j}$	$F_{1,j}$	$F_{0,j}F_{1,j}L_j$	$F_{1,j}^2 L_j$	$F_{a,j}$
AB	L	0	−0.71	0	0.5 L	+0.40P
BC	L	0	−0.71	0	0.5 L	+0.40P
CD	L	−P	−0.71	0.71PL	0.5 L	−0.60P
BD	1.41 L	—	1.0	—	1.41 L	−0.56P
AC	1.41 L	1.41P	1.0	2.0PL	1.41 L	+0.85P
AD	L	0	−0.71	0	0.5 L	+0.40P
				$\sum = 2.71\,PL$	$\sum = 4.82\,L$	

We note in the preceding that Δ_{BD} is positive, which means that Δ_{BD} is in the direction of the unit loads; that is, B approaches D and the diagonal BD in the released structure decreases in length. Therefore, in the complete structure, the member BD, which prevents this shortening, must be in compression as shown; also a_{BD} will always be positive, since it contains the term $F_{1,j}2$. Finally, we note that the cut member BD is included in the calculation of the displacements in the released structure, since its deformation, under a unit load, contributes to a_{BD}. ▄

Example 5.16

Calculate the forces in the members of the truss shown in Fig. 5.21(a). All members have the same cross-sectional area A and Young's modulus E. See Ex. 1.1.

By inspection we see that the truss is both internally and externally statically indeterminate, since it would remain stable and in equilibrium if one of the diagonals, AD or BD, and the support at C were removed; the degree of indeterminacy is therefore 2. Unlike the truss in Example 5.15, we could not remove *any* member, since, if BC or

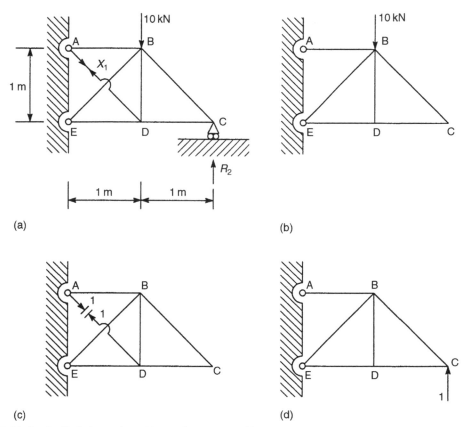

(a)

(b)

(c)

(d)

FIGURE 5.21 Statically Indeterminate Truss of Example 5.16

CD were removed, the outer half of the truss would become a mechanism while the portion ABDE would remain statically indeterminate. Therefore, we select AD and the support at C as the releases, giving the statically determinate truss shown in Fig. 5.21(b); we designate the force in the member AD as X_1 and the vertical reaction at C as R_2.

In this case, we shall have two compatibility conditions, one for the diagonal AD and one for the support at C. We therefore need to investigate three loading cases: one in which the actual loads are applied to the released statically determinate truss in Fig. 5.21(b), a second in which unit loads are applied to the cut member AD (Fig. 5.21(c)), and a third in which a unit load is applied at C in the direction of R_2 (Fig. 5.21(d)). By comparison with the previous example, the compatibility conditions are

$$\Delta_{AD} + a_{11}X_1 + a_{12}R_2 = 0 \tag{i}$$

$$v_C + a_{21}X_1 + a_{22}R_2 = 0 \tag{ii}$$

in which Δ_{AD} and v_C are, respectively, the change in length of the diagonal AD and the vertical displacement of C due to the actual loads acting on the released truss, while a_{11}, a_{12}, and so forth, are flexibility coefficients, which we have previously defined. The calculations are similar to those carried out in Example 5.15 and are shown in Table 5.7. From Table 5.7,

$$\Delta_{AD} = \sum_{j=1}^{n} \frac{F_{0,j}F_{1,j}(X_1)L_j}{AE} = \frac{-27.1}{AE} \quad \text{(i.e., AD increases in length)}$$

$$v_C = \sum_{j=1}^{n} \frac{F_{0,j}F_{1,j}(R_2)L_j}{AE} = \frac{-48.11}{AE} \quad \text{(i.e., C is displaced downward)}$$

$$a_{11} = \sum_{j=1}^{n} \frac{F^2_{1,j}(X_1)L_j}{AE} = \frac{4.32}{AE}$$

$$a_{22} = \sum_{j=1}^{n} \frac{F^2_{1,j}(R_2)L_j}{AE} = \frac{11.62}{AE}$$

$$a_{12} = a_{21} \sum_{j=1}^{n} \frac{F_{1,j}(X_1)F_{1,j}(R_2)L_j}{AE} = \frac{2.7}{AE}$$

Table 5.7 Example 5.16

Member	L_j	F_{0j}	$F_{1,j}$ (X_1)	$F_{1,j}$ (R_2)	$F_{0,j}F_{1,j}$ $(X_1)\,L_j$	$F_{0,j}F_{1,j}$ $(R_2)\,L_j$	$F_{1,j}^2$ $(X_1)L_j$	$F_{1,j}^2$ $(R_2)L_j$	$F_{1,j}\,(X_1)$ $F_{1,j}\,(R_2)\,L_j$	F_{aj}
AB	1	10.0	−0.71	−2.0	−7.1	−20.0	0.5	4.0	1.41	0.67
BC	1.41	0	0	−1.41	0	0	0	2.81	0	−4.45
CD	1	0	0	1.0	0	0	0	1.0	0	3.15
DE	1	0	−0.71	1.0	0	0	0.5	1.0	−0.71	0.12
AD	1.41	0	1.0	0	0	0	1.41	0	0	4.28
BE	1.41	−14.14	1.0	1.41	−20.0	−28.11	1.41	2.81	2.0	−5.4
BD	1	0	0	−0.71	0	0	0.5	0	0	−3.03
					$\sum = -27.1$	$\sum = -48.11$	$\sum = 4.32$	$\sum = 11.62$	$\sum = 2.7$	

Substituting in Eqs. (i) and (ii) and multiplying through by AE, we have

$$-27.1 + 4.32X_1 + 2.7R_2 = 0 \tag{iii}$$

$$-48.11 + 2.7X_1 + 11.62R_2 = 0 \tag{iv}$$

Solving Eqs. (iii) and (iv), we obtain

$$X_1 = 4.28\text{kN}, R_2 = 3.15\text{kN}$$

The actual forces, $F_{a,j}$, in the members of the complete truss are now calculated by the method of joints and are listed in the final column of Table 5.7.

5.6.1 Self-straining trusses

Statically indeterminate trusses, unlike the statically determinate type, may be subjected to self-straining, in which internal forces are present before external loads are applied. Such a situation may be caused by a local temperature change or by an initial lack of fit of a member. In cases such as these, the term on the right-hand side of the compatibility equations, Eq. (ii) in Example 5.15 and Eqs. (i) and (ii) in Example 5.16, would not be zero.

Example 5.17

The truss shown in Fig. 5.22(a) is unstressed when the temperature of each member is the same, but due to local conditions, the temperature in the member BC is increased by 30°C. If the cross-sectional area of each member is 200 mm² and the coefficient of linear expansion of the members is $7 \times 10^{-6}/°\text{C}$, calculate the resulting forces in the members; Young's modulus $E = 200,000$ N/mm². See Ex. 1.1.

Due to the temperature rise, the increase in length of the member BC is $3 \times 10^3 \times 30 \times 7 \times 10^{-6} = 0.63$ mm. The truss has a degree of internal statical indeterminacy equal to 1 (by inspection). We therefore release the truss by cutting the member BC, which has experienced the temperature rise, as shown in Fig. 5.22(b); we suppose that the force in BC is X_1. Since there are no external loads on the truss, Δ_{BC} is zero and the compatibility condition becomes

$$a_{11}X_1 = -0.63 \tag{i}$$

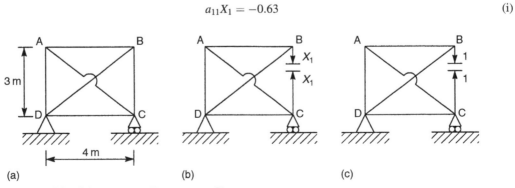

(a)   (b)   (c)

FIGURE 5.22 Self-Straining Due to a Temperature Change

Table 5.8 Example 5.17

Member	L_j (mm)	$F_{1,j}$	$F_{1,j}^2 L_j$	$F_{a,j}$ (N)
AB	4,000	1.33	7,111.1	−700
BC	3,000	1.0	3,000.0	−525
CD	4,000	1.33	7,111.1	−700
DA	3,000	1.0	3,000.0	−525
AC	5,000	−1.67	13,888.9	875
DB	5,000	−1.67	13,888.9	875
			$\sum = 48,000.0$	

in which, as before,

$$a_{11} = \sum_{j=1}^{n} \frac{F_{1,j}^2 L_j}{AE}$$

Note that the extension of BC is negative, since it is opposite in direction to X_1. The solution is now completed in Table 5.8. Hence,

$$a_{11} = \frac{48,000}{200 \times 200,000} = 1.2 \times 10^{-3}$$

Then, from Eq. (i),

$$X_1 = -525 \text{ N}$$

The forces, $F_{a,j}$, in the members of the complete truss are given in the final column of Table 5.8. Compare the preceding with the solution of Example 5.9.

5.7 TOTAL POTENTIAL ENERGY

In the spring–mass system shown in its unstrained position in Fig. 5.23(a), we normally define the *potential energy* of the mass as the product of its weight, Mg, and its height, h, above some arbitrarily fixed datum. In other words, it possesses energy by virtue of its position. After deflection to an

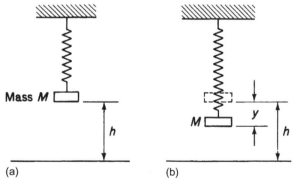

Mass M

h

(a)

M

y

h

(b)

FIGURE 5.23 (a) Potential Energy of a Spring–Mass System; (b) Loss in Potential Energy Due to a Change in Position

equilibrium state (Fig. 5.23(b)), the mass has lost an amount of potential energy equal to Mgy. Therefore, we may associate deflection with a loss of potential energy. Alternatively, we may argue that the gravitational force acting on the mass does work during its displacement, resulting in a loss of energy. Applying this reasoning to the elastic system of Fig. 5.1(a) and assuming that the potential energy of the system is zero in the unloaded state, the *loss* of potential energy of the load P as it produces a deflection y is Py. Thus, the potential energy V of P in the deflected equilibrium state is given by

$$V = -Py$$

We now define the *total potential energy* (TPE) of a system in its deflected equilibrium state as the sum of its internal or strain energy and the potential energy of the applied external forces. Hence, for the single member–force configuration of Fig. 5.1(a),

$$\text{TPE} = U + V = \int_0^y P \, dy - Py$$

For a general system consisting of loads $P_1, P_2, \ldots, P_n$ producing *corresponding displacements* (i.e., displacements in the directions of the loads; see Section 5.10), $\Delta_1, \Delta_2, \ldots, \Delta_n$ the potential energy of all the loads is

$$V = \sum_{r=1}^{n} V_r = \sum_{r=1}^{n} (-P_r \Delta_r)$$

and the total potential energy of the system is given by

$$\text{TPE} = U + V = U + \sum_{r=1}^{n} (-P_r \Delta_r) \tag{5.16}$$

5.8 PRINCIPLE OF THE STATIONARY VALUE OF THE TOTAL POTENTIAL ENERGY

Let us now consider an elastic body in equilibrium under a series of external loads, $P_1, P_2, \ldots, P_n$, and suppose that we impose small virtual displacements $\delta\Delta_1, \delta\Delta_2, \ldots, \delta\Delta_n$ in the directions of the loads. The virtual work done by the loads is

$$\sum_{r=1}^{n} P_r \delta\Delta_r$$

This work is accompanied by an increment of strain energy δU in the elastic body, since by specifying virtual displacements of the loads, we automatically impose virtual displacements on the particles of the body itself, as the body is continuous and is assumed to remain so. This increment in strain energy may be regarded as negative virtual work done by the particles so that the total work done during the virtual displacement is

$$-\delta U + \sum_{r=1}^{n} P_r \delta\Delta_r$$

The body is in equilibrium under the applied loads, so that by the principle of virtual work, this expression must be equal to zero. Hence,

$$\delta U - \sum_{r=1}^{n} P_r \delta \Delta_r = 0 \qquad (5.17)$$

The loads P_r remain constant during the virtual displacement; therefore, Eq. (5.17) may be written

$$\delta U - \delta \sum_{r=1}^{n} P_r \Delta_r = 0$$

or, from Eq. (5.16),

$$\delta(U + V) = 0 \qquad (5.18)$$

Thus, the total potential energy of an elastic system has a stationary value for all small displacements if the system is in equilibrium. It may also be shown that, if the stationary value is a minimum, the equilibrium is stable. A qualitative demonstration of this fact is sufficient for our purposes, although mathematical proofs exist[1]. In Fig. 5.24, the positions A, B, and C of a particle correspond to different equilibrium states. The total potential energy of the particle in each of its three positions is proportional to its height h above some arbitrary datum, since we are considering a single particle for which the strain energy is zero. Clearly, at each position, the first-order variation, $\partial(U + V)/\partial u$, is zero (indicating equilibrium), but only at B, where the total potential energy is a minimum is the equilibrium stable. At A and C, we have unstable and neutral equilibrium, respectively.

To summarize, the principle of the stationary value of the total potential energy may be stated as follows:

The total potential energy of an elastic system has a stationary value for all small displacements when the system is in equilibrium; further, the equilibrium is stable if the stationary value is a minimum.

This principle is often used in the approximate analysis of structures where an exact analysis does not exist. We illustrate the application of the principle in Example 5.18, where we suppose that the displaced form of the beam is unknown and must be assumed; this approach is called the *Rayleigh–Ritz* method.

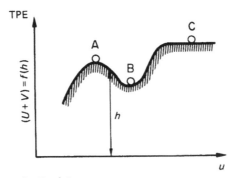

FIGURE 5.24 States of Equilibrium of a Particle

Example 5.18

Determine the deflection of the mid-span point of the linearly elastic, simply supported beam shown in Fig. 5.25; the flexural rigidity of the beam is EI. See Ex. 1.1.

The assumed displaced shape of the beam must satisfy the boundary conditions for the beam. Generally, trigonometric or polynomial functions have been found to be the most convenient, where, however, the simpler the function, the less accurate is the solution. Let us suppose that the displaced shape of the beam is given by

$$v = v_B \sin \frac{\pi z}{L} \tag{i}$$

in which v_B is the displacement at the mid-span point. From Eq. (i), we see that $v = 0$ when $z = 0$ and $z = L$ and that $v = v_B$ when $z = L/2$. Also $dv/dz = 0$ when $z = L/2$, so that the displacement function satisfies the boundary conditions of the beam.

The strain energy, U, due to bending of the beam, is given by[3]

$$U = \int_L \frac{M^2}{2EI} \, dz \tag{ii}$$

Also,

$$M = -EI \frac{d^2 v}{dz^2} \tag{iii}$$

(see Chapter 15). Substituting in Eq. (iii) for v from Eq. (i) and for M in Eq. (ii) from (iii),

$$U = \frac{EI}{2} \int_0^L \frac{v_B^2 \pi^4}{L^4} \sin^2 \frac{\pi z}{L} \, dz$$

which gives

$$U = \frac{\pi^4 EI v_B^2}{4L^3}$$

The total potential energy of the beam is then given by

$$\text{TPE} = U + V = \frac{\pi^4 EI v_B^2}{4L^3} - W v_B$$

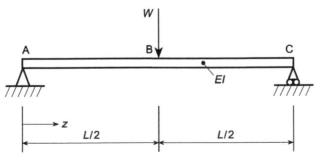

FIGURE 5.25 Approximate Determination of Beam Deflection Using Total Potential Energy

Then, from the principle of the stationary value of the total potential energy,

$$\frac{\partial(U + V)}{\partial v_B} = \frac{\pi^4 E I v_B^2}{2L^3} - W = 0$$

from which

$$v_B = \frac{2WL^3}{\pi^4 EI} = 0.02053 \frac{WL^3}{EI} \tag{iv}$$

The exact expression for the mid-span displacement is[3]

$$v_B = \frac{WL^3}{48EI} = 0.02083 \frac{WL^3}{EI} \tag{v}$$

Comparing the exact (Eq. (v)) and approximate results (Eq. (iv)), we see that the difference is less than 2 percent. Further, the approximate displacement is less than the exact displacement, since, by assuming a displaced shape, we have, in effect, forced the beam into taking that shape by imposing restraint; the beam is therefore stiffer.

5.9 PRINCIPLE OF SUPERPOSITION

An extremely useful principle employed in the analysis of linearly elastic structures is that of superposition. The principle states that, if the displacements at all points in an elastic body are proportional to the forces producing them, that is, the body is linearly elastic, the effect on such a body of a number of forces is the sum of the effects of the forces applied separately. We make immediate use of the principle in the derivation of the reciprocal theorem in the following section.

5.10 RECIPROCAL THEOREM

The reciprocal theorem is an exceptionally powerful method of analysis of linearly elastic structures and is accredited in turn to Maxwell, Betti, and Rayleigh. However, before we establish the theorem, we consider a useful property of linearly elastic systems resulting from the principle of superposition. This principle enables us to express the deflection of any point in a structure in terms of a constant coefficient and the applied loads. For example, a load P_1 applied at a point 1 in a linearly elastic body produces a deflection Δ_1 at the point given by

$$\Delta_1 = a_{11}P_1$$

in which the *influence* or *flexibility* coefficient a_{11} is defined as the deflection at the point 1 in the direction of P_1, produced by a unit load at the point 1 applied in the direction of P_1. Clearly, if the body supports a system of loads such as those shown in Fig. 5.26, each of the loads $P_1, P_2, \ldots, P_n$ contributes to the deflection at the point 1. Thus, the *corresponding deflection* Δ_1 at the point 1 (i.e., the total deflection in the direction of P_1 produced by all the loads) is

$$\Delta_1 = a_{11}P_1 + a_{12}P_2 + \cdots + a_{1n}P_n$$

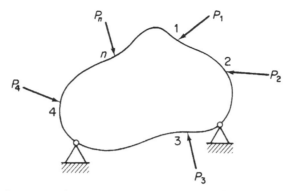

FIGURE 5.26 Linearly Elastic Body Subjected to Loads P_1, P_2, P_3, . . ., P_n

where a_{12} is the deflection at the point 1 in the direction of P_1 produced by a unit load at the point 2 in the direction of the load P_2, and so on. The corresponding deflections at the points of application of the complete system of loads are

$$\left.\begin{array}{l}
\Delta_1 = a_{11}P_1 + a_{12}P_2 + a_{13}P_3 + \cdots + a_{1n}P_n \\
\Delta_2 = a_{21}P_1 + a_{22}P_2 + a_{23}P_3 + \cdots + a_{2n}P_n \\
\Delta_3 = a_{31}P_1 + a_{32}P_2 + a_{33}P_3 + \cdots + a_{3n}P_n \\
\vdots \\
\Delta_n = a_{n1}P_1 + a_{n2}P_2 + a_{n3}P_3 + \cdots + a_{nn}P_n
\end{array}\right\} \tag{5.19}$$

or, in matrix form,

$$\begin{Bmatrix} \Delta_1 \\ \Delta_2 \\ \Delta_3 \\ \vdots \\ \Delta_n \end{Bmatrix} = \begin{bmatrix} a_{11} & a_{12} & a_{13} & \cdots & a_{1n} \\ a_{21} & a_{22} & a_{23} & \cdots & a_{2n} \\ a_{31} & a_{32} & a_{33} & \cdots & a_{3n} \\ \vdots & \vdots & \vdots & \vdots & \vdots \\ a_{n1} & a_{n2} & a_{n3} & \cdots & a_{nn} \end{bmatrix} \begin{Bmatrix} P_1 \\ P_2 \\ P_3 \\ \vdots \\ P_n \end{Bmatrix}$$

which may be written in shorthand matrix notation as

$$\{\Delta\} = [A]\{P\}$$

Suppose now that an elastic body is subjected to a gradually applied force P_1 at a point 1 and then, while P_1 remains in position, a force P_2 is gradually applied at another point 2. The total strain energy U of the body is given by

$$U_1 = \frac{P_1}{2}(a_{11}P_1) + \frac{P_2}{2}(a_{22}P_2) + P_1(a_{12}P_2) \tag{5.20}$$

The third term on the right-hand side of Eq. (5.20) results from the additional work done by P_1 as it is displaced through a further distance $a_{12}P_2$ by the action of P_2. If we now remove the loads and apply P_2 followed by P_1, we have

$$U_2 = \frac{P_2}{2}(a_{22}P_2) + \frac{P_1}{2}(a_{11}P_1) + P_2(a_{21}P_1) \tag{5.21}$$

By the principle of superposition, the strain energy stored is independent of the order in which the loads are applied. Hence,

$$U_1 = U_2$$

and it follows that

$$a_{12} = a_{21} \tag{5.22}$$

Thus, in its simplest form, the reciprocal theorem states the following:

The deflection at point 1 in a given direction due to a unit load at point 2 in a second direction is equal to the deflection at point 2 in the second direction due to a unit load at point 1 in the first direction.

In a similar manner, we derive the relationship between moments and rotations:

The rotation at a point 1 due to a unit moment at a point 2 is equal to the rotation at point 2 produced by a unit moment at point 1.

Finally, we have

The rotation at point 1 due to a unit load at point 2 is numerically equal to the deflection at point 2 in the direction of the unit load due to a unit moment at point 1.

Example 5.19

A cantilever 800 mm long with a prop 500 mm from the wall deflects in accordance with the following observations when a point load of 40 N is applied to its end:

Distance (mm)	0	100	200	300	400	500	600	700	800
Deflection (mm)	0	−0.3	−1.4	−2.5	−1.9	0	2.3	4.8	10.6

What is the angular rotation of the beam at the prop due to a 30 N load applied 200 mm from the wall, together with a 10 N load applied 350 mm from the wall? See Ex. 1.1.

The initial deflected shape of the cantilever is plotted as shown in Fig. 5.27(a) and the deflections at D and E produced by the 40 N load determined. The solution then proceeds as follows.

Deflection at D due to 40 N load at C = −1.4 mm.

Hence, from the reciprocal theorem, the deflection at C due to a 40 N load at D = −1.4 mm.

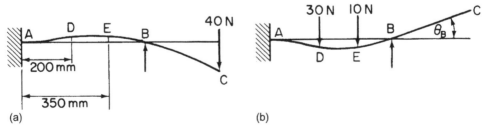

FIGURE 5.27 (a) Given Deflected Shape of Propped Cantilever; (b) Determination of the Deflection of C

It follows that the deflection at C due to a 30 N load at D $= -3/4 \times 1.4 = -1.05$ mm.

Similarly, the deflection at C due to a 10 N load at E $= -1/4 \times 2.4 = -0.6$ mm.

Therefore, the total deflection at C, produced by the 30 and 10 N loads acting simultaneously (Fig. 5.27(b)), is $-1.05 - 0.6 = -1.65$ mm, from which the angular rotation of the beam at B, θ_B, is given by

$$\theta_B = \tan^{-1}\frac{1.65}{300} = \tan^{-1}0.0055$$

or

$$\theta_B = 0°19'$$

Example 5.20

An elastic member is pinned to a drawing board at its ends A and B. When a moment M is applied at A, A rotates θ_A, B rotates θ_B, and the center deflects δ_1. The same moment M applied to B rotates B, θ_C, and deflects the center through δ_2. Find the moment induced at A when a load W is applied to the center in the direction of the measured deflections, both A and B being restrained against rotation. See Ex. 1.1.

The three load conditions and the relevant displacements are shown in Fig. 5.28. From Figs. 5.28(a) and (b), the rotation at A due to M at B is, from the reciprocal theorem, equal to the rotation at B due to M at A. Hence,

$$\theta_{A(b)} = \theta_B$$

It follows that the rotation at A due to M_B at B is

$$\theta_{A(c),1} = M_B/M\theta_B \tag{i}$$

Also the rotation at A due to unit load at C is equal to the deflection at C due to unit moment at A. Therefore,

$$\frac{\theta_{A(c),2}}{W} = \frac{\delta_1}{M}$$

or

$$\theta_{A(c),2} = \frac{W}{M}\delta_1 \tag{ii}$$

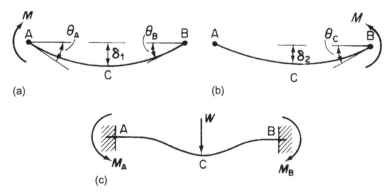

(a)   (b)

(c)

FIGURE 5.28 Model Analysis of a Fixed Beam

where $\theta_{A(c),2}$ is the rotation at A due to W at C. Finally, the rotation at A due to M_A at A is, from Fig. 5.28(a) and (c),

$$\theta_{A(c),3} = \frac{M_A}{M}\theta_A \tag{iii}$$

The total rotation at A produced by M_A at A, W at C, and M_B at B is, from Eqs. (i), (ii), and (iii),

$$\theta_{A(c),1} + \theta_{A(c),2} + \theta_{A(c),3} = \frac{M_B}{M}\theta_B + \frac{W}{M}\delta_1 + \frac{M_A}{M}\theta_A = 0 \tag{iv}$$

since the end A is restrained from rotation. Similarly, the rotation at B is given by

$$\frac{M_B}{M}\theta_C + \frac{W}{M}\delta_2 + \frac{M_A}{M}\theta_B = 0 \tag{v}$$

Solving Eqs. (iv) and (v) for M_A gives

$$M_A = W\left(\frac{\delta_2\theta_B - \delta_1\theta_C}{\theta_A\theta_C - \theta_B^2}\right)$$

The fact that the arbitrary moment M does not appear in the expression for the restraining moment at A (similarly it does not appear in M_B), produced by the load W, indicates an extremely useful application of the reciprocal theorem, namely, the model analysis of statically indeterminate structures. For example, the fixed beam of Fig. 5.28(c) could be a full-scale bridge girder. It is then only necessary to construct a model, say of Perspex, having the same flexural rigidity EI as the full-scale beam and measure rotations and displacements produced by an arbitrary moment M to obtain fixing moments in the full-scale beam supporting a full-scale load.

5.11 TEMPERATURE EFFECTS

A uniform temperature applied across a beam section produces an expansion of the beam, as shown in Fig. 5.29, provided there are no constraints. However, a linear temperature gradient across the beam section causes the upper fibers of the beam to expand more than the lower ones, producing a bending strain, as shown in Fig. 5.30, without the associated bending stresses, again provided no constraints are present.

Consider an element of the beam of depth h and length δz subjected to a linear temperature gradient over its depth, as shown in Fig. 5.31(a). The upper surface of the element increases in length to $\delta z(1 + \alpha t)$ (see Section 1.15.1), where α is the coefficient of linear expansion of the material of the beam. Thus, from Fig. 5.31(b),

$$\frac{R}{\delta z} = \frac{R+h}{\delta z(1+\alpha t)}$$

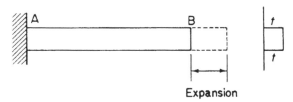

FIGURE 5.29 Expansion of Beam Due to Uniform Temperature

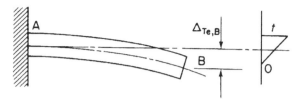

FIGURE 5.30 Bending of Beam Due to Linear Temperature Gradient

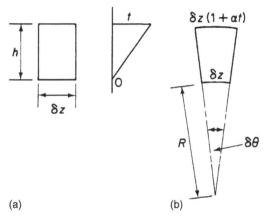

(a)   (b)

FIGURE 5.31 (a) Linear Temperature Gradient Applied to Beam Element; (b) Bending of Beam Element Due to Temperature Gradient

giving

$$R = h/\alpha t \tag{5.23}$$

Also.

$$\delta\theta = \delta z / R$$

so that, from Eq. (5.23).

$$\delta\theta = \frac{\delta z \alpha t}{h} \tag{5.24}$$

We may now apply the principle of the stationary value of the total complementary energy in conjunction with the unit load method to determine the deflection Δ_{Te}, due to the temperature of any point of the beam shown in Fig. 5.30. We have seen that this principle is equivalent to the application of the principle of virtual work, where virtual forces act through real displacements. Therefore, we may specify that the displacements are those produced by the temperature gradient, while the virtual force system is the unit load. Thus, the deflection $\Delta_{Te,B}$ of the tip of the beam is found by writing down the increment in total complementary energy caused by the application of a virtual unit load at B and equating the resulting expression to zero (see Eqs. (5.7) and Eq. (i) of Example 5.2). Thus,

$$\delta C = \int_L M_1 \, d\theta - 1 \, \Delta_{\text{Te,B}} = 0$$

or

$$\Delta_{\text{Te,B}} = \int_L M_1 \, d\theta \tag{5.25}$$

where M_1 is the bending moment at any section due to the unit load. Substituting for $d\theta$ from Eq. (5.24), we have

$$\Delta_{\text{Te,B}} = \int_L M_1 \frac{\alpha t}{h} \, dz \tag{5.26}$$

where t can vary arbitrarily along the span of the beam but only linearly with depth. For a beam supporting some form of external loading, the total deflection is given by the superposition of the temperature deflection from Eq. (5.26) and the bending deflection from Eq. (5.15); thus,

$$\Delta = \int_L M_1 \left(\frac{M_0}{EI} + \frac{\alpha t}{h} \right) dz \tag{5.27}$$

Example 5.21

Determine the deflection of the tip of the cantilever in Fig. 5.32 with the temperature gradient shown. See Ex. 1.1.

Applying a unit load vertically downward at B, $M_1 = 1 \times z$. Also the temperature t at a section z is $t_0 \, (l - z)/l$. Substituting in Eq. (5.26) gives

$$\Delta_{\text{Te,B}} = \int_0^l z \frac{\alpha}{h} \frac{t_0}{l} (l - z) dz \tag{i}$$

Integrating Eq. (i) gives

$$\Delta_{\text{Te,B}} = \frac{\alpha t_0 l^2}{6h} \quad (\text{i.e., downward})$$

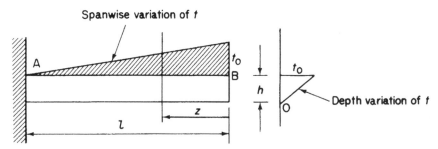

Spanwise variation of t

Depth variation of t

FIGURE 5.32 Beam of Example 5.21

References

[1] Charlton TM. Energy principles in applied statics. London: Blackie; 1959.
[2] Gregory MS. Introduction to extremum principles. London: Butterworths; 1969.
[3] Megson THG. Structural and stress analysis. 2nd ed. Oxford: Elsevier; 2005.

Further reading

Argyris JH, Kelsey S. Energy theorems and structural analysis. London: Butterworths; 1960.
Hoff NJ. The analysis of structures. New York: John Wiley and Sons; 1956.
Timoshenko SP, Gere JM. Theory of elastic stability. New York: McGraw-Hill; 1961.

PROBLEMS

P.5.1. Find the magnitude and the direction of the movement of the joint C of the plane pin-jointed frame loaded as shown in Fig. P.5.1. The value of L/AE for each member is 1/20 mm/N.

Answer: 5.24 mm at 14.7° to left of vertical

P.5.2. A rigid triangular plate is suspended from a horizontal plane by three vertical wires attached to its corners. The wires are each 1 mm diameter, 1440 mm long, with a modulus of elasticity of 196,000 N/mm². The ratio of the lengths of the sides of the plate is 3:4:5. Calculate the deflection at the point of application due to a 100 N load placed at a point equidistant from the three sides of the plate.

Answer: 0.33 mm

P.5.3. The pin-jointed space frame shown in Fig. P.5.3 is attached to rigid supports at points 0, 4, 5, and 9 and is loaded by a force P in the x direction and a force $3P$ in the negative y direction at the point 7. Find the rotation of member 27 about the z axis due to this loading. Note that the plane frames 01234 and 56789 are identical. All members have the same cross-sectional area A and Young's modulus E.

Answer: $382P/9AE$

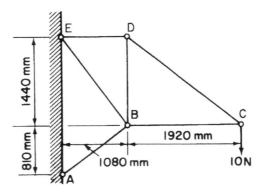

FIGURE P.5.1

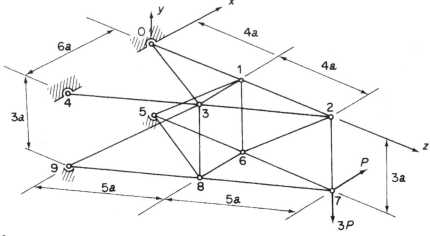

FIGURE P.5.3

P.5.4. A horizontal beam is of uniform material throughout but has a second moment of area of I for the central half of the span L and $I/2$ for each section in both outer quarters of the span. The beam carries a single central concentrated load P.

(a) Derive a formula for the central deflection of the beam, due to P, when simply supported at each end of the span.

(b) If both ends of the span are encastré determine the magnitude of the fixed end moments.

Answer: $3PL^3/128EI$, $5PL/48$ (hogging)

P.5.5. The tubular steel post shown in Fig. P.5.5 supports a load of 250 N at the free end C. The outside diameter of the tube is 100 mm and the wall thickness is 3 mm. Neglecting the weight of the tube find the horizontal deflection at C. The modulus of elasticity is 206,000 N/mm^2.

Answer: 53.3 mm

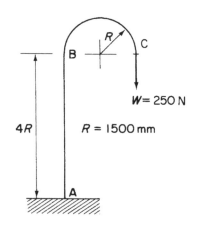

FIGURE P.5.5

P.5.6. A simply supported beam AB of span L and uniform section carries a distributed load of intensity varying from zero at A to w_0/unit length at B according to the law

$$w = \frac{2w_0 z}{L}\left(1 - \frac{z}{2L}\right)$$

per unit length. If the deflected shape of the beam is given approximately by the expression

$$v = a_1 \sin\frac{\pi z}{L} + a_2 \sin\frac{2\pi z}{L}$$

evaluate the coefficients a_1 and a_2 and find the deflection of the beam at mid-span.

   Answer: $a_1 = 2w_0 L^4(\pi^2 + 4)/EI\pi^7$, $a_2 = -w_0 L^4/16EI\pi^5$, $0.00918\, w_0 L^4/EI$.

P.5.7. A uniform simply supported beam, span L, carries a distributed loading which varies according to a parabolic law across the span. The load intensity is zero at both ends of the beam and w_0 at its mid-point. The loading is normal to a principal axis of the beam cross-section and the relevant flexural rigidity is EI. Assuming that the deflected shape of the beam can be represented by the series

$$v = \sum_{i=1}^{\infty} a_i \sin\frac{i\pi z}{L}$$

find the coefficients a_i and the deflection at the mid-span of the beam using only the first term in this series.

   Answer: $a_i = 32w_0 L^4/EI\pi^7 i^7\,(i\ \text{odd})$, $w_0 L^4/94.4EI$.

P.5.8. Figure P.5.8 shows a plane pin-jointed framework pinned to a rigid foundation. All its members are made of the same material and have equal cross-sectional area A, except member 12, which has area $A\sqrt{2}$. Under some system of loading, member 14 carries a tensile stress of 0.7 N/mm^2. Calculate the change in temperature which, if applied to member 14 only, reduces the stress in that member to zero. Take the coefficient of linear expansion as $\alpha = 24 \times 10^{-6}$/°C and Young's modulus $E = 70,000$ N/mm^2.

   Answer: 5.6°C

P.5.8. MATLAB Use the Symbolic Math Toolbox in MATLAB to repeat Problem P.5.8, assuming that member 14 carries tensile stresses of 0.5, 0.7, 0.9, and 1.1 N/mm^2.

   Answer: (i)   4.0°C
                (ii) 5.6°C
                (iii) 7.2°C
                (iv) 8.8°C

P.5.9. The plane, pin-jointed rectangular framework shown in Fig. P.5.9(a) has one member (24) which is loosely attached at joint 2, so that relative movement between the end of the member and the joint may occur when the framework is loaded. This movement is a maximum of 0.25 mm and takes place only in the direction 24. Figure P.5.9(b) shows joint 2 in detail when the framework is not loaded. Find the value of the load P at which member 24 just becomes an effective part of the structure and also the loads in all the members when P is 10,000 N. All bars are of the same material ($E = 70,000$ N/mm^2) and have a cross-sectional area of 300 mm^2.

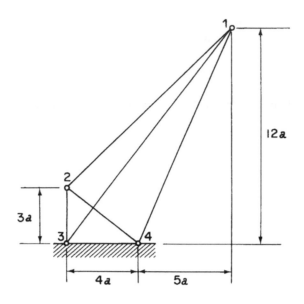

FIGURE P.5.8

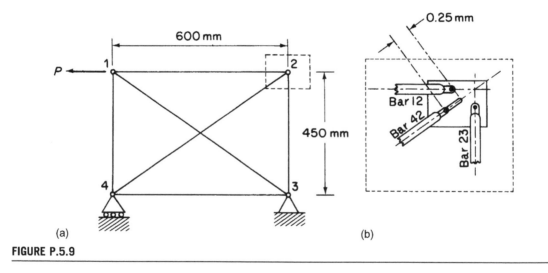

(a)   (b)

FIGURE P.5.9

Answer: $P = 294$ N, $F_{12} = 2481.6$ N(T), $F_{23} = 1861.2$ N(T), $F_{34} = 2481.6$ N(T),
$F_{41} = 5638.9$ N(C), $F_{13} = 9398.1$ N(T), $F_{24} = 3102.0$ N(C).

P.5.10. The plane frame ABCD of Fig. P.5.10 consists of three straight members with rigid joints at B and C, freely hinged to rigid supports at A and D. The flexural rigidity of AB and CD is twice that of BC. A distributed load is applied to AB, varying linearly in intensity from zero at A to w per unit length

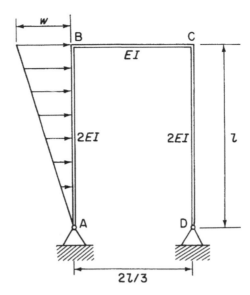

FIGURE P.5.10

at B. Determine the distribution of bending moment in the frame, illustrating your results with a sketch showing the principal values.

Answer: $M_B = 7\,wl^2/45$, $M_C = 8\,wl^2/45$,

Cubic distribution on AB, linear on BC and CD

P.5.11. A bracket BAC is composed of a circular tube AB, whose second moment of area is $1.5I$, and a beam AC, whose second moment of area is I and which has negligible resistance to torsion. The two members are rigidly connected together at A and built into a rigid abutment at B and C, as shown in Fig. P.5.11. A load P is applied at A in a direction normal to the plane of the figure. Determine the fraction of the load supported at C. Both members are of the same material, for which $G = 0.38E$.

Answer: $0.72P$.

P.5.12. In the plane pin-jointed framework shown in Fig. P.5.12, bars 25, 35, 15, and 45 are linearly elastic with modulus of elasticity E. The remaining three bars obey a nonlinear elastic stress–strain law given by

$$\varepsilon = \frac{\tau}{E}\left[1 + \left(\frac{\tau}{\tau_0}\right)^n\right]$$

where τ is the stress corresponding to strain ε. Each of bars 15, 45, and 23 has a cross-sectional area A, and each of the remainder has an area of $A/\sqrt{3}$. The length of member 12 is equal to the length of member $34 = 2\,L$. If a vertical load P_0 is applied at joint 5, as shown, show that the force in the member 23, that is, F_{23}, is given by the equation

$$\alpha^n x^{n+1} + 3.5x + 0.8 = 0$$

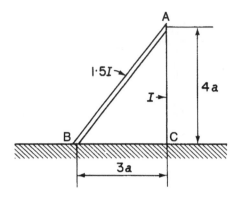

FIGURE P.5.11

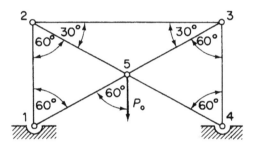

FIGURE P.5.12

where

$$x = F_{23}/P_0 \quad \text{and} \quad \alpha = P_0/A\tau_0$$

P.5.13. Figure P.5.13 shows a plan view of two beams, AB 9150 mm long and DE 6100 mm long. The simply supported beam AB carries a vertical load of 100,000 N applied at F, a distance one third of the span from B. This beam is supported at C on the encastré beam DE. The beams are of uniform cross-section and have the same second moment of area 83.5×10^6 mm^4. $E = 200,000$ N/mm^2. Calculate the deflection of C.

Answer: 5.6 mm

P.5.14. The plane structure shown in Fig. P.5.14 consists of a uniform continuous beam ABC pinned to a fixture at A and supported by a framework of pin-jointed members. All members other than ABC have the same cross-sectional area A. For ABC, the area is $4A$ and the second moment of area for bending is $Aa^2/16$. The material is the same throughout. Find (in terms of w, A, a, and Young's modulus E) the vertical displacement of point D under the vertical loading shown. Ignore shearing strains in the beam ABC.

Answer: 30,232 $wa^2/3AE$

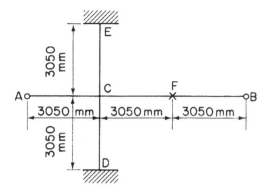

FIGURE P.5.13

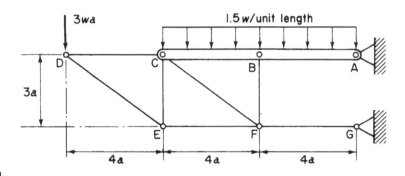

FIGURE P.5.14

P.5.15. The fuselage frame shown in Fig. P.5.15 consists of two parts, ACB and ADB, with frictionless pin joints at A and B. The bending stiffness is constant in each part, with value EI for ACB and xEI for ADB. Find x so that the maximum bending moment in ADB is one half of that in ACB. Assume that the deflections are due to bending strains only.

Answer: 0.092

P.5.16. A transverse frame in a circular section fuel tank is of radius r and constant bending stiffness EI. The loading on the frame consists of the hydrostatic pressure due to the fuel and the vertical support reaction P, which is equal to the weight of fuel carried by the frame, shown in Fig. P.5.16. Taking into account only strains due to bending, calculate the distribution of bending moment around the frame in terms of the force P, the frame radius r, and the angle θ.

Answer: $M = Pr(0.160 - 0.080 \cos\theta - 0.159\theta \sin\theta)$

P.5.17. The frame shown in Fig. P.5.17 consists of a semi-circular arc, center B, radius a, of constant flexural rigidity EI jointed rigidly to a beam of constant flexural rigidity $2EI$. The frame is subjected to an outward loading, as shown, arising from an internal pressure p_0. Find the bending moment at points

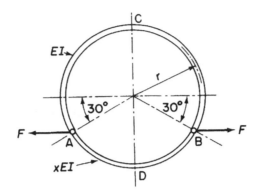

FIGURE P.5.15

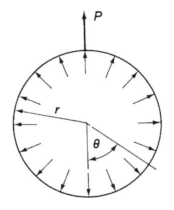

FIGURE P.5.16

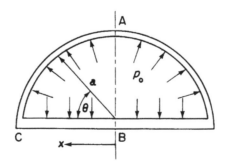

FIGURE P.5.17

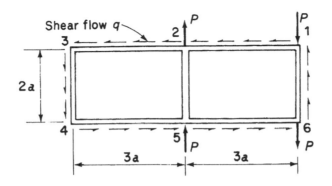

FIGURE P.5.18

A, B, and C and locate any points of contraflexure. A is the mid-point of the arc. Neglect deformations of the frame due to shear and normal forces.

Answer: $M_A = -0.057p_0a^2$, $M_B = -0.292p_0a^2$, $M_C = 0.208p_0a^2$

Points of contraflexure: in AC, at 51.7° from horizontal; in BC, 0.764a from B

P.5.18. The rectangular frame shown in Fig. P.5.18 consists of two horizontal members 123 and 456 rigidly joined to three vertical members 16, 25, and 34. All five members have the same bending stiffness EI. The frame is loaded in its own plane by a system of point loads P, which are balanced by a constant shear flow q around the outside. Determine the distribution of the bending moment in the frame and sketch the bending moment diagram. In the analysis, take into account only bending deformations.

Answer: Shears only at mid-points of vertical members. On the lower half of the frame, $S_{43} = 0.27P$ to right, $S_{52} = 0.69P$ to left, $S_{61} = 1.08P$ to left; the bending moment diagram follows in Figure P.5.19.

P.5.19. A circular fuselage frame, shown in Fig. P.5.19, of radius r and constant bending stiffness EI, has a straight floor beam of length $r\sqrt{2}$, bending stiffness EI, rigidly fixed to the frame at either end. The frame is loaded by a couple T applied at its lowest point and a constant equilibrating shear flow q

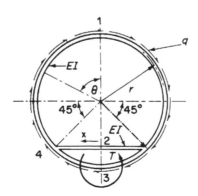

FIGURE P.5.19

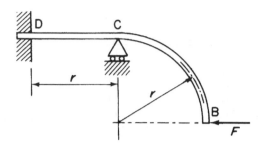

FIGURE P.5.20

around its periphery. Determine the distribution of the bending moment in the frame, illustrating your answer by means of a sketch. In the analysis, deformations due to shear and end load may be considered negligible. The depth of the frame cross-section in comparison with the radius r may also be neglected.

Answer: $M_{14} = T(0.29 \sin\theta - 0.16\theta),$ $M_{24} = 0.30Tx/r,$ $M_{43} = T(0.59 \sin\theta - 0.16\theta).$

P.5.20. A thin-walled member BCD is rigidly built-in at D and simply supported at the same level at C, as shown in Fig. P.5.20. Find the horizontal deflection at B due to the horizontal force F. Full account must be taken of deformations due to shear and direct strains, as well as to bending. The member is of uniform cross-section, of area A, relevant second moment of area in bending $I = Ar^2/400$, and "reduced: effective area in shearing $A' = A/4$. Poisson's ratio for the material is $v = 1/3$. Give the answer in terms of F, r, A, and Young's modulus E.

Answer: 448 Fr/EA.

P.5.21. Figure P.5.21 shows two cantilevers, the end of one being vertically above the other and connected to it by a spring AB. Initially, the system is unstrained. A weight W placed at A causes a vertical deflection at A of δ_1 and a vertical deflection at B of δ_2. When the spring is removed, the weight W at A causes a deflection at A of δ_3. Find the extension of the spring when it is replaced and the weight W is transferred to B.

Answer: $\delta_2 (\delta_1 - \delta_2)/(\delta_3 - \delta_1)$

P.5.22. A beam 2400 mm long is supported at two points A and B, which are 1440 mm apart; point A is 360 mm from the left-hand end of the beam and point B is 600 mm from the right-hand end; the value of EI for the beam is 240×10^8 N mm^2. Find the slope at the supports due to a load of 2000 N applied at the mid-point of AB. Use the reciprocal theorem in conjunction with this result to find the deflection at the mid-point of AB due to loads of 3000 N applied at each of the extreme ends of the beam.

Answer: 0.011, 15.8 mm

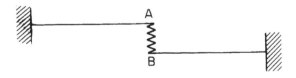

FIGURE P.5.21

P.5.22. MATLAB Use the Symbolic Math Toolbox in MATLAB to repeat Problem P.5.22. In addition, use the reciprocal theorem to calculate the displacement at the mid-point of AB due to the following combinations of loads at C and D:

	(i)	(ii)	(iii)	(iv)	(v)
C:	2000 N	2500 N	3000 N	3500 N	4000 N
D:	4000 N	3500 N	3000 N	2500 N	2000 N

Answer: 0.011 rad

(i)	(ii)	(iii)	(iv)	(v)
17.16 mm	16.5 mm	15.84 mm	15.18 mm	14.52 mm

P.5.23. Figure P.5.23 shows a cantilever beam subjected to linearly varying temperature gradients along its length and through its depth. Calculate the deflection at the free end of the beam.

Answer: $2\alpha t_0 L^2/3h$

P.5.24. Figure P.5.24 shows a frame pinned to its support at A and B. The frame center-line is a circular arc and the section is uniform, of bending stiffness EI and depth d. Find an expression for the maximum stress produced by a uniform temperature gradient through the depth, the temperatures on the outer and inner surfaces being, respectively, raised and lowered by amount T. The points A and B are unaltered in position.

Answer: $1.30ET\alpha$

P.5.25. A uniform, semi-circular fuselage frame is pin-jointed to a rigid portion of the structure and is subjected to a given temperature distribution on the inside, as shown in Fig. P.5.25. The temperature falls linearly across the section of the frame to zero on the outer surface. Find the values of the reactions at the pin-joints and show that the distribution of the bending moment in the frame is

$$M = \frac{0.59\ EI\alpha\theta_0 \cos \psi}{h}$$

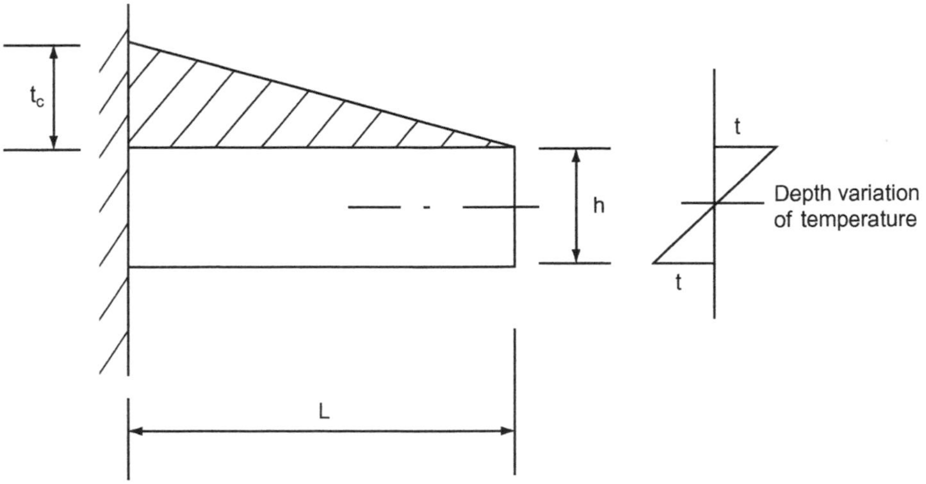

FIGURE P.5.23

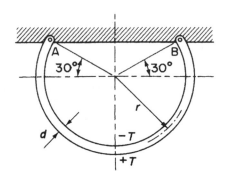

FIGURE P.5.24

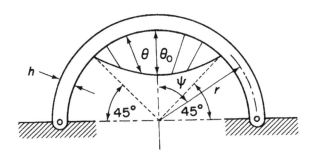

FIGURE P.5.25

given that

(a) The temperature distribution is

$$\theta = \theta_0 \cos 2\psi, \quad \text{for} - \pi/4 < \Psi < \pi/4$$
$$\theta = 0, \qquad\quad \text{for} - \pi/4 > \Psi > \pi/4$$

(b) Only bending deformations are to be taken into account:

$$\alpha = \text{coefficient of linear expansion of frame material}$$
$$EI = \text{bending rigidity of frame}$$
$$h = \text{depth of cross-section}$$
$$r = \text{mean radius of frame.}$$

Matrix methods

6

Actual aircraft structures consist of numerous components generally arranged in an irregular manner. These components are usually continuous and therefore, theoretically, possess an infinite number of degrees of freedom and redundancies. Analysis is possible only if the actual structure is replaced by an idealized approximation or model. This procedure is discussed to some extent in Chapter 19, where we note that the greater the simplification introduced by the idealization, the less complex but more inaccurate becomes the analysis. In aircraft design, where structural weight is of paramount importance, an accurate knowledge of component loads and stresses is essential, so that at some stage in the design these must be calculated as accurately as possible. This accuracy is achieved only by considering an idealized structure which closely represents the actual structure. Standard methods of structural analysis are inadequate for coping with the necessary degree of complexity in such idealized structures. This situation led, in the late 1940s and early 1950s, to the development of *matrix methods* of analysis and at the same time to the emergence of high-speed, electronic, digital computers. Conveniently, matrix methods are ideally suited for expressing structural theory and expressing that theory in a form suitable for numerical solution by computer.

A structural problem may be formulated in either of two ways. One approach proceeds with the displacements of the structure as the unknowns, the internal forces then follow from the determination of these displacements; in the alternative approach, forces are treated as being initially unknown. In the language of matrix methods, these two approaches are known as the *stiffness* (or *displacement*) method and the *flexibility* (or *force*) method, respectively. The most widely used of these two methods is the stiffness method, and for this reason, we shall concentrate on this particular approach. Argyris and Kelsey[1], however, showed that complete duality exists between the two methods, in that the form of the governing equations is the same whether they are expressed in terms of displacements or forces.

Generally, actual structures must be idealized to some extent before they become amenable to analysis. Examples of some simple idealizations and their effect on structural analysis are presented in Chapter 19 for aircraft structures. Outside the realms of aeronautical engineering, the representation of a truss girder by a pin-jointed framework is a well-known example of the idealization of what are known as "skeletal" structures. Such structures are assumed to consist of a number of elements joined at points called *nodes*. The behavior of each element may be determined by basic methods of structural analysis and hence the behavior of the complete structure is obtained by superposition. Operations such as this are easily carried out by matrix methods, as we shall see later in this chapter.

A more difficult type of structure to idealize is the continuum structure; in this category are dams, plates, shells, and obviously, aircraft fuselage and wing skins. A method, extending the matrix technique for skeletal structures, of representing continua by any desired number of elements connected at their nodes was developed by Clough et al[2]. at the Boeing Aircraft Company and the University of

Berkeley in California. The elements may be of any desired shape but the simplest, used in plane stress problems, are the triangular and quadrilateral elements. We shall discuss the *finite element method*, as it is known, in greater detail later.

Initially, we shall develop the matrix stiffness method of solution for simple skeletal and beam structures. The fundamentals of matrix algebra are assumed.

6.1 **NOTATION**

Generally, we shall consider structures subjected to forces, $F_{x,1}, F_{y,1}, F_{z,1}, F_{x,2}, F_{y,2}, F_{z,2}, \ldots, F_{x,n}, F_{y,n}, F_{z,n}$, at nodes 1, 2, ..., n, at which the displacements are $u_1, v_1, w_1, u_2, v_2, w_2, \ldots, u_n, v_n, w_n$. The numerical suffixes specify nodes, while the algebraic suffixes relate the direction of the forces to an arbitrary set of axes, x, y, z. Nodal displacements u, v, w represent displacements in the positive directions of the x, y, and z axes, respectively. The forces and nodal displacements are written as column matrices (alternatively known as *column vectors*)

$$
\begin{Bmatrix} F_{x,1} \\ F_{y,1} \\ F_{z,1} \\ F_{x,2} \\ F_{y,2} \\ F_{z,2} \\ \vdots \\ F_{x,n} \\ F_{y,n} \\ F_{z,n} \end{Bmatrix}
\qquad
\begin{Bmatrix} u_1 \\ v_1 \\ w_1 \\ u_2 \\ v_2 \\ w_2 \\ \vdots \\ u_n \\ v_n \\ w_n \end{Bmatrix}
$$

which, when once established for a particular problem, may be abbreviated to

$$\{F\} \quad \{\delta\}$$

The generalized force system $\{F\}$ can contain moments M and torques T in addition to direct forces, in which case, $\{\delta\}$ includes rotations θ. Therefore, in referring simply to a nodal force system, we imply the possible presence of direct forces, moments, and torques, while the corresponding nodal displacements can be translations and rotations.

For a complete structure, the nodal forces and nodal displacements are related through a *stiffness matrix* $[K]$. We shall see that, in general,

$$\{F\} = [K]\{\delta\} \tag{6.1}$$

where $[K]$ is a symmetric matrix of the form

$$
[K] = \begin{bmatrix} k_{11} & k_{12} & \cdots & k_{1n} \\ k_{21} & k_{22} & \cdots & k_{2n} \\ \cdots & \cdots & \cdots & \cdots \\ k_{n1} & k_{n2} & \cdots & k_{nn} \end{bmatrix} \tag{6.2}
$$

The element k_{ij} (that is, the element located on row i and in column j) is known as the *stiffness influence coefficient* (note $k_{ij} = k_{ji}$). Once the stiffness matrix $[K]$ has been formed, the complete solution to a problem follows from routine numerical calculations that are carried out, in most practical cases, by computer.

6.2 STIFFNESS MATRIX FOR AN ELASTIC SPRING

The formation of the stiffness matrix $[K]$ is the most crucial step in the matrix solution of any structural problem. We shall show in the subsequent work how the stiffness matrix for a complete structure may be built up from a consideration of the stiffness of its individual elements. First, however, we shall investigate the formation of $[K]$ for a simple spring element, which exhibits many of the characteristics of an actual structural member.

The spring of stiffness k shown in Fig. 6.1 is aligned with the x axis and supports forces $F_{x,1}$ and $F_{x,2}$ at its nodes 1 and 2, where the displacements are u_1 and u_2. We build up the stiffness matrix for this simple case by examining different states of nodal displacement. First, we assume that node 2 is prevented from moving, such that $u_1 = u_1$ and $u_2 = 0$. Hence,

$$F_{x,1} = ku_1$$

and, from equilibrium, we see that

$$F_{x,2} = -F_{x,1} = -ku_1 \tag{6.3}$$

which indicates that $F_{x,2}$ has become a reactive force in the opposite direction to $F_{x,1}$. Secondly, we take the reverse case, where $u_1 = 0$ and $u_2 = u_2$, and obtain

$$F_{x,2} = ku_2 = -F_{x,1} \tag{6.4}$$

By superposition of these two conditions, we obtain relationships between the applied forces and the nodal displacements for the state when $u_1 = u_1$ and $u_2 = u_2$. Thus,

$$\left. \begin{array}{l} F_{x,1} = ku_1 - ku_2 \\ F_{x,2} = -ku_1 + ku_2 \end{array} \right\} \tag{6.5}$$

Writing Eq. (6.5) in matrix form, we have

$$\left\{ \begin{array}{c} F_{x,1} \\ F_{x,2} \end{array} \right\} = \left[\begin{array}{cc} k & -k \\ -k & k \end{array} \right] \left\{ \begin{array}{c} u_1 \\ u_2 \end{array} \right\} \tag{6.6}$$

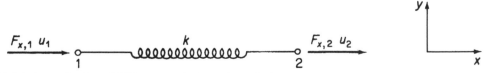

FIGURE 6.1 Determination of Stiffness Matrix for a Single Spring

and by comparison with Eq. (6.1), we see that the stiffness matrix for this spring element is

$$[K] = \begin{bmatrix} k & -k \\ -k & k \end{bmatrix} \tag{6.7}$$

which is a symmetric matrix of order 2×2.

6.3 STIFFNESS MATRIX FOR TWO ELASTIC SPRINGS IN LINE

Bearing in mind the results of the previous section, we shall now proceed, initially by a similar process, to obtain the stiffness matrix of the composite two-spring system shown in Fig. 6.2. The notation and sign convention for the forces and nodal displacements are identical to those specified in section 6.1.

First, let us suppose that $u_1 = u_1$ and $u_2 = u_3 = 0$. By comparison with the single spring case, we have

$$F_{x,1} = k_a u_1 = -F_{x,2} \tag{6.8}$$

but, in addition, $F_{x,3} = 0$, since $u_2 = u_3 = 0$.

Secondly, we put $u_1 = u_3 = 0$ and $u_2 = u_2$. Clearly, in this case, the movement of node 2 takes place against the combined spring stiffnesses k_a and k_b. Hence,

$$\left.\begin{aligned} F_{x,2} &= (k_a + k_b)u_2 \\ F_{x,1} &= -k_a u_2 + F_{x,3} = -k_b u_2 \end{aligned}\right\} \tag{6.9}$$

Thus, the reactive force $F_{x,1}$ $(= -k_a u_2)$ is not directly affected by the fact that node 2 is connected to node 3 but is determined solely by the displacement of node 2. Similar conclusions are drawn for the reactive force $F_{x,3}$.

Finally, we set $u_1 = u_2 = 0$, $u_3 = u_3$ and obtain

$$\left.\begin{aligned} F_{x,3} &= k_b u_3 = -F_{x,2} \\ F_{x,1} &= 0 \end{aligned}\right\} \tag{6.10}$$

Superimposing these three displacement states, we have, for the condition $u_1 = u_1$, $u_2 = u_2$, $u_3 = u_3$,

$$\left.\begin{aligned} F_{x,1} &= k_a u_1 - k_a u_2 \\ F_{x,2} &= -k_a u_1 + (k_a + k_b)u_2 - k_b u_3 \\ F_{x,3} &= -k_b u_2 + k_b u_3 \end{aligned}\right\} \tag{6.11}$$

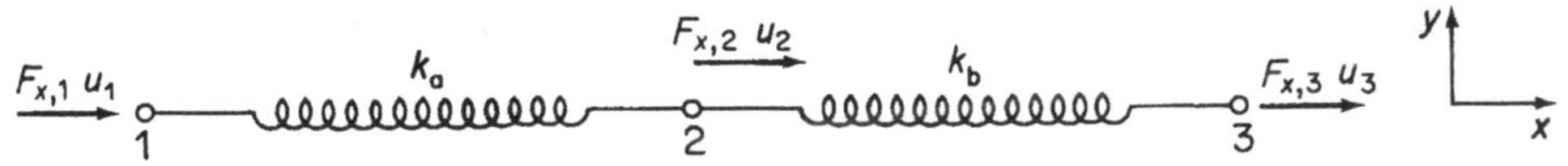

FIGURE 6.2 Stiffness Matrix for a Two-Spring System

Writing Eqs. (6.11) in matrix form gives

$$\begin{Bmatrix} F_{x,1} \\ F_{x,2} \\ F_{x,3} \end{Bmatrix} = \begin{bmatrix} k_a & -k_a & 0 \\ -k_a & k_a + k_b & -k_b \\ 0 & -k_b & k_b \end{bmatrix} \begin{Bmatrix} u_1 \\ u_2 \\ u_3 \end{Bmatrix} \tag{6.12}$$

Comparison of Eqs. (6.12) with Eq. (6.1) shows that the stiffness matrix $[K]$ of this two-spring system is

$$[K] = \begin{bmatrix} k_a & -k_a & 0 \\ -k_a & k_a + k_b & -k_b \\ 0 & -k_b & k_b \end{bmatrix} \tag{6.13}$$

Equation (6.13) is a symmetric matrix of order 3×3.

It is important to note that the order of a stiffness matrix may be predicted from a knowledge of the number of nodal forces and displacements. For example, Eq. (6.7) is a 2×2 matrix connecting *two* nodal forces with *two* nodal displacements; Eq. (6.13) is a 3×3 matrix relating *three* nodal forces to *three* nodal displacements. We deduce that a stiffness matrix for a structure in which n nodal forces relate to n nodal displacements is of order $n \times n$. The order of the stiffness matrix does not, however, bear a direct relation to the number of nodes in a structure, since it is possible for more than one force to be acting at any one node.

So far, we built up the stiffness matrices for the single- and two-spring assemblies by considering various states of displacement in each case. Such a process clearly becomes tedious for more complex assemblies involving a large number of springs, so that a shorter, alternative procedure is desirable. From our remarks in the preceding paragraph and by reference to Eq. (6.2), we could have deduced at the outset of the analysis that the stiffness matrix for the two-spring assembly would be of the form

$$[K] = \begin{bmatrix} k_{11} & k_{12} & k_{13} \\ k_{21} & k_{22} & k_{23} \\ k_{31} & k_{32} & k_{33} \end{bmatrix} \tag{6.14}$$

The element k_{11} of this matrix relates the force at node 1 to the displacement at node 1 and so on. Hence, remembering the stiffness matrix for the single spring (Eq. (6.7)), we may write down the stiffness matrix for an elastic element connecting nodes 1 and 2 in a structure as

$$[K_{12}] = \begin{bmatrix} k_{11} & k_{12} \\ k_{21} & k_{22} \end{bmatrix} \tag{6.15}$$

and for the element connecting nodes 2 and 3 as

$$[K_{23}] = \begin{bmatrix} k_{22} & k_{23} \\ k_{32} & k_{33} \end{bmatrix} \tag{6.16}$$

In our two-spring system, the stiffness of the spring joining nodes 1 and 2 is k_a and that of the spring joining nodes 2 and 3 is k_b. Therefore, by comparison with Eq. (6.7), we rewrite Eqs. (6.15) and (6.16) as

$$[K_{12}] = \begin{bmatrix} k_a & -k_a \\ -k_a & k_a \end{bmatrix}, \quad [K_{23}] = \begin{bmatrix} k_b & -k_b \\ -k_b & k_b \end{bmatrix} \tag{6.17}$$

Substituting in Eq. (6.14) gives

$$[K] = \begin{bmatrix} k_a & -k_a & 0 \\ -k_a & k_a + k_b & -k_b \\ 0 & -k_b & k_b \end{bmatrix}$$

which is identical to Eq. (6.13). We see that only the k_{22} term (linking the force at node 2 to the displacement at node 2) receives contributions from both springs. This results from the fact that node 2 is directly connected to both nodes 1 and 3, while nodes 1 and 3 are each joined directly only to node 2. Also, the elements k_{13} and k_{31} of $[K]$ are zero since nodes 1 and 3 are not directly connected and are therefore not affected by each other's displacement.

The formation of a stiffness matrix for a complete structure thus becomes a relatively simple matter of the superposition of individual or element stiffness matrices. The procedure may be summarized as follows: terms of the form k_{ii} on the main diagonal consist of the sum of the stiffnesses of all the structural elements meeting at node i while off-diagonal terms of the form k_{ij} consist of the sum of the stiffnesses of all the elements connecting node i to node j.

An examination of the stiffness matrix reveals that it possesses certain properties. For example, the sum of the elements in any column is zero, indicating that the conditions of equilibrium are satisfied. Also, the nonzero terms are concentrated near the leading diagonal, while all the terms in the leading diagonal are positive; the latter property derives from the physical behavior of any actual structure in which positive nodal forces produce positive nodal displacements.

Further inspection of Eq. (6.13) shows that its determinant vanishes. As a result, the stiffness matrix $[K]$ is singular and its inverse does not exist. We shall see that this means that the associated set of simultaneous equations for the unknown nodal displacements cannot be solved, for the simple reason that we have placed no limitation on any of the displacements u_1, u_2, or u_3. Thus, the application of external loads results in the system moving as a rigid body. Sufficient boundary conditions must therefore be specified to enable the system to remain stable under load. In this particular problem, we shall demonstrate the solution procedure by assuming that node 1 is fixed; thst is, $u_1 = 0$.

The first step is to rewrite Eq. (6.13) in partitioned form as

$$\begin{Bmatrix} F_{x,1} \\ F_{x,2} \\ F_{x,3} \end{Bmatrix} = \begin{bmatrix} k_a & \vdots & -k_a & 0 \\ \cdots & & \cdots & \\ -k_a & \vdots & k_a + k_b & -k_b \\ 0 & \vdots & -k_b & k_b \end{bmatrix} \begin{Bmatrix} u_1 = 0 \\ u_2 \\ u_3 \end{Bmatrix} \tag{6.18}$$

In Eq. (6.18), $F_{x,1}$ is the unknown reaction at node 1, u_1 and u_2 are unknown nodal displacements, while $F_{x,2}$ and $F_{x,3}$ are known applied loads. Expanding Eq. (6.18) by matrix multiplication, we obtain

$$\{F_{x,1}\} = \begin{bmatrix} -k_a & 0 \end{bmatrix} \begin{Bmatrix} u_2 \\ u_3 \end{Bmatrix} \qquad \begin{Bmatrix} F_{x,2} \\ F_{x,3} \end{Bmatrix} = \begin{bmatrix} k_a + k_b & -k_b \\ -k_b & k_b \end{bmatrix} \begin{Bmatrix} u_2 \\ u_3 \end{Bmatrix} \tag{6.19}$$

Inversion of the second of Eqs. (6.19) gives u_2 and u_3 in terms of $F_{x,2}$ and $F_{x,3}$. Substitution of these values in the first equation then yields $F_{x,1}$.

Thus,

$$\begin{Bmatrix} u_2 \\ u_3 \end{Bmatrix} = \begin{bmatrix} k_a + k_b & -k_b \\ -k_b & k_b \end{bmatrix}^{-1} \begin{Bmatrix} F_{x,2} \\ F_{x,3} \end{Bmatrix}$$

or

$$\left\{ \begin{array}{c} u_2 \\ u_3 \end{array} \right\} = \begin{bmatrix} 1/k_a & 1/k_a \\ 1/k_a & 1/k_b + 1/k_a \end{bmatrix} \left\{ \begin{array}{c} F_{x,2} \\ F_{x,3} \end{array} \right\}$$

Hence,

$$\left\{ F_{x,1} \right\} = \begin{bmatrix} -k_a & 0 \end{bmatrix} \begin{bmatrix} 1/k_a & 1/k_a \\ 1/k_a & 1/k_b + 1/k_a \end{bmatrix} \left\{ \begin{array}{c} F_{x,2} \\ F_{x,3} \end{array} \right\}$$

which gives

$$F_{x,1} = -F_{x,2} - F_{x,3}$$

as would be expected from equilibrium considerations. In problems where reactions are not required, equations relating known applied forces to unknown nodal displacements may be obtained by deleting the rows and columns of $[K]$ corresponding to zero displacements. This procedure eliminates the necessity of rearranging rows and columns in the original stiffness matrix when the fixed nodes are not conveniently grouped together.

Finally, the internal forces in the springs may be determined from the force–displacement relationship of each spring. Thus, if S_a is the force in the spring joining nodes 1 and 2, then

$$S_a = k_a(u_2 - u_1)$$

Similarly for the spring between nodes 2 and 3,

$$S_b = k_b(u_3 - u_2)$$

6.4 MATRIX ANALYSIS OF PIN-JOINTED FRAMEWORKS

The formation of stiffness matrices for pin-jointed frameworks and the subsequent determination of nodal displacements follow a similar pattern to that described for a spring assembly. A member in such a framework is assumed to be capable of carrying axial forces only and obeys a unique force–deformation relationship given by

$$F = \frac{AE}{L} \delta$$

where F is the force in the member, δ its change in length, A its cross-sectional area, L its unstrained length, and E its modulus of elasticity. This expression is seen to be equivalent to the spring–displacement relationships of Eqs. (6.3) and (6.4), so that we immediately write down the stiffness matrix for a member by replacing k by AE/L in Eq. (6.7):

$$[K] = \begin{bmatrix} AE/L & -AE/L \\ -AE/L & AE/L \end{bmatrix}$$

or

$$[K] = \frac{AE}{L} \begin{bmatrix} 1 & -1 \\ -1 & 1 \end{bmatrix} \tag{6.20}$$

so that, for a member aligned with the x axis, joining nodes i and j subjected to nodal forces $F_{x,i}$ and $F_{x,j}$, we have

$$\left\{ \begin{array}{c} F_{x,i} \\ F_{x,j} \end{array} \right\} = \frac{AE}{L} \begin{bmatrix} 1 & -1 \\ -1 & 1 \end{bmatrix} \left\{ \begin{array}{c} u_i \\ u_j \end{array} \right\} \tag{6.21}$$

The solution proceeds in a similar manner to that given in the previous section for a spring or spring assembly. However, some modification is necessary, since frameworks consist of members set at various angles to one another. Figure 6.3 shows a member of a framework inclined at an angle θ to a set of arbitrary reference axes x, y. We shall refer every member of the framework to this *global coordinate* system, as it is known, when we consider the complete structure, but we shall use a *member* or *local* coordinate system $\overline{x}, \overline{y}$ when considering individual members. Nodal forces and displacements referred to local coordinates are written as $\overline{F}, \overline{u}$, etc., so that Eq. (6.21) becomes, in terms of local coordinates,

$$\left\{ \begin{array}{c} \overline{F_{x,i}} \\ \overline{F_{x,j}} \end{array} \right\} = \frac{AE}{L} \begin{bmatrix} 1 & -1 \\ -1 & 1 \end{bmatrix} \left\{ \begin{array}{c} \overline{u_i} \\ \overline{u_j} \end{array} \right\} \tag{6.22}$$

where the element stiffness matrix is written $[\overline{K_{ij}}]$.

In Fig. 6.3, external forces $\overline{F_{x,i}}$ and $\overline{F_{x,j}}$ are applied to nodes i and j. Note that $\overline{F_{y,i}}$ and $\overline{F_{y,j}}$ do not exist, since the member can only support axial forces. However, $\overline{F_{x,i}}$ and $\overline{F_{x,j}}$ have components $F_{x,i}, F_{y,i}$ and $F_{x,j}, F_{y,j}$, respectively, so that, whereas only two force components appear for the member in terms of local coordinates, four components are present when global coordinates are used. Therefore, if we are to transfer from local to global coordinates, Eq. (6.22) must be expanded to an order consistent with the use of global coordinates; that is,

$$\left\{ \begin{array}{c} \overline{F_{x,i}} \\ \overline{F_{y,i}} \\ \overline{F_{x,j}} \\ \overline{F_{y,j}} \end{array} \right\} = \frac{AE}{L} \begin{bmatrix} 1 & 0 & -1 & 0 \\ 0 & 0 & 0 & 0 \\ -1 & 0 & 1 & 0 \\ 0 & 0 & 0 & 0 \end{bmatrix} \left\{ \begin{array}{c} \overline{u_i} \\ \overline{v_i} \\ \overline{u_j} \\ \overline{v_j} \end{array} \right\} \tag{6.23}$$

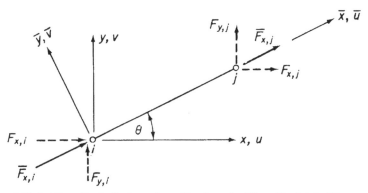

FIGURE 6.3 Local and Global Coordinate Systems for a Member of a Plane Pin-Jointed Framework

Equation (6.23) does not change the basic relationship between $\overline{F}_{x,i}$, $\overline{F}_{x,j}$ and $\overline{u}_i$, $\overline{u}_j$ as defined in Eq. (6.22).

From Fig. 6.3, we see that

$$\overline{F}_{x,i} = F_{x,i}\cos\theta + F_{y,i}\sin\theta$$
$$\overline{F}_{y,i} = -F_{x,i}\sin\theta + F_{y,i}\cos\theta$$

and

$$\overline{F}_{x,j} = F_{x,j}\cos\theta + F_{y,j}\sin\theta$$
$$\overline{F}_{y,j} = -F_{x,j}\sin\theta + F_{y,j}\cos\theta$$

Writing λ for $\cos\theta$ and μ for $\sin\theta$, we express the preceding equations in matrix form as

$$\begin{Bmatrix} \overline{F}_{x,i} \\ \overline{F}_{y,i} \\ \overline{F}_{x,j} \\ \overline{F}_{y,j} \end{Bmatrix} = \begin{bmatrix} \lambda & \mu & 0 & 0 \\ -\mu & \lambda & 0 & 0 \\ 0 & 0 & \lambda & \mu \\ 0 & 0 & -\mu & \lambda \end{bmatrix} \begin{Bmatrix} F_{x,i} \\ F_{y,i} \\ F_{x,j} \\ F_{y,j} \end{Bmatrix} \tag{6.24}$$

or, in abbreviated form,

$$\{\overline{F}\} = [T]\{F\} \tag{6.25}$$

where $[T]$ is known as the *transformation matrix*. A similar relationship exists between the sets of nodal displacements. Thus, again using our shorthand notation,

$$\{\overline{\delta}\} = [T]\{\delta\} \tag{6.26}$$

Substituting now for $\{\overline{F}\}$ and $\{\overline{\delta}\}$ in Eq. (6.23) from Eqs. (6.25) and (6.26), we have

$$[T]\{F\} = [\overline{K}_{ij}][T]\{\delta\}$$

Hence,

$$\{F\} = [T^{-1}][\overline{K}_{ij}][T]\{\delta\} \tag{6.27}$$

It may be shown that the inverse of the transformation matrix is its transpose; that is,

$$[T^{-1}] = [T]^{\mathrm{T}}$$

Therefore, we rewrite Eq. (6.27) as

$$\{F\} = [T]^{\mathrm{T}}[\overline{K}_{ij}][T]\{\delta\} \tag{6.28}$$

The nodal force system referred to global coordinates, $\{F\}$, is related to the corresponding nodal displacements by

$$\{F\} = [K_{ij}]\{\delta\} \tag{6.29}$$

where $[K_{ij}]$ is the member stiffness matrix referred to global coordinates. Comparison of Eqs. (6.28) and (6.29) shows that

$$[K_{ij}] = [T]^{\text{T}}[\overline{K_{ij}}][T]$$

Substituting for $[T]$ from Eq. (6.24) and $[\overline{K_{ij}}]$ from Eq. (6.23), we obtain

$$[K_{ij}] = \frac{AE}{L} \begin{bmatrix} \lambda^2 & \lambda\mu & -\lambda^2 & -\lambda\mu \\ \lambda\mu & \mu^2 & -\lambda\mu & -\mu^2 \\ -\lambda^2 & -\lambda\mu & \lambda^2 & \lambda\mu \\ -\lambda\mu & -\mu^2 & \lambda\mu & \mu^2 \end{bmatrix} \tag{6.30}$$

By evaluating $\lambda(= \cos\theta)$ and $\mu(= \sin\theta)$ for each member and substituting in Eq. (6.30), we obtain the stiffness matrix, referred to global coordinates, for each member of the framework.

In Section 6.3, we determined the internal force in a spring from the nodal displacements. Applying similar reasoning to the framework member, we may write down an expression for the internal force S_{ij} in terms of the local coordinates:

$$S_{ij} = \frac{AE}{L}(\overline{u}_j - \overline{u}_i) \tag{6.31}$$

Now,

$$\overline{u}_j = \lambda u_j + \mu v_j$$

$$\overline{u}_i = \lambda u_i + \mu v_i$$

Hence,

$$\overline{u}_j - \overline{u}_i = \lambda(u_j - u_i) + \mu(v_j - v_i)$$

Substituting in Eq. (6.31) and rewriting in matrix form, we have

$$S_{ij} = \frac{AE}{L} \begin{bmatrix} \lambda & \mu \\ & ij \end{bmatrix} \begin{Bmatrix} u_j - u_i \\ v_j - v_i \end{Bmatrix} \tag{6.32}$$

Example 6.1

Determine the horizontal and vertical components of the deflection of node 2 and the forces in the members of the pin-jointed framework shown in Fig. 6.4. The product AE is constant for all members. See Ex. 1.1.

We see in this problem that nodes 1 and 3 are pinned to a fixed foundation and are therefore not displaced. Hence, with the global coordinate system shown

$$u_1 = v_1 = u_3 = v_3 = 0$$

The external forces are applied at node 2 such that $F_{x,2} = 0$, $F_{y,2} = -W$; the nodal forces at 1 and 3 are then unknown reactions.

The first step in the solution is to assemble the stiffness matrix for the complete framework by writing down the member stiffness matrices referred to the global coordinate system using Eq. (6.30). The direction cosines λ and μ take different values for each of the three members; therefore, remembering that the angle θ is measured counter-clockwise from the positive direction of the x axis, we have Table 6.1.

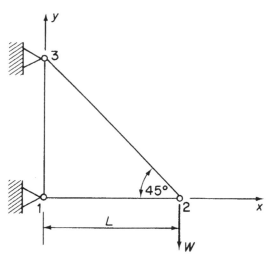

FIGURE 6.4 Pin-Jointed Framework of Example 6.1

Table 6.1 Example 6.1			
Member	**θ**	**λ**	**μ**
1–2	0	1	0
1–3	90	0	1
2–3	135	$-1/\sqrt{2}$	$1/\sqrt{2}$

The member stiffness matrices are therefore

$$[K_{12}] = \frac{AE}{L} \begin{bmatrix} 1 & 0 & -1 & 0 \\ 0 & 0 & 0 & 0 \\ -1 & 0 & 1 & 0 \\ 0 & 0 & 0 & 0 \end{bmatrix} \qquad [K_{13}] = \frac{AE}{L} \begin{bmatrix} 0 & 0 & 0 & 0 \\ 0 & 1 & 0 & -1 \\ 0 & 0 & 0 & 0 \\ 0 & -1 & 0 & 1 \end{bmatrix}$$

$$[K_{23}] = \frac{AE}{\sqrt{2}L} \begin{bmatrix} \frac{1}{2} & -\frac{1}{2} & -\frac{1}{2} & \frac{1}{2} \\ -\frac{1}{2} & \frac{1}{2} & \frac{1}{2} & -\frac{1}{2} \\ -\frac{1}{2} & \frac{1}{2} & \frac{1}{2} & -\frac{1}{2} \\ \frac{1}{2} & -\frac{1}{2} & -\frac{1}{2} & \frac{1}{2} \end{bmatrix} \qquad (i)$$

The next stage is to add the member stiffness matrices to obtain the stiffness matrix for the complete framework. Since there are six possible nodal forces producing six possible nodal displacements, the complete stiffness matrix is of the order 6×6. Although the addition is not difficult in this simple problem, care must be taken when solving more complex structures to ensure that the matrix elements are placed in the correct position in the complete stiffness matrix. This may be achieved by expanding each member stiffness matrix to the order of the complete stiffness matrix by inserting appropriate rows and columns of zeros. Such a method is, however, time and space consuming. An alternative procedure is suggested here. The complete stiffness matrix is of the form shown in Eq. (ii):

$$
\begin{Bmatrix} F_{x,1} \\ F_{y,1} \\ F_{x,2} \\ F_{y,2} \\ F_{x,3} \\ F_{y,3} \end{Bmatrix} = \begin{bmatrix} [k_{11}] & [k_{12}] & [k_{13}] \\ [k_{21}] & [k_{22}] & [k_{23}] \\ [k_{31}] & [k_{32}] & [k_{33}] \end{bmatrix} \begin{Bmatrix} u_1 \\ v_1 \\ u_2 \\ v_2 \\ u_3 \\ v_3 \end{Bmatrix} \tag{ii}
$$

The complete stiffness matrix has been divided into a number of submatrices in which $[k_{11}]$ is a 2×2 matrix relating the nodal forces $F_{x,1}$, $F_{y,1}$ to the nodal displacements u_1, v_1, and so on. It is a simple matter to divide each member stiffness matrix into submatrices of the form $[k_{11}]$, as shown in Eqs. (iii). All that remains is to insert each submatrix into its correct position in Eq. (ii), adding the matrix elements where they overlap; for example, the $[k_{11}]$ submatrix in Eq. (ii) receives contributions from $[K_{12}]$ and $[K_{13}]$. The complete stiffness matrix is then of the form shown in Eq. (iv). It is sometimes helpful, when considering the stiffness matrix separately, to write the nodal displacement above the appropriate column (see Eq. (iv)). We note that $[K]$ is symmetrical, all the diagonal terms are positive, and the sum of each row and column is zero

$$
[K_{12}] = \frac{AE}{L} \begin{bmatrix} \begin{matrix} 1 & 0 \\ & k_{11} \\ 0 & 0 \end{matrix} & \begin{matrix} -1 & 0 \\ & k_{12} \\ 0 & 0 \end{matrix} \\ \begin{matrix} -1 & 0 \\ & k_{21} \\ 0 & 0 \end{matrix} & \begin{matrix} 1 & 0 \\ & k_{22} \\ 0 & 0 \end{matrix} \end{bmatrix}
$$

$$
[K_{13}] = \frac{AE}{L} \begin{bmatrix} \begin{matrix} 0 & 0 \\ & k_{11} \\ 0 & 1 \end{matrix} & \begin{matrix} 0 & 0 \\ & k_{13} \\ 0 & -1 \end{matrix} \\ \begin{matrix} 0 & 0 \\ & k_{31} \\ 0 & -1 \end{matrix} & \begin{matrix} 0 & 0 \\ & k_{33} \\ 0 & 1 \end{matrix} \end{bmatrix} \tag{iii}
$$

$$
[K_{23}] = \frac{AE}{\sqrt{2}L} \begin{bmatrix} \begin{matrix} \frac{1}{2} & -\frac{1}{2} \\ & k_{22} \\ -\frac{1}{2} & \frac{1}{2} \end{matrix} & \begin{matrix} -\frac{1}{2} & \frac{1}{2} \\ & k_{23} \\ \frac{1}{2} & -\frac{1}{2} \end{matrix} \\ \begin{matrix} -\frac{1}{2} & \frac{1}{2} \\ & k_{32} \\ \frac{1}{2} & -\frac{1}{2} \end{matrix} & \begin{matrix} \frac{1}{2} & -\frac{1}{2} \\ & k_{33} \\ -\frac{1}{2} & \frac{1}{2} \end{matrix} \end{bmatrix}
$$

$$
\begin{Bmatrix} F_{x,1} \\ F_{y,1} \\ F_{x,2} \\ F_{y,2} \\ F_{x,3} \\ F_{y,3} \end{Bmatrix} = \frac{AE}{L}
\begin{array}{c}
\begin{array}{cccccc} u_1 & v_1 & u_2 & v_2 & u_3 & v_3 \end{array} \\
\begin{bmatrix}
1 & 0 & -1 & 0 & 0 & 0 \\
0 & 1 & 0 & 0 & 0 & -1 \\
-1 & 0 & 1+\dfrac{1}{2\sqrt{2}} & -\dfrac{1}{2\sqrt{2}} & -\dfrac{1}{2\sqrt{2}} & \dfrac{1}{2\sqrt{2}} \\
0 & 0 & -\dfrac{1}{2\sqrt{2}} & \dfrac{1}{2\sqrt{2}} & \dfrac{1}{2\sqrt{2}} & -\dfrac{1}{2\sqrt{2}} \\
0 & 0 & -\dfrac{1}{2\sqrt{2}} & \dfrac{1}{2\sqrt{2}} & \dfrac{1}{2\sqrt{2}} & -\dfrac{1}{2\sqrt{2}} \\
0 & -1 & \dfrac{1}{2\sqrt{2}} & -\dfrac{1}{2\sqrt{2}} & -\dfrac{1}{2\sqrt{2}} & 1+\dfrac{1}{2\sqrt{2}}
\end{bmatrix}
\end{array}
\begin{Bmatrix} u_1 = 0 \\ v_1 = 0 \\ u_2 \\ v_2 \\ u_3 = 0 \\ v_3 = 0 \end{Bmatrix}
\qquad \text{(iv)}
$$

If we now delete rows and columns in the stiffness matrix corresponding to zero displacements, we obtain the unknown nodal displacements u_2 and v_2 in terms of the applied loads $F_{x,2}\ (=0)$ and $F_{y,2}\ (=-W)$:

$$
\begin{Bmatrix} F_{x,2} \\ F_{y,2} \end{Bmatrix} = \frac{AE}{L}
\begin{bmatrix}
1+\dfrac{1}{2\sqrt{2}} & -\dfrac{1}{2\sqrt{2}} \\
-\dfrac{1}{2\sqrt{2}} & \dfrac{1}{2\sqrt{2}}
\end{bmatrix}
\begin{Bmatrix} u_2 \\ v_2 \end{Bmatrix}
\qquad \text{(v)}
$$

Inverting Eq. (v) gives

$$
\begin{Bmatrix} u_2 \\ v_2 \end{Bmatrix} = \frac{L}{AE}
\begin{bmatrix} 1 & 1 \\ 1 & 1+2\sqrt{2} \end{bmatrix}
\begin{Bmatrix} F_{x,2} \\ F_{y,2} \end{Bmatrix}
\qquad \text{(vi)}
$$

from which

$$
u_2 = \frac{L}{AE}\left(F_{x,2}+F_{y,2}\right) = -\frac{WL}{AE}
\qquad \text{(vii)}
$$

$$
v_2 = \frac{L}{AE}\left[F_{x,2}+(1+2\sqrt{2})F_{y,2}\right] = -\frac{WL}{AE}(1+2\sqrt{2})
\qquad \text{(viii)}
$$

The reactions at nodes 1 and 3 are now obtained by substituting for u_2 and v_2 from Eq. (vi) into Eq. (iv):

$$
\begin{Bmatrix} F_{x,1} \\ F_{y,1} \\ F_{x,3} \\ F_{y,3} \end{Bmatrix} =
\begin{bmatrix}
-1 & 0 \\
0 & 0 \\
-\dfrac{1}{2\sqrt{2}} & \dfrac{1}{2\sqrt{2}} \\
\dfrac{1}{2\sqrt{2}} & -\dfrac{1}{2\sqrt{2}}
\end{bmatrix}
\begin{bmatrix} 1 & 1 \\ 1 & 1+2\sqrt{2} \end{bmatrix}
\begin{Bmatrix} F_{x,2} \\ F_{y,2} \end{Bmatrix}
$$

$$
=
\begin{bmatrix}
-1 & -1 \\
0 & 0 \\
0 & 1 \\
0 & -1
\end{bmatrix}
\begin{Bmatrix} F_{x,2} \\ F_{y,2} \end{Bmatrix}
$$

giving

$$
\begin{aligned}
F_{x,1} &= -F_{x,2} - F_{y,2} = W \\
F_{y,1} &= 0 \\
F_{x,3} &= F_{y,2} = -W \\
F_{y,3} &= W
\end{aligned}
$$

Finally, the forces in the members are found from Eqs. (6.32), (vii), and (viii):

$$
S_{12} = \frac{AE}{L} [1 \quad 0] \left\{ \begin{array}{c} u_2 - u_1 \\ v_2 - v_1 \end{array} \right\} = -W \quad \text{(compression)}
$$

$$
S_{13} = \frac{AE}{L} [0 \quad 1] \left\{ \begin{array}{c} u_3 - u_1 \\ v_3 - v_1 \end{array} \right\} = - 0 \ \text{(as expected)}
$$

$$
S_{23} = \frac{AE}{\sqrt{2}L} \left[-\frac{1}{\sqrt{2}} \quad \frac{1}{\sqrt{2}} \right] \left\{ \begin{array}{c} u_3 - u_2 \\ v_3 - v_2 \end{array} \right\} = \sqrt{2}W \ \text{(tension)}
$$

6.5 APPLICATION TO STATICALLY INDETERMINATE FRAMEWORKS

The matrix method of solution described in the previous sections for spring and pin-jointed framework assemblies is completely general and therefore applicable to any structural problem. We observe that at no stage in Example 6.1 did the question of the degree of indeterminacy of the framework arise. It follows that problems involving statically indeterminate frameworks (and other structures) are solved in an identical manner to that presented in Example 6.1, the stiffness matrices for the redundant members being included in the complete stiffness matrix as before.

6.6 MATRIX ANALYSIS OF SPACE FRAMES

The procedure for the matrix analysis of space frames is similar to that for plane pin-jointed frameworks. The main difference lies in the transformation of the member stiffness matrices from local to global coordinates, since, as we see from Fig. 6.5, axial nodal forces $\overline{F_{x,i}}$ and $\overline{F_{x,j}}$ each now has three global components $F_{x,i}, F_{y,i}, F_{z,i}$ and $F_{x,j}, F_{y,j}, F_{z,j}$, respectively. The member stiffness matrix referred to global coordinates is therefore of the order of 6×6 so that $[K_{ij}]$ of Eq. (6.22) must be expanded to the same order to allow for this. Hence,

$$
[\overline{K_{ij}}] = \frac{AE}{L}
\begin{array}{c}
\begin{array}{cccccc} \overline{u}_i & \overline{v}_i & \overline{w}_i & \overline{u}_j & \overline{v}_j & \overline{w}_j \end{array} \\
\left[
\begin{array}{cccccc}
1 & 0 & 0 & -1 & 0 & 0 \\
0 & 0 & 0 & 0 & 0 & 0 \\
0 & 0 & 0 & 0 & 0 & 0 \\
-1 & 0 & 0 & 1 & 0 & 0 \\
0 & 0 & 0 & 0 & 0 & 0 \\
0 & 0 & 0 & 0 & 0 & 0
\end{array}
\right]
\end{array}
\tag{6.33}
$$

In Fig. 6.5, the member ij is of length L, cross-sectional area A, and modulus of elasticity E. Global and local coordinate systems are designated as for the two-dimensional case. Further, we suppose that

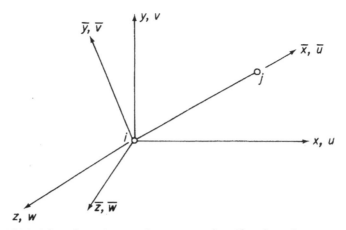

FIGURE 6.5 Local and Global Coordinate Systems for a Member in a Pin-Jointed Space Frame

$$\theta_{x\bar{x}} = \text{angle between } x \text{ and } \bar{x}$$
$$\theta_{x\bar{y}} = \text{angle between } x \text{ and } \bar{y}$$
$$\vdots$$
$$\theta_{z\bar{y}} = \text{angle between } z \text{ and } \bar{y}$$
$$\vdots$$

Therefore, nodal forces referred to the two systems of axes are related as follows:

$$\left.\begin{aligned}
\overline{F}_x &= F_x \cos\theta_{x\bar{x}} + F_y \cos\theta_{x\bar{y}} + F_z \cos\theta_{x\bar{z}} \\
\overline{F}_y &= F_x \cos\theta_{y\bar{x}} + F_y \cos\theta_{y\bar{y}} + F_z \cos\theta_{y\bar{z}} \\
\overline{F}_z &= F_x \cos\theta_{z\bar{x}} + F_y \cos\theta_{z\bar{y}} + F_z \cos\theta_{z\bar{z}}
\end{aligned}\right\} \tag{6.34}$$

Writing

$$\left.\begin{aligned}
\lambda_{\bar{x}} &= \cos\theta_{x\bar{x}}, & \lambda_{\bar{y}} &= \cos\theta_{x\bar{y}}, & \lambda_{\bar{z}} &= \cos\theta_{x\bar{z}} \\
\mu_{\bar{x}} &= \cos\theta_{y\bar{x}}, & \mu_{\bar{y}} &= \cos\theta_{y\bar{y}}, & \mu_{\bar{z}} &= \cos\theta_{y\bar{z}} \\
\nu_{\bar{x}} &= \cos\theta_{z\bar{x}}, & \nu_{\bar{y}} &= \cos\theta_{z\bar{y}}, & \nu_{\bar{z}} &= \cos\theta_{z\bar{z}}
\end{aligned}\right\} \tag{6.35}$$

we may express Eq. (6.34) for nodes i and j in matrix form as

$$
\begin{Bmatrix}
\overline{F}_{x,i} \\
\overline{F}_{y,i} \\
\overline{F}_{z,i} \\
\overline{F}_{x,j} \\
\overline{F}_{y,j} \\
\overline{F}_{z,j}
\end{Bmatrix}
=
\begin{bmatrix}
\lambda_{\bar{x}} & \mu_{\bar{x}} & \nu_{\bar{x}} & 0 & 0 & 0 \\
\lambda_{\bar{y}} & \mu_{\bar{y}} & \nu_{\bar{y}} & 0 & 0 & 0 \\
\lambda_{\bar{z}} & \mu_{\bar{z}} & \nu_{\bar{z}} & 0 & 0 & 0 \\
0 & 0 & 0 & \lambda_{\bar{x}} & \mu_{\bar{x}} & \nu_{\bar{x}} \\
0 & 0 & 0 & \lambda_{\bar{y}} & \mu_{\bar{y}} & \nu_{\bar{y}} \\
0 & 0 & 0 & \lambda_{\bar{z}} & \mu_{\bar{z}} & \nu_{\bar{z}}
\end{bmatrix}
\begin{Bmatrix}
F_{x,i} \\
F_{y,i} \\
F_{z,i} \\
F_{x,j} \\
F_{y,j} \\
F_{z,j}
\end{Bmatrix}
\tag{6.36}
$$

or in abbreviated form

$$\{\overline{F}\} = [T]\{F\}$$

The derivation of $[K_{ij}]$ for a member of a space frame proceeds on identical lines to that for the plane frame member. Thus, as before,

$$[K_{ij}] = [T]^{\mathrm{T}}[\overline{K_{ij}}][T]$$

Substituting for $[T]$ and $[\overline{K_{ij}}]$ from Eqs. (6.36) and (6.33) gives

$$[K_{ij}] = \frac{AE}{L}\begin{bmatrix}
\lambda_{\bar{x}}^2 & \lambda_{\bar{x}}\mu_{\bar{x}} & \lambda_{\bar{x}}\nu_{\bar{x}} & -\lambda_{\bar{x}}^2 & -\lambda_{\bar{x}}\mu_{\bar{x}} & -\lambda_{\bar{x}}\nu_{\bar{x}} \\
\lambda_{\bar{x}}\mu_{\bar{x}} & \mu_{\bar{x}}^2 & \mu_{\bar{x}}\nu_{\bar{x}} & -\lambda_{\bar{x}}\mu_{\bar{x}} & -\mu_{\bar{x}}^2 & -\mu_{\bar{x}}\nu_{\bar{x}} \\
\lambda_{\bar{x}}\nu_{\bar{x}} & \mu_{\bar{x}}\nu_{\bar{x}} & \nu_{\bar{x}}^2 & -\lambda_{\bar{x}}\nu_{\bar{x}} & -\mu_{\bar{x}}\nu_{\bar{x}} & -\nu_{\bar{x}}^2 \\
-\lambda_{\bar{x}}^2 & -\lambda_{\bar{x}}\mu_{\bar{x}} & -\lambda_{\bar{x}}\nu_{\bar{x}} & \lambda_{\bar{x}}^2 & \lambda_{\bar{x}}\mu_{\bar{x}} & \lambda_{\bar{x}}\nu_{\bar{x}} \\
-\lambda_{\bar{x}}\mu_{\bar{x}} & -\mu_{\bar{x}}^2 & -\mu_{\bar{x}}\nu_{\bar{x}} & \lambda_{\bar{x}}\mu_{\bar{x}} & \mu_{\bar{x}}^2 & \mu_{\bar{x}}\nu_{\bar{x}} \\
-\lambda_{\bar{x}}\nu_{\bar{x}} & -\mu_{\bar{x}}\nu_{\bar{x}} & -\nu_{\bar{x}}^2 & \lambda_{\bar{x}}\nu_{\bar{x}} & \mu_{\bar{x}}\nu_{\bar{x}} & \nu_{\bar{x}}^2
\end{bmatrix} \tag{6.37}$$

All the suffixes in Eq. (6.37) are $\bar{x}$, so that we may rewrite the equation in simpler form, namely,

$$[K_{ij}] = \frac{AE}{L}\begin{bmatrix}
\lambda^2 & & & \vdots & & \mathrm{SYM} \\
\lambda\mu & \mu^2 & & \vdots & & \\
\lambda\nu & \mu\nu & \nu^2 & \vdots & & \\
\cdots & \cdots & \cdots & \cdots & \cdots & \cdots \\
-\lambda^2 & -\lambda\mu & -\lambda\nu & \vdots & \lambda^2 & \\
-\lambda\mu & -\mu^2 & -\mu\nu & \vdots & \lambda\mu & \mu^2 \\
-\lambda\nu & -\mu\nu & -\nu^2 & \vdots & \lambda\nu & \mu\nu & \nu^2
\end{bmatrix} \tag{6.38}$$

where λ, μ, and ν are the direction cosines between the x, y, z and $\bar{x}$ axes, respectively.

The complete stiffness matrix for a space frame is assembled from the member stiffness matrices in a similar manner to that for the plane frame and the solution completed as before.

6.7 STIFFNESS MATRIX FOR A UNIFORM BEAM

Our discussion so far has been restricted to structures comprising members capable of resisting axial loads only. Many structures, however, consist of beam assemblies in which the individual members resist shear and bending forces in addition to axial loads. We shall now derive the stiffness matrix for a uniform beam and consider the solution of rigid jointed frameworks formed by an assembly of beams, or beam elements, as they are sometimes called.

Figure 6.6 shows a uniform beam ij of flexural rigidity EI and length L subjected to nodal forces $F_{y,i}$, $F_{y,j}$ and nodal moments M_i, M_j in the xy plane. The beam suffers nodal displacements and rotations v_i, v_j and θ_i, θ_j. We do not include axial forces here, since their effects have already been determined in our investigation of pin-jointed frameworks.

The stiffness matrix $[K_{ij}]$ may be built up by considering various deflected states for the beam and superimposing the results, as we did initially for the spring assemblies of Figs 6.1 and 6.2 or,

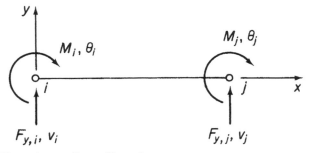

FIGURE 6.6 Forces and Moments on a Beam–Element

alternatively, it may be written down directly from the well-known beam slope–deflection equations.[3] We shall adopt the latter procedure. From slope–deflection theory, we have

$$M_i = -\frac{6EI}{L^2}v_i + \frac{4EI}{L}\theta_i + \frac{6EI}{L^2}v_j + \frac{2EI}{L}\theta_j \qquad (6.39)$$

and

$$M_j = -\frac{6EI}{L^2}v_i + \frac{2EI}{L}\theta_i + \frac{6EI}{L^2}v_j + \frac{2EI}{L}\theta_j \qquad (6.40)$$

Also, considering vertical equilibrium, we obtain

$$F_{y,i} + F_{y,j} = 0 \qquad (6.41)$$

and, from moment equilibrium about node j, we have

$$F_{y,i}L + M_i + M_j = 0 \qquad (6.42)$$

Hence, the solution of Eqs. (6.39)–(6.42) gives

$$-F_{y,i} = F_{y,j} = -\frac{12EI}{L^3}v_i + \frac{6EI}{L^2}\theta_i + \frac{12EI}{L^3}v_j + \frac{6EI}{L^2}\theta_j \qquad (6.43)$$

Expressing Eqs. (6.39), (6.40), and (6.43) in matrix form yields

$$\begin{Bmatrix} F_{y,i} \\ M_i \\ F_{y,j} \\ M_j \end{Bmatrix} = EI \begin{bmatrix} 12/L^3 & -6/L^2 & -12/L^3 & -6/L^2 \\ -6/L^2 & 4/L & 6/L^2 & 2/L \\ -12/L^3 & 6/L^2 & 12/L^3 & 6/L^2 \\ 6/L^2 & 2/L & 6/L^2 & 4/L \end{bmatrix} \begin{Bmatrix} v_i \\ \theta_i \\ v_j \\ \theta_j \end{Bmatrix} \qquad (6.44)$$

which is of the form

$$\{F\} = [K_{ij}]\{\delta\}$$

where $[K_{ij}]$ is the stiffness matrix for the beam.

It is possible to write Eq. (6.44) in an alternative form such that the elements of $[K_{ij}]$ are pure numbers:

$$\begin{Bmatrix} F_{y,i} \\ M_i/L \\ F_{y,j} \\ M_j/L \end{Bmatrix} = \frac{EI}{L^3} \begin{bmatrix} 12 & -6 & -12 & -6 \\ -6 & 4 & 6 & 2 \\ -12 & 6 & 12 & 6 \\ -6 & 2 & 6 & 4 \end{bmatrix} \begin{Bmatrix} v_i \\ \theta_i L \\ v_j \\ \theta_j L \end{Bmatrix}$$

This form of Eq. (6.44) is particularly useful in numerical calculations for an assemblage of beams in which EI/L^3 is constant.

Equation (6.44) is derived for a beam whose axis is aligned with the x axis so that the stiffness matrix defined by Eq. (6.44) is actually $[\overline{K_{ij}}]$, the stiffness matrix referred to a local coordinate system. If the beam is positioned in the xy plane with its axis arbitrarily inclined to the x axis, then the x and y axes form a global coordinate system and it becomes necessary to transform Eq. (6.44) to allow for this. The procedure is similar to that for the pin-jointed framework member of Section 6.4, in that $[\overline{K_{ij}}]$ must be expanded to allow for the fact that nodal displacements $\overline{u}_i$ and $\overline{u}_j$, which are irrelevant for the beam in local coordinates, have components u_i, v_i and u_j, v_j in global coordinates:

$$[\overline{K_{ij}}] = EI \begin{matrix} \begin{matrix} u_i & \quad v_i & \quad \theta_i & \quad u_j & \quad v_j & \quad \theta_j \end{matrix} \\ \begin{bmatrix} 0 & 0 & 0 & 0 & 0 & 0 \\ 0 & 12/L^3 & -6/L^2 & 0 & -12/L^3 & -6/L^2 \\ 0 & -6/L^2 & 4/L & 0 & 6/L^2 & 2/L \\ 0 & 0 & 0 & 0 & 0 & 0 \\ 0 & -12/L^3 & 6/L^2 & 0 & 12/L^3 & 6/L^2 \\ 0 & -6/L^2 & 2/L & 0 & 6/L^2 & 4/L \end{bmatrix} \end{matrix} \quad (6.45)$$

We may deduce the transformation matrix $[T]$ from Eq. (6.24) if we remember that, although u and v transform in exactly the same way as in the case of a pin-jointed member, the rotations θ remain the same in either local or global coordinates.

Hence,

$$[T] = \begin{bmatrix} \lambda & \mu & 0 & 0 & 0 & 0 \\ -\mu & \lambda & 0 & 0 & 0 & 0 \\ 0 & 0 & 1 & 0 & 0 & 0 \\ 0 & 0 & 0 & \lambda & \mu & 0 \\ 0 & 0 & 0 & -\mu & \lambda & 0 \\ 0 & 0 & 0 & 0 & 0 & 1 \end{bmatrix} \quad (6.46)$$

where λ and μ have previously been defined. Thus, since

$$[K_{ij}] = [T]^{\mathrm{T}} [\overline{K_{ij}}][T] \qquad \text{(see Section 6.4)}$$

we have, from Eqs. (6.45) and (6.46),

$$
[K_{ij}] = EI
\begin{bmatrix}
12\mu^2/L^3 & & & & & \text{SYM} \\
-12\lambda\mu/L^3 & 12\lambda^2/L^3 & & & & \\
6\mu/L^2 & -6\lambda/L^2 & 4/L & & & \\
-12\mu^2/L^3 & 12\lambda\mu/L^3 & -6u/L^2 & 12\mu^2/L^3 & & \\
12\lambda\mu/L^3 & -12\lambda^2/L^3 & 6\lambda/L^2 & -12\lambda\mu/L^3 & 12\lambda^2/L^3 & \\
6\mu/L^2 & -6\lambda/L^2 & 2/L & 6\mu/L^2 & 6\lambda/L^2 & 4\lambda/L
\end{bmatrix}
\tag{6.47}
$$

Again, the stiffness matrix for the complete structure is assembled from the member stiffness matrices, the boundary conditions are applied, and the resulting set of equations solved for the unknown nodal displacements and forces.

The internal shear forces and bending moments in a beam may be obtained in terms of the calculated nodal displacements. For a beam joining nodes i and j, we shall have obtained the unknown values of v_i, θ_i and v_j, θ_j. The nodal forces $F_{y,i}$ and M_i are then obtained from Eq. (6.44) if the beam is aligned with the x axis. Hence,

$$
\left.
\begin{aligned}
F_{y,i} &= EI\left(\frac{12}{L^3}v_i - \frac{6}{L^2}\theta_i - \frac{12}{L^3}v_j - \frac{6}{L^2}\theta_j\right) \\[2mm]
M_i &= EI\left(-\frac{6}{L^2}v_i + \frac{4}{L}\theta_i + \frac{6}{L^2}v_j - \frac{2}{L}\theta_j\right)
\end{aligned}
\right\}
\tag{6.48}
$$

Similar expressions are obtained for the forces at node j. From Fig. 6.6, we see that the shear force S_y and bending moment M in the beam are given by

$$
\left.
\begin{aligned}
S_y &= F_{y,i} \\
M &= F_{y,i}x + M_i
\end{aligned}
\right\}
\tag{6.49}
$$

Substituting Eq. (6.48) into Eq. (6.49) and expressing in matrix form yields

$$
\begin{Bmatrix} S_y \\ M \end{Bmatrix} = EI
\begin{bmatrix}
\dfrac{12}{L^3} & -\dfrac{6}{L^2} & -\dfrac{12}{L^3} & -\dfrac{6}{L^2} \\[3mm]
\dfrac{12}{L^3}x - \dfrac{6}{L^2} & -\dfrac{6}{L^2}x + \dfrac{4}{L} & \dfrac{12}{L^3}x + \dfrac{6}{L^2} & -\dfrac{6}{L^2}x + \dfrac{2}{L}
\end{bmatrix}
\begin{Bmatrix} v_i \\ \theta_i \\ v_j \\ \theta_j \end{Bmatrix}
\tag{6.50}
$$

The matrix analysis of the beam in Fig. 6.6 is based on the condition that no external forces are applied between the nodes. Obviously, in a practical case, a beam supports a variety of loads along its length and therefore such beams must be idealized into a number of *beam–elements* for which this condition holds. The idealization is accomplished by merely specifying nodes at points along the beam such that any element lying between adjacent nodes carries, at the most, a uniform shear and a linearly varying bending moment. For example, the beam of Fig. 6.7 is idealized into beam–elements 1–2, 2–3, and 3–4, for which the unknown nodal displacements are v_2, θ_2, θ_3, v_4 and θ_4 ($v_1 = \theta_1 = v_3 = 0$).

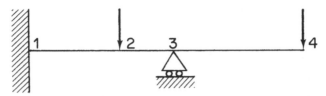

FIGURE 6.7 Idealization of a Beam into Beam–Elements

Beams supporting distributed loads require special treatment, in that the distributed load is replaced by a series of statically equivalent point loads at a selected number of nodes. Clearly, the greater the number of nodes chosen, the more accurate but more complicated and therefore time consuming is the analysis. Figure 6.8 shows a typical idealization of a beam supporting a uniformly distributed load. Details of the analysis of such beams may be found in Martin.[4]

Many simple beam problems may be idealized into a combination of two beam–elements and three nodes. A few examples of such beams are shown in Fig. 6.9. If we therefore assemble a stiffness matrix for the general case of a two beam–element system, we may use it to solve a variety of problems simply by inserting the appropriate loading and support conditions. Consider the assemblage of two

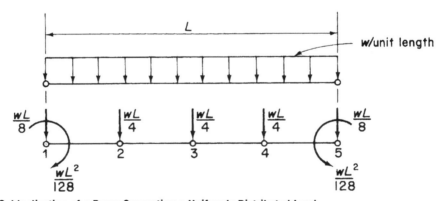

FIGURE 6.8 Idealization of a Beam Supporting a Uniformly Distributed Load

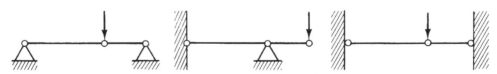

FIGURE 6.9 Idealization of Beams into Beam–Elements

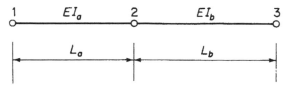

FIGURE 6.10 Assemblage of Two Beam–Elements

beam–elements shown in Fig. 6.10. The stiffness matrices for the beam–elements 1–2 and 2–3 are obtained from Eq. (6.44):

$$
[K_{12}] = EI_a
\begin{bmatrix}
\begin{array}{cc}
12/L_a^3 & -6/L_a^2 \\
-6/L_a^2 & 4/L_a
\end{array} \;\; k_{11} &
\begin{array}{cc}
-12/L_a^3 & -6/L_a^2 \\
6/L_a^2 & 2/L_a
\end{array} \;\; k_{12} \\[14pt]
\begin{array}{cc}
-12/L_a^3 & 6/L_a^2 \\
-6/L_a^2 & 2/L_a
\end{array} \;\; k_{21} &
\begin{array}{cc}
12/L_a^3 & 6/L_a^2 \\
6/L_a^2 & 4/L_a
\end{array} \;\; k_{22}
\end{bmatrix}
\qquad (6.51)
$$

(column headings: $v_1 \quad \theta_1 \quad v_2 \quad \theta_2$)

$$
[K_{23}] = EI_b
\begin{bmatrix}
\begin{array}{cc}
12/L_b^3 & -6/L_b^2 \\
-6/L_b^2 & 4/L_b
\end{array} \;\; k_{22} &
\begin{array}{cc}
-12/L_b^3 & -6/L_b^2 \\
6/L_b^2 & 2/L_b
\end{array} \;\; k_{23} \\[14pt]
\begin{array}{cc}
-12/L_b^3 & 6/L_b^2 \\
-6/L_b^2 & 2/L_b
\end{array} \;\; k_{32} &
\begin{array}{cc}
12/L_b^3 & 6/L_b^2 \\
6/L_b^2 & 4/L_b
\end{array} \;\; k_{33}
\end{bmatrix}
\qquad (6.52)
$$

(column headings: $v_2 \quad \theta_2 \quad v_3 \quad \theta_3$)

The complete stiffness matrix is formed by superimposing $[K_{12}]$ and $[K_{23}]$ as described in Example 6.1. Hence,

$$
[K] = E
\begin{bmatrix}
\dfrac{12I_a}{L_a^3} & -\dfrac{6I_a}{L_a^2} & -\dfrac{12I_a}{L_a^3} & \dfrac{6I_a}{L_a^2} & 0 & 0 \\[10pt]
-\dfrac{6I_a}{L_a^2} & \dfrac{4I_a}{L_a} & \dfrac{6I_a}{L_a^2} & \dfrac{2I_a}{L_a} & 0 & 0 \\[10pt]
-\dfrac{12I_a}{L_a^3} & \dfrac{6I_a}{L_a^2} & 12\left(\dfrac{I_a}{L_a^3}+\dfrac{I_b}{L_b^3}\right) & 6\left(\dfrac{I_a}{L_a^2}-\dfrac{I_b}{L_b^2}\right) & -\dfrac{12I_b}{L_b^3} & -\dfrac{6I_b}{L_b^2} \\[10pt]
-\dfrac{6I_a}{L_a^2} & \dfrac{2I_a}{L_a} & 6\left(\dfrac{I_a}{L_a^2}-\dfrac{I_b}{L_b^2}\right) & 4\left(\dfrac{I_a}{L_a}+\dfrac{I_b}{L_b}\right) & \dfrac{6I_b}{L_b^2} & \dfrac{2I_b}{L_b} \\[10pt]
0 & 0 & -\dfrac{12I_b}{L_b^3} & \dfrac{6I_b}{L_b^2} & \dfrac{12I_b}{L_b^3} & \dfrac{6I_b}{L_b^2} \\[10pt]
0 & 0 & -\dfrac{6I_b}{L_b^2} & \dfrac{2I_b}{L_b} & \dfrac{6I_b}{L_b^2} & \dfrac{4I_b}{L_b}
\end{bmatrix}
\qquad (6.53)
$$

Example 6.2

Determine the unknown nodal displacements and forces in the beam shown in Fig. 6.11. The beam is of uniform section throughout. See Ex. 1.1.

The beam may be idealized into two beam–elements, 1–2 and 2–3. From Fig. 6.11, we see that $v_1 = v_3 = 0$, $F_{y,2} = -W$, $M_2 = +M$. Therefore, eliminating rows and columns corresponding to zero displacements from Eq. (6.53), we obtain

$$
\begin{Bmatrix} F_{y,2} = -W \\ M_2 = M \\ M_1 = 0 \\ M_3 = 0 \end{Bmatrix} = EI \begin{bmatrix} 27/2L^3 & 9/2L^2 & 6/L^2 & -3/2L^2 \\ 9/2L^2 & 6/L & 2/L & 1/L \\ 6/L^2 & 2/L & 4/L & 0 \\ -3/2L^2 & 1/L & 0 & 2/L \end{bmatrix} \begin{Bmatrix} v_2 \\ \theta_2 \\ \theta_1 \\ \theta_3 \end{Bmatrix} \tag{i}
$$

Equation (i) may be written such that the elements of $[K]$ are pure numbers:

$$
\begin{Bmatrix} F_{y,2} = -W \\ M_2/L = M/L \\ M_1/L = 0 \\ M_3/L = 0 \end{Bmatrix} = \frac{EI}{2L^3} \begin{bmatrix} 27 & 9 & 112 & -3 \\ 9 & 12 & 4 & 2 \\ 12 & 4 & 8 & 0 \\ -3 & 2 & 0 & 4 \end{bmatrix} \begin{Bmatrix} v_2 \\ \theta_2 L \\ \theta_1 L \\ \theta_3 L \end{Bmatrix} \tag{ii}
$$

Expanding Eq. (ii) by matrix multiplication, we have

$$
\begin{Bmatrix} -W \\ M/L \end{Bmatrix} = \frac{EI}{2L^3} \left(\begin{bmatrix} 27 & 9 \\ 9 & 12 \end{bmatrix} \begin{Bmatrix} v_2 \\ \theta_2 L \end{Bmatrix} + \begin{bmatrix} 12 & -3 \\ 4 & 2 \end{bmatrix} \begin{Bmatrix} \theta_1 L \\ \theta_3 L \end{Bmatrix} \right) \tag{iii}
$$

and

$$
\begin{Bmatrix} 0 \\ 0 \end{Bmatrix} = \frac{EI}{2L^3} \left(\begin{bmatrix} 12 & 4 \\ -3 & 2 \end{bmatrix} \begin{Bmatrix} v_2 \\ \theta_2 L \end{Bmatrix} + \begin{bmatrix} 8 & 0 \\ 0 & 4 \end{bmatrix} \begin{Bmatrix} \theta_1 L \\ \theta_3 L \end{Bmatrix} \right) \tag{iv}
$$

Equation (iv) gives

$$
\begin{Bmatrix} \theta_1 L \\ \theta_3 L \end{Bmatrix} = \begin{bmatrix} -\dfrac{3}{2} & -\dfrac{1}{2} \\ -\dfrac{3}{4} & -\dfrac{1}{2} \end{bmatrix} \begin{Bmatrix} v_2 \\ \theta_2 L \end{Bmatrix} \tag{v}
$$

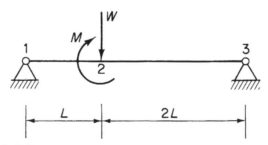

FIGURE 6.11 Beam of Example 6.2

Substituting Eq. (v) in Eq. (iii), we obtain

$$\left\{ \begin{array}{c} v_2 \\ \theta_2 L \end{array} \right\} = \frac{L^3}{9EI} \begin{bmatrix} -4 & -2 \\ -2 & 3 \end{bmatrix} \left\{ \begin{array}{c} -W \\ M/L \end{array} \right\} \tag{vi}$$

from which the unknown displacements at node 2 are

$$v_2 = -\frac{4}{9}\frac{WL^3}{EI} - \frac{2}{9}\frac{ML^2}{EI}$$

$$\theta_2 = \frac{2}{9}\frac{WL^2}{EI} + \frac{1}{3}\frac{ML}{EI}$$

In addition, from Eq. (v), we find that

$$\theta_1 = \frac{5}{9}\frac{WL^3}{EI} + \frac{1}{6}\frac{ML}{EI}$$

$$\theta_3 = -\frac{4}{9}\frac{WL^2}{EI} - \frac{1}{3}\frac{ML}{EI}$$

Note that the solution has been obtained by inverting two 2×2 matrices rather than the 4×4 matrix of Eq. (ii). This simplification has been brought about by the fact that $M_1 = M_3 = 0$.

The internal shear forces and bending moments can now be found using Eq. (6.50). For the beam–element 1–2, we have

$$S_{y,12} = EI\left(\frac{12}{L^3}v_1 - \frac{6}{L^2}\theta_1 - \frac{12}{L^3}v_2 - \frac{6}{L^2}\theta_2 \right)$$

or

$$S_{y,12} = \frac{2}{3}W - \frac{1}{3}\frac{M}{L}$$

and

$$M_{12} = EI\left[\left(\frac{12}{L^3}x - \frac{6}{L^2} \right)v_1 + \left(-\frac{6}{L^2}x + \frac{4}{L} \right)\theta_1 + \left(-\frac{12}{L^3}x + \frac{6}{L^2} \right)v_2 + \left(-\frac{6}{L^2}x + \frac{2}{L} \right)\theta_2 \right]$$

which reduces to

$$M_{12} = \left(\frac{2}{3}W - \frac{1}{3}\frac{M}{L} \right)x$$

■

6.8 FINITE ELEMENT METHOD FOR CONTINUUM STRUCTURES

In the previous sections, we have discussed the matrix method of solution of structures composed of elements connected only at nodal points. For skeletal structures consisting of arrangements of beams, these nodal points fall naturally at joints and at positions of concentrated loading. Continuum structures, such as flat plates, aircraft skins, shells, etc., possess no such natural subdivisions and must therefore be artificially idealized into a number of elements before matrix methods can be used. These *finite elements*, as they are known, may be two- or three-dimensional but the most commonly used are two-dimensional triangular and quadrilateral shaped elements. The idealization may be carried out in any number of different ways, depending on such factors as the type of problem, the accuracy of the solution required, and the time and money available. For example, a *coarse* idealization involving a small number of large elements provides a comparatively rapid but very approximate solution, while a *fine* idealization of small elements produces more accurate results but takes longer and consequently costs more. Frequently, *graded meshes* are used, in which small elements are placed in regions where high stress concentrations are expected, for example, around cut-outs and loading points. The principle is illustrated in Fig. 6.12, where a graded system of triangular elements is used to examine the stress concentration around a circular hole in a flat plate.

Although the elements are connected at an infinite number of points around their boundaries it is assumed that they are interconnected only at their corners or nodes. Thus, compatibility of displacement is ensured only at the nodal points. However, in the finite element method, a displacement pattern is chosen for each element which may satisfy some, if not all, of the compatibility requirements along the sides of adjacent elements.

Since we are employing matrix methods of solution, we are concerned initially with the determination of nodal forces and displacements. Therefore, the system of loads on the structure must be replaced by an equivalent system of nodal forces. Where these loads are concentrated, the elements are chosen such that a node occurs at the point of application of the load. In the case of distributed loads, equivalent nodal concentrated loads must be calculated.[4]

The solution procedure is identical in outline to that described in the previous sections for skeletal structures; the differences lie in the idealization of the structure into finite elements and the calculation of the stiffness matrix for each element. The latter procedure, which in general terms is applicable to all finite elements, may be specified in a number of distinct steps. We shall illustrate the method by establishing the stiffness matrix for the simple one-dimensional beam–element of Fig. 6.6, for which we have already derived the stiffness matrix using slope–deflection.

FIGURE 6.12 Finite Element Idealization of a Flat Plate with a Central Hole

6.8.1 Stiffness matrix for a beam–element

The first step is to choose a suitable coordinate and node numbering system for the element and define its nodal displacement vector $\{\delta^e\}$ and nodal load vector $\{F^e\}$. Use is made here of the superscript e to denote element vectors, since, in general, a finite element possesses more than two nodes. Again, we are not concerned with axial or shear displacements, so that, for the beam–element of Fig. 6.6, we have

$$\{\delta^e\} = \begin{Bmatrix} v_i \\ \theta_i \\ v_j \\ \theta_j \end{Bmatrix}, \{F^e\} = \begin{Bmatrix} F_{y,i} \\ M_i \\ F_{y,j} \\ M_j \end{Bmatrix}$$

Since each of these vectors contains four terms the element stiffness matrix $[K^e]$ is of order of 4×4.

In the second step, we select a displacement function which uniquely defines the displacement of all points in the beam–element in terms of the nodal displacements. This displacement function may be taken as a polynomial which must include four arbitrary constants, corresponding to the four nodal degrees of freedom of the element:

$$v(x) = \alpha_1 + \alpha_2 x + \alpha_3 x^2 + \alpha_4 x^3 \tag{6.54}$$

Equation (6.54) is of the same form as that derived from elementary bending theory for a beam subjected to concentrated loads and moments and may be written in matrix form as

$$\{v(x)\} = [1 \ \ x \ \ x^2 \ \ x^3] \begin{Bmatrix} \alpha_1 \\ \alpha_2 \\ \alpha_3 \\ \alpha_4 \end{Bmatrix}$$

or in abbreviated form as

$$\{v(x)\} = [f(x)]\{\alpha\} \tag{6.55}$$

The rotation θ at any section of the beam–element is given by $\partial v/\partial x$; therefore,

$$\theta = \alpha_2 + 2\alpha_3 x + 3\alpha_4 x^2 \tag{6.56}$$

From Eqs. (6.54) and (6.56), we can write expressions for the nodal displacements v_i, θ_i and v_j, θ_j at $x = 0$ and $x = L$, respectively. Hence,

$$\left.\begin{aligned} v_i &= \alpha_1 \\ \theta_i &= \alpha_2 \\ v_j &= \alpha_1 + \alpha_2 L + \alpha_3 L^2 + \alpha_4 L^3 \\ \theta_j &= \alpha_2 + 2\alpha_3 L + 3\alpha_4 L^2 \end{aligned}\right\} \tag{6.57}$$

Writing Eqs. (6.57) in matrix form gives

$$\begin{Bmatrix} v_i \\ \theta_i \\ v_j \\ \theta_j \end{Bmatrix} = \begin{bmatrix} 1 & 0 & 0 & 0 \\ 0 & 1 & 0 & 0 \\ 1 & L & L^2 & L^3 \\ 0 & 1 & 2L & 3L^2 \end{bmatrix} \begin{Bmatrix} \alpha_1 \\ \alpha_2 \\ \alpha_3 \\ \alpha_4 \end{Bmatrix} \tag{6.58}$$

or

$$\{\delta^e\} = [A]\{\alpha\} \tag{6.59}$$

The third step follows directly from Eqs. (6.58) and (6.55), in that we express the displacement at any point in the beam–element in terms of the nodal displacements. Using Eq. (6.59), we obtain

$$\{\alpha\} = [A^{-1}]\{\delta^e\} \tag{6.60}$$

Substituting in Eq. (6.55) gives

$$\{v(x)\} = [f(x)][A^{-1}]\{\delta^e\} \tag{6.61}$$

where $[A^{-1}]$ is obtained by inverting $[A]$ in Eq. (6.58) and may be shown to be given by

$$[A^{-1}] = \begin{bmatrix} 1 & 0 & 0 & 0 \\ 0 & 1 & 0 & 0 \\ -3/L^2 & -2/L & 3/L^2 & -1/L \\ 2/L^3 & 1/L^2 & -2/L^3 & 1/L^2 \end{bmatrix} \tag{6.62}$$

In step four, we relate the strain $\{\varepsilon(x)\}$ at any point x in the element to the displacement $\{v(x)\}$ and hence to the nodal displacements $\{\delta^e\}$. Since we are concerned here with bending deformations only, we may represent the strain by the curvature $\partial^2 v/\partial x^2$. Hence, from Eq. (6.54),

$$\frac{\partial^2 v}{\partial x^2} = 2\alpha_3 + 6\alpha_4 x \tag{6.63}$$

or in matrix form

$$\{\varepsilon\} = [0 \ 0 \ 2 \ 6x] \begin{Bmatrix} \alpha_1 \\ \alpha_2 \\ \alpha_3 \\ \alpha_4 \end{Bmatrix} \tag{6.64}$$

which we write as

$$\{\varepsilon\} = [C]\{\alpha^e\} \tag{6.65}$$

Substituting for $\{\alpha\}$ in Eq. (6.65) from Eq. (6.60), we have

$$\{\varepsilon\} = [C][A^{-1}]\{\delta^e\} \tag{6.66}$$

Step five relates the internal stresses in the element to the strain $\{\varepsilon\}$ and hence, using Eq. (6.66), to the nodal displacements $\{\delta^e\}$. In our beam–element, the stress distribution at any section depends entirely on the value of the bending moment M at that section. Therefore, we may represent a "state of stress" $\{\sigma\}$ at any section by the bending moment M, which, from simple beam theory, is given by

$$M = EI\frac{\partial^2 v}{\partial x^2}$$

or

$$\{\sigma\} = [EI]\{\varepsilon\} \tag{6.67}$$

which we write as

$$\{\sigma\} = [D]\{\varepsilon\} \tag{6.68}$$

The matrix $[D]$ in Eq. (6.68) is the "elasticity" matrix relating "stress" and "strain." In this case, $[D]$ consists of a single term, the flexural rigidity EI of the beam. Generally, however, $[D]$ is of a higher order. If we now substitute for $\{\varepsilon\}$ in Eq. (6.68) from Eq. (6.66), we obtain the "stress" in terms of the nodal displacements; that is,

$$\{\sigma\} = [D][C][A^{-1}]\{\delta^e\} \tag{6.69}$$

The element stiffness matrix is finally obtained in step six, in which we replace the internal "stresses" $\{\sigma\}$ by a statically equivalent nodal load system $\{F^e\}$, thereby relating nodal loads to nodal displacements (from Eq. (6.69)) and defining the element stiffness matrix $[K^e]$. This is achieved by employing the principle of the stationary value of the total potential energy of the beam (see Section 5.8), which comprises the internal strain energy U and the potential energy V of the nodal loads. Thus,

$$U + V = \frac{1}{2}\int_{\text{vol}} \{\varepsilon\}^T\{\sigma\}\mathrm{d}(\text{vol}) - \{\delta^e\}^T\{F^e\} \tag{6.70}$$

Substituting in Eq. (6.70) for $\{\varepsilon\}$ from Eq. (6.66) and $\{\sigma\}$ from Eq. (6.69), we have

$$U + V = \frac{1}{2}\int_{\text{vol}} \{\delta^e\}^T[A^{-1}]^T[C]^T[D][C][A^{-1}]\{\delta^e\}\mathrm{d}(\text{vol}) - \{\delta^e\}^T\{F^e\} \tag{6.71}$$

The total potential energy of the beam has a stationary value with respect to the nodal displacements $\{\delta^e\}^T$; hence, from Eq. (6.71),

$$\frac{\partial(U + V)^T}{\partial(\delta^e)^T} = \int_{\text{vol}} [A^{-1}]^T[C]^T[D][C][A^{-1}]\{\delta^e\}\mathrm{d}(\text{vol}) - \{F^e\} = 0 \tag{6.72}$$

from which

$$\{F^e\} = \left[\int_{\text{vol}} [C]^T[A^{-1}]^T[D][C][A^{-1}]\mathrm{d}(\text{vol})\right]\{\delta^e\} \tag{6.73}$$

or, writing $[C][A^{-1}]$ as $[B]$, we obtain

$$\{F^e\} = \left[\int_{\text{vol}} [B]^T[D][B]\mathrm{d}(\text{vol})\right]\{\delta^e\} \tag{6.74}$$

from which the element stiffness matrix is clearly

$$\{K^e\} = \left[\int_{\text{vol}} [B]^T[D][B]\mathrm{d}(\text{vol})\right] \tag{6.75}$$

From Eqs. (6.62) and (6.64) we have

$$[B] = [C][A^{-1}] = \begin{bmatrix} 0 & 0 & 2 & 6x \end{bmatrix}\begin{bmatrix} 1 & 0 & 0 & 0 \\ 0 & 1 & 0 & 0 \\ -3/L^2 & -2/L & 3/L^2 & -1/L \\ 2/L^3 & 1/L^2 & -2/L^3 & 1/L^2 \end{bmatrix}$$

or

$$[B]^{\mathrm{T}} = \begin{bmatrix} -\dfrac{6}{L^2} + \dfrac{12x}{L^3} \\[2mm] -\dfrac{4}{L} + \dfrac{6x}{L^2} \\[2mm] \dfrac{6}{L^2} - \dfrac{12x}{L^3} \\[2mm] -\dfrac{2}{L} + \dfrac{6x}{L^2} \end{bmatrix} \tag{6.76}$$

Hence,

$$[K^{\mathrm{e}}] = \int_0^L \begin{bmatrix} -\dfrac{6}{L^2} + \dfrac{12x}{L^3} \\[2mm] -\dfrac{4}{L} + \dfrac{6x}{L^2} \\[2mm] \dfrac{6}{L^2} - \dfrac{12x}{L^3} \\[2mm] -\dfrac{2}{L} + \dfrac{6x}{L^2} \end{bmatrix} [EI] \begin{bmatrix} -\dfrac{6}{L^2} + \dfrac{12x}{L^3} & -\dfrac{4}{L} + \dfrac{6x}{L^2} & \dfrac{6}{L^2} - \dfrac{12x}{L^3} & -\dfrac{2}{L} + \dfrac{6x}{L^2} \end{bmatrix} \mathrm{d}x$$

which gives

$$[K^{\mathrm{e}}] = \frac{EI}{L^3} \begin{bmatrix} 12 & -6L & -12 & -6L \\ -6L & 4L^2 & 6L & 2L^2 \\ -12 & 6L & 12 & 6L \\ -6L & 2L^2 & 6L & 4L^2 \end{bmatrix} \tag{6.77}$$

Equation (6.77) is identical to the stiffness matrix (see Eq. (6.44)) for the uniform beam of Fig. 6.6.

Finally, in step seven, we relate the internal "stresses," $\{\sigma\}$, in the element to the nodal displacements $\{\delta^{\mathrm{e}}\}$. This has in fact been achieved to some extent in Eq. (6.69), namely,

$$\{\sigma\} = [D][C]\left[A^{-1}\right]\{\delta^{\mathrm{e}}\}$$

or, from the preceding,

$$\{\sigma\} = [D][B]\{\delta^{\mathrm{e}}\} \tag{6.78}$$

Equation (6.78) is usually written

$$\{\sigma\} = [H]\{\delta^{\mathrm{e}}\} \tag{6.79}$$

in which $[H] = [D][B]$ is the stress–displacement matrix. For this particular beam–element, $[D] = EI$ and $[B]$ is defined in Eq. (6.76). Thus,

$$[H] = [EI] \begin{bmatrix} -\dfrac{6}{L^2} + \dfrac{12x}{L^3} & -\dfrac{4}{L} + \dfrac{6x}{L^2} & \dfrac{6}{L^2} - \dfrac{12x}{L^3} & -\dfrac{2}{L} + \dfrac{6x}{L^2} \end{bmatrix} \tag{6.80}$$

6.8.2 **Stiffness matrix for a triangular finite element**

Triangular finite elements are used in the solution of plane stress and plane strain problems. Their advantage over other shaped elements lies in their ability to represent irregular shapes and boundaries with relative simplicity.

In the derivation of the stiffness matrix, we shall adopt the step-by-step procedure of the previous example. Initially, therefore, we choose a suitable coordinate and node numbering system for the element and define its nodal displacement and nodal force vectors. Figure 6.13 shows a triangular element referred to axes Oxy and having nodes i, j, and k lettered counterclockwise. It may be shown that the inverse of the $[A]$ matrix for a triangular element contains terms giving the actual area of the element; this area is positive if the preceding node lettering or numbering system is adopted. The element is to be used for plane elasticity problems and has therefore two degrees of freedom per node, giving a total of six degrees of freedom for the element, which results in a 6×6 element stiffness matrix $[K^e]$. The nodal forces and displacements are shown, and the complete displacement and force vectors are

$$\{\delta^e\} = \begin{Bmatrix} u_i \\ v_i \\ u_j \\ v_j \\ u_k \\ v_k \end{Bmatrix} \quad \{F^e\} = \begin{Bmatrix} F_{x,i} \\ F_{y,i} \\ F_{x,j} \\ F_{y,j} \\ F_{x,k} \\ F_{y,k} \end{Bmatrix} \tag{6.81}$$

We now select a displacement function which must satisfy the boundary conditions of the element, that is, the condition that each node possesses two degrees of freedom. Generally, for computational purposes, a polynomial is preferable to, say, a trigonometric series, since the terms in a polynomial can be calculated much more rapidly by a digital computer. Furthermore, the total number of degrees of freedom is six, so that only six coefficients in the polynomial can be obtained. Suppose that the displacement function is

$$\begin{aligned} u(x, y) &= \alpha_1 + \alpha_2 x + \alpha_3 y \\ v(x, y) &= \alpha_4 + \alpha_5 x + \alpha_6 y \end{aligned} \Bigg\} \tag{6.82}$$

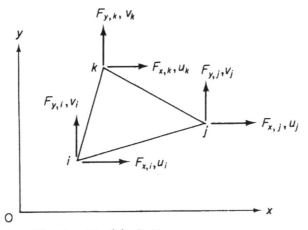

FIGURE 6.13 Triangular Element for Plane Elasticity Problems

The constant terms, α_1 and α_4, are required to represent any in-plane rigid body motion, that is, motion without strain, while the linear terms enable states of constant strain to be specified; Eqs. (6.82) ensure compatibility of displacement along the edges of adjacent elements. Writing Eqs. (6.82) in matrix form gives

$$\left\{ \begin{array}{c} u(x,y) \\ v(x,y) \end{array} \right\} = \begin{bmatrix} 1 & x & y & 0 & 0 & 0 \\ 0 & 0 & 0 & 1 & x & y \end{bmatrix} \left\{ \begin{array}{c} \alpha_1 \\ \alpha_2 \\ \alpha_3 \\ \alpha_4 \\ \alpha_5 \\ \alpha_6 \end{array} \right\} \tag{6.83}$$

Comparing Eq. (6.83) with Eq. (6.55), we see that it is of the form

$$\left\{ \begin{array}{c} u(x,y) \\ v(x,y) \end{array} \right\} = [f(x,y)]\{\alpha\} \tag{6.84}$$

Substituting values of displacement and coordinates at each node in Eq. (6.84), we have, for node i,

$$\left\{ \begin{array}{c} u_i \\ v_i \end{array} \right\} = \begin{bmatrix} 1 & x_i & y_i & 0 & 0 & 0 \\ 0 & 0 & 0 & 1 & x_i & y_i \end{bmatrix} \{\alpha\}$$

Similar expressions are obtained for nodes j and k, so that, for the complete element, we obtain

$$\left\{ \begin{array}{c} u_i \\ v_i \\ u_j \\ v_j \\ u_k \\ v_k \end{array} \right\} = \begin{bmatrix} 1 & x_i & y_i & 0 & 0 & 0 \\ 0 & 0 & 0 & 1 & x_i & y_i \\ 1 & x_j & y_j & 0 & 0 & 0 \\ 0 & 0 & 0 & 1 & x_j & y_j \\ 1 & x_k & y_k & 0 & 0 & 0 \\ 0 & 0 & 0 & 1 & x_k & y_k \end{bmatrix} \left\{ \begin{array}{c} \alpha_1 \\ \alpha_2 \\ \alpha_3 \\ \alpha_4 \\ \alpha_5 \\ \alpha_6 \end{array} \right\} \tag{6.85}$$

From Eq. (6.81) and by comparison with Eqs. (6.58) and (6.59), we see that Eq. (6.85) takes the form

$$\{\delta^e\} = [A]\{\alpha\}$$

Hence (step 3), we obtain

$$\{\alpha\} = [A^{-1}]\{\delta^e\} \quad \text{(compare with Eq. (6.60))}$$

The inversion of $[A]$, defined in Eq. (6.85), may be achieved algebraically, as illustrated in Example 6.3. Alternatively, the inversion may be carried out numerically for a particular element by computer. Substituting for $\{\alpha\}$ from the preceding into Eq. (6.84) gives

$$\left\{ \begin{array}{c} u(x,y) \\ v(x,y) \end{array} \right\} = [f(x,y)][A^{-1}]\{\delta^e\} \quad \text{(compare with Eq. (6.61))} \tag{6.86}$$

The strains in the element are

$$\{\varepsilon\} = \left\{ \begin{array}{c} \varepsilon_x \\ \varepsilon_y \\ \gamma_{xy} \end{array} \right\} \tag{6.87}$$

From Eqs. (1.18) and (1.20), we see that

$$\varepsilon_x = \frac{\partial u}{\partial x}, \quad \varepsilon_y = \frac{\partial v}{\partial y}, \quad \gamma_{xy} = \frac{\partial u}{\partial y} + \frac{\partial v}{\partial x} \tag{6.88}$$

Substituting for u and v in Eqs. (6.88) from Eqs. (6.82) gives

$$\varepsilon_x = \alpha_2$$
$$\varepsilon_y = \alpha_6$$
$$\gamma_{xy} = \alpha_3 + \alpha_5$$

or in matrix form

$$\{\varepsilon\} = \begin{bmatrix} 0 & 1 & 0 & 0 & 0 & 0 \\ 0 & 0 & 0 & 0 & 0 & 1 \\ 0 & 0 & 1 & 0 & 1 & 0 \end{bmatrix} \begin{Bmatrix} \alpha_1 \\ \alpha_2 \\ \alpha_3 \\ \alpha_4 \\ \alpha_5 \\ \alpha_6 \end{Bmatrix} \tag{6.89}$$

which is of the form

$$\{\varepsilon\} = [C]\{\alpha\} \quad \text{(see Eqs. (6.64) and (6.65))}$$

Substituting for $\{\alpha\}(= [A^{-1}] \{\delta^e\})$, we obtain

$$\{\varepsilon\} = [C][A^{-1}]\{\delta^e\} \quad \text{(compare with Eq. (6.66))}$$

or

$$\{\varepsilon\} = [B]\{\delta^e\} \quad \text{(see Eq. 6.76)}$$

where $[C]$ is defined in Eq. (6.89).

In step five, we relate the internal stresses $\{\sigma\}$ to the strain $\{\varepsilon\}$ and hence, using step four, to the nodal displacements $\{\delta^e\}$. For plane stress problems,

$$\{\sigma\} = \begin{Bmatrix} \sigma_x \\ \sigma_y \\ \tau_{xy} \end{Bmatrix} \tag{6.90}$$

and

$$\left. \begin{aligned} \varepsilon_x &= \frac{\sigma_x}{E} - \frac{\nu\sigma_y}{E} \\ \varepsilon_y &= \frac{\sigma_y}{E} - \frac{\nu\sigma_x}{E} \\ \gamma_{xy} &= \frac{\tau_{xy}}{G} = \frac{2(1+\nu)}{E}\tau_{xy} \end{aligned} \right\} \quad \text{(see Chapter 1)}$$

Thus, in matrix form,

$$\{\varepsilon\} = \begin{Bmatrix} \varepsilon_x \\ \varepsilon_y \\ \gamma_{xy} \end{Bmatrix} = \frac{1}{E} \begin{bmatrix} 1 & -\nu & 0 \\ -\nu & 1 & 0 \\ 0 & 0 & 2(1+\nu) \end{bmatrix} \begin{Bmatrix} \sigma_x \\ \sigma_y \\ \tau_{xy} \end{Bmatrix} \tag{6.91}$$

It may be shown that (see Chapter 1)

$$\{\sigma\} = \left\{ \begin{array}{c} \sigma_x \\ \sigma_y \\ \tau_{xy} \end{array} \right\} = \frac{E}{1-v^2} \begin{bmatrix} 1 & v & 0 \\ v & 1 & 0 \\ 0 & 0 & \frac{1}{2}(1-v) \end{bmatrix} \left\{ \begin{array}{c} \varepsilon_x \\ \varepsilon_y \\ \gamma_{xy} \end{array} \right\} \tag{6.92}$$

which has the form of Eq. (6.68); that is,

$$\{\sigma\} = [D]\{\varepsilon\}$$

Substituting for $\{\varepsilon\}$ in terms of the nodal displacements $\{\delta^e\}$, we obtain

$$\{\sigma\} = [D][B]\{\delta^e\} \quad \text{(see Eq. (6.69))}$$

In the case of plane strain, the elasticity matrix $[D]$ takes a different form to that defined in Eq. (6.92). For this type of problem,

$$\varepsilon_x = \frac{\sigma_x}{E} - \frac{v\sigma_y}{E} - \frac{v\sigma_z}{E}$$

$$\varepsilon_y = \frac{\sigma_y}{E} - \frac{v\sigma_x}{E} - \frac{v\sigma_z}{E}$$

$$\varepsilon_z = \frac{\sigma_z}{E} - \frac{v\sigma_x}{E} - \frac{v\sigma_y}{E} = 0$$

$$\gamma_{xy} = \frac{\tau_{xy}}{G} = \frac{2(1+v)}{E}\tau_{xy}$$

Eliminating σ_z and solving for σ_x, σ_y, and τ_{xy}, gives

$$\{\sigma\} = \left\{ \begin{array}{c} \sigma_x \\ \sigma_y \\ \tau_{xy} \end{array} \right\} = \frac{E(1-v)}{(1+v)(1-2v)} \begin{bmatrix} 1 & \frac{v}{1-v} & 0 \\ \frac{v}{1-v} & 1 & 0 \\ 0 & 0 & \frac{(1-2v)}{2(1-v)} \end{bmatrix} \left\{ \begin{array}{c} \varepsilon_x \\ \varepsilon_y \\ \gamma_{xy} \end{array} \right\} \tag{6.93}$$

which again takes the form

$$\{\sigma\} = [D]\{\varepsilon\}$$

Step six, in which the internal stresses $\{\sigma\}$ are replaced by the statically equivalent nodal forces $\{F^e\}$, proceeds in an identical manner to that described for the beam–element. Thus,

$$\{F^e\} = \left[\int_{\text{vol}} [B]^T [D][B] \, \mathrm{d}(\text{vol}) \right] \{\delta^e\}$$

as in Eq. (6.74), whence

$$\{K^e\} = \left[\int_{\text{vol}} [B]^T [D][B] \, \mathrm{d}(\text{vol}) \right]$$

In this expression $[B] = [C][A^{-1}]$, where $[A]$ is defined in Eq. (6.85) and $[C]$ in Eq. (6.89). The elasticity matrix $[D]$ is defined in Eq. (6.92) for plane stress problems or in Eq. (6.93) for plane strain problems. We note that the $[C]$, $[A]$ (therefore $[B]$), and $[D]$ matrices contain only constant terms and may therefore be taken outside the integration in the expression for $[K^e]$, leaving only $\int d(\text{vol})$, which is simply the area A of the triangle times its thickness t. Thus,

$$[K^e] = [[B]^T[D][B]At] \tag{6.94}$$

Finally, the element stresses follow from Eq. (6.79); that is,

$$\{\sigma\} = [H]\{\delta^e\}$$

where $[H] = [D][B]$ and $[D]$ and $[B]$ have previously been defined. It is usually found convenient to plot the stresses at the centroid of the element.

Of all the finite elements in use, the triangular element is probably the most versatile. It may be used to solve a variety of problems ranging from two-dimensional flat plate structures to three-dimensional folded plates and shells. For three-dimensional applications, the element stiffness matrix $[K^e]$ is transformed from an in-plane xy coordinate system to a three-dimensional system of global coordinates by the use of a transformation matrix similar to those developed for the matrix analysis of skeletal structures. In addition to the above, triangular elements may be adapted for use in plate flexure problems and for the analysis of bodies of revolution.

Example 6.3

A constant strain triangular element has corners 1(0, 0), 2(4, 0), and 3(2, 2) referred to a Cartesian Oxy axes system and is 1 unit thick. If the elasticity matrix $[D]$ has elements $D_{11} = D_{22} = a$, $D_{12} = D_{21} = b$, $D_{13} = D_{23} = D_{31} = D_{32} = 0$ and $D_{33} = c$, derive the stiffness matrix for the element. See Ex. 1.1.

From Eq. (6.82),

$$u_1 = \alpha_1 + \alpha_2(0) + \alpha_3(0)$$

that is,

$$u_1 = \alpha_1 \tag{i}$$

$$u_2 = \alpha_1 + \alpha_2(4) + \alpha_3(0)$$

that is,

$$u_2 = \alpha_1 + 4\alpha_2 \tag{ii}$$

$$u_3 = \alpha_1 + \alpha_2(2) + \alpha_3(2)$$

that is,

$$u_3 = \alpha_1 + 2\alpha_2 + 2\alpha_3 \tag{iii}$$

From Eq. (i),

$$\alpha_1 = u_1 \tag{iv}$$

and, from Eqs (ii) and (iv),

$$\alpha_2 = \frac{u_2 - u_1}{4} \tag{v}$$

Then, from Eqs (iii) to (v),

$$\alpha_3 = \frac{2u_3 - u_1 - u_2}{4} \tag{vi}$$

Substituting for α_1, α_2, and α_3 in the first of Eqs. (6.82) gives

$$u = u_1 + \left(\frac{u_2 - u_1}{4}\right)x + \left(\frac{2u_3 - u_1 - u_2}{4}\right)y$$

or

$$u = \left(1 - \frac{x}{4} - \frac{y}{4}\right)u_1 + \left(\frac{x}{4} - \frac{y}{4}\right)u_2 + \frac{y}{2}u_3 \tag{vii}$$

Similarly,

$$v = \left(1 - \frac{x}{4} - \frac{y}{4}\right)v_1 + \left(\frac{x}{4} - \frac{y}{4}\right)v_2 + \frac{y}{2}v_3 \tag{viii}$$

Now, from Eq. (6.88),

$$\varepsilon_x = \frac{\partial u}{\partial x} = -\frac{u_1}{4} + \frac{u_2}{4}$$

$$\varepsilon_y = \frac{\partial v}{\partial y} = -\frac{v_1}{4} + \frac{v_2}{4} + \frac{v_3}{2}$$

and

$$\gamma_{xy} = \frac{\partial u}{\partial y} + \frac{\partial v}{\partial x} = -\frac{u_1}{4} - \frac{u_2}{4} + \frac{v_1}{4} + \frac{v_2}{4}$$

Hence,

$$[B]\{\delta^e\} = \begin{bmatrix} \dfrac{\partial u}{\partial x} \\[2mm] \dfrac{\partial v}{\partial y} \\[2mm] \dfrac{\partial u}{\partial y} + \dfrac{\partial v}{\partial x} \end{bmatrix} = \frac{1}{4} \begin{bmatrix} -1 & 0 & 1 & 0 & 0 & 0 \\ 0 & -1 & 0 & -1 & 0 & 2 \\ -1 & -1 & -1 & 1 & 2 & 0 \end{bmatrix} \begin{Bmatrix} u_1 \\ v_1 \\ u_2 \\ v_2 \\ u_3 \\ v_3 \end{Bmatrix} \tag{ix}$$

Also,

$$[D] = \begin{bmatrix} a & b & 0 \\ b & a & 0 \\ 0 & 0 & c \end{bmatrix}$$

Hence,

$$[D][B] = \frac{1}{4} \begin{bmatrix} -a & -b & a & -b & 0 & 2b \\ -b & -a & b & -a & 0 & 2a \\ -c & -c & -c & c & 2c & 0 \end{bmatrix}$$

and

$$[B]^{\mathrm{T}}[D][B] = \frac{1}{16} \begin{bmatrix} a+c & b+c & -a+c & b-c & -2c & -2b \\ b+c & a+c & -b+c & a-c & -2c & -2a \\ -a+c & -b+c & a+c & -b-c & -2c & 2b \\ b-c & a-c & -b-c & a+c & 2c & -2a \\ -2c & -2c & -2c & 2c & 4c & 0 \\ -2b & -2a & 2b & -2a & 0 & 4a \end{bmatrix}$$

Then, from Eq. (6.94),

$$[K^{\mathrm{e}}] = \frac{1}{4} \begin{bmatrix} a+c & b+c & -a+c & b-c & -2c & -2b \\ b+c & a+c & -b+c & a-c & -2c & -2a \\ -a+c & -b+c & a+c & -b-c & -2c & 2b \\ b-c & a-c & -b-c & a+c & 2c & -2a \\ -2c & -2c & -2c & 2c & 4c & 0 \\ -2b & -2a & 2b & -2a & 0 & 4a \end{bmatrix}$$

6.8.3 Stiffness matrix for a quadrilateral element

Quadrilateral elements are frequently used in combination with triangular elements to build up particular geometrical shapes.

Figure 6.14 shows a quadrilateral element referred to axes Oxy and having corner nodes, i, j, k, and l; the nodal forces and displacements are also shown and the displacement and force vectors are

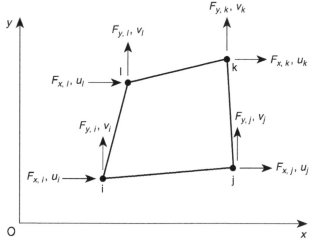

FIGURE 6.14 Quadrilateral element Subjected to Nodal In-Plane Forces and Displacements

$$\{\delta^e\} = \begin{Bmatrix} u_i \\ v_i \\ u_j \\ v_j \\ u_k \\ v_k \\ u_l \\ v_l \end{Bmatrix}, \{F^e\} = \begin{Bmatrix} F_{x,i} \\ F_{y,i} \\ F_{x,j} \\ F_{y,j} \\ F_{x,k} \\ F_{y,k} \\ F_{x,l} \\ F_{y,l} \end{Bmatrix} \tag{6.95}$$

As in the case of the triangular element, we select a displacement function that satisfies the total of eight degrees of freedom of the nodes of the element; again this displacement function is in the form of a polynomial with a maximum of eight coefficients:

$$\left. \begin{aligned} u(x,y) &= \alpha_1 + \alpha_2 x + \alpha_3 y + \alpha_4 xy \\ v(x,y) &= \alpha_5 + \alpha_6 x + \alpha_7 y + \alpha_8 xy \end{aligned} \right\} \tag{6.96}$$

The constant terms, α_1 and α_5, are required, as before, to represent the in-plane rigid body motion of the element, while the two pairs of linear terms enable states of constant strain to be represented throughout the element. Further, the inclusion of the xy terms results in both the $u\,(x, y)$ and $v\,(x, y)$ displacements having the same algebraic form, so that the element behaves in exactly the same way in the x direction as it does in the y direction.

Writing Eqs. (6.96) in matrix form gives

$$\begin{Bmatrix} u(x,y) \\ v(x,y) \end{Bmatrix} = \begin{bmatrix} 1 & x & y & xy & 0 & 0 & 0 & 0 \\ 0 & 0 & 0 & 0 & 1 & x & y & xy \end{bmatrix} \begin{Bmatrix} \alpha_1 \\ \alpha_2 \\ \alpha_3 \\ \alpha_4 \\ \alpha_5 \\ \alpha_6 \\ \alpha_7 \\ \alpha_8 \end{Bmatrix} \tag{6.97}$$

or

$$\begin{Bmatrix} u(x,y) \\ v(x,y) \end{Bmatrix} = [f(x,y)]\{\alpha\} \tag{6.98}$$

Now, substituting the coordinates and values of displacement at each node, we obtain

$$\begin{Bmatrix} u_i \\ v_i \\ u_j \\ v_j \\ u_k \\ v_k \\ u_l \\ v_l \end{Bmatrix} = \begin{bmatrix} 1 & x_i & y_i & x_iy_i & 0 & 0 & 0 & 0 \\ 0 & 0 & 0 & 0 & 1 & x_i & y_i & x_iy_i \\ 1 & x_j & y_j & x_jy_j & 0 & 0 & 0 & 0 \\ 0 & 0 & 0 & 0 & 1 & x_j & y_j & x_jy_j \\ 1 & x_k & y_k & x_ky_k & 0 & 0 & 0 & 0 \\ 0 & 0 & 0 & 0 & 1 & x_k & y_k & x_ky_k \\ 1 & x_l & y_l & x_ly_l & 0 & 0 & 0 & 0 \\ 0 & 0 & 0 & 0 & 1 & x_l & y_l & x_ly_l \end{bmatrix} \begin{Bmatrix} \alpha_1 \\ \alpha_2 \\ \alpha_3 \\ \alpha_4 \\ \alpha_5 \\ \alpha_6 \\ \alpha_7 \\ \alpha_8 \end{Bmatrix} \tag{6.99}$$

which is of the form

$$\{\delta^e\} = [A]\{\alpha\}$$

Then,

$$\{\alpha\} = [A^{-1}]\{\delta^e\} \tag{6.100}$$

The inversion of $[A]$ is illustrated in Example 6.4 but, as in the case of the triangular element, is most easily carried out by means of a computer. The remaining analysis is identical to that for the triangular element except that the $\{\varepsilon\}$–$\{\alpha\}$ relationship (see Eq. (6.89)) becomes

$$\{\varepsilon\} = \begin{bmatrix} 0 & 1 & 0 & y & 0 & 0 & 0 & 0 \\ 0 & 0 & 0 & 0 & 0 & 0 & 1 & x \\ 0 & 0 & 1 & x & 0 & 1 & 0 & y \end{bmatrix} \begin{Bmatrix} \alpha_1 \\ \alpha_2 \\ \alpha_3 \\ \alpha_4 \\ \alpha_5 \\ \alpha_6 \\ \alpha_7 \\ \alpha_8 \end{Bmatrix} \tag{6.101}$$

Example 6.4

A rectangular element used in a plane stress analysis has corners whose coordinates (in metres), referred to an Oxy axes system, are 1(–2, –1), 2(2, –1), 3(2, 1), and 4(–2, 1); the displacements (also in metres) of the corners are

$$u_1 = 0.001, \quad u_2 = 0.003, \quad u_3 = -0.003, \quad u_4 = 0$$
$$v_1 = -0.004, \quad v_2 = -0.002, \quad v_3 = 0.001, \quad v_4 = 0.001$$

If Young's modulus $E = 200,000$ N/mm^2 and Poisson's ratio $v = 0.3$, calculate the stresses at the center of the element. See Ex. 1.1.

From the first of Eqs. (6.96),

$$u_1 = \alpha_1 - 2\alpha_2 - \alpha_3 + 2\alpha_4 = 0.001 \tag{i}$$

$$u_2 = \alpha_1 + 2\alpha_2 - \alpha_3 - 2\alpha_4 = 0.003 \tag{ii}$$

$$u_3 = \alpha_1 + 2\alpha_2 + \alpha_3 + 2\alpha_4 = -0.003 \tag{iii}$$

$$u_4 = \alpha_1 - 2\alpha_2 + \alpha_3 - 2\alpha_4 = 0 \tag{iv}$$

Subtracting Eq. (ii) from Eq. (i),

$$\alpha_2 - \alpha_4 = 0.0005 \tag{v}$$

Now, subtracting Eq. (iv) from Eq. (iii),

$$\alpha_2 + \alpha_4 = -0.00075 \tag{vi}$$

Then, subtracting Eq. (vi) from Eq. (v),

$$\alpha_4 = -0.000625 \tag{vii}$$

whence, from either Eq. (v) or (vi),

$$\alpha_2 = -0.000125 \tag{viii}$$

Adding Eqs. (i) and (ii),

$$\alpha_1 - \alpha_3 = 0.002 \qquad\qquad\qquad\qquad \text{(ix)}$$

Adding Eqs. (iii) and (iv),

$$\alpha_1 + \alpha_3 = -0.0015 \qquad\qquad\qquad\qquad \text{(x)}$$

Then, adding Eqs. (ix) and (x),

$$\alpha_1 = 0.00025 \qquad\qquad\qquad\qquad \text{(xi)}$$

and, from either Eq. (ix) or (x),

$$\alpha_3 = -0.00175 \qquad\qquad\qquad\qquad \text{(xii)}$$

The second of Eqs. (6.96) is used to determine α_5, α_6, α_7, α_8 in an identical manner to the preceding. Thus,

$$\alpha_5 = -0.001$$
$$\alpha_6 = 0.00025$$
$$\alpha_7 = 0.002$$
$$\alpha_8 = -0.00025$$

Now, substituting for $\alpha_1, \alpha_2, \ldots, \alpha_8$ in Eqs. (6.96),

$$u_i = 0.00025 - 0.000125x - 0.00175y - 0.000625xy$$

and

$$v_i = -0.001 + 0.00025x + 0.002y - 0.00025xy$$

Then, from Eqs. (6.88),

$$\varepsilon_x = \frac{\partial u}{\partial x} = -0.000125 - 0.000625y$$

$$\varepsilon_y = \frac{\partial v}{\partial y} = 0.002 - 0.00025x$$

$$\gamma_{xy} = \frac{\partial u}{\partial y} + \frac{\partial v}{\partial x} = -0.0015 - 0.000625x - 0.00025y$$

Therefore, at the center of the element ($x = 0$, $y = 0$),

$$\varepsilon_x = -0.000125$$
$$\varepsilon_y = 0.002$$
$$\gamma_{xy} = -0.0015$$

so that, from Eqs. (6.92),

$$\sigma_x = \frac{E}{1 - v^2}(\varepsilon_x + v\varepsilon_y) = \frac{200,000}{1 - 0.3^2}[-0.000125 + (0.3 \times 0.002)]$$

that is,

$$\sigma_x = 104.4 \text{ N/mm}^2$$

$$\sigma_y = \frac{E}{1 - v^2}(\varepsilon_y + v\varepsilon_x) = \frac{200,000}{1 - 0.3^2}[0.002 + (0.3 \times 0.000125)]$$

that is,

$$\sigma_y = 431.3 \, \text{N/mm}^2$$

and

$$\tau_{xy} = \frac{E}{1 - v^2} \times \frac{1}{2} (1 - v)\gamma_{xy} = \frac{E}{2(1 + v)} \gamma_{xy}$$

Thus,

$$\tau_{xy} = \frac{200{,}000}{2(1 + 0.3)} \times (-0.0015)$$

that is,

$$\tau_{xy} = -115.4 \, \text{N/mm}^2$$

The application of the finite element method to three-dimensional solid bodies is a straightforward extension of the analysis of two-dimensional structures. The basic three-dimensional elements are the tetrahedron and the rectangular prism, both shown in Fig. 6.15. The tetrahedron has four nodes, each possessing 3 degrees of freedom, a total of 12 degrees of freedom for the element, while the prism has eight nodes and therefore a total of 24 degrees of freedom. Displacement functions for each element require polynomials in x, y, and z; for the tetrahedron, the displacement function is of the first degree with 12 constant coefficients, while that for the prism may be of a higher order to accommodate the 24 degrees of freedom. A development in the solution of three-dimensional problems has been the introduction of curvilinear coordinates. This enables the tetrahedron and prism to be distorted into arbitrary shapes that are better suited for fitting actual boundaries. For more detailed discussions of the finite element method, reference should be made to the work of Jenkins,[5] Zienkiewicz and Cheung,[6] and to the many research papers published on the method.

New elements and new applications of the finite element method are still being developed, some of which lie outside the field of structural analysis. These fields include soil mechanics, heat transfer, fluid and seepage flow, magnetism, and electricity.

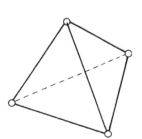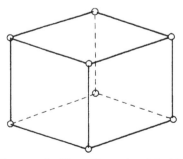

FIGURE 6.15 Tetrahedron and rectangular Prism Finite Elements for Three-Dimensional Problems

---

Example 6.4 MATLAB

Repeat Example 6.4 using MATLAB. Use matrix methods when possible. See Ex. 1.1.

Values for the stresses in the element, rounded to the first decimal place, are obtained through the following MATLAB file:

```
% Declare any needed variables
E = 200000; % Youngs Modulus
v_p = 0.3; % Poisson's ratio
x = [-2 2 2 -2]'; % x-location of corners
y = [-1 -1 1 1]'; % y-location of corners
u = [0.001 0.003 -0.003 0]; % x-displacement of corners
v = [-0.004 -0.002 0.001 0.001]; % y-displacement of corners

% Calculate alpha values using Eqs (6.99) and (6.100)
A = zeros(8,8);
delta = zeros(8,1);
for i=1:1:4
 A((i-1)*2+1,:) = [1 x(i) y(i) x(i)*y(i) 0 0 0 0];
 A((i-1)*2+2,:) = [0 0 0 0 1 x(i) y(i) x(i)*y(i)];
 delta((i-1)*2+1) = u(i);
 delta((i-1)*2+2) = v(i);
end
alpha = A\delta;

% Calculate the strains at the center (x=0,y=0) of the element using Eq. (6.101)
x = 0;
y = 0;
B = [0 1 0 y 0 0 0 0;
 0 0 0 0 0 0 1 x;
 0 0 1 x 0 1 0 y];
strain = B*alpha;

% Calculate the stresses using Eq. (6.92)
C = [1 v_p 0;
 v_p 1 0;
 0 0 (1-v_p)/2];
sig = E/(1-v_p^2)*(C*strain);

% Output stresses rounded to the first decimal to the Command Window
disp(['sig_x =' num2str(round(sig(1)*10)/10) 'N/mm^2'])
disp(['sig_y =' num2str(round(sig(2)*10)/10) 'N/mm^2'])
disp(['tau_xy =' num2str(round(sig(3)*10)/10) 'N/mm^2'])
```

The Command Window outputs resulting from this MATLAB file are as follows:

```
sig_x = 104.4 N/mm^2
sig_y = 431.3 N/mm^2
tau_xy = -115.4 N/mm^2
```

References

[1] Argyris JH, Kelsey S. Energy theorems and structural analysis. London: Butterworth scientific publications; 1960.

[2] Clough RW, Turner MJ, Martin HC, Topp LJ. Stiffness and deflection analysis of complex structures. J Aero Sciences 1956;23(9).

[3] Megson THG. Structural and stress analysis. 2nd ed. Oxford: Elsevier; 2005.

[4] Martin HC. Introduction to matrix methods of structural analysis. New York: McGraw-Hill; 1966.

[5] Jenkins WM. Matrix and digital computer methods in structural analysis. London: McGraw-Hill; 1969.

[6] Zienkiewicz OC, Cheung YK. The finite element method in structural and continuum mechanics. London: McGraw-Hill; 1967.

Further reading

Zienkiewicz OC, Holister GS. Stress Analysis. London: John Wiley and Sons; 1965.

PROBLEMS

P.6.1. Figure P.6.1 shows a square symmetrical pin-jointed truss 1234, pinned to rigid supports at 2 and 4 and loaded with a vertical load at 1. The axial rigidity EA is the same for all members. Use the stiffness method to find the displacements at nodes 1 and 3 and solve for all the internal member forces and support reactions.

Answer: $v_1 = -PL/\sqrt{2}AE$, $\quad v_3 = -0.293PL/AE$, $\quad S_{12} = P/2 = S_{14}$,
$S_{23} = -0.207P = S_{43}$, $\quad S_{13} = 0.293P$
$F_{x,2} = -F_{x,4} = 0.207P$, $\quad F_{y,2} = F_{y,4} = P/2$.

P.6.2. Use the stiffness method to find the ratio H/P for which the displacement of node 4 of the plane pin-jointed frame shown loaded in Fig. P.6.2 is zero and, for that case, give the displacements of nodes 2 and 3. All members have equal axial rigidity EA.

Answer: $H/P = 0.449$, $\quad v_2 = -4Pl/(9 + 2\sqrt{3})AE$,
$v_3 = -6PL/(9 + 2\sqrt{3})AE$.

P.6.3. Form the matrices required to solve completely the plane truss shown in Fig. P.6.3 and determine the force in member 24. All members have equal axial rigidity.

Answer: $S_{24} = 0$.

P.6.4. The symmetrical plane rigid jointed frame 1234567, shown in Fig. P.6.4, is fixed to rigid supports at 1 and 5 and supported by rollers inclined at 45° to the horizontal at nodes 3 and 7. It carries a vertical point load P at node 4 and a uniformly distributed load w per unit length on the span 26. Assuming the same flexural rigidity EI for all members, set up the stiffness equations which, when solved, give the nodal displacements of the frame. Explain how the member forces can be obtained.

P.6.5. The frame shown in Fig. P.6.5 has the planes xz and yz as planes of symmetry. The nodal coordinates of one quarter of the frame are given in Table P.6.5(i). In this structure, the deformation of each member is due to a single effect, this being axial, bending, or torsional. The mode of deformation

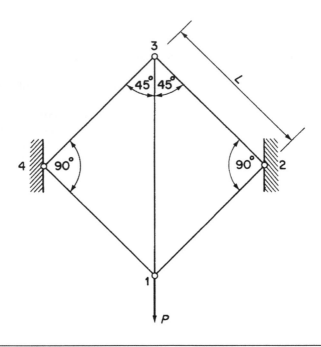

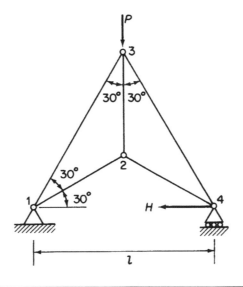

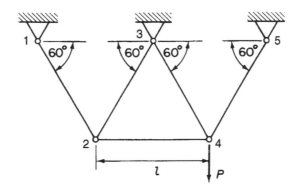

FIGURE P.6.3

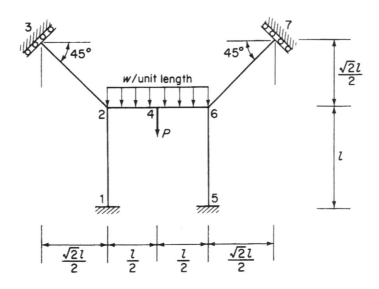

FIGURE P.6.4

of each member is given in Table P.6.5(ii), together with the relevant rigidity. Use the *direct stiffness* method to find all the displacements and hence calculate the forces in all the members. For member 123, plot the shear force and bending moment diagrams. Briefly outline the sequence of operations in a typical computer program suitable for linear frame analysis.

Answer: $S_{29} = S_{28} = \sqrt{2}P/6$ (tension)
$M_3 = -M_1 = PL/9$ (hogging), $M_2 = 2PL/9$ (sagging)
$SF_{12} = -SF_{23} = P/3$

Twisting moment in 37, $PL/18$ (counterclockwise).

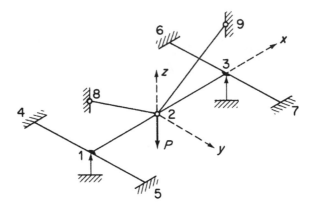

FIGURE P.6.5

Table P.6.5(i) Nodal Coordinates

Node	x	y	z
2	0	0	0
3	L	0	0
7	L	$0.8L$	0
9	L	0	L

Table P.6.5(ii) Mode of Defprmation

		Effect	
Member	**Axial**	**Bending**	**Torsional**
23	—	EI	—
37	—	—	$GJ = 0.8EI$
29	$EA = 6\sqrt{2}\dfrac{EI}{L^2}$	—	—

P.6.6. Given that the force–displacement (stiffness) relationship for the beam–element shown in Fig. P.6.6(a) may be expressed in the following form:

$$\begin{Bmatrix} F_{y,1} \\ M_1/L \\ F_{y,2} \\ M_2/L \end{Bmatrix} = \frac{EI}{L^3} \begin{bmatrix} 12 & -6 & -12 & -6 \\ -6 & 4 & 6 & 2 \\ -12 & 6 & 12 & 6 \\ 6 & 2 & 6 & 4 \end{bmatrix} \begin{Bmatrix} v_1 \\ \theta_1 L \\ v_2 \\ \theta_2 L \end{Bmatrix}$$

Obtain the force–displacement (stiffness) relationship for the variable section beam (Fig. P.6.6(b)), composed of elements 12, 23, and 34. Such a beam is loaded and supported symmetrically, as shown in Fig. P.6.6(c). Both ends are rigidly fixed and the ties FB, CH have a cross-section area a_1 and the ties

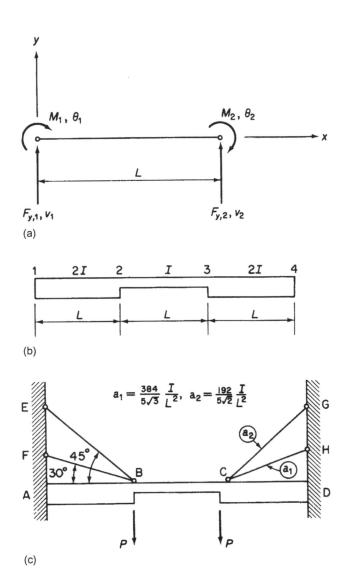

FIGURE P.6.6

EB, CG a cross-section area a_2. Calculate the deflections under the loads, the forces in the ties, and all other information necessary for sketching the bending moment and shear force diagrams for the beam. Neglect axial effects in the beam. The ties are made from the same material as the beam.

Answer: $v_B = v_C = -5PL^3/144EI, \theta_B = -\theta_C = PL^2/24EI,$
         $S_1 = 2P/3, S_2 = \sqrt{2}\,P/3,$
         $F_{y,A} = P/3, MA = -PL/4.$

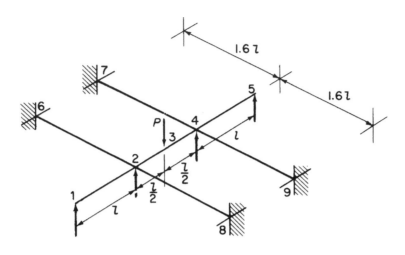

FIGURE P.6.7

P.6.7. The symmetrical rigid jointed grillage shown in Fig. P.6.7 is encastré at 6, 7, 8, and 9 and rests on simple supports at 1, 2, 4, and 5. It is loaded with a vertical point load P at 3. Use the stiffness method to find the displacements of the structure and calculate the support reactions and the forces in all the members. Plot the bending moment diagram for 123. All members have the same section properties and $GJ = 0.8EI$.

Answer: $F_{y,1} = F_{y,5} = -P/16$
$F_{y,2} = F_{y,4} = 9P/16$
$M_{21} = M_{45} = -Pl/16$ (hogging)
$M_{23} = M_{43} = -Pl/12$ (hogging) Twisting moment in 62, 82, 74, and 94 is $Pl/96$.

P.6.8. It is required to form the stiffness matrix of a triangular element 123 with coordinates $(0, 0)$, $(a, 0)$, and $(0, a)$, respectively, to be used for "plane stress" problems.
    (a) Form the $[B]$ matrix.
    (b) Obtain the stiffness matrix $[K^e]$.
Why, in general, is a finite element solution not an exact solution?

P.6.9. It is required to form the stiffness matrix of a triangular element 123 for use in stress analysis problems. The coordinates of the element are $(1, 1)$, $(2, 1)$, and $(2, 2)$, respectively.
    (a) Assume a suitable displacement field explaining the reasons for your choice.
    (b) Form the $[B]$ matrix.
    (c) Form the matrix that gives, when multiplied by the element nodal displacements, the stresses in the element. Assume a general $[D]$ matrix.

P.6.10. It is required to form the stiffness matrix for a rectangular element of side $2a \times 2b$ and thickness t for use in "plane stress" problems.
    (a) Assume a suitable displacement field.
    (b) Form the $[C]$ matrix.
    (c) Obtain $\int_{\text{vol}} [C]^T [D][C] \mathrm{d}V$.

Note that the stiffness matrix may be expressed as

$$[K^e] = [A^{-1}]^T \left[\int_{vol} [C]^T [D][C] dV \right] [A^{-1}]$$

P.6.11. A square element 1234, whose corners have coordinates x, y (in metres) of $(-1, -1)$, $(1, -1)$, $(1, 1)$, and $(-1, 1)$, respectively, was used in a plane stress finite element analysis. The following nodal displacements (mm) were obtained:

$$u_1 = 0.1 \quad u_2 = 0.3 \quad u_3 = 0.6 \quad u_4 = 0.1$$
$$v_1 = 0.1 \quad v_2 = 0.3 \quad v_3 = 0.7 \quad v_4 = 0.5$$

If Young's modulus $E = 200{,}000$ N/mm² and Poisson's ratio $v = 0.3$, calculate the stresses at the center of the element.

Answer: $\sigma_x = 51.65 \text{ N/mm}^2$, $\sigma_y = 55.49 \text{ N/mm}^2$, $\tau_{xy} = 13.46 \text{ N/mm}^2$

P.6.12. A rectangular element used in plane stress analysis has corners whose coordinates in metres referred to an Oxy axes system are $1(-2, -1)$, $2(2, -1)$, $3(2, 1)$, $4(-2, 1)$. The displacements of the corners (in metres) are

$$u_1 = 0.001 \quad u_2 = 0.003 \quad u_3 = -0.003 \quad u_4 = 0$$
$$v_1 = -0.004 \quad v_2 = -0.002 \quad v_3 = 0.001 \quad v_4 = 0.001$$

If Young's modulus is $200{,}000$ N/mm² and Poisson's ratio is 0.3, calculate the strains at the center of the element.

Answer: $\varepsilon_x = -0.000125$, $\varepsilon_y = 0.002$, $\gamma_{xy} = -0.0015$.

P.6.12 MATLAB Use MATLAB to repeat Problem P.6.12. In addition, calculate the strains at the following (x,y) locations in the element:

$$(-1, 0.5) \quad (0, 0.5) \quad (1, 0.5)$$
$$(-1, 0) \quad (0, 0) \quad (1, 0)$$
$$(-1, -0.5) \quad (0, -0.5) \quad (1, -0.5)$$

Answer:
(i) $(x, y) = (-1, 0.5), \epsilon_x = -0.0004375, \epsilon_y = 0.0023, \gamma_{xy} = -0.0010$
(ii) $(x, y) = (0, 0.5), \varepsilon_x = -0.0004375, \varepsilon_y = 0.0020, \gamma_{xy} = -0.0016$
(iii) $(x, y) = (1, 0.5), \varepsilon_x = -0.0004375, \varepsilon_y = 0.0018, \gamma_{xy} = -0.0023$
(iv) $(x, y) = (-1, 0), \varepsilon_x = -0.000125, \varepsilon_y = 0.0023, \gamma_{xy} = -0.000875$
(v) $(x, y) = (0, 0), \varepsilon_x = -0.000125, \varepsilon_y = 0.0020, \gamma_{xy} = -0.0015$
(vi) $(x, y) = (1, 0), \varepsilon_x = -0.000125, \varepsilon_y = 0.0018, \gamma_{xy} = -0.0021$
(vii) $(x, y) = (-1, -0.5), \varepsilon_x = 0.0001875, \varepsilon_y = 0.0023, \gamma_{xy} = -0.00075$
(viii) $(x, y) = (0, -0.5), \varepsilon_x = 0.0001875, \varepsilon_y = 0.002, \gamma_{xy} = -0.0014$
(ix) $(x, y) = (1, -0.5), \varepsilon_x = 0.0001875, \varepsilon_y = 0.0018, \gamma_{xy} = -0.002$

P.6.13. A constant strain triangular element has corners $1(0,0)$, $2(4,0)$, and $3(2,2)$ and is 1 unit thick. If the elasticity matrix $[D]$ has elements $D_{11} = D_{22} = a$, $D_{12} = D_{21} = b$, $D_{13} = D_{23} = D_{31} = D_{32} = 0$ and $D_{33} = c$, derive the stiffness matrix for the element.

Answer:

$$[K^e] = \frac{1}{4}\begin{bmatrix} a+c & & & & & \\ b+c & a+c & & & & \\ -a+c & -b+c & a+c & & & \\ b-c & a-c & -b-c & a+c & & \\ -2c & -2c & -2c & 2c & 4c & \\ -2b & -2a & 2b & -2a & 0 & 4a \end{bmatrix}$$

P.6.14. The following interpolation formula is suggested as a displacement function for deriving the stiffness of a plane stress rectangular element of uniform thickness t shown in Fig. P.6.14.

$$u = \frac{1}{4ab}[(a-x)(b-y)u_1 + (a+x)(b-y)u_2 + (a+x)(b+y)u_3 + (a-x)(b+y)u_1]$$

Form the strain matrix and obtain the stiffness coefficients K_{11} and K_{12} in terms of the material constants c, d, and e defined next. In the elasticity matrix $[D]$,

$$D_{11} = D_{22} = c \quad D_{12} = d \quad D_{33} = e \quad \text{and} \quad D_{13} = D_{23} = 0$$

Answer: $K_{11} = t(4c+e)/6, \quad K_{12} = t(d+e)/4$

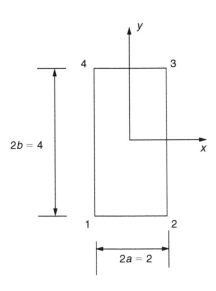

FIGURE P.6.14

Thin plate theory A3

Bending of thin plates

Generally, we define a thin plate as a sheet of material whose thickness is small compared with its other dimensions but which is capable of resisting bending in addition to membrane forces. Such a plate forms a basic part of an aircraft structure, being, for example, the area of stressed skin bounded by adjacent stringers and ribs in a wing structure or by adjacent stringers and frames in a fuselage.

In this chapter, we investigate the effect of a variety of loading and support conditions on the small deflection of rectangular plates. Two approaches are presented: an "exact" theory based on the solution of a differential equation and an energy method relying on the principle of the stationary value of the total potential energy of the plate and its applied loading. The latter theory is used in Chapter 9 to determine buckling loads for unstiffened and stiffened panels.

7.1 PURE BENDING OF THIN PLATES

The thin rectangular plate of Fig. 7.1 is subjected to pure bending moments of intensity M_x and M_y per unit length uniformly distributed along its edges. The former bending moment is applied along the edges parallel to the y axis, the latter along the edges parallel to the x axis. We assume that these bending moments are positive when they produce compression at the upper surface of the plate and tension at the lower.

If we further assume that the displacement of the plate in a direction parallel to the z axis is small compared with its thickness t and sections which are plane before bending remain plane after bending, then, as in the case of simple beam theory, the middle plane of the plate does not deform during the bending and is therefore a *neutral plane*. We take the neutral plane as the reference plane for our system of axes.

Let us consider an element of the plate of side $\delta x\delta y$ and having a depth equal to the thickness t of the plate, as shown in Fig. 7.2(a). Suppose that the radii of curvature of the neutral plane n are ρ_x and ρ_y in the xz and yz planes, respectively (Fig. 7.2(b)). Positive curvature of the plate corresponds to the positive bending moments which produce displacements in the positive direction of the z or downward axis. Again, as in simple beam theory, the direct strains ε_x and ε_y corresponding to direct stresses σ_x and σ_y of an elemental lamina of thickness δz a distance z below the neutral plane are given by

$$\varepsilon_x = \frac{z}{\rho_x}, \quad \varepsilon_y = \frac{z}{\rho_y} \tag{7.1}$$

Referring to Eqs. (1.52), we have

$$\varepsilon_x = \frac{1}{E}(\sigma_x - v\sigma_y), \quad \varepsilon_y = \frac{1}{E}(\sigma_y - v\sigma_x) \tag{7.2}$$

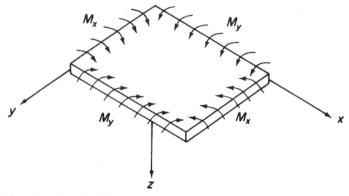

FIGURE 7.1 Plate Subjected to Pure Bending

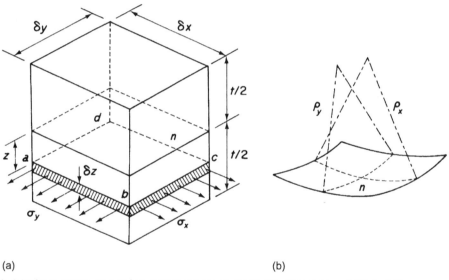

(a)   (b)

FIGURE 7.2 (a) Direct Stress on Lamina of Plate Element; (b) Radii of Curvature of Neutral Plane

Substituting for ε_x and ε_y from Eqs. (7.1) into (7.2) and rearranging gives

$$\left.\begin{array}{l} \sigma_x = \dfrac{Ez}{1-v^2}\left(\dfrac{1}{\rho_x}+\dfrac{v}{\rho_y}\right) \\[4mm] \sigma_y = \dfrac{Ez}{1-v^2}\left(\dfrac{1}{\rho_y}+\dfrac{v}{\rho_x}\right) \end{array}\right\} \qquad (7.3)$$

As would be expected from our assumption of plane sections remaining plane, the direct stresses vary linearly across the thickness of the plate, their magnitudes depending on the curvatures (i.e., bending moments) of the plate. The internal direct stress distribution on each vertical surface of the element must be in equilibrium with the applied bending moments. Thus,

$$M_x \delta y = \int_{-t/2}^{t/2} \sigma_x z \delta y \, dz$$

and

$$M_y \delta x = \int_{-t/2}^{t/2} \sigma_y z \delta x \, dz$$

Substituting for σ_x and σ_y from Eqs. (7.3) gives

$$M_x = \int_{-t/2}^{t/2} \frac{Ez^2}{1-v^2} \left(\frac{1}{\rho_x} + \frac{v}{\rho_y} \right) dz$$

$$M_y = \int_{-t/2}^{t/2} \frac{Ez^2}{1-v^2} \left(\frac{1}{\rho_y} + \frac{v}{\rho_x} \right) dz$$

Let

$$D = \int_{-t/2}^{t/2} \frac{Ez^2}{1-v^2} \, dz = \frac{Et^3}{12(1-v^2)} \tag{7.4}$$

Then,

$$M_x = D \left(\frac{1}{\rho_x} + \frac{v}{\rho_y} \right) \tag{7.5}$$

$$M_y = D \left(\frac{1}{\rho_y} + \frac{v}{\rho_x} \right) \tag{7.6}$$

in which D is known as the *flexural rigidity* of the plate.

If w is the deflection of any point on the plate in the z direction, then we may relate w to the curvature of the plate in the same manner as the well-known expression for beam curvature:

$$\frac{1}{\rho_x} = -\frac{\partial^2 w}{\partial x^2} \quad \frac{1}{\rho_y} = -\frac{\partial^2 w}{\partial y^2}$$

the negative signs resulting from the fact that the centers of curvature occur above the plate, in which region z is negative. Equations (7.5) and (7.6) then become

$$M_x = -D \left(\frac{\partial^2 w}{\partial x^2} + v \frac{\partial^2 w}{\partial y^2} \right) \tag{7.7}$$

$$M_y = -D \left(\frac{\partial^2 w}{\partial y^2} + v \frac{\partial^2 w}{\partial x^2} \right) \tag{7.8}$$

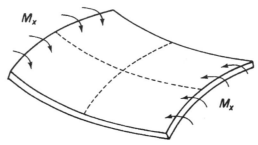

FIGURE 7.3 Anticlastic Bending

Equations (7.7) and (7.8) define the deflected shape of the plate provided that M_x and M_y are known. If either M_x or M_y is zero, then

$$\frac{\partial^2 w}{\partial x^2} = -\nu \frac{\partial^2 w}{\partial y^2} \quad \text{or} \quad \frac{\partial^2 w}{\partial y^2} = -\nu \frac{\partial^2 w}{\partial x^2}$$

and the plate has curvatures of opposite signs. The case of $M_y = 0$ is illustrated in Fig. 7.3. A surface possessing two curvatures of opposite sign is known as an *anticlastic surface*, as opposed to a *synclastic surface*, which has curvatures of the same sign. Further, if $M_x = M_y = M$, then from Eqs. (7.5) and (7.6),

$$\frac{1}{\rho_x} = \frac{1}{\rho_y} = \frac{1}{\rho}$$

Therefore, the deformed shape of the plate is spherical and of curvature

$$\frac{1}{\rho} = \frac{M}{D(1 + \nu)} \tag{7.9}$$

7.2 PLATES SUBJECTED TO BENDING AND TWISTING

In general, the bending moments applied to the plate will not be in planes perpendicular to its edges. Such bending moments, however, may be resolved in the normal manner into tangential and perpendicular components, as shown in Fig. 7.4. The perpendicular components are seen to be M_x and M_y as before, while the tangential components M_{xy} and M_{yx} (again, these are moments per unit length)

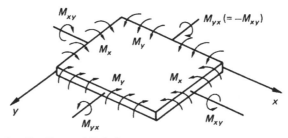

FIGURE 7.4 Plate Subjected to Bending and Twisting

produce twisting of the plate about axes parallel to the x and y axes. The system of suffixes and the sign convention for these twisting moments must be clearly understood to avoid confusion. M_{xy} is a twisting moment intensity in a vertical x plane parallel to the y axis, while M_{yx} is a twisting moment intensity in a vertical y plane parallel to the x axis. Note that the first suffix gives the direction of the axis of the twisting moment. We also define positive twisting moments as being clockwise when viewed along their axes in directions parallel to the positive directions of the corresponding x or y axis. In Fig. 7.4, therefore, all moment intensities are positive.

Since the twisting moments are tangential moments or torques they are resisted by a system of horizontal shear stresses τ_{xy}, as shown in Fig. 7.6. From a consideration of complementary shear stresses (see Fig. 7.6), $M_{xy} = -M_{yx}$, so that we may represent a general moment application to the plate in terms of M_x, M_y, and M_{xy} as shown in Fig. 7.5(a). These moments produce tangential and normal moments, M_t and M_n, on an arbitrarily chosen diagonal plane FD. We may express these moment intensities (in an analogous fashion to the complex stress systems of Section 1.6) in terms of M_x, M_y and M_{xy}. Thus, for equilibrium of the triangular element ABC of Fig. 7.5(b) in a plane perpendicular to AC

$$M_n AC = M_x AB \cos\alpha + M_y BC \sin\alpha - M_{xy} AB \sin\alpha - M_{xy} BC \cos\alpha$$

giving

$$M_n = M_x \cos^2\alpha + M_y \sin^2\alpha - M_{xy} \sin 2\alpha \tag{7.10}$$

Similarly, for equilibrium in a plane parallel to CA,

$$M_t AC = M_x AB \sin\alpha - M_y BC \cos\alpha + M_{xy} AB \cos\alpha - M_{xy} BC \sin\alpha$$

or

$$M_t = \frac{(M_x - M_y)}{2} \sin 2\alpha + M_{xy} \cos 2\alpha \tag{7.11}$$

(compare Eqs. (7.10) and (7.11) with Eqs. (1.8) and (1.9)). We observe, from Eq. (7.11), that there are two values of α, differing by 90° and given by

$$\tan 2\alpha = -\frac{2M_{xy}}{M_x - M_y}$$

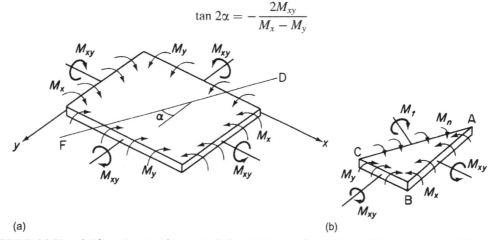

(a)   (b)

FIGURE 7.5 (a) Plate Subjected to Bending and Twisting; (b) Tangential and Normal Moments on an Arbitrary Plane

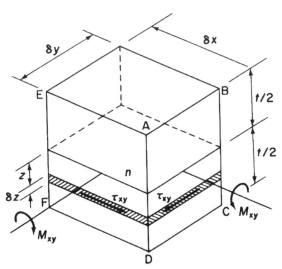

FIGURE 7.6 Complementary Shear Stresses Due to Twisting Moments M_{xy}

for which $M_t = 0$, leaving normal moments of intensity M_n on two mutually perpendicular planes. These moments are termed *principal moments* and their corresponding curvatures *principal curvatures*. For a plate subjected to pure bending and twisting in which M_x, M_y, and M_{xy} are invariable throughout the plate, the principal moments are the algebraically greatest and least moments in the plate. It follows that there are no shear stresses on these planes and that the corresponding direct stresses, for a given value of z and moment intensity, are the algebraically greatest and least values of direct stress in the plate.

Let us now return to the loaded plate of Fig. 7.5(a). We have established, in Eqs. (7.7) and (7.8), the relationships between the bending moment intensities M_x and M_y and the deflection w of the plate. The next step is to relate the twisting moment M_{xy} to w. From the principle of superposition, we may consider M_{xy} acting separately from M_x and M_y. As stated previously, M_{xy} is resisted by a system of horizontal complementary shear stresses on the vertical faces of sections taken throughout the thickness of the plate parallel to the x and y axes. Consider an element of the plate formed by such sections, as shown in Fig. 7.6. The complementary shear stresses on a lamina of the element a distance z below the neutral plane are, in accordance with the sign convention of Section 1.2, τ_{xy}. Therefore, on the face ABCD,

$$M_{xy}\delta y = -\int_{-t/2}^{t/2} \tau_{xy}\delta yz \, dz$$

and, on the face ADFE,

$$M_{xy}\delta x = -\int_{-t/2}^{t/2} \tau_{xy}\delta xz \, dz$$

giving

$$M_{xy} = -\int_{-t/2}^{t/2} \tau_{xy} z \, dz$$

or, in terms of the shear strain γ_{xy} and modulus of rigidity G,

$$M_{xy} = -G\int_{-t/2}^{t/2} \gamma_{xy} z \, dz \qquad (7.12)$$

Referring to Eqs. (1.20), the shear strain γ_{xy} is given by

$$\gamma_{xy} = \frac{\partial v}{\partial x} + \frac{\partial u}{\partial y}$$

We require, of course, to express γ_{xy} in terms of the deflection w of the plate; this may be accomplished as follows. An element taken through the thickness of the plate suffers rotations equal to $\partial w/\partial x$ and $\partial w/\partial y$ in the xz and yz planes, respectively. Considering the rotation of such an element in the xz plane, as shown in Fig. 7.7, we see that the displacement u in the x direction of a point a distance z below the neutral plane is

$$u = -\frac{\partial w}{\partial x} z$$

Similarly, the displacement v in the y direction is

$$v = -\frac{\partial w}{\partial y} z$$

Hence, substituting for u and v in the expression for γ_{xy}, we have

$$\gamma_{xy} = -2z\frac{\partial^2 w}{\partial x \partial y} \qquad (7.13)$$

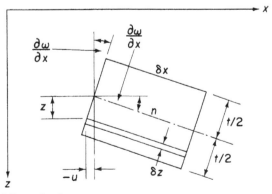

FIGURE 7.7 Determination of Shear Strain γ_{xy}

whence from Eq. (7.12),

$$M_{xy} = G \int_{-t/2}^{t/2} 2z^2 \frac{\partial^2 w}{\partial x \partial y} \, dz$$

or

$$M_{xy} = \frac{Gt^3}{6} \frac{\partial^2 w}{\partial x \partial y}$$

Replacing G by the expression $E/2 \, (1 + v)$ established in Eq. (1.50) gives

$$M_{xy} = \frac{Et^3}{12(1 + v)} \frac{\partial^2 w}{\partial x \partial y}$$

Multiplying the numerator and denominator of this equation by the factor $(1 - v)$ yields

$$M_{xy} = D(1 - v) \frac{\partial^2 w}{\partial x \partial y} \tag{7.14}$$

Equations (7.7), (7.8), and (7.14) relate the bending and twisting moments to the plate deflection and are analogous to the bending moment–curvature relationship for a simple beam.

7.3 PLATES SUBJECTED TO A DISTRIBUTED TRANSVERSE LOAD

The relationships between bending and twisting moments, and plate deflection are now employed in establishing the general differential equation for the solution of a thin rectangular plate, supporting a distributed transverse load of intensity q per unit area (see Fig. 7.8). The distributed load may, in general, vary over the surface of the plate and is therefore a function of x and y. We assume, as in the preceding analysis, that the middle plane of the plate is the neutral plane and that the plate deforms such that plane sections remain plane after bending. This latter assumption introduces an apparent inconsistency in the theory. For plane sections to remain plane, the shear strains γ_{xz} and γ_{yz} must be

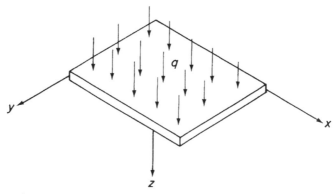

FIGURE 7.8 Plate Supporting a Distributed Transverse Load

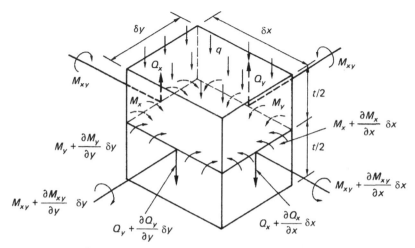

FIGURE 7.9 Plate Element Subjected to Bending, Twisting, and Transverse Loads

zero. However, the transverse load produces transverse shear forces (and therefore stresses), as shown in Fig. 7.9. We therefore assume that, although $\gamma_{xz} = \tau_{xz}/G$ and $\gamma_{yz} = \tau_{yz}/G$ are negligible, the corresponding shear forces are of the same order of magnitude as the applied load q and the moments M_x, M_y, and M_{xy}. This assumption is analogous to that made in a slender beam theory, in which shear strains are ignored.

The element of plate shown in Fig. 7.9 supports bending and twisting moments as previously described and, in addition, vertical shear forces Q_x and Q_y per unit length on faces perpendicular to the x and y axes, respectively. The variation of shear stresses τ_{xz} and τ_{yz} along the small edges δx, δy of the element is neglected and the resultant shear forces $Q_x \delta y$ and $Q_y \delta x$ are assumed to act through the centroid of the faces of the element. From the previous sections,

$$M_x = \int_{-t/2}^{t/2} \sigma_x z \, dz, \quad M_y = \int_{-t/2}^{t/2} \sigma_y z \, dz \quad M_{xy} = (-M_{yx}) = -\int_{-t/2}^{t/2} \tau_{xy} z \, dz$$

In a similar fashion,

$$Q_x = \int_{-t/2}^{t/2} \tau_{xz} \, dz, \quad Q_y = \int_{-t/2}^{t/2} \tau_{yz} \, dz \tag{7.15}$$

For equilibrium of the element parallel to Oz and assuming that the weight of the plate is included in q,

$$\left(Q_x + \frac{\partial Q_x}{\partial x} \delta x \right) \delta y - Q_x \delta y + \left(Q_y + \frac{\partial Q_y}{\partial y} \delta y \right) \delta x - Q_y \delta x + q \delta x \delta y = 0$$

or, after simplification,

$$\frac{\partial Q_x}{\partial x} + \frac{\partial Q_y}{\partial y} + q = 0 \tag{7.16}$$

Taking moments about the x axis,

$$M_{xy}\delta y - \left(M_{xy} + \frac{\partial M_{xy}}{\partial x}\delta x \right)\delta y - M_y\delta x + \left(M_y + \frac{\partial M_y}{\partial y}\delta y \right)\delta x$$
$$- \left(Q_y + \frac{\partial Q_y}{\partial y}\delta y \right)\delta x\delta y + Q_x\frac{\delta y^2}{2} - \left(Q_x + \frac{\partial Q_x}{\partial x}\delta x \right)\frac{\delta y^2}{2} - q\delta x\frac{\delta y^2}{2} = 0$$

Simplifying this equation and neglecting small quantities of a higher order than those retained gives

$$\frac{\partial M_{xy}}{\partial x} - \frac{\partial M_y}{\partial y} + Q_y = 0 \tag{7.17}$$

Similarly, taking moments about the y axis, we have

$$\frac{\partial M_{xy}}{\partial y} - \frac{\partial M_x}{\partial x} + Q_x = 0 \tag{7.18}$$

Substituting in Eq. (7.16) for Q_x and Q_y from Eqs. (7.18) and (7.17), we obtain

$$\frac{\partial^2 M_x}{\partial x^2} - \frac{\partial^2 M_{xy}}{\partial x\partial y} + \frac{\partial^2 M_y}{\partial y^2} - \frac{\partial^2 M_{xy}}{\partial x\partial y} = -q$$

or

$$\frac{\partial^2 M_x}{\partial x^2} - 2\frac{\partial^2 M_{xy}}{\partial x\partial y} + \frac{\partial^2 M_y}{\partial y^2} = -q \tag{7.19}$$

Replacing M_x, M_{xy}, and M_y in Eq. (7.19) from Eqs. (7.7), (7.14), and (7.8) gives

$$\frac{\partial^4 w}{\partial x^4} + 2\frac{\partial^4 w}{\partial x^2\partial y^2} + \frac{\partial^4 w}{\partial y^4} = \frac{q}{D} \tag{7.20}$$

This equation may also be written

$$\left(\frac{\partial^2}{\partial x^2} + \frac{\partial^2}{\partial y^2} \right) \left(\frac{\partial^2 w}{\partial x^2} + \frac{\partial^2 w}{\partial y^2} \right) = \frac{q}{D}$$

or

$$\left(\frac{\partial^2}{\partial x^2} + \frac{\partial^2}{\partial y^2} \right)^2 w = \frac{q}{D}$$

The operator $(\partial^2/\partial x^2 + \partial^2/\partial y^2)$ is the well-known Laplace operator in two dimensions and is sometimes written as ∇^2. Therefore,

$$(\nabla^2)^2 w = \frac{q}{D}$$

Generally, the transverse distributed load q is a function of x and y, so that the determination of the deflected form of the plate reduces to obtaining a solution to Eq. (7.20), which satisfies the known boundary conditions of the problem. The bending and twisting moments follow from Eqs. (7.7), (7.8), and (7.14), and the shear forces per unit length Q_x and Q_y are found from Eqs. (7.17) and (7.18) by substitution for M_x, M_y, and M_{xy} in terms of the deflection w of the plate; thus,

$$Q_x = \frac{\partial M_x}{\partial x} - \frac{\partial M_{xy}}{\partial y} = -D\frac{\partial}{\partial x}\left(\frac{\partial^2 w}{\partial x^2} + \frac{\partial^2 w}{\partial y^2}\right) \tag{7.21}$$

$$Q_y = \frac{\partial M_y}{\partial y} - \frac{\partial M_{xy}}{\partial x} = -D\frac{\partial}{\partial y}\left(\frac{\partial^2 w}{\partial x^2} + \frac{\partial^2 w}{\partial y^2}\right) \tag{7.22}$$

Direct and shear stresses are then calculated from the relevant expressions relating them to M_x, M_y, M_{xy}, Q_x, and Q_y.

Before discussing the solution to Eq. (7.20) for particular cases, we shall establish boundary conditions for various types of edge support.

7.3.1 The simply supported edge

Let us suppose that the edge $x = 0$ of the thin plate shown in Fig. 7.10 is free to rotate but not to deflect. The edge is then said to be *simply supported*. The bending moment along this edge must be zero and also the deflection $w = 0$. Thus,

$$(w)_{x=0} = 0 \quad \text{and} \quad (M_x)_{x=0} = -D\left(\frac{\partial^2 w}{\partial x^2} + v\frac{\partial^2 w}{\partial y^2}\right)_{x=0} = 0$$

The condition that $w = 0$ along the edge $x = 0$ also means that

$$\frac{\partial w}{\partial y} = \frac{\partial^2 w}{\partial y^2} = 0$$

along this edge. These boundary conditions therefore reduce to

$$(w)_{x=0} = 0, \quad \left(\frac{\partial^2 w}{\partial x^2}\right)_{x=0} = 0 \tag{7.23}$$

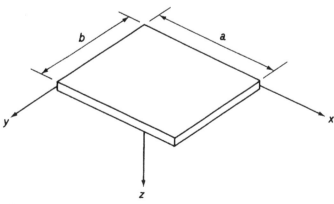

FIGURE 7.10 Plate of Dimensions *a* × *b*

7.3.2 The built-in edge

If the edge $x = 0$ is built-in or firmly clamped so that it can neither rotate nor deflect, then, in addition to w, the slope of the middle plane of the plate normal to this edge must be zero. That is,

$$(w)_{x=0} = 0, \quad \left(\frac{\partial w}{\partial x}\right)_{x=0} = 0 \tag{7.24}$$

7.3.3 The free edge

Along a free edge, there are no bending moments, twisting moments, or vertical shearing forces, so that, if $x = 0$ is the free edge, then

$$(M_x)_{x=0} = 0, \quad (M_{xy})_{x=0} = 0, \quad (Q_x)_{x=0} = 0$$

giving, in this instance, three boundary conditions. However, Kirchhoff (1850) showed that only two boundary conditions are necessary to obtain a solution to Eq. (7.20) and that the reduction is obtained by replacing the two requirements of zero twisting moment and zero shear force by a single equivalent condition. Thomson and Tait (1883) gave a physical explanation of how this reduction may be effected. They pointed out that the horizontal force system equilibrating the twisting moment M_{xy} may be replaced along the edge of the plate by a vertical force system.

Consider two adjacent elements δy_1 and δy_2 along the edge of the thin plate of Fig. 7.11. The twisting moment $M_{xy}\delta y_1$ on the element δy_1 may be replaced by *forces* M_{xy} a distance δy_1 apart. Note that M_{xy}, being a twisting moment per unit length, has the dimensions of force. The twisting moment on the adjacent element δy_2 is $[M_{xy} + (\partial M_{xy}/\partial y)\delta y]\delta y_2$. Again, this may be replaced by forces $M_{xy} + (\partial M_{xy}/\partial y)\delta y$. At the common surface of the two adjacent elements there is now a resultant force $(\partial M_{xy}/\partial y)\delta y$ or a vertical force per unit length of $\partial M_{xy}/\partial y$. For the sign convention for Q_x shown in

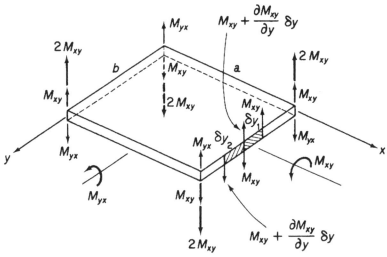

FIGURE 7.11 Equivalent Vertical Force System

Fig. 7.9, we have a statically equivalent vertical force per unit length of $(Q_x - \partial M_{xy}/\partial y)$. The separate conditions for a free edge of $(M_{xy})_{x=0} = 0$ and $(Q_x)_{x=0} = 0$ are therefore replaced by the equivalent condition

$$\left(Q_x - \frac{\partial M_{xy}}{\partial y}\right)_{x=0} = 0$$

or, in terms of deflection,

$$\left[\frac{\partial^3 w}{\partial x^3} + (2 - v)\frac{\partial^3 w}{\partial x \partial y^2}\right]_{x=0} = 0 \tag{7.25}$$

Also, for the bending moment along the free edge to be zero,

$$(M_x)_{x=0} = \left(\frac{\partial^2 w}{\partial x^2} + v\frac{\partial^2 w}{\partial y^2}\right)_{x=0} = 0 \tag{7.26}$$

The replacement of the twisting moment M_{xy} along the edges $x = 0$ and $x = a$ of a thin plate by a vertical force distribution results in leftover concentrated forces at the corners of M_{xy}, as shown in Fig. 7.11. By the same argument concentrated forces M_{yx} are produced by the replacement of the twisting moment M_{yx}. Since $M_{xy} = -M_{yx}$, the resultant forces $2M_{xy}$ act at each corner as shown and must be provided by external supports if the corners of the plate are not to move. The directions of these forces are easily obtained if the deflected shape of the plate is known. For example, a thin plate simply supported along all four edges and uniformly loaded has $\partial w/\partial x$ positive and numerically increasing, with increasing y near the corner $x = 0$, $y = 0$. Hence, $\partial^2 w/\partial x \partial y$ is positive at this point and, from Eq. (7.14), we see that M_{xy} is positive and M_{yx} negative; the resultant force $2M_{xy}$ is therefore downward. From symmetry, the force at each remaining corner is also $2M_{xy}$ downward, so that the tendency is for the corners of the plate to rise.

Having discussed various types of boundary conditions, we shall proceed to obtain the solution for the relatively simple case of a thin rectangular plate of dimensions $a \times b$, simply supported along each of its four edges and carrying a distributed load $q(x, y)$. We have shown that the deflected form of the plate must satisfy the differential equation

$$\frac{\partial^4 w}{\partial x^4} + 2\frac{\partial^4 w}{\partial x^2 \partial y^2} + \frac{\partial^4 w}{\partial y^4} = \frac{q(x, y)}{D}$$

with the boundary conditions

$$(w)_{x=0,a} = 0, \qquad \left(\frac{\partial^2 w}{\partial x^2}\right)_{x=0,a} = 0$$

$$(w)_{y=0,b} = 0, \qquad \left(\frac{\partial^2 w}{\partial y^2}\right)_{x=0,b} = 0$$

Navier (1820) showed that these conditions are satisfied by representing the deflection w as an infinite trigonometrical or Fourier series:

$$w = \sum_{m=1}^{\infty}\sum_{n=1}^{\infty} A_{mn} \sin\frac{m\pi x}{a} \sin\frac{n\pi y}{b} \tag{7.27}$$

in which m represents the number of half waves in the x direction and n the corresponding number in the y direction. Further, A_{mn} are unknown coefficients which must satisfy the preceding differential equation and may be determined as follows.

We may also represent the load $q(x, y)$ by a Fourier series:

$$q(x, y,) = \sum_{m=1}^{\infty} \sum_{n=1}^{\infty} A_{mn} \sin \frac{m\pi x}{a} \sin \frac{n\pi y}{b} \tag{7.28}$$

A particular coefficient $a_{m'n'}$ is calculated by first multiplying both sides of Eq. (7.28) by $\sin(m'\pi x/a) \sin(n'\pi y/b)$ and integrating with respect to x from 0 to a and with respect to y from 0 to b. Thus,

$$\int_0^a \int_0^b q(x, y) \sin \frac{m'\pi x}{a} \sin \frac{n'\pi y}{b} \, dxdy$$

$$= \sum_{m=1}^{\infty} \sum_{n=1}^{\infty} \int_0^a \int_0^b a_{mn} \sin \frac{m\pi x}{a} \sin \frac{m'\pi x}{a} \sin \frac{n\pi y}{b} \sin \frac{n'\pi y}{b} \, dxdy$$

$$= \frac{ab}{4} a_{m'n'}$$

since

$$\int_0^a \sin \frac{m\pi x}{a} \sin \frac{m'\pi x}{a} \, dx = 0, \quad \text{when } m \neq m'$$

$$= \frac{a}{2}, \quad \text{when } m = m'$$

and

$$\int_0^b \sin \frac{n\pi y}{b} \sin \frac{n'\pi y}{b} \, dy = 0, \quad \text{when } n \neq n'$$

$$= \frac{b}{2}, \quad \text{when } n = n'$$

It follows that

$$a_{m'n'} = \frac{4}{ab} \int_0^a \int_0^b q(x, y) \sin \frac{m'\pi x}{a} \sin \frac{n'\pi y}{b} \, dxdy \tag{7.29}$$

Substituting now for w and $q(x, y)$ from Eqs. (7.27) and (7.28) into the differential equation for w, we have

$$\sum_{m=1}^{\infty} \sum_{n=1}^{\infty} \left\{ A_{mn} \left[\left(\frac{m\pi}{a}\right)^4 + 2\left(\frac{m\pi}{a}\right)^2 \left(\frac{n\pi}{b}\right)^2 + \left(\frac{n\pi}{b}\right)^4 \right] - \frac{a_{mn}}{D} \right\} \sin \frac{m\pi x}{a} \sin \frac{n\pi y}{b} = 0$$

This equation is valid for all values of x and y, so that

$$A_{mn} \left[\left(\frac{m\pi}{a}\right)^4 + 2\left(\frac{m\pi}{a}\right)^2 \left(\frac{n\pi}{b}\right)^2 + \left(\frac{n\pi}{b}\right)^4 \right] - \frac{a_{mn}}{D} = 0$$

or in alternative form

$$A_{mn}\pi^4 \left(\frac{m^2}{a^2} + \frac{n^2}{b^2}\right)^2 - \frac{a_{mn}}{D} = 0$$

giving

$$A_{mn} = \frac{1}{\pi^4 D} \frac{a_{mn}}{[(m^2/a^2) + (n^2/b^2)]^2}$$

Hence,

$$w = \frac{1}{\pi^4 D} \sum_{m=1}^{\infty} \sum_{n=1}^{\infty} \frac{a_{mn}}{[(m^2/a^2) + (n^2/b^2)]^2} \sin\frac{m\pi x}{a} \sin\frac{n\pi y}{b} \qquad (7.30)$$

in which a_{mn} is obtained from Eq. (7.29). Equation (7.30) is the general solution for a thin rectangular plate under a transverse load $q(x, y)$.

Example 7.1

A thin rectangular plate $a \times b$ is simply supported along its edges and carries a uniformly distributed load of intensity q_0. Determine the deflected form of the plate and the distribution of bending moment. See Ex. 1.1.

Since $q(x, y) = q_0$, we find from Eq. (7.29) that,

$$a_{mn} = \frac{4q_0}{ab} \int_0^a \int_0^b \sin\frac{m\pi x}{a} \sin\frac{n\pi y}{b} \, dx\, dy = \frac{16q_0}{\pi^2 mn}$$

where m and n are odd integers. For m or n even, $a_{mn} = 0$. Hence, from Eq. (7.30),

$$w = \frac{16q_0}{\pi^6 D} \sum_{m=1,3,5}^{\infty} \sum_{n=1,3,5}^{\infty} \frac{\sin(m\pi x/a)\sin(n\pi y/b)}{mn[(m^2/a^2) + (n^2/b^2)]^2} \qquad (i)$$

The maximum deflection occurs at the center of the plate, where $x = a/2$, $y = b/2$. Thus,

$$w_{max} = \frac{16q_0}{\pi^6 D} \sum_{m=1,3,5}^{\infty} \sum_{n=1,3,5}^{\infty} \frac{\sin(m\pi/2)\sin(n\pi/2)}{mn[(m^2/a^2) + (n^2/b^2)]^2} \qquad (ii)$$

This series is found to converge rapidly, the first few terms giving a satisfactory answer. For a square plate, taking $\nu = 0.3$, summation of the first four terms of the series gives

$$w_{max} = 0.0443q_0 \frac{a^4}{Et^3}$$

Substitution for w from Eq. (i) into the expressions for bending moment, Eqs. (7.7) and (7.8), yields

$$M_x = \frac{16q_0}{\pi^4} \sum_{m=1,3,5}^{\infty} \sum_{n=1,3,5}^{\infty} \frac{[(m^2/a^2) + \nu(n^2/b^2)]}{mn[(m^2/a^2) + (n^2/b^2)]^2} \sin\frac{m\pi x}{a} \sin\frac{n\pi y}{b} \qquad (iii)$$

$$M_y = \frac{16q_0}{\pi^4} \sum_{m=1,3,5}^{\infty} \sum_{n=1,3,5}^{\infty} \frac{[\nu(m^2/a^2) + (n^2/b^2)]}{mn[(m^2/a^2) + (n^2/b^2)]^2} \sin\frac{m\pi x}{a} \sin\frac{n\pi y}{b} \qquad (iv)$$

Maximum values occur at the center of the plate. For a square plate $a = b$ and the first five terms give

$$m_{x,\text{max}} = m_{y,\text{max}} = 0.0479 q_0 a^2$$

Comparing Eqs. (7.3) with Eqs. (7.5) and (7.6), we observe that

$$\sigma_x = \frac{12 M_x z}{t^3}, \quad \sigma_y = \frac{12 M_y z}{t^3}$$

Again the maximum values of these stresses occur at the center of the plate at $z = \pm t/2$, so that

$$\sigma_{x,\text{max}} = \frac{6 M_x}{t^2}, \quad \sigma_{y,\text{max}} = \frac{6 M_y}{t^2}$$

For the square plate,

$$\sigma_{x,\text{max}} = \sigma_{y,\text{max}} = 0.287 q_0 \frac{a^2}{t^2}$$

The twisting moment and shear stress distributions follow in a similar manner.

The infinite series (Eq. (7.27)) assumed for the deflected shape of a plate gives an exact solution for displacements and stresses. However, a more rapid, but approximate, solution may be obtained by assuming a displacement function in the form of a polynomial. The polynomial must, of course, satisfy the governing differential equation (Eq. (7.20)) and the boundary conditions of the specific problem. The "guessed" form of the deflected shape of a plate is the basis for the energy method of solution described in Section 7.6.

Example 7.2

Show that the deflection function

$$w = A(x^2 y^2 - bx^2 y - axy^2 + abxy)$$

is valid for a rectangular plate of sides a and b, built in on all four edges and subjected to a uniformly distributed load of intensity q. If the material of the plate has a Young's modulus E and is of thickness t, determine the distributions of bending moment along the edges of the plate. See Ex. 1.1.

Differentiating the deflection function gives

$$\frac{\partial^4 w}{\partial x^4} = 0, \quad \frac{\partial^4 w}{\partial y^4} = 0, \quad \frac{\partial^4 w}{\partial x^2 \partial y^2} = 4A$$

Substituting in Eq. (7.20), we have

$$0 + 2 \times 4A + 0 = \text{constant} = \frac{q}{D}$$

The deflection function is therefore valid and

$$A = \frac{q}{8D}$$

The bending moment distributions are given by Eqs. (7.7) and (7.8); that is,

$$M_x = -\frac{q}{4}[y^2 - by + \nu(x^2 - ax)] \tag{i}$$

$$M_y = -\frac{q}{4}[x^2 - ax + v(y^2 - by)] \tag{ii}$$

For the edges $x = 0$ and $x = a$,

$$M_x = -\frac{q}{4}(y^2 - by), \quad M_y = -\frac{vq}{4}(y^2 - by)$$

For the edges $y = 0$ and $y = b$

$$M_x = -\frac{vq}{4}(x^2 - ax), \quad M_y = -\frac{q}{4}(x^2 - ax)$$

Example 7.2 MATLAB

Repeat Example 7.2 using the Symbolic Math Toolbox in MATLAB. See Ex. 1.1.

Expressions for M_x and M_y along the edges of the plate are obtained through the following MATLAB file:

```
% Declare any needed variables
syms w A x y b a q M_x M_y v D

% Define the given deflection function
w = A*(x^2*y^2 - b*x^2*y - a*x*y^2 + a*b*x*y);

% Check Eq. (7.20)
check = diff(w,x,4) + 2*diff(diff(w,x,2),y,2) + diff(w,y,4) - q/D;
if diff(check,x) == sym(0) && diff(check,y) == sym(0) && check ~= -q/D
 disp('The deflection function is valid')
 disp(' ')

% Solve check for the constant A and substitute back into w
A_val = solve(check,A);
w = subs(w,A,A_val);

% Calculate the bending moment distributions using Eqs (7.7) and (7.8)
eqI = -D*(diff(w,x,2) + v*diff(w,y,2)); % M_x
eqII = -D*(diff(w,y,2) + v*diff(w,x,2)); % M_y

% Output expressions of M_x, M_y for the edge x=0 to the Command Window
disp('For the edge x = 0:')
disp(['M_x =' char(simplify(subs(eqI,x,0)))])
disp(['M_y =' char(simplify(subs(eqII,x,0)))])
disp(' ')

% Output expressions of M_x, M_y for the edge x=a to the Command Window
disp('For the edge x = a:')
disp(['M_x =' char(simplify(subs(eqI,x,a)))])
disp(['M_y =' char(simplify(subs(eqII,x,a)))])
disp(' ')
```

```
% Output expressions of M_x, M_y for the edge y=0 to the Command Window
disp('For the edge y = 0:')
disp(['M_x =' char(simplify(subs(eqI,y,0)))])
disp(['M_y =' char(simplify(subs(eqII,y,0)))])
disp(' ')

% Output expressions of M_x, M_y for the edge y=b to the Command Window
disp('For the edge y = b:')
disp(['M_x =' char(simplify(subs(eqI,y,b)))])
disp(['M_y =' char(simplify(subs(eqII,y,b)))])
disp(' ')
else
disp('The deflection function does not satisfy Equation (7.20)')
disp(' ')
end
```

The Command Window outputs resulting from this MATLAB file are as follows. The deflection function is valid

```
For the edge x = 0:
M_x = (q*y*(b - y))/4
M_y = (q*v*y*(b - y))/4

For the edge x = a:
M_x = (q*y*(b - y))/4
M_y = (q*v*y*(b - y))/4

For the edge y = 0:
M_x = (q*v*x*(a - x))/4
M_y = (q*x*(a - x))/4

For the edge y = b:
M_x = (q*v*x*(a - x))/4
M_y = (q*x*(a - x))/4
```

7.4 COMBINED BENDING AND IN-PLANE LOADING OF A THIN RECTANGULAR PLATE

So far our discussion has been limited to small deflections of thin plates produced by different forms of transverse loading. In these cases, we assumed that the middle or neutral plane of the plate remained unstressed. Additional in-plane tensile, compressive, or shear loads produce stresses in the middle plane, and these, if of sufficient magnitude, affect the bending of the plate. Where the in-plane stresses are small compared with the critical buckling stresses, it is sufficient to consider the two systems separately; the total stresses are then obtained by superposition. On the other hand, if the in-plane stresses are not small, then their effect on the bending of the plate must be considered.

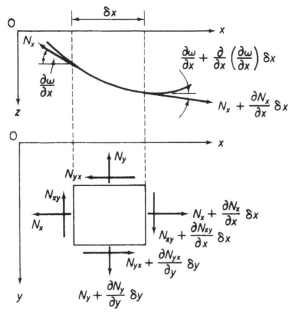

FIGURE 7.12 In-Plane Forces on Plate Element

The elevation and plan of a small element $\delta x \delta y$ of the middle plane of a thin deflected plate are shown in Fig. 7.12. Direct and shear forces per unit length produced by the in-plane loads are given the notation N_x, N_y, and N_{xy} and are assumed to be acting in positive senses in the directions shown. Since there are no resultant forces in the x or y directions from the transverse loads (see Fig. 7.9), we need only include the in-plane loads shown in Fig. 7.12 when considering the equilibrium of the element in these directions. For equilibrium parallel to Ox,

$$\left(N_x + \frac{\partial N_x}{\partial x}\delta x\right)\delta y \cos\left(\frac{\partial w}{\partial x} + \frac{\partial^2 w}{\partial x^2}\delta x\right) - N_x\delta y \cos\frac{\partial w}{\partial x} + \left(N_{yx} + \frac{\partial N_{yx}}{\partial y}\delta y\right)\delta x - N_{yx}\delta x = 0$$

For small deflections, $\partial w/\partial x$ and $(\partial w/\partial x) + (\partial^2 w/\partial x^2)\delta x$ are small and the cosines of these angles are therefore approximately equal to one. The equilibrium equation thus simplifies to

$$\frac{\partial N_x}{\partial x} + \frac{\partial N_{yx}}{\partial y} = 0 \tag{7.31}$$

Similarly, for equilibrium in the y direction, we have

$$\frac{\partial N_y}{\partial y} + \frac{\partial N_{xy}}{\partial x} = 0 \tag{7.32}$$

Note that the components of the in-plane shear loads per unit length are, to a first order of approximation, the value of the shear load multiplied by the projection of the element on the relevant axis.

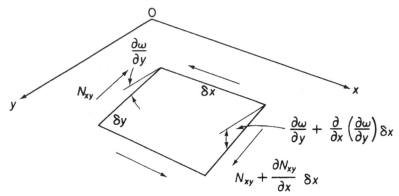

FIGURE 7.13 Component of Shear Loads in the z Direction

The determination of the contribution of the shear loads to the equilibrium of the element in the z direction is complicated by the fact that the element possesses curvature in both the xz and yz planes. Therefore, from Fig. 7.13, the component in the z direction due to the N_{xy} shear loads only is

$$\left(N_{xy} + \frac{\partial N_{xy}}{\partial x}\,\delta x\right)\delta y\left(\frac{\partial w}{\partial y} + \frac{\partial^2 w}{\partial x \partial y}\,\delta x\right) - N_{xy}\delta y\,\frac{\partial w}{\partial y}$$

or

$$N_{xy}\frac{\partial^2 w}{\partial x \partial y}\,\delta x \delta y + \frac{\partial N_{xy}}{\partial x}\frac{\partial w}{\partial y}\,\delta x \delta y$$

neglecting terms of a lower order. Similarly, the contribution of N_{yx} is

$$N_{yx}\frac{\partial^2 w}{\partial x \partial y}\,\delta x \delta y + \frac{\partial N_{yx}}{\partial y}\frac{\partial w}{\partial x}\,\delta x \delta y$$

The components arising from the direct forces per unit length are readily obtained from Fig. 7.12, namely,

$$\left(N_x + \frac{\partial N_x}{\partial x}\,\delta x\right)\delta y\left(\frac{\partial w}{\partial x} + \frac{\partial^2 w}{\partial x^2}\,\delta x\right) - N_x\delta y\,\frac{\partial w}{\partial x}$$

or

$$N_x\frac{\partial^2 w}{\partial x^2}\,\delta x \delta y + \frac{\partial N_x}{\partial x}\frac{\partial w}{\partial x}\,\delta x \delta y$$

and, similarly,

$$N_y\frac{\partial^2 w}{\partial y^2}\,\delta x \delta y + \frac{\partial N_y}{\partial y}\frac{\partial w}{\partial y}\,\delta x \delta y$$

The total force in the z direction is found from the summation of these expressions and is

$$N_x \frac{\partial^2 w}{\partial x^2} \delta x \delta y + \frac{\partial N_x}{\partial x} \frac{\partial w}{\partial x} \delta x \delta y + N_y \frac{\partial^2 w}{\partial y^2} \delta x \delta y + \frac{\partial N_y}{\partial y} \frac{\partial w}{\partial y} \delta x \delta y$$

$$+ \frac{\partial N_{xy}}{\partial x} \frac{\partial w}{\partial y} \delta x \delta y + 2 N_{xy} \frac{\partial^2 w}{\partial x \partial y} \delta x \delta y + \frac{\partial N_{xy}}{\partial y} \frac{\partial w}{\partial x} \delta x \delta y$$

in which N_{yx} is equal to and is replaced by N_{xy}. Using Eqs. (7.31) and (7.32), we reduce this expression to

$$\left(N_x \frac{\partial^2 w}{\partial x^2} + N_y \frac{\partial^2 w}{\partial y^2} + 2 N_{xy} \frac{\partial^2 w}{\partial x \partial y} \right) \delta x \delta y$$

Since the in-plane forces do not produce moments along the edges of the element, Eqs. (7.17) and (7.18) remain unaffected. Further, Eq. (7.16) may be modified simply by the addition of the preceding vertical component of the in-plane loads to $q \delta x \delta y$. Therefore, the governing differential equation for a thin plate supporting transverse and in-plane loads is, from Eq. (7.20),

$$\frac{\partial^4 w}{\partial x^4} + 2 \frac{\partial^4 w}{\partial x^2 \partial y^2} + \frac{\partial^4 w}{\partial y^4} = \frac{1}{D} \left(q + N_x \frac{\partial^2 w}{\partial x^2} + N_y \frac{\partial^2 w}{\partial y^2} + 2 N_{xy} \frac{\partial^2 w}{\partial x \partial y} \right) \tag{7.33}$$

Example 7.3

Determine the deflected form of the thin rectangular plate of Example 7.1 if, in addition to a uniformly distributed transverse load of intensity q_0, it supports an in-plane tensile force N_x per unit length. See Ex. 1.1.

The uniform transverse load may be expressed as a Fourier series (see Eq. (7.28) and Example 7.1); that is,

$$q = \frac{16 q_0}{\pi^2} \sum_{m=1,3,5}^{\infty} \sum_{n=1,3,5}^{\infty} \frac{1}{mn} \sin \frac{m \pi x}{a} \sin \frac{n \pi y}{b}$$

Equation (7.33) then becomes, on substituting for q,

$$\frac{\partial^4 w}{\partial x^4} + 2 \frac{\partial^4 w}{\partial x^2 \partial y^2} + \frac{\partial^4 w}{\partial y^4} - \frac{N_x}{D} \frac{\partial^2 w}{\partial x^2} = \frac{16 q_0}{\pi^2 D} \sum_{m=1,3,5}^{\infty} \sum_{n=1,3,5}^{\infty} \frac{1}{mn} \sin \frac{m \pi x}{a} \sin \frac{n \pi y}{b} \tag{i}$$

The appropriate boundary conditions are

$$w = \frac{\partial^2 w}{\partial x^2} = 0, \text{ at } x = 0 \text{ and } a$$

$$w = \frac{\partial^2 w}{\partial y^2} = 0, \text{ at } y = 0 \text{ and } b$$

These conditions may be satisfied by the assumption of a deflected form of the plate given by

$$w = \sum_{m=1}^{\infty} \sum_{n=1}^{\infty} A_{mn} \sin \frac{m \pi x}{a} \sin \frac{n \pi y}{b}$$

Substituting this expression into Eq. (i) gives

$$A_{mn} = \frac{16q_0}{\pi^6 Dmn\left[\left(\dfrac{m^2}{a^2} + \dfrac{n^2}{b^2}\right)^2 + \dfrac{N_x m^2}{\pi^2 D a^2}\right]}, \quad \text{for odd } m \text{ and } n$$

$$A_{mn} = 0, \text{ for even } m \text{ and } n$$

Therefore,

$$w = \frac{16q_0}{\pi^6 D} \sum_{m=1,3,5}^{\infty} \sum_{n=1,3,5}^{\infty} \frac{1}{mn\left[\left(\dfrac{m^2}{a^2} + \dfrac{n^2}{b^2}\right)^2 + \dfrac{N_x m^2}{\pi^2 D a^2}\right]} \sin\frac{m\pi x}{a} \sin\frac{n\pi y}{b} \tag{ii}$$

Comparing Eq. (ii) with Eq. (i) of Example 7.1, we see that, as a physical inspection would indicate, the presence of a tensile in-plane force decreases deflection. Conversely, a compressive in-plane force would increase the deflection.

7.5 BENDING OF THIN PLATES HAVING A SMALL INITIAL CURVATURE

Suppose that a thin plate has an initial curvature so that the deflection of any point in its middle plane is w_0. We assume that w_0 is small compared with the thickness of the plate. The application of transverse and in-plane loads causes the plate to deflect a further amount w_1, so that the total deflection is then $w = w_0 + w_1$. However, in the derivation of Eq. (7.33), we note that the left-hand side is obtained from expressions for bending moments, which themselves depend on the change of curvature. We therefore use the deflection w_1 on the left-hand side, not w. The effect on bending of the in-plane forces depends on the total deflection w so that we write Eq. (7.33)

$$\frac{\partial^4 w_1}{\partial x^4} + 2\frac{\partial^4 w_1}{\partial x^2 \partial y^2} + \frac{\partial^4 w_1}{\partial y^4}$$

$$= \frac{1}{D}\left[q + N_x\frac{\partial^2(w_0 + w_1)}{\partial x^2} + N_y\frac{\partial^2(w_0 + w_1)}{\partial y^2} + 2N_{xy}\frac{\partial^2(w_0 + w_1)}{\partial x \partial y}\right] \tag{7.34}$$

The effect of an initial curvature on deflection is therefore equivalent to the application of a transverse load of intensity

$$N_x\frac{\partial^2 w_0}{\partial x^2} + N_y\frac{\partial^2 w_0}{\partial y^2} + 2N_{xy}\frac{\partial^2 w_0}{\partial x \partial y}$$

Thus, in-plane loads alone produce bending, provided there is an initial curvature.

Assuming that the initial form of the deflected plate is

$$w_0 = \sum_{m=1}^{\infty} \sum_{n=1}^{\infty} A_{mn} \sin\frac{m\pi x}{a} \sin\frac{n\pi y}{b} \tag{7.35}$$

then, by substitution in Eq. (7.34), we find that, if N_x is compressive and $N_y = N_{xy} = 0$,

$$w_1 = \sum_{m=1}^{\infty} \sum_{n=1}^{\infty} B_{mn} \sin\frac{m\pi x}{a} \sin\frac{n\pi y}{b} \tag{7.36}$$

where

$$B_{mn} = \frac{A_{mn}N_x}{(\pi^2 D/a^2)[m + (n^2a^2/mb^2)]^2 - N_x}$$

We shall return to the consideration of initially curved plates in the discussion of the experimental determination of buckling loads of flat plates in Chapter 9.

7.6 ENERGY METHOD FOR THE BENDING OF THIN PLATES

Two types of solution are obtainable for thin-plate bending problems by the application of the principle of the stationary value of the total potential energy of the plate and its external loading. The first, in which the form of the deflected shape of the plate is known, produces an exact solution; the second, the *Rayleigh–Ritz* method, assumes an approximate deflected shape in the form of a series having a finite number of terms chosen to satisfy the boundary conditions of the problem and also to give the kind of deflection pattern expected.

In Chapter 5, we saw that the total potential energy of a structural system comprised the internal or strain energy of the structural member plus the potential energy of the applied loading. We now proceed to derive expressions for these quantities for the loading cases considered in the preceding sections.

7.6.1 Strain energy produced by bending and twisting

In thin-plate analysis, we are concerned with deflections normal to the loaded surface of the plate. These, as in the case of slender beams, are assumed to be primarily due to bending action, so that the effects of shear strain and shortening or stretching of the middle plane of the plate are ignored. Therefore, it is sufficient for us to calculate the strain energy produced by bending and twisting only, as this will be applicable, for the reason of the previous assumption, to all loading cases. It must be remembered that we are only neglecting the contributions of only shear and direct *strains* on the deflection of the plate; the stresses producing them must not be ignored.

Consider the element $\delta x \times \delta y$ of a thin plate $a \times b$ shown in elevation in the xz plane in Fig. 7.14(a). Bending moments M_x per unit length applied to its δy edge produce a change in slope between its ends equal to $(\partial^2 w/\partial x^2)\delta x$. However, since we regard the moments M_x as positive in the sense shown, then this change in slope, or relative rotation, of the ends of the element is negative, as the slope decreases with increasing x. The bending strain energy due to M_x is then

$$\frac{1}{2}M_x\delta y\left(-\frac{\partial^2 w}{\partial x^2}\delta x\right)$$

Similarly, in the yz plane the contribution of M_y to the bending strain energy is

$$\frac{1}{2}M_y\delta x\left(-\frac{\partial^2 w}{\partial y^2}\delta y\right)$$

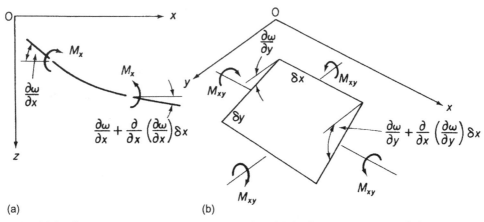

FIGURE 7.14 (a) Strain Energy of an Element Due to Bending; (b) Strain Energy Due to Twisting

The strain energy due to the twisting moment per unit length, M_{xy}, applied to the δy edges of the element, is obtained from Fig. 7.14(b). The relative rotation of the δy edges is $(\partial^2 w/\partial x \partial y)\delta x$, so that the corresponding strain energy is

$$\frac{1}{2}M_{xy}\delta y\frac{\partial^2 w}{\partial x \partial y}\delta x$$

Finally, the contribution of the twisting moment M_{xy} on the δx edges is, in a similar fashion,

$$\frac{1}{2}M_{xy}\delta x\frac{\partial^2 w}{\partial x \partial y}\delta y$$

The total strain energy of the element from bending and twisting is thus

$$\frac{1}{2}\left(-M_x\frac{\partial^2 w}{\partial x^2} - M_y\frac{\partial^2 w}{\partial y^2} + 2M_{xy}\frac{\partial^2 w}{\partial x \partial y}\right)\delta x \delta y$$

Substitution for M_x, M_y, and M_{xy} from Eqs. (7.7), (7.8), and (7.14) gives the total strain energy of the element as

$$\frac{D}{2}\left[\left(\frac{\partial^2 w}{\partial x^2}\right)^2 + \left(\frac{\partial^2 w}{\partial y^2}\right)^2 + 2v\frac{\partial^2 w}{\partial x^2}\frac{\partial^2 w}{\partial y^2} + 2(1-v)\left(\frac{\partial^2 w}{\partial x \partial y}\right)^2\right]\delta x \delta y$$

which on rearranging becomes

$$\frac{D}{2}\left\{\left(\frac{\partial^2 w}{\partial x^2} + \frac{\partial^2 w}{\partial y^2}\right)^2 - 2(1-v)\left[\frac{\partial^2 w}{\partial x^2}\frac{\partial^2 w}{\partial y^2} - \left(\frac{\partial^2 w}{\partial x \partial y}\right)^2\right]\right\}\delta x \delta y$$

Hence, the total strain energy U of the rectangular plate $a \times b$ is

$$U = \frac{D}{2}\int_0^a \int_0^b \left\{\left(\frac{\partial^2 w}{\partial x^2} + \frac{\partial^2 w}{\partial y^2}\right)^2 - 2(1-v)\left[\frac{\partial^2 w}{\partial x^2}\frac{\partial^2 w}{\partial y^2} - \left(\frac{\partial^2 w}{\partial x \partial y}\right)^2\right]\right\}dx\,dy \qquad (7.37)$$

Note that, if the plate is subject to pure bending only, then $M_{xy} = 0$ and, from Eq. (7.14), $\partial^2 w/\partial x \partial y = 0$, so that Eq. (7.37) simplifies to

$$U = \frac{D}{2} \int_0^a \int_0^b \left[\left(\frac{\partial^2 w}{\partial x^2} \right)^2 + \left(\frac{\partial^2 w}{\partial y^2} \right)^2 + 2v \frac{\partial^2 w}{\partial x^2} \frac{\partial^2 w}{\partial y^2} \right] dx\, dy \qquad (7.38)$$

7.6.2 Potential energy of a transverse load

An element $\delta x \times \delta y$ of the transversely loaded plate of Fig. 7.8 supports a load $q \delta x \delta y$. If the displacement of the element normal to the plate is w, then the potential energy δV of the load on the element referred to the undeflected plate position is

$$\delta V = -wq \delta x \delta y \quad \text{(see Section 5.7)}$$

Therefore, the potential energy V of the total load on the plate is given by

$$V = - \int_0^a \int_0^b wq \ dx\, dy \qquad (7.39)$$

7.6.3 Potential energy of in-plane loads

We may consider each load N_x, N_y, and N_{xy} in turn, then use the principle of superposition to determine the potential energy of the loading system when they act simultaneously. Consider an elemental strip of width δy along the length a of the plate in Fig. 7.15(a). The compressive load on this strip is $N_x \delta y$, and due to the bending of the plate, the horizontal length of the strip decreases by an amount λ, as shown in Fig. 7.15(b). The potential energy δV_x of the load $N_x \delta y$, referred to the undeflected position of the plate as the datum, is then

$$\delta V_x = -N_x \lambda \delta y \qquad (7.40)$$

From Fig. 7.15(b), the length of a small element δa of the strip is

$$\delta a = \left(\delta x^2 + \delta w^2 \right)^{\frac{1}{2}}$$

and, since $\partial w/\partial x$ is small,

$$\delta a \approx \delta x \left[1 + \frac{1}{2} \left(\frac{\partial w}{\partial x} \right)^2 \right]$$

Hence,

$$a = \int_0^{a'} \left[1 + \frac{1}{2} \left(\frac{\partial w}{\partial x} \right)^2 \right] dx$$

giving

$$a = a' + \int_0^{a'} \frac{1}{2} \left(\frac{\partial w}{\partial x} \right)^2 dx$$

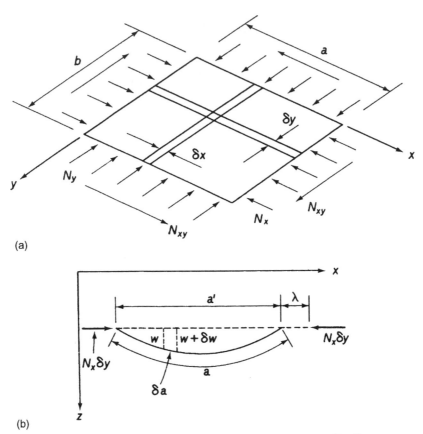

FIGURE 7.15 (a) In-Plane Loads on a Plate; (b) Shortening of an Element Due to Bending

and

$$\lambda = a - a' = \int_0^{a'} \frac{1}{2} \left(\frac{\partial w}{\partial x} \right)^2 dx$$

Since

$$\int_0^{a'} \frac{1}{2} \left(\frac{\partial w}{\partial x} \right)^2 dx \quad \text{only differs from} \quad \int_0^{a} \frac{1}{2} \left(\frac{\partial w}{\partial x} \right)^2 dx$$

by a term of negligible order, we write

$$\lambda = \int_0^{a} \frac{1}{2} \left(\frac{\partial w}{\partial x} \right)^2 dx \tag{7.41}$$

The potential energy V_x of the N_x loading follows from Eqs. (7.40) and (7.41), thus

$$V_x = -\frac{1}{2} \int_0^{a} \int_0^{b} N_x \left(\frac{\partial w}{\partial x} \right)^2 dx \ dy \tag{7.42}$$

Similarly,

$$V_y = -\frac{1}{2}\int_0^a \int_0^b N_y \left(\frac{\partial w}{\partial y}\right)^2 dx\ dy \qquad (7.43)$$

The potential energy of the in-plane shear load N_{xy} may be found by considering the work done by N_{xy} during the shear distortion corresponding to the deflection w of an element. This shear strain is the reduction in the right angle C_2AB_1 to the angle C_1AB_1 of the element in Fig. 7.16 or, rotating C_2A with respect to AB_1 to AD in the plane C_1AB_1, the angle DAC_1. The displacement C_2D is equal to $(\partial w/\partial y)\delta y$ and the angle DC_2C_1 is $\partial w/\partial x$. Thus C_1D is equal to

$$\frac{\partial w}{\partial x}\frac{\partial w}{\partial y}\delta y$$

and the angle DAC_1, representing the shear strain corresponding to the bending displacement w, is

$$\frac{\partial w}{\partial x}\frac{\partial w}{\partial y}$$

so that the work done on the element by the shear force $N_{xy}\delta x$ is

$$\frac{1}{2}N_{xy}\delta x \frac{\partial w}{\partial x}\frac{\partial w}{\partial y}$$

Similarly, the work done by the shear force $N_{xy}\delta y$ is

$$\frac{1}{2}N_{xy}\delta y \frac{\partial w}{\partial x}\frac{\partial w}{\partial y}$$

and the total work done, taken over the complete plate, is

$$\frac{1}{2}\int_0^a \int_b^b 2N_{xy}\frac{\partial w}{\partial x}\frac{\partial w}{\partial y}dx\ dy$$

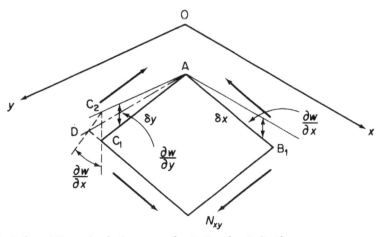

FIGURE 7.16 Calculation of Shear Strain Corresponding to Bending Deflection

It follows immediately that the potential energy of the N_{xy} loads is

$$V_{xy} = -\frac{1}{2}\int_0^a\int_0^b 2N_{xy}\frac{\partial w}{\partial x}\frac{\partial w}{\partial y}\ dx\,dy \qquad (7.44)$$

and, for the complete in-plane loading system, we have, from Eqs. (7.42), (7.43), and (7.44), a potential energy of

$$V = -\frac{1}{2}\int_0^a\int_0^b\left[N_x\left(\frac{\partial w}{\partial x}\right)^2 + N_y\left(\frac{\partial w}{\partial y}\right)^2 + 2N_{xy}\frac{\partial w}{\partial x}\frac{\partial w}{\partial y}\right]\ dx\,dy \qquad (7.45)$$

We are now in a position to solve a wide range of thin-plate problems, provided that the deflections are small, obtaining exact solutions if the deflected form is known or approximate solutions if the deflected shape has to be guessed.

Considering the rectangular plate of Section 7.3, simply supported along all four edges and subjected to a uniformly distributed transverse load of intensity q_0, we know that its deflected shape is given by Eq. (7.27), namely,

$$w = \sum_{m=1}^{\infty}\sum_{n=1}^{\infty} A_{mn}\sin\frac{m\pi x}{a}\sin\frac{n\pi y}{b}$$

The total potential energy of the plate is, from Eqs. (7.37) and (7.39),

$$U + V = \int_0^a\int_0^b\left\{\frac{D}{2}\left[\left(\frac{\partial^2 w}{\partial x^2} + \frac{\partial^2 w}{\partial y^2}\right)^2 - 2(1-v)\left\{\frac{\partial^2 w}{\partial x^2}\frac{\partial^2 w}{\partial y^2} - \left(\frac{\partial^2 w}{\partial x\partial y}\right)^2\right\}\right] - wq_0\right\}\ dx\,dy \quad (7.46)$$

Substituting in Eq. (7.46) for w and realizing that "cross-product" terms integrate to zero, we have

$$U + V = \int_0^a\int_0^b\left\{\frac{D}{2}\sum_{m=1}^{\infty}\sum_{n=1}^{\infty}A_{mn}^2\left[\pi^4\left(\frac{m^2}{a^2}+\frac{n^2}{b^2}\right)^2\sin^2\frac{m\pi x}{a}\sin^2\frac{n\pi y}{b}\right.\right.$$

$$-2(1-v)\frac{m^2n^2\pi^4}{a^2b^2}\left(\sin^2\frac{m\pi x}{a}\sin^2\frac{n\pi y}{b} - \cos^2\frac{m\pi x}{a}\cos^2\frac{n\pi y}{b}\right)\right]$$

$$\left.-q_0\sum_{m=1}^{\infty}\sum_{n=1}^{\infty}A_{mn}\sin\frac{m\pi x}{a}\sin\frac{n\pi y}{b}\right\}\ dx\,dy$$

The term multiplied by $2(1-v)$ integrates to zero and the mean value of $\sin^2$ or $\cos^2$ over a complete number of half waves is $\frac{1}{2}$, thus integration of the above expression yields

$$U + V = \frac{D}{2}\sum_{m=1,3,5}^{\infty}\sum_{n=1,3,5}^{\infty}A_{mn}^2\frac{\pi^4ab}{4}\left(\frac{m^2}{a^2}+\frac{n^2}{b^2}\right)^2 - q_0\sum_{m=1,3,5}^{\infty}\sum_{n=1,3,5}^{\infty}A_{mn}\frac{4ab}{\pi^2mn} \qquad (7.47)$$

From the principle of the stationary value of the total potential energy, we have

$$\frac{\partial(U+V)}{\partial A_{mn}} = \frac{D}{2}2A_{mn}\frac{\pi^4ab}{4}\left(\frac{m^2}{a^2}+\frac{n^2}{b^2}\right)^2 - q_0\frac{4ab}{\pi^2mn} = 0$$

so that

$$A_{mn} = \frac{16q_0}{\pi^6 Dmn[(m^2/a^2) + (n^2/b^2)]^2}$$

giving a deflected form

$$w = \frac{16q_0}{\pi^6 D} \sum_{m=1,3,5}^{\infty} \sum_{n=1,3,5}^{\infty} \frac{\sin(m\pi x/a)\sin(n\pi y/b)}{mn[(m^2/a^2) + (n^2/b^2)]^2}$$

which is the result obtained in Eq. (i) of Example 7.1.

This solution is exact, since we know the true deflected shape of the plate in the form of an infinite series for w. Frequently, the appropriate infinite series is not known, so that only an approximate solution may be obtained. The method of solution, known as the *Rayleigh–Ritz* method, involves the selection of a series for w containing a finite number of functions of x and y. These functions are chosen to satisfy the boundary conditions of the problem as far as possible and also to give the type of deflection pattern expected. Naturally, the more representative the guessed functions are, the more accurate the solution becomes.

Suppose that the guessed series for w in a particular problem contains three functions of x and y. Thus;

$$w = A_1 f_1(x, y) + A_2 f_2(x, y) + A_3 f_3(x, y)$$

where A_1, A_2, and A_3 are unknown coefficients. We now substitute for w in the appropriate expression for the total potential energy of the system and assign stationary values with respect to A_1, A_2, and A_3 in turn. Thus,

$$\frac{\partial(U + V)}{\partial A_1} = 0, \quad \frac{\partial(U + V)}{\partial A_2} = 0, \quad \frac{\partial(U + V)}{\partial A_3} = 0$$

giving three equations which are solved for A_1, A_2, and A_3.

Example 7.4

A rectangular plate $a \times b$, is simply supported along each edge and carries a uniformly distributed load of intensity q_0. Assuming a deflected shape given by

$$w = A_{11} \sin\frac{\pi x}{a} \sin\frac{\pi y}{b}$$

determine the value of the coefficient A_{11} and hence find the maximum value of deflection. See Ex. 1.1.

The expression satisfies the boundary conditions of zero deflection and zero curvature (i.e., zero bending moment) along each edge of the plate. Substituting for w in Eq. (7.46), we have

$$U + V = \int_0^a \int_0^b \left[\frac{DA_{11}^2}{2} \left\{ \frac{\pi^4}{(a^2 b^2)^2}(a^2 + b^2)^2 \sin^2\frac{\pi x}{a} \sin^2\frac{\pi y}{b} - 2(1 - v) \right. \right.$$

$$\left. \times \left[\frac{\pi^4}{a^2 b^2} \sin^2\frac{\pi x}{a} \sin^2\frac{\pi y}{b} - \frac{\pi^4}{a^2 b^2} \cos^2\frac{\pi x}{a} \cos^2\frac{\pi y}{b} \right] \right\}$$

$$\left. -q_0 A_{11} \sin\frac{\pi x}{a} \sin\frac{\pi y}{b} \right] dx\,dy$$

from which

$$U + V = \frac{DA_{11}^2}{2}\frac{\pi^4}{4a^3b^3}\left(a^2 + b^2\right)^2 - q_0 A_{11}\frac{4ab}{\pi^2}$$

so that

$$\frac{\partial(U + V)}{\partial A_{11}} = \frac{DA_{11}\pi^4}{4a^3b^3}\left(a^2 + b^2\right)^2 - q_0\frac{4ab}{\pi^2} = 0$$

and

$$A_{11} = \frac{16q_0 a^4 b^4}{\pi^6 D(a^2 + b^2)^2}$$

giving

$$w = \frac{16q_0 a^4 b^4}{\pi^6 D(a^2 + b^2)^2}\sin\frac{\pi x}{a}\sin\frac{\pi y}{b}$$

At the center of the plate, w is a maximum and

$$w_{max} = \frac{16q_0 a^4 b^4}{\pi^6 D(a^2 + b^2)^2}$$

For a square plate and assuming $v = 0.3$,

$$w_{max} = 0.0455 q_0\frac{a^4}{Et^3}$$

which compares favorably with the result of Example 7.1.

In this chapter, we have dealt exclusively with small deflections of thin plates. For a plate subjected to large deflections, the middle plane is stretched due to *bending*, so that Eq. (7.33) requires modification. The relevant theory is outside the scope of this book but may be found in a variety of references.

Example 7.4 MATLAB

Repeat Example 7.4 using the Symbolic Math Toolbox in MATLAB. See Ex. 1.1.

Expressions for A_{11} and the maximum deflection are obtained through the following MATLAB file:

```
% Declare any needed variables
syms w A_11 x y b a q_0 UV v D

% Define the given deflection function
w = A_11*sin(pi*x/a)*sin(pi*y/b);

% Evaluate Eq. (7.46) to calculate U+V (UV)
w_xx = diff(w,x,2);
w_yy = diff(w,y,2);
w_xy = diff(diff(w,x),y);

UV = int(int((D/2*((w_xx+w_yy)^2 - 2*(1-v)*(w_xx*w_yy-w_xy^2)) - w*q_0),y,0,b),x,0,a);
```

```
% Differentiate UV with respect to A_11, set equal to 0, and solve for A_11
A_11val = solve(diff(UV,A_11),A_11);

% Substitute A_11 back into w
w = subs(w,A_11,A_11val);

% Due to the boundary conditions and the form of w, there will only be one maximum value.
% Therefore, w_max will occur where the gradient of w is 0.
x_max = solve(diff(w,x),x);
y_max = solve(diff(w,y),y);

% Substitute x_max and y_max into w
w_max = subs(subs(w,x,x_max),y,y_max);

% Output expressions for A_11 and w_max to the Command Window
disp(['A_11 =' char(A_11val)])
disp(['w_max =' char(w_max)])
```

The Command Window outputs resulting from this MATLAB file are as follows:

```
A_11 = (16*a^4*b^4*q_0)/(D*pi^6*(a^2 + b^2)^2)
w_max = (16*a^4*b^4*q_0)/(D*pi^6*(a^2 + b^2)^2)
```

Further reading

Jaeger JC. Elementary theory of elastic plates. New York: Pergamon press; 1964.
Timoshenko SP, Woinowsky-Krieger S. Theory of plates and shells. 2nd ed. New York: McGraw-Hill; 1959.
Timoshenko SP, Gere JM. Theory of elastic stability. 2nd ed. New York: McGraw-Hill; 1961.
Wang CT. Applied elasticity. New York: McGraw-Hill; 1953.

PROBLEMS

P.7.1. A plate 10 mm thick is subjected to bending moments M_x equal to 10 Nm/mm and M_y equal to 5 Nm/mm. Calculate the maximum direct stresses in the plate.

Answer: $\sigma_{x,max} = \pm 600 \text{N/mm}^2$, $\sigma_{y,max} = \pm 300 \text{N/mm}^2$.

P.7.2. For the plate and loading of problem P.7.1, find the maximum twisting moment per unit length in the plate and the direction of the planes on which this occurs.

Answer: 2.5 Nm/mm at 45° to the x and y axes.

P.7.3. The plate of the previous two problems is subjected to a twisting moment of 5 Nm/mm along each edge in addition to the bending moments of $M_x = 10$ Nm/mm and $M_y = 5$ N m/mm. Determine the principal moments in the plate, the planes on which they act, and the corresponding principal stresses.

Answer: 13.1 Nm/mm,   1.9 Nm/mm, $\alpha = -31.7°$, $\alpha = +58.3°$, $\pm 786 \text{ N/mm}^2$,
          $\pm 114 \text{ N/mm}^2$.

P.7.3 MATLAB Use MATLAB to repeat Problem P.7.3 for the following combinations of M_x, M_y, and M_{xy}. Do not calculate the principal stresses.

	(i)	(ii)	(iii)	(iv)	(v)
M_x	6	8	10	12	14
M_x	7	6	5	4	3
M_x	3	4	5	6	7

Answer:
(i) $M_I = 9.5$ N m/mm, $M_{II} = 3.5$ N m/mm, $\alpha = -49.7°$ or $40.3°$
(ii) $M_I = 11.1$ N m/mm, $M_{II} = 2.9$ N m/mm, $\alpha = -38°$ or $52°$
(iii) $M_I = 13.1$ N m/mm, $M_{II} = 1.9$ N m/mm, $\alpha = -31.7°$ or $58.3°$
(iv) $M_I = 15.2$ N m/mm, $M_{II} = 0.8$ N m/mm, $\alpha = -28.2°$ or $61.8°$
(v) $M_I = 17.4$ N m/mm, $M_{II} = 0.4$ N m/mm, $\alpha = -25.9°$ or $64.1°$

P.7.4. A thin rectangular plate of length a and width $2a$ is simply supported along the edges $x = 0$, $x = a$, $y = -a$, and $y = +a$. The plate has a flexural rigidity D, a Poisson's ratio of 0.3, and carries a load distribution given by $q(x, y) = q_0 \sin(\pi x/a)$. If the deflection of the plate is represented by the expression

$$w = \frac{qa^4}{D\pi^4}\left(1 + A\cosh\frac{\pi y}{a} + B\frac{\pi y}{a}\sinh\frac{\pi y}{a}\right)\sin\frac{\pi x}{a}$$

determine the values of the constants A and B.

Answer: $A = -0.2213$, $B = 0.0431$.

P.7.5. A thin, elastic square plate of side a is simply supported on all four sides and supports a uniformly distributed load q. If the origin of axes coincides with the center of the plate show that the deflection of the plate can be represented by the expression

$$w = \frac{q}{96(1-v)D}[2(x^4 + y^4) - 3a^2(1-v)(x^2 + y^2) - 12vx^2y^2 + A]$$

where D is the flexural rigidity, v is Poisson's ratio and A is a constant. Calculate the value of A and hence the central deflection of the plate.

Answer: $A = a^4(5 - 3v)/4$, Cen. def. $= qa^4 (5 - 3v)/384D(1 - v)$

P.7.6. The deflection of a square plate of side a which supports a lateral load represented by the function $q(x, y)$ is given by

$$w(x, y) = w_0\cos\frac{\pi x}{a}\cos\frac{3\pi y}{a}$$

where x and y are referred to axes whose origin coincides with the center of the plate and w_0 is the deflection at the center. If the flexural rigidity of the plate is D and Poisson's ratio is v, determine the loading function q, the support conditions of the plate, the reactions at the plate corners, and the bending moments at the center of the plate.

Answer: $q(x, y) = w_0 D 100\frac{\pi^4}{a^4}\cos\frac{\pi x}{a}\cos\frac{3\pi y}{a}$

The plate is simply supported on all edges. Reactions: $-6w_0D\left(\frac{\pi}{a}\right)^2(1-v)$

$$M_x = w_0D\left(\frac{\pi}{a}\right)^2(1+9v), \quad M_y = w_0D\left(\frac{\pi}{a}\right)^2(9+v)$$

P.7.7. A simply supported square plate $a \times a$ carries a distributed load according to the formula

$$q(x,y) = q_0\frac{x}{a}$$

where q_0 is its intensity at the edge $x = a$. Determine the deflected shape of the plate.

Answer: $w = \frac{8q_0a^4}{\pi^6D}\sum_{m=1,2,3}^{\infty}\sum_{n=1,3,5}^{\infty}\frac{(-1)^{m+1}}{mn(m^2+n^2)^2}\sin\frac{m\pi x}{a}\sin\frac{n\pi y}{a}$

P.7.8. An elliptic plate of major and minor axes $2a$ and $2b$ and of small thickness t is clamped along its boundary and is subjected to a uniform pressure difference p between the two faces. Show that the usual differential equation for normal displacements of a thin flat plate subject to lateral loading is satisfied by the solution

$$w = w_0\left(1 - \frac{x^2}{a^2} - \frac{y^2}{b^2}\right)^2$$

where w_0 is the deflection at the center, which is taken as the origin.

Determine w_0 in terms of p and the relevant material properties of the plate and hence expressions for the greatest stresses due to bending at the center and at the ends of the minor axis.

Answer: $w_0 = \dfrac{3p(1-v^2)}{2Et^3\left(\dfrac{3}{a^4} + \dfrac{2}{a^2b^2} + \dfrac{3}{b^4}\right)}$

Center,

$$\sigma_{x,max} = \frac{\pm 3pa^2b^2(b^2+va^2)}{t^2(3b^4+2a^2b^2+3a^4)}, \quad \sigma_{y,max} = \frac{\pm 3pa^2b^2(a^2+vb^2)}{t^2(3b^4+2a^2b^2+3a^4)}$$

Ends of minor axis,

$$\sigma_{x,max} = \frac{\pm 6pa^4b^2}{t^2(3b^4+2a^2b^2+3a^4)}, \quad \sigma_{y,max} = \frac{\pm 6pb^4a^2}{t^2(3b^4+2a^2b^2+3a^4)}$$

P.7.9. Use the energy method to determine the deflected shape of a rectangular plate $a \times b$, simply supported along each edge and carrying a concentrated load W at a position (ξ, η) referred to axes through a corner of the plate. The deflected shape of the plate can be represented by the series

$$w = \sum_{m=1}^{\infty}\sum_{n=1}^{\infty}A_{mn}\sin\frac{m\pi x}{a}\sin\frac{n\pi y}{b}$$

Answer: $A_{mn} = \dfrac{4W \sin \dfrac{m\pi\xi}{a} \sin \dfrac{n\pi\eta}{b}}{\pi^4 Dab[(m^2/a^2) + (n^2/b^2)]^2}$

P.7.10. If, in addition to the point load W, the plate of problem P.7.9 supports an in-plane compressive load of N_x per unit length on the edges $x = 0$ and $x = a$, calculate the resulting deflected shape.

Answer: $A_{mn} = \dfrac{4W \sin \dfrac{m\pi\xi}{a} \sin \dfrac{n\pi\eta}{b}}{abD\pi^4 \left[\left(\dfrac{m^2}{a^2} + \dfrac{n^2}{b^2} \right)^2 - \dfrac{m^2 N_x}{\pi^2 a^2 D} \right]}$

P.7.11. A square plate of side a is simply supported along all four sides and is subjected to a transverse uniformly distributed load of intensity q_0. It is proposed to determine the deflected shape of the plate by the Rayleigh–Ritz method employing a "guessed" form for the deflection of

$$w = A_{11} \left(1 - \frac{4x^2}{a^2} \right) \left(1 - \frac{4y^2}{a^2} \right)$$

in which the origin is taken at the center of the plate.

Comment on the degree to which the boundary conditions are satisfied and find the central deflection assuming $v = 0.3$.

Answer: $0.0389 q_0 a^4 / Et^3$

P.7.11 MATLAB Use the Symbolic Math Toolbox in MATLAB to repeat Problem P.7.11. In addition, calculate the deflections at the following (x,y) locations in the plate.

$$(a/4, a/4), (0,0), \quad (a/3, a/3)$$

Answer: (i) $(x, y) = (a/4, a/4), w = 0.0218 a^4 q0 / Et^3$
  (ii) $(x, y) = (0, 0), w = 0.0388 a^4 q0 / Et^3$
  (iii) $(x, y) = (a/3, a/3), w = 0.012 a^4 q0 / Et^3$

P.7.12. A rectangular plate $a \times b$, simply supported along each edge, possesses a small initial curvature in its unloaded state given by

$$w_0 = A_{11} \sin \frac{\pi x}{a} \sin \frac{\pi y}{b}$$

Determine, using the energy method, its final deflected shape when it is subjected to a compressive load N_x per unit length along the edges $x = 0$, $x = a$.

Answer: $w = \dfrac{A_{11}}{\left[1 - \dfrac{N_x a^2}{\pi^2 D} \Big/ \left(1 + \dfrac{a^2}{b^2} \right)^2 \right]} \sin \dfrac{\pi x}{a} \sin \dfrac{\pi y}{b}$

Structural instability

A4

Columns

8

A large proportion of an aircraft's structure consists of thin webs stiffened by slender longerons or stringers. Both are susceptible to failure by buckling at a buckling stress or critical stress, which is frequently below the limit of proportionality and seldom appreciably above the yield stress of the material. Clearly, for this type of structure, buckling is the most critical mode of failure, so that the prediction of buckling loads of columns, thin plates, and stiffened panels is extremely important in aircraft design. In this chapter, we consider the buckling failure of all these structural elements and also the flexural–torsional failure of thin-walled open tubes of low torsional rigidity.

Two types of structural instability arise: *primary* and *secondary*. The former involves the complete element, there being no change in cross-sectional area while the wavelength of the buckle is of the same order as the length of the element. Generally, solid and thick-walled columns experience this type of failure. In the latter mode, changes in cross-sectional area occur and the wavelength of the buckle is on the order of the cross-sectional dimensions of the element. Thin-walled columns and stiffened plates may fail in this manner.

8.1 EULER BUCKLING OF COLUMNS

The first significant contribution to the theory of the buckling of columns was made as early as 1744 by Euler. His classical approach is still valid, and likely to remain so, for slender columns possessing a variety of end restraints. Our initial discussion is therefore a presentation of the Euler theory for the small elastic deflection of perfect columns. However, we investigate first the nature of buckling and the difference between theory and practice.

It is common experience that, if an increasing axial compressive load is applied to a slender column, there is a value of the load at which the column will suddenly bow or buckle in some unpredetermined direction. This load is patently the buckling load of the column or something very close to the buckling load. Clearly, this displacement implies a degree of asymmetry in the plane of the buckle caused by geometrical or material imperfections of the column and its load. However, in our theoretical stipulation of a perfect column in which the load is applied precisely along the perfectly straight centroidal axis, there is perfect symmetry, so that, theoretically, there can be no sudden bowing or buckling. We therefore require a precise definition of the buckling load that may be used in our analysis of the perfect column.

If the perfect column of Fig. 8.1 is subjected to a compressive load P, only shortening of the column occurs, no matter what the value of P. However, if the column is displaced a small amount by a lateral load F, then, at values of P below the critical or buckling load, P_{CR}, removal of F results in a return of the column to its undisturbed position, indicating a state of stable equilibrium. At the critical load, the displacement does not disappear and, in fact, the column remains in *any* displaced position as long as

Introduction to Aircraft Structural Analysis, Second Edition

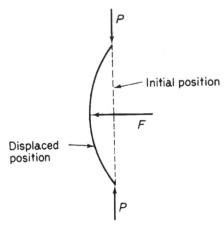

FIGURE 8.1 Definition of Buckling Load for a Perfect Column

the displacement is small. Thus, the buckling load P_{CR} is associated with a state of *neutral equilibrium*. For $P > P_{CR}$, enforced lateral displacements increase and the column is unstable.

Consider the pin-ended column AB of Fig. 8.2. We assume that it is in the displaced state of neutral equilibrium associated with buckling, so that the compressive load P has attained the critical value P_{CR}. Simple bending theory (see Chapter 15) gives

$$EI\frac{d^2v}{dz^2} = -M$$

or

$$EI\frac{d^2_v}{dz^2} = -P_{CR}v \tag{8.1}$$

so that the differential equation of bending of the column is

$$\frac{d^2v}{dz^2} + \frac{P_{CR}}{EI}v = 0 \tag{8.2}$$

The well-known solution of Eq. (8.2) is

$$v = A\cos\mu z + B\sin\mu z \tag{8.3}$$

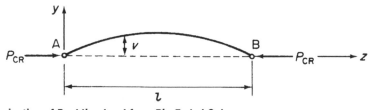

FIGURE 8.2 Determination of Buckling Load for a Pin-Ended Column

where $\mu^2 = P_{CR}/EI$ and A and B are unknown constants. The boundary conditions for this particular case are $v = 0$ at $z = 0$ and l. Therefore, $A = 0$ and

$$B \sin \mu l = 0$$

For a non-trivial solution (i.e., $v \neq 0$),

$$\sin \mu l = 0 \quad \text{or} \quad \mu l = n\pi, \quad \text{where } n = 1, 2, 3, \ldots$$

giving

$$\frac{P_{CR} l^2}{EI} = n^2 \pi^2$$

or

$$P_{CR} = \frac{n^2 \pi^2 EI}{l^2} \tag{8.4}$$

Note that Eq. (8.3) cannot be solved for v no matter how many of the available boundary conditions are inserted. This is to be expected, since the neutral state of equilibrium means that v is indeterminate.

The smallest value of buckling load, in other words the smallest value of P that can maintain the column in a neutral equilibrium state, is obtained by substituting $n = 1$ in Eq. (8.4). Hence,

$$P_{CR} = \frac{\pi^2 EI}{l^2} \tag{8.5}$$

Other values of P_{CR} corresponding to $n = 2, 3, \ldots$, are

$$P_{CR} = \frac{4\pi^2 EI}{l^2}, \frac{9\pi^2 EI}{l^2}, \ldots$$

These higher values of buckling load cause more complex modes of buckling such as those shown in Fig. 8.3. The different shapes may be produced by applying external restraints to a very slender column at the points of contraflexure to prevent lateral movement. If no restraints are provided then these forms of buckling are unstable and have little practical meaning.

The critical stress, σ_{CR}, corresponding to P_{CR}, is, from Eq. (8.5)

$$\sigma_{CR} = \frac{\pi^2 E}{(l/r)^2} \tag{8.6}$$

where r is the radius of gyration of the cross-sectional area of the column. The term l/r is known as the *slenderness ratio* of the column. For a column that is not doubly symmetrical, r is the least radius of

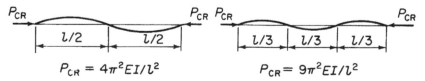

FIGURE 8.3 Buckling Loads for Different Buckling Modes of a Pin-Ended Column

gyration of the cross-section since the column bends about an axis about which the flexural rigidity EI is least. Alternatively, if buckling is prevented in all but one plane, then EI is the flexural rigidity in that plane.

Equations (8.5) and (8.6) may be written in the form

$$P_{CR} = \frac{\pi^2 EI}{l_e^2} \tag{8.7}$$

and

$$\sigma_{CR} = \frac{\pi^2 E}{(l_e/r)^2} \tag{8.8}$$

where l_e is the *effective length* of the column. This is the length of a *pin-ended column* that has the same critical load as that of a column of length l but with different end conditions. The determination of critical load and stress is carried out in an identical manner to that for the pin-ended column, except that the boundary conditions are different in each case. Table 8.1 gives the solution in terms of effective length for columns having a variety of end conditions. In addition, the boundary conditions referred to the coordinate axes of Fig. 8.2 are quoted. The last case in Table 8.1 involves the solution of a transcendental equation; this is most readily accomplished by a graphical method.

Let us now examine the buckling of the perfect pin-ended column of Fig. 8.2 in greater detail. We showed, in Eq. (8.4), that the column buckles at *discrete* values of axial load and that associated with each value of buckling load is a particular buckling mode (Fig. 8.3). These discrete values of buckling load are called *eigenvalues*, their associated functions (in this case $v = B \sin n\pi z/l$) are called *eigenfunctions* and the problem itself is called an *eigenvalue problem*.

Further, suppose that the lateral load F in Fig. 8.1 is removed. Since the column is perfectly straight, homogeneous, and loaded exactly along its axis, it suffers only axial compression as P is increased. This situation, theoretically, continues until yielding of the material of the column occurs. However, as we have seen, for values of P below P_{CR}, the column is in stable equilibrium, whereas for $P > P_{CR}$, the column is unstable. A plot of load against lateral deflection at mid-height therefore has the form shown in Fig. 8.4, where, at the point $P = P_{CR}$, it is theoretically possible for the column to take one of three deflection paths. Thus, if the column remains undisturbed, the deflection at mid-height continues to be zero but unstable (i.e., the trivial solution of Eq. (8.3), $v = 0$), or if disturbed, the column buckles in either of two lateral directions; the point at which this possible branching occurs is called a *bifurcation point*; further bifurcation points occur at the higher values of $P_{CR}(4\pi^2 EI/l^2, 9\pi^2 EI/l^2, \ldots)$.

Table 8.1 Column Length Solutions

Ends	l_e/l	Boundary conditions
Both pinned	1.0	$v = 0$ at $z = 0$ and l
Both fixed	0.5	$v = 0$ at $z = 0$ and $z = l$, $dv/dz = 0$ at $z = l$
One fixed, the other free	2.0	$v = 0$ and $dv/dz = 0$ at $z = 0$
One fixed, the other pinned	0.6998	$dv/dz = 0$ at $z = 0$, $v = 0$ at $z = l$ and $z = 0$

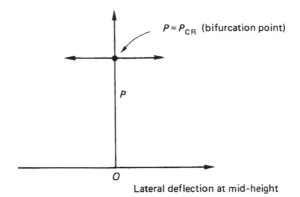

FIGURE 8.4 Behavior of a Perfect Pin-Ended Column

Example 8.1

A uniform column of length L and flexural stiffness EI is simply supported at its ends and has an additional elastic support at mid-span. This support is such that, if a lateral displacement v_c occurs at this point, a restoring force kv_c is generated at the point. Derive an equation giving the buckling load of the column. If the buckling load is $4\pi^2EI/L^2$, find the value of k. Also, if the elastic support is infinitely stiff, show that the buckling load is given by the equation $\tan \lambda L/2 = \lambda L/2$, where $\lambda = \sqrt{P/EI}$. See Ex. 1.1.

The column is shown in its displaced position in Fig. 8.5. The bending moment at any section of the column is given by

$$M = Pv - \frac{kv_c}{2}z$$

so that, by comparison with Eq. (8.1),

$$EI\frac{d^2v}{dz^2} = -Pv + \frac{kv_c}{2}z$$

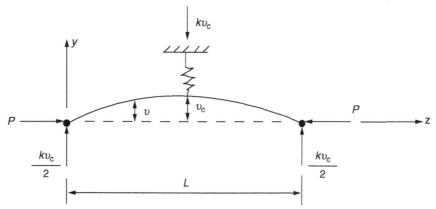

FIGURE 8.5 Column of Example 8.1

giving

$$\frac{d^2v}{dz^2} + \lambda^2 v = \frac{kv_c}{2EI}z \tag{i}$$

The solution of Eq. (i) is of standard form and is

$$v = A\cos\lambda z + B\sin\lambda z + \frac{kv_c}{2P}z$$

The constants A and B are found using the boundary conditions of the column, which are $v = 0$ when $z = 0$, $v = v_c$, when $z = L/2$ and $(dv/dz) = 0$ when $z = L/2$.

From the first of these, $A = 0$, while from the second,

$$B = v_c(1 - kL/4P)/\sin(\lambda L/2)$$

The third boundary condition gives, since $v_c \neq 0$, the required equation; that is,

$$\left(1 - \frac{kL}{4P}\right)\cos\frac{\lambda L}{2} + \frac{k}{2P\lambda}\sin\frac{\lambda L}{2} = 0$$

Rearranging,

$$P = \frac{kL}{4}\left(1 - \frac{\tan(\lambda L/2)}{\lambda L/2}\right)$$

If P (buckling load) $= 4\pi^2 EI/L^2$, then $\lambda L/2 = \pi$, so that $k = 4P/L$. Finally, if $k \to \infty$,

$$\tan\frac{\lambda L}{2} = \frac{\lambda L}{2} \tag{ii}$$

Note that Eq. (ii) is the transcendental equation which would be derived when determining the buckling load of a column of length $L/2$, built in at one end, and pinned at the other.

Example 8.1 MATLAB
Repeat the derivation of the column buckling load in Example 8.1 using MATLAB. See Ex. 1.1.

The expression for the column buckling load (P) is obtained through the following MATLAB file:

```
% Declare any needed variables
syms M P v k v_c z EI lambda L A B C PI
P_buck = EI/(L^2)*4*PI^2;
lambda_sq = P_buck/EI;

% Define the bending moment equation at any section in the column
M = P*v - k*v_c*z/2;

% Substituting M intoEq.(8.1)and solving the second order differential
% equation results in the following general solution for v
C = solve(EI*diff(C,2)+subs(M,v,C),C); %Equation 8.1
v = A*cos(lambda*z) + B*sin(lambda*z) + C; % General solution ofEq.(8.1)
```

```
% Check boundary conditions to solve for A and B and P
% Boundary Condition #1: v = 0 when z = 0
A_val = solve(subs(v,z,0),A);
v = subs(v,A,A_val);

% Boundary Condition #2: v = v_c when z = L/2
B_val = solve(subs(v,z,L/2)-v_c,B);
v = subs(v,B,B_val);

% Boundary Condition #3: v_z = 0 when z = L/2
P = solve(subs(diff(v,z),z,L/2),P);

% Output a simplified expression of P to the Command Window
disp(['P =' char(simplify(P))])
```

The Command Window output resulting from this MATLAB file is as follows:

```
P = (L*k)/4 - (k*tan((L*lambda)/2))/(2*lambda)
```

8.2 INELASTIC BUCKLING

We have shown that the critical stress, Eq. (8.8), depends only on the elastic modulus of the material of the column and the slenderness ratio l/r. For a given material, the critical stress increases as the slenderness ratio decreases, that is, as the column becomes shorter and thicker. A point is reached when the critical stress is greater than the yield stress of the material, so that Eq. (8.8) is no longer applicable. For mild steel, this point occurs at a slenderness ratio of approximately 100, as shown in Fig. 8.6. We therefore require some alternative means of predicting column behavior at low values of the slenderness ratio.

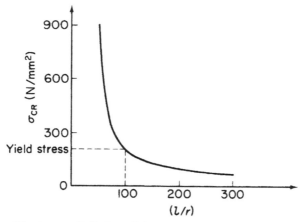

FIGURE 8.6 Critical Stress–Slenderness Ratio for a Column

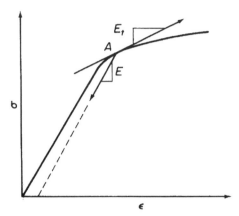

FIGURE 8.7 Elastic Moduli for a Material Stressed above the Elastic Limit

It was assumed in the derivation of Eq. (8.8) that the stresses in the column remain within the elastic range of the material, so that the modulus of elasticity $E(= d\sigma/d\varepsilon)$ was constant. Above the elastic limit, $d\sigma/d\varepsilon$ depends upon the value of stress and whether the stress is increasing or decreasing. Thus, in Fig. 8.7, the elastic modulus at the point A is the *tangent modulus* E_t if the stress is increasing but E if the stress is decreasing.

Consider a column having a plane of symmetry and subjected to a compressive load P such that the direct stress in the column P/A is above the elastic limit. If the column is given a small deflection, v, in its plane of symmetry, then the stress on the concave side increases while the stress on the convex side decreases. Thus, in the cross-section of the column shown in Fig. 8.8(a) the compressive stress decreases in the area A_1 and increases in the area A_2, while the stress on the line nn is unchanged. Since these changes take place outside the elastic limit of the material, we see, from our remarks in

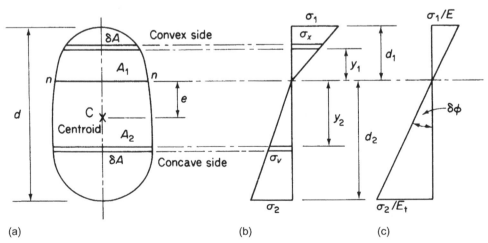

FIGURE 8.8 Determination of Reduced Elastic Modulus

the previous paragraph, that the modulus of elasticity of the material in the area A_1 is E while that in A_2 is E_t. The homogeneous column now behaves as if it were nonhomogeneous, with the result that the stress distribution is changed to the form shown in Fig. 8.8(b); the linearity of the distribution follows from an assumption that plane sections remain plane.

As the axial load is unchanged by the disturbance,

$$\int_0^{d_1} \sigma_x dA = \int_0^{d_2} \sigma_v \, dA \tag{8.9}$$

Also, P is applied through the centroid of each end section a distance e from nn, so that

$$\int_0^{d_1} \sigma_x(y_1 + e)dA + \int_0^{d_2} \sigma_v(y_2 - e) \, dA = -Pv \tag{8.10}$$

From Fig. 8.8(b),

$$\sigma_x = \frac{\sigma_1}{d_1}y_1, \quad \sigma_v = \frac{\sigma_2}{d_2}y_2 \tag{8.11}$$

The angle between two close, initially parallel, sections of the column is equal to the change in slope d^2v/dz^2 of the column between the two sections. This, in turn, must be equal to the angle $\delta\phi$ in the strain diagram of Fig. 8.8(c). Hence,

$$\frac{d^2v}{dz^2} = \frac{\sigma_1}{Ed_1} = \frac{\sigma_2}{E_t d_2} \tag{8.12}$$

and Eq. (8.9) becomes, from Eqs. (8.11) and (8.12),

$$E\frac{d^2v}{dz^2}\int_0^{d_1} y_1 dA - E_t\frac{d^2v}{dz^2}\int_0^{d_2} y_2 \, dA = 0 \tag{8.13}$$

Further, in a similar manner, from Eq. (8.10),

$$\frac{d^2v}{dz^2}\left(E\int_0^{d_1} y_1^2 dA + E_t\int_0^{d_2} y_2^2 dA\right) + e\frac{d^2v}{dz^2}\left(E\int_0^{d_1} y_1 \, dA - E_t\int_0^{d_2} y_2 \, dA\right) = -Pv \tag{8.14}$$

The second term on the left-hand side of Eq. (8.14) is zero, from Eq. (8.13). Therefore, we have

$$\frac{d^2v}{dz^2}(EI_1 + E_t I_2) = -Pv \tag{8.15}$$

in which

$$I_1 = \int_0^{d_1} y_1^2 dA \quad \text{and} \quad I_2 = \int_0^{d_2} y_2^2 \, dA$$

the second moments of area about nn of the convex and concave sides of the column, respectively. Putting

$$E_r I = EI_1 + E_t I_2$$

or

$$E_r = E\frac{I_1}{I} + E_t\frac{I_2}{I} \tag{8.16}$$

where E_r, known as the *reduced modulus*, gives

$$E_rI\frac{d^2v}{dz^2} + Pv = 0$$

Comparing this with Eq. (8.2), we see that, if P is the critical load P_{CR}, then

$$P_{CR} = \frac{\pi^2 E_r I}{l_e^2} \tag{8.17}$$

and

$$\sigma_{CR} = \frac{\pi^2 E_r}{(l_e/r)^2} \tag{8.18}$$

This method for predicting critical loads and stresses outside the elastic range is known as the *reduced modulus theory*. From Eq. (8.13), we have

$$E\int_0^{d_1} y_1\,dA - E_t\int_0^{d_2} y_2\,dA = 0 \tag{8.19}$$

which, together with the relationship $d = d_1 + d_2$, enables the position of *nn* to be found.

It is possible that the axial load P is increased at the time of the lateral disturbance of the column such that no strain reversal occurs on its convex side. The compressive stress therefore increases over the complete section so that the tangent modulus applies over the whole cross-section. The analysis is then the same as that for column buckling within the elastic limit except that E_t is substituted for E. Hence, the *tangent modulus theory* gives

$$P_{CR} = \frac{\pi^2 E_t I}{l_e^2} \tag{8.20}$$

and

$$\sigma_{CR} = \frac{\pi^2 E_t}{(l_e/r)^2} \tag{8.21}$$

By a similar argument, a reduction in P could result in a decrease in stress over the whole cross-section. The elastic modulus applies in this case and the critical load and stress are given by the standard Euler theory; namely, Eqs. (8.7) and (8.8).

In Eq. (8.16), I_1 and I_2 are together greater than I while E is greater than E_t. It follows that the reduced modulus E_r is greater than the tangent modulus E_t. Consequently, buckling loads predicted by the reduced modulus theory are greater than buckling loads derived from the tangent modulus theory, so that, although we specified theoretical loading situations where the different theories apply, there remains the difficulty of deciding which should be used for design purposes.

Extensive experiments carried out on aluminum alloy columns by the aircraft industry in the 1940s showed that the actual buckling load was approximately equal to the tangent modulus load. Shanley (1947) explained that, for columns with small imperfections, an increase of axial load and bending occur simultaneously. He then showed analytically that, after the tangent modulus load is reached, the strain on the concave side of the column increases rapidly while that on the convex side decreases slowly. The large deflection corresponding to the rapid strain increase on the concave side, which occurs soon after the tangent modulus load is passed, means that it is possible to exceed the tangent modulus load by only a small amount. It follows that the buckling load of columns is given most accurately for practical purposes by the tangent modulus theory.

Empirical formulae have been used extensively to predict buckling loads, although in view of the close agreement between experiment and the tangent modulus theory, they would appear unnecessary. Several formulae are in use; for example, the *Rankine, Straight-line*, and *Johnson's parabolic* formulae are given in many books on elastic stability.[1]

8.3 EFFECT OF INITIAL IMPERFECTIONS

Obviously, it is impossible in practice to obtain a perfectly straight homogeneous column and to ensure that it is exactly axially loaded. An actual column may be bent with some eccentricity of load. Such imperfections influence to a large degree the behavior of the column, which, unlike the perfect column, begins to bend immediately the axial load is applied.

Let us suppose that a column, initially bent, is subjected to an increasing axial load P, as shown in Fig. 8.9. In this case, the bending moment at any point is proportional to the change in curvature of the column from its initial bent position. Thus,

$$EI\frac{d^2v}{dz^2} - EI\frac{d^2v_0}{dz^2} = Pv \tag{8.22}$$

which, on rearranging, becomes

$$\frac{d^2v}{dz^2} + \lambda^2 v = \frac{d^2v_0}{dz^2} \tag{8.23}$$

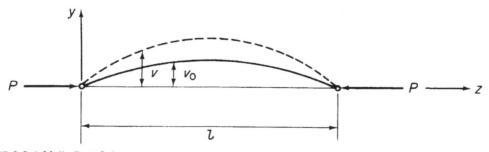

FIGURE 8.9 Initially Bent Column

where $\lambda^2 = P/EI$. The final deflected shape, v, of the column depends upon the form of its unloaded shape, v_0. Assuming that

$$v_0 = \sum_{n=1}^{\infty} A_n \sin \frac{n\pi z}{l} \tag{8.24}$$

and, substituting in Eq. (8.23), we have

$$\frac{d^2v}{dz^2} + \lambda^2 v = -\frac{\pi^2}{l^2} \sum_{n=1}^{\infty} n^2 A_n \sin \frac{n\pi z}{l}$$

The general solution to this equation is

$$v = B \cos \lambda z + D \sin \lambda z + \sum_{n=1}^{\infty} \frac{n^2 A_n}{n^2 - \alpha} \sin \frac{n\pi z}{l}$$

where B and D are constants of integration and $\alpha = \lambda^2 l^2/\pi^2$. The boundary conditions are $v = 0$ at $z = 0$ and l, giving $B = D = 0$, from which

$$v = \sum_{n=1}^{\infty} \frac{n^2 A_n}{n^2 - \alpha} \sin \frac{n\pi z}{l} \tag{8.25}$$

Note that, in contrast to the perfect column, we are able to obtain a non-trivial solution for deflection. This is to be expected, since the column is in stable equilibrium in its bent position at all values of P.

An alternative form for α is

$$\alpha = \frac{Pl^2}{\pi^2 EI} = \frac{P}{P_{CR}}$$

(see Eq. (8.5)). Thus, α is always less than 1 and approaches unity when P approaches P_{CR}, so that the first term in Eq. (8.25) usually dominates the series. A good approximation, therefore, for deflection when the axial load is in the region of the critical load, is

$$v = \frac{A_1}{1 - \alpha} \sin \frac{\pi z}{l} \tag{8.26}$$

or, at the center of the column, where $z = l/2$,

$$v = \frac{A_1}{1 - P/P_{CR}} \tag{8.27}$$

in which A_1 is seen to be the initial central deflection. If central deflections $\delta (= v - A_1)$ are measured from the initially bowed position of the column, then from Eq. (8.27), we obtain

$$\frac{A_1}{1 - P/P_{CR}} - A_1 = \delta$$

which gives, on rearranging,

$$\delta = P_{CR} \frac{\delta}{P} - A_1 \tag{8.28}$$

and we see that a graph of δ plotted against δ/P has a slope, in the region of the critical load, equal to P_{CR} and an intercept equal to the initial central deflection. This is the well-known *Southwell plot* for the experimental determination of the elastic buckling load of an imperfect column.

Timoshenko and Gere[1] also showed that Eq. (8.27) may be used for a perfectly straight column with small eccentricities of column load.

Example 8.2

The pin-jointed column shown in Fig. 8.10 carries a compressive load P applied eccentrically at a distance e from the axis of the column. Determine the maximum bending moment in the column. See Ex. 1.1.

The bending moment at any section of the column is given by

$$M = P(e + v)$$

Then, by comparison with Eq. (8.1),

$$EI\frac{d^2v}{dz^2} = -P(e + v)$$

giving

$$\frac{d^2v}{dz^2} + \mu^2 v = -\frac{Pe}{EI} \quad (\mu^2 = P/EI) \tag{i}$$

The solution of Eq. (i) is of standard form and is

$$v = A\cos \mu z + B \sin \mu z - e$$

The boundary conditions are $v = 0$ when $z = 0$ and $(dv/dz) = 0$ when $z = L/2$. From the first of these, $A = e$, while from the second,

$$B = e \tan \frac{\mu L}{2}$$

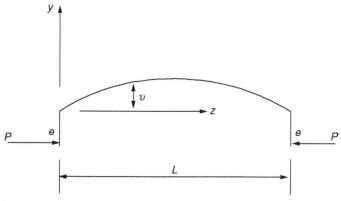

FIGURE 8.10 Eccentrically Loaded Column of Example 8.2.

The equation for the deflected shape of the column is then

$$v = e\left[\frac{\cos\mu(z - L/2)}{\cos\mu L/2} - 1\right]$$

The maximum value of v occurs at mid-span, where $z = L/2$; that is,

$$v_{\max} = e\left(\sec\frac{\mu L}{2} - 1\right)$$

The maximum bending moment is given by

$$M(\max) = Pe + Pv_{\max}$$

so that

$$M(\max) = Pe\sec\frac{\mu L}{2}$$

■

8.4 STABILITY OF BEAMS UNDER TRANSVERSE AND AXIAL LOADS

Stresses and deflections in a linearly elastic beam subjected to transverse loads, as predicted by simple beam theory, are directly proportional to the applied loads. This relationship is valid if the deflections are small, such that the slight change in geometry produced in the loaded beam has an insignificant effect on the loads themselves. This situation changes drastically when axial loads act simultaneously with the transverse loads. The internal moments, shear forces, stresses, and deflections then become dependent upon the magnitude of the deflections as well as the magnitude of the external loads. They are also sensitive, as we observed in the previous section, to beam imperfections, such as initial curvature and eccentricity of axial load. Beams supporting both axial and transverse loads are sometimes known as *beam-columns* or simply as *transversely loaded columns*.

We consider first the case of a pin-ended beam carrying a uniformly distributed load of intensity w per unit length and an axial load P, as shown in Fig. 8.11. The bending moment at any section of the beam is

$$M = Pv + \frac{wlz}{2} - \frac{wz^2}{2} = -EI\frac{d^2v}{dz^2}$$

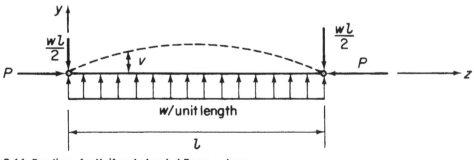

FIGURE 8.11 Bending of a Uniformly Loaded Beam-column

giving

$$\frac{d^2 v}{dz^2} + \frac{P}{EI} v = \frac{w}{2EI}(z^2 - lz) \tag{8.29}$$

The standard solution to Eq. (8.29) is

$$v = A \cos \lambda z + B \sin \lambda z \frac{w}{2P}\left(z^2 - lz - \frac{2}{\lambda^2}\right)$$

where A and B are unknown constants and $\lambda^2 = P/EI$. Substituting the boundary conditions $v = 0$ at $z = 0$ and l gives

$$A = \frac{w}{\lambda^2 P}, \quad B = \frac{w}{\lambda^2 P \sin \lambda l}(l - \cos \lambda l)$$

so that the deflection is determinate for any value of w and P and is given by

$$v = \frac{w}{\lambda^2 P}\left[\cos \lambda z + \left(\frac{1 - \cos \lambda l}{\sin \lambda l}\right)\sin \lambda z\right] + \frac{w}{2P}\left(z^2 - lz - \frac{2}{\lambda^2}\right) \tag{8.30}$$

In beam-columns, as in beams, we are primarily interested in maximum values of stress and deflection. For this particular case, the maximum deflection occurs at the center of the beam and is, after some transformation of Eq. (8.30),

$$v_{\max} = \frac{w}{\lambda^2 P}\left(\sec\frac{\lambda l}{2} - 1\right) - \frac{wl^2}{8P} \tag{8.31}$$

The corresponding maximum bending moment is

$$M_{\max} = -Pv_{\max} - \frac{wl^2}{8}$$

or, from Eq. (8.31),

$$M_{\max} = \frac{w}{\lambda^2}\left(1 - \sec\frac{\lambda l}{2}\right) \tag{8.32}$$

We may rewrite Eq. (8.32) in terms of the Euler buckling load $P_{CR} = \pi^2 EI/l^2$ for a pin-ended column, hence

$$M_{\max} = \frac{wl^2}{\pi^2}\frac{P_{CR}}{P}\left(1 - \sec\frac{\pi}{2}\sqrt{\frac{P}{P_{CR}}}\right) \tag{8.33}$$

As P approaches P_{CR}, the bending moment (and deflection) becomes infinite. However, this theory is based on the assumption of small deflections (otherwise, $d^2 v/dz^2$ is not a close approximation for curvature), so that such a deduction is invalid. The indication is, though, that large deflections are produced by the presence of a compressive axial load no matter how small the transverse load might be.

Let us consider now the beam-column of Fig. 8.12, with hinged ends carrying a concentrated load W at a distance a from the right-hand support. For

$$z \le l - a, \quad EI\frac{d^2 v}{dz^2} = -M = -Pv - \frac{Waz}{l} \tag{8.34}$$

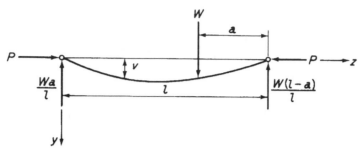

FIGURE 8.12 Beam-Column Supporting A Point Load

and, for

$$z \geq l - a, \quad EI\frac{d^2v}{dz^2} = -M = -Pv - \frac{W}{l}(l-a)(l-z) \tag{8.35}$$

Writing

$$\lambda^2 = \frac{P}{EI}$$

Eq. (8.34) becomes

$$\frac{d^2v}{dz^2} + \lambda^2 v = -\frac{Wa}{EIl}z$$

the general solution of which is

$$v = A\cos\lambda z + B\sin\lambda z - \frac{Wa}{Pl}z \tag{8.36}$$

Similarly, the general solution of Eq. (8.35) is

$$v = C\cos\lambda z + D\sin\lambda z - \frac{W}{Pl}(l-a)(l-z) \tag{8.37}$$

where A, B, C, and D are constants, which are found from the boundary conditions as follows.

When $z = 0$, $v = 0$, therefore, from Eq. (8.36), $A = 0$. At $z = l$, $v = 0$, giving, from Eq. (8.37), $C = -D\tan\lambda l$. At the point of application of the load, the deflection and slope of the beam given by Eqs (8.36) and (8.37) must be the same. Hence, equating deflections,

$$B\sin\lambda(l-a) - \frac{Wa}{Pl}(l-a) = D[\sin\lambda(l-a) - \tan\lambda l\cos\lambda(l-a)] - \frac{Wa}{Pl}(l-a)$$

and, equating slopes,

$$B\lambda\cos\lambda(l-a) - \frac{Wa}{Pl} = D\lambda[\cos\lambda(l-a) - \tan\lambda l\sin\lambda(l-a)] + \frac{W}{Pl}(l-a)$$

Solving these equations for B and D and substituting for A, B, C, and D in Eqs. (8.36) and (8.37), we have

$$v = \frac{W \sin \lambda a}{P\lambda \sin \lambda l} \sin \lambda z - \frac{Wa}{Pl} z, \quad \text{for } z \le l - a \tag{8.38}$$

$$v = \frac{W \sin \lambda(l-a)}{P\lambda \sin \lambda l} \sin \lambda(l-z) - \frac{W}{Pl}(l-a)(l-z), \quad \text{for } z \ge l - a \tag{8.39}$$

These equations for the beam-column deflection enable the bending moment and resulting bending stresses to be found at all sections.

A particular case arises when the load is applied at the center of the span. The deflection curve is then symmetrical with a maximum deflection under the load of

$$v_{max} = \frac{W}{2P\lambda} \tan \frac{\lambda l}{2} - \frac{Wl}{4P}$$

Finally, we consider a beam-column subjected to end moments M_A and M_B in addition to an axial load P (Fig. 8.13). The deflected form of the beam-column may be found by using the principle of superposition and the results of the previous case. First, we imagine that M_B acts alone with the axial load P. If we assume that the point load W moves towards B and simultaneously increases so that the product $Wa = \text{constant} = M_B$, then, in the limit as a tends to zero, we have the moment M_B applied at B. The deflection curve is then obtained from Eq. (8.38) by substituting λa for $\sin \lambda a$ (since λa is now very small) and M_B for Wa:

$$v = \frac{M_B}{P} \left(\frac{\sin \lambda z}{\sin \lambda l} - \frac{z}{l} \right) \tag{8.40}$$

In a similar way, we find the deflection curve corresponding to M_A acting alone. Suppose that W moves toward A such that the product $W(l-a) = \text{constant} = M_A$. Then, as $(l-a)$ tends to zero, we have $\sin \lambda(l-a) = \lambda(l-a)$ and Eq. (8.39) becomes

$$v = \frac{M_A}{P} \left[\frac{\sin \lambda(l-z)}{\sin \lambda l} - \frac{(l-z)}{l} \right] \tag{8.41}$$

The effect of the two moments acting simultaneously is obtained by superposition of the results of Eqs. (8.40) and (8.41). Hence, for the beam-column of Fig. 8.13,

$$v = \frac{M_B}{P} \left(\frac{\sin \lambda z}{\sin \lambda l} - \frac{z}{l} \right) + \frac{M_A}{P} \left[\frac{\sin \lambda(l-z)}{\sin \lambda l} - \frac{(l-z)}{l} \right] \tag{8.42}$$

FIGURE 8.13 Beam-Column Supporting End Moments

Equation (8.42) is also the deflected form of a beam-column supporting eccentrically applied end loads at A and B. For example, if e_A and e_B are the eccentricities of P at the ends A and B, respectively, then $M_A = Pe_A$, $M_B = Pe_B$, giving a deflected form of

$$v = e_B \left(\frac{\sin \lambda z}{\sin \lambda l} - \frac{z}{l} \right) + e_A \left[\frac{\sin \lambda (l - z)}{\sin \lambda l} - \frac{(l - z)}{l} \right] \tag{8.43}$$

Other beam-column configurations featuring a variety of end conditions and loading regimes may be analyzed by a similar procedure.

8.5 ENERGY METHOD FOR THE CALCULATION OF BUCKLING LOADS IN COLUMNS

The fact that the total potential energy of an elastic body possesses a stationary value in an equilibrium state may be used to investigate the neutral equilibrium of a buckled column. In particular, the energy method is extremely useful when the deflected form of the buckled column is unknown and has to be guessed.

First, we consider the pin-ended column shown in its buckled position in Fig. 8.14. The internal or strain energy U of the column is assumed to be produced by bending action alone and is given by the well-known expression

$$U = \int_0^l \frac{M^2}{2EI} \, dz \tag{8.44}$$

or alternatively, since $EI \, d^2v/dz^2 = -M$,

$$U = \frac{EI}{2} \int_0^l \left(\frac{d^2v}{dz^2} \right)^2 dz \tag{8.45}$$

The potential energy V of the buckling load P_{CR}, referred to the straight position of the column as the datum, is then

$$V = -P_{CR}\delta$$

where δ is the axial movement of P_{CR} caused by the bending of the column from its initially straight position. By reference to Fig. 7.15(b) and Eq. (7.41), we see that

$$\delta = \frac{1}{2} \int_0^l \left(\frac{dv}{dz} \right)^2 dz$$

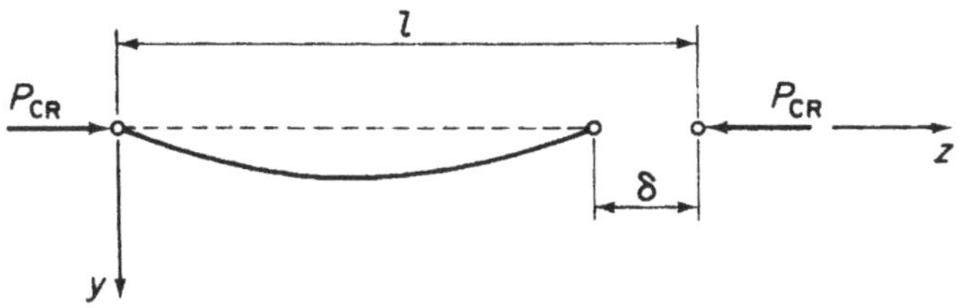

FIGURE 8.14 Shortening of a Column Due to Buckling

giving

$$V = -\frac{P_{CR}}{2} \int_0^l \left(\frac{dv}{dz}\right)^2 dz \tag{8.46}$$

The total potential energy of the column in the neutral equilibrium of its buckled state is therefore

$$U + V = \int_0^l \frac{M^2}{2EI} \, dz - \frac{P_{CR}}{2} \int_0^l \left(\frac{dv}{dz}\right)^2 dz \tag{8.47}$$

or, using the alternative form of U from Eq. (8.45),

$$U + V = \frac{EI}{2} \int_0^l \left(\frac{d^2v}{dz^2}\right)^2 dz - \frac{P_{CR}}{2} \int_0^l \left(\frac{dv}{dz}\right)^2 dz \tag{8.48}$$

We saw in Chapter 7 that exact solutions of plate bending problems are obtainable by energy methods when the deflected shape of the plate is known. An identical situation exists in the determination of critical loads for column and thin-plate buckling modes. For the pin-ended column under discussion, a deflected form of

$$v = \sum_{n=1}^{\infty} A_n \sin \frac{n\pi z}{l} \tag{8.49}$$

satisfies the boundary conditions of

$$(v)_{z=0} = (v)_{z=l} = 0, \quad \left(\frac{d^2v}{dz^2}\right)_{z=0} = \left(\frac{d^2v}{dz^2}\right)_{z=l} = 0$$

and is capable, within the limits for which it is valid and if suitable values for the constant coefficients A_n are chosen, of representing any continuous curve. We are therefore in a position to find P_{CR} exactly. Substituting Eq. (8.49) into Eq. (8.48) gives

$$U + V = \frac{EI}{2} \int_0^l \left(\frac{\pi}{l}\right)^4 \left(\sum_{n=1}^{\infty} n^2 A_n \sin \frac{n\pi z}{l}\right)^2 dz - \frac{P_{CR}}{2} \int_0^l \left(\frac{\pi}{l}\right)^2 \left(\sum_{n=1}^{\infty} n A_n \cos \frac{n\pi z}{l}\right)^2 dz \tag{8.50}$$

The product terms in both integrals of Eq. (8.50) disappear on integration, leaving only integrated values of the squared terms. Thus,

$$U + V = \frac{\pi^4 EI}{4l^3} \sum_{n=1}^{\infty} n^4 A_n^2 - \frac{\pi^2 P_{CR}}{4l} \sum_{n=1}^{\infty} n^2 A_n^2 \tag{8.51}$$

Assigning a stationary value to the total potential energy of Eq. (8.51) with respect to each coefficient A_n in turn, then taking A_n as being typical, we have

$$\frac{\partial (U + V)}{\partial A_n} = \frac{\pi^4 EI n^4 A_n}{2l^3} - \frac{\pi^2 P_{CR} n^2 A_n}{2l} = 0$$

from which

$$P_{CR} = \frac{\pi^2 EI n^2}{l^2}$$

as before.

We see that each term in Eq. (8.49) represents a particular deflected shape with a corresponding critical load. Hence, the first term represents the deflection of the column shown in Fig. 8.14, with $P_{CR} = \pi^2 EI/l^2$. The second and third terms correspond to the shapes shown in Fig. 8.3, having critical loads of $4\pi^2 EI/l^2$ and $9\pi^2 EI/l^2$ and so on. Clearly, the column must be constrained to buckle into these more complex forms. In other words, the column is being forced into an unnatural shape, is consequently stiffer, and offers greater resistance to buckling, as we observe from the higher values of critical load. Such buckling modes, as stated in Section 8.1, are unstable and are generally of academic interest only.

If the deflected shape of the column is known, it is immaterial which of Eqs. (8.47) or (8.48) is used for the total potential energy. However, when only an approximate solution is possible, Eq. (8.47) is preferable, since the integral involving bending moment depends upon the accuracy of the assumed form of v, whereas the corresponding term in Eq. (8.48) depends upon the accuracy of d^2v/dz^2. Generally, for an assumed deflection curve, v is obtained much more accurately than d^2v/dz^2.

Suppose that the deflection curve of a particular column is unknown or extremely complicated. We then assume a reasonable shape which satisfies, as far as possible, the end conditions of the column and the pattern of the deflected shape (Rayleigh–Ritz method). Generally, the assumed shape is in the form of a finite series involving a series of unknown constants and assumed functions of z. Let us suppose that v is given by

$$v = A_1 f_1(z) + A_2 f_2(z) + A_3 f_3(z)$$

Substitution in Eq. (8.47) results in an expression for total potential energy in terms of the critical load and the coefficients A_1, A_2, and A_3 as the unknowns. Assigning stationary values to the total potential energy with respect to A_1, A_2, and A_3 in turn produces three simultaneous equations from which the ratios A_1/A_2, A_1/A_3, and the critical load are determined. Absolute values of the coefficients are unobtainable, since the deflections of the column in its buckled state of neutral equilibrium are indeterminate.

As a simple illustration, consider the column shown in its buckled state in Fig. 8.15. An approximate shape may be deduced from the deflected shape of a tip-loaded cantilever. Thus,

$$v = \frac{v_0 z^2}{2l^3}(3l - z)$$

This expression satisfies the end conditions of deflection, that is, $v = 0$ at $z = 0$ and $v = v_0$ at $z = l$. In addition, it satisfies the conditions that the slope of the column is zero at the built-in end and that

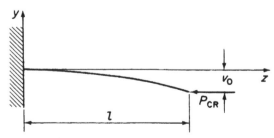

FIGURE 8.15 Buckling Load for a Built-in Column by the Energy Method

the bending moment, that is, d^2v/dz^2, is zero at the free end. The bending moment at any section is $M = P_{CR}(v_0 - v)$, so that substitution for M and v in Eq. (8.47) gives

$$U + V = \frac{P_{CR}^2 v_0^2}{2EI} \int_0^l \left(1 - \frac{3z^2}{2l^2} + \frac{z^3}{2l^3}\right)^2 dz - \frac{P_{CR}}{2} \int_0^l \left(\frac{3v_0}{2l^3}\right)^3 z^2(2l - z)^2 dz$$

Integrating and substituting the limits, we have

$$U + V = \frac{17}{35}\frac{P_{CR}^2 v_0^2 l}{2EI} - \frac{3}{5}P_{CR}\frac{v_0^2}{l} = 0$$

Hence,

$$\frac{\partial(U + V)}{\partial v_0} = \frac{17}{35}\frac{P_{CR}^2 v_0 l}{EI} - \frac{6P_{CR}v_0}{5l} = 0$$

from which

$$P_{CR} = \frac{42EI}{17l^2} = 2.471\frac{EI}{l^2}$$

This value of the critical load compares with the exact value (see Table 8.1) of $\pi^2 EI/4l^2 = 2.467EI/l^2$; the error, in this case, is seen to be extremely small. Approximate values of the critical load obtained by the energy method are always greater than the correct values. The explanation lies in the fact that an assumed deflected shape implies the application of constraints to force the column to take up an artificial shape. This, as we have seen, has the effect of stiffening the column with a consequent increase in critical load.

It will be observed that the solution for this example may be obtained by simply equating the increase in internal energy (U) to the work done by the external critical load ($-V$). This is always the case when the assumed deflected shape contains a single unknown coefficient, such as v_0 in the above example.

8.6 FLEXURAL–TORSIONAL BUCKLING OF THIN-WALLED COLUMNS

It is recommended that the reading of this section be delayed until after Chapter 27 has been studied.

In some instances, thin-walled columns of open cross-section do not buckle in bending as predicted by the Euler theory but twist without bending or bend and twist simultaneously, producing flexural–torsional buckling. The solution to this type of problem relies on the theory presented in Chapter 27 for the torsion of open section beams subjected to warping (axial) restraint. Initially, however, we shall establish a useful analogy between the bending of a beam and the behavior of a pin-ended column.

The bending equation for a simply supported beam carrying a uniformly distributed load of intensity w_y and having Cx and Cy as principal centroidal axes is

$$EI_{xx}\frac{d^4v}{dz^4} = w_y \quad \text{(see Chapter 15)} \tag{8.52}$$

Also, the equation for the buckling of a pin-ended column about the Cx axis is (see Eq. (8.1))

$$EI_{xx}\frac{d^2v}{dz^2} = -P_{CR}v \tag{8.53}$$

Differentiating Eq. (8.53) twice with respect to z gives

$$EI_{xx}\frac{d^4v}{dz^4} = -P_{CR}\frac{d^2v}{dz^2} \tag{8.54}$$

Comparing Eqs. (8.52) and (8.54), we see that the behavior of the column may be obtained by considering it as a simply supported beam carrying a uniformly distributed load of intensity w_y given by

$$w_y = -P_{CR}\frac{d^2v}{dz^2} \tag{8.55}$$

Similarly, for buckling about the Cy axis,

$$w_x = -P_{CR}\frac{d^2u}{dz^2} \tag{8.56}$$

Consider now a thin-walled column having the cross-section shown in Fig. 8.16 and suppose that the centroidal axes Cxy are principal axes (see Chapter 15); $S(x_S, y_S)$ is the shear center of the column (see Chapter 16) and its cross-sectional area is A. Due to the flexural–torsional buckling produced, say, by a compressive axial load P, the cross-section suffers translations u and v parallel to Cx and Cy, respectively, and a rotation θ, positive counterclockwise, about the shear center S. Thus, due to translation, C and S move to C' and S' and, due to rotation about S', C' moves to C''. The total movement of C, u_C, in the x direction is given by

$$u_c = u + C'D = u + C'C'' \sin \alpha \; (S'\hat{C}'C'' \simeq 90°)$$

But

$$C'C'' = C'S'\theta = CS\theta$$

Hence.

$$u_C = u + \theta\, CS \sin \alpha = u + y_S\theta \tag{8.57}$$

Also, the total movement of C in the y direction is

$$v_C = v - DC'' = v - C'C'' \cos \alpha = v - \theta CS \cos \alpha$$

so that

$$v_C = v - x_s\theta \tag{8.58}$$

Since, at this particular cross-section of the column, the centroidal axis has been displaced, the axial load P produces bending moments about the displaced x and y axes given, respectively, by

$$M_x = Pv_C = P(v - x_S\theta) \tag{8.59}$$

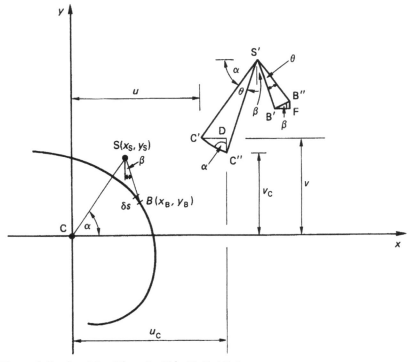

FIGURE 8.16 Flexural–Torsional Buckling of a Thin-Walled Column

and

$$M_y = Pu_C = P(u + y_S\theta) \tag{8.60}$$

From simple beam theory (Chapter 15),

$$EI_{xx}\frac{\mathrm{d}^2 v}{\mathrm{d}z^2} = -M_x = -P(v - x_S\theta) \tag{8.61}$$

and

$$EI_{yy}\frac{\mathrm{d}^2 u}{\mathrm{d}z^2} = -M_y = -P(u + y_S\theta) \tag{8.62}$$

where I_{xx} and I_{yy} are the second moments of area of the cross-section of the column about the principal centroidal axes, E is Young's modulus for the material of the column, and z is measured along the centroidal longitudinal axis.

The axial load P on the column at any cross-section, is distributed as a uniform direct stress σ. Thus, the direct load on any element of length δs at a point $B(x_B, y_B)$ is $\sigma t\,\mathrm{d}s$ acting in a direction parallel to the longitudinal axis of the column. In a similar manner to the movement of C to C″, the point B is displaced to B″. The horizontal movement of B in the x direction is then

$$u_B = u + B'F = u + B'B'' \cos \beta$$

But

$$B'B'' = S'B'\theta = SB\theta$$

Hence,

$$u_B = u + \theta SB \ \cos \beta$$

or

$$u_B = u + (y_S - y_B)\theta \tag{8.63}$$

Similarly, the movement of B in the y direction is

$$v_B = v - (x_S - x_B)\theta \tag{8.64}$$

Therefore, from Eqs. (8.63) and (8.64) and referring to Eqs. (8.55) and (8.56), we see that the compressive load on the element δs at B, $\sigma t \delta s$, is equivalent to lateral loads

$$-\sigma t \delta s \frac{d^2}{dz^2}[u + (y_S - y_B)\theta] \text{ in the } x \text{ direction}$$

and

$$-\sigma t \delta s \frac{d^2}{dz^2}[v - (x_S - x_B)\theta] \text{ in the } y \text{ direction}$$

The lines of action of these equivalent lateral loads do not pass through the displaced position S′ of the shear center and therefore produce a torque about S′ leading to the rotation θ. Suppose that the element δs at B is of unit length in the longitudinal z direction. The torque per unit length of the column $\delta T(z)$ acting on the element at B is then given by

$$\delta T(z) = -\sigma t \delta s \frac{d^2}{dz^2}[u + (y_S - y_B)\theta](y_S - y_B)$$

$$+ \sigma t \delta s \frac{d^2}{dz^2}[v - (x_S - x_B)\theta](x_S - x_B) \tag{8.65}$$

Integrating Eq. (8.65) over the complete cross-section of the column gives the torque per unit length acting on the column; that is,

$$T(z) = -\int_{\text{Sect}} \sigma t \frac{d^2 u}{dz^2}(y_S - y_B)ds - \int_{\text{Sect}} \sigma t (y_S - y_B)^2 \frac{d^2\theta}{dz^2} \ ds$$

$$+\int_{\text{Sect}} \sigma t \frac{d^2 v}{dz^2}(x_S - x_B)ds - \int_{\text{Sect}} \sigma t (x_S - x_B)^2 \frac{d^2\theta}{dz^2} \ ds \tag{8.66}$$

Expanding Eq. (8.66) and noting that σ is constant over the cross-section, we obtain

$$
\begin{aligned}
T(z) = & -\sigma\frac{d^2u}{dz^2}y_S\int_{\text{Sect}}t\,ds + \sigma\frac{d^2u}{dz^2}\int_{\text{Sect}}ty_B\,ds - \sigma\frac{d^2\theta}{dz^2}y_S^2\int_{\text{Sect}}t\,ds \\
& +\sigma\frac{d^2\theta}{dz^2}2y_S\int_{\text{Sect}}ty_B\,ds - \sigma\frac{d^2\theta}{dz^2}\int_{\text{Sect}}ty_B^2\,ds + \sigma\frac{d^2v}{dz^2}x_S\int_{\text{Sect}}t\,ds \\
& -\sigma\frac{d^2v}{dz^2}\int_{\text{Sect}}tx_B\,ds - \sigma\frac{d^2\theta}{dz^2}x_S^2\int_{\text{Sect}}t\,ds + \sigma\frac{d^2\theta}{dz^2}2x_S\int_{\text{Sect}}tx_B\,ds \\
& -\sigma\frac{d^2\theta}{dz^2}\int_{\text{Sect}}tx_B^2\,ds
\end{aligned}
\tag{8.67}
$$

Equation (8.67) may be rewritten

$$
T(z) = P\left(x_S\frac{d^2v}{dz^2} - y_S\frac{d^2u}{dz^2}\right) - \frac{P}{A}\frac{d^2\theta}{dz^2}(Ay_S^2 + I_{xx} + Ax_S^2 + I_{yy})
\tag{8.68}
$$

In Eq. (8.68), the term $I_{xx} + I_{yy} + A(x_S^2 + y_S^2))$ is the polar second moment of area I_0 of the column about the shear center S. Thus, Eq. (8.68) becomes

$$
T(z) = P\left(x_S\frac{d^2v}{dz^2} - y_S\frac{d^2u}{dz^2}\right) - I_0\frac{P}{A}\frac{d^2\theta}{dz^2}
\tag{8.69}
$$

Substituting for $T(z)$ from Eq. (8.69) in Eq. (27.11), the general equation for the torsion of a thin-walled beam, we have

$$
E\Gamma\frac{d^4\theta}{dz^4} - \left(GJ - I_0\frac{P}{A}\right)\frac{d^2\theta}{dz^2} - Px_S\frac{d^2v}{dz^2} + Py_S\frac{d^2u}{dz^2} = 0
\tag{8.70}
$$

Equations (8.61), (8.62), and (8.70) form three simultaneous equations which may be solved to determine the flexural–torsional buckling loads.

As an example, consider the case of a column of length L in which the ends are restrained against rotation about the z axis and against deflection in the x and y directions; the ends are also free to rotate about the x and y axes and are free to warp. Thus, $u = v = \theta = 0$ at $z = 0$ and $z = L$. Also, since the column is free to rotate about the x and y axes at its ends, $M_x = M_y = 0$ at $z = 0$ and $z = L$, and from Eqs. (8.61) and (8.62),

$$
\frac{d^2v}{dz^2} = \frac{d^2u}{dz^2} = 0 \text{ at } z = 0 \text{ and } z = L
$$

Further, the ends of the column are free to warp so that

$$
\frac{d^2\theta}{dz^2} = 0 \text{ at } z = 0 \text{ and } z = L \quad \text{(see Eq. (27.1))}
$$

An assumed buckled shape given by

$$
u = A_1\sin\frac{\pi z}{L}, \quad v = A_2\sin\frac{\pi z}{L} \quad \theta = A_3\sin\frac{\pi z}{L}
\tag{8.71}
$$

in which A_1, A_2, and A_3 are unknown constants, satisfies the preceding boundary conditions. Substituting for u, v, and θ from Eqs. (8.71) into Eqs. (8.61), (8.62), and (8.70), we have

$$\left.\begin{array}{l} \left(P - \dfrac{\pi^2 EI_{xx}}{L^2}\right) A_2 - Px_S A_3 = 0 \\[4mm] \left(P - \dfrac{\pi^2 EI_{yy}}{L^2}\right) A_1 - Py_S A_3 = 0 \\[4mm] Py_S A_1 - Px_S A_2 - \left(\dfrac{\pi^2 E\Gamma}{L^2} + GJ - \dfrac{I_0}{A}P\right) A_3 = 0 \end{array}\right\} \tag{8.72}$$

For nonzero values of A_1, A_2, and A_3, the determinant of Eqs. (8.72) must equal zero that is,

$$\begin{vmatrix} 0 & P - \pi^2 EI_{xx}/L^2 & -Px_S \\ P - \pi^2 EI_{yy}/L^2 & 0 & Py_S \\ Py_S & -Px_S & I_0 P/A - \pi^2 E\Gamma/L^2 - GJ \end{vmatrix} = 0 \tag{8.73}$$

The roots of the cubic equation formed by the expansion of the determinant give the critical loads for the flexural–torsional buckling of the column; clearly, the lowest value is significant.

In the case where the shear center of the column and the centroid of area coincide, that is, the column has a doubly symmetrical cross-section, $x_S = y_S = 0$ and Eqs. (8.61), (8.62), and (8.70) reduce, respectively, to

$$EI_{xx}\frac{d^2 v}{dz^2} = -Pv \tag{8.74}$$

$$EI_{yy}\frac{d^2 u}{dz^2} = -Pu \tag{8.75}$$

$$E\Gamma\frac{d^4\theta}{dz^4} = \left(GJ - I_0\frac{P}{A}\right)\frac{d^2\theta}{dz^2} = 0 \tag{8.76}$$

Equations (8.74), (8.75), and (8.76), unlike Eqs. (8.61), (8.62), and (8.70), are uncoupled and provide three separate values of buckling load. Thus, Eqs. (8.74) and (8.75) give values for the Euler buckling loads about the x and y axes, respectively, while Eq. (8.76) gives the axial load that produces pure torsional buckling; clearly the buckling load of the column is the lowest of these values. For the column whose buckled shape is defined by Eqs. (8.71), substitution for v, u, and θ in Eqs. (8.74), (8.75), and (8.76), respectively, gives

$$P_{CR(xx)} = \frac{\pi^2 EI_{xx}}{L^2} \quad P_{CR(yy)} = \frac{\pi^2 EI_{yy}}{L^2} \quad P_{CR(\theta)} = \frac{A}{I_0}\left(GJ + \frac{\pi^2 E\Gamma}{L^2}\right) \tag{8.77}$$

Example 8.3

A thin-walled pin-ended column is 2 m long and has the cross-section shown in Fig. 8.17. If the ends of the column are free to warp, determine the lowest value of axial load which causes buckling and specify the buckling mode. Take $E = 75,000$ N/mm^2 and $G = 21,000$ N/mm^2. See Ex. 1.1.

Since the cross-section of the column is doubly symmetrical, the shear center coincides with the centroid of area and $x_S = y_S = 0$; Eqs. (8.74), (8.75), and (8.76) therefore apply. Further, the boundary conditions are those of the column whose buckled shape is defined by Eqs. (8.71), so that the buckling load of the column is the lowest of the three values given by Eqs. (8.77).

The cross-sectional area A of the column is

$$A = 2.5(2 \times 37.5 + 75) = 375 \text{ mm}^2$$

The second moments of area of the cross-section about the centroidal axes Cxy are (see Chapter 15), respectively,

$$I_{xx} = 2 \times 37.5 \times 2.5 \times 37.5^2 + 2.5 \times 75^3/12 = 3.52 \times 10^5 \text{ mm}^4$$
$$I_{yy} = 2 \times 2.5 \times 37.5^3/12 = 0.22 \times 10^5 \text{ mm}^4$$

The polar second moment of area I_0 is

$$I_0 = I_{xx} + I_{yy} + A(x_S^2 + y_S^2) \quad \text{(see the derivation of Eq. (8.69))}$$

that is

$$I_0 = 3.52 \times 10^5 + 0.22 \times 10^5 = 3.74 \times 10^5 \text{ mm}^4$$

The torsion constant J is obtained using Eq. (18.11), which gives

$$J = 2 \times 37.5 \times 2.5^3/3 + 75 \times 2.5^3/3 = 781.3 \text{ mm}^4$$

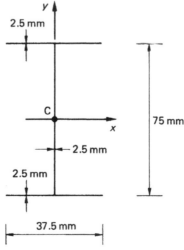

FIGURE 8.17 Column Section of Example 8.3

Finally, Γ is found using the method of Section 27.2 and is

$$\Gamma = 2.5 \times 37.5^3 \times 75^2/24 = 30.9 \times 10^6 \text{ mm}^6$$

Substituting these values in Eqs. (8.77), we obtain

$$P_{CR(xx)} = 6.5 \times 10^4 \text{N}, \quad P_{CR(yy)} = 0.41 \times 10^4 \text{N}, \quad P_{CR(\theta)} = 2.22 \times 10^4 \text{N}$$

The column therefore buckles in bending about the Cy axis when subjected to an axial load of 0.41×10^4 N.

Equation (8.73) for the column whose buckled shape is defined by Eqs. (8.71) may be rewritten in terms of the three separate buckling loads given by Eqs. (8.77):

$$\begin{vmatrix} 0 & P - P_{CR(xx)} & -Px_S \\ P - P_{CR(yy)} & 0 & Py_S \\ Py_S & -Px_S & I_0(P - P_{CR(\theta)})/A \end{vmatrix} = 0 \tag{8.78}$$

If the column has, say, Cx as an axis of symmetry, then the shear center lies on this axis and $y_S = 0$. Equation (i) thereby reduces to

$$\begin{vmatrix} P - P_{CR(xx)} & -Px_S \\ -Px_S & I_0 - (P - P_{CR(\theta)})/A \end{vmatrix} = 0 \tag{8.79}$$

The roots of the quadratic equation formed by expanding Eq. (8.79) are the values of axial load which produce flexural–torsional buckling about the longitudinal and x axes. If $P_{CR(yy)}$ is less than the smallest of these roots the column will buckle in pure bending about the y axis.

Example 8.4

A column of length 1 m has the cross-section shown in Fig. 8.18. If the ends of the column are pinned and free to warp, calculate its buckling load; $E = 70,000 \text{ N/mm}^2$, $G = 30,000 \text{ N/mm}^2$. See Ex. 1.1.

In this case, the shear center S is positioned on the Cx axis, so that $y_S = 0$ and Eq. (8.79) applies. The distance $\bar{x}$ of the centroid of area C from the web of the section is found by taking first moments of area about the web:

$$2(100 + 100 + 100)\bar{x} = 2 \times 2 \times 100 \times 50$$

which gives

$$\bar{x} = 33.3 \text{ mm}$$

The position of the shear center S is found using the method of Example 17.1; this gives $x_S = -76.2$ mm. The remaining section properties are found by the methods specified in Example 8.3 and follow:

$$A = 600 \text{ mm}^2, \quad I_{xx} = 1.17 \times 10^6 \text{ mm}^4, \quad I_{yy} = 0.67 \times 10^6 \text{ mm}^4,$$
$$I_0 = 5.35 \times 10^6 \text{mm}^4, \quad J = 800 \text{ mm}^4, \quad \Gamma = 2488 \times 10^6 \text{mm}^6$$

From Eq. (8.77),

$$P_{CR(yy)} = 4.63 \times 10^5 \text{N}, \quad P_{CR(xx)} = 8.08 \times 10^5 \text{N}, \quad P_{CR(\theta)} = 1.97 \times 10^5 \text{N}$$

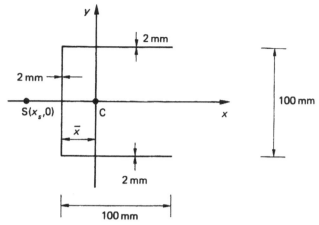

FIGURE 8.18 Column Section of Example 8.4

Expanding Eq. (8.79),

$$(P - P_{CR(xx)})(P - P_{CR(\theta)})I_0/A - P^2 x_S^2 = 0 \qquad \text{(i)}$$

Rearranging Eq. (i),

$$P^2(1 - A x_S^2/I_0) - P(P_{CR(xx)} + P_{CR(\theta)}) + P_{CR(xx)}P_{CR(\theta)} = 0 \qquad \text{(ii)}$$

Substituting the values of the constant terms in Eq. (ii), we obtain

$$P^2 - 29.13 \times 10^5 P + 46.14 \times 10^{10} = 0 \qquad \text{(iii)}$$

The roots of Eq. (iii) give two values of critical load, the lowest of which is

$$P = 1.68 \times 10^5 \text{N}$$

It can be seen that this value of flexural–torsional buckling load is lower than any of the uncoupled buckling loads $P_{CR(xx)}$, $P_{CR(yy)}$, or $P_{CR(\theta)}$; the reduction is due to the interaction of the bending and torsional buckling modes.

Example 8.4 MATLAB

Repeat Example 8.4 using the MATLAB and the calculated section properties. See Ex. 1.1.

The value of the critical buckling load is obtained through the following MATLAB file:

```
% Declare any needed variables
syms P
L = 1000;
t = 2;
E = 70000;
G = 30000;
```

```
y_s = 0;
x_s = -76.2;
A = 600;
I_xx = 1.17e6;
I_yy = 0.67e6;
I_0 = 5.32e6;
J = 800;
T = 2488e6;

% Evaluate Eq.(8.77)
P_CRyy = pi^2*E*I_yy/L^2;
P_CRxx = pi^2*E*I_xx/L^2;
P_CRtheta = A*(G*J + pi^2*E*T/L^2)/I_0;

% Substitute results of Eq. (8.77) into Eq.(8.79)
eq_I = det([P-P_CRxx -P*x_s; -P*x_s I_0*(P-P_CRtheta)/A]);

% Solve eq_I for the critical buckling load (P)
P = solve(eq_I,P);

% Output the minimum value of P to the Command Window
disp(['P =' num2str(min(double(P))) 'N'])
```

The Command Window output resulting from this MATLAB file is as follows:

```
P = 167785.096 N
```

Example 8.5

A thin-walled column has the cross-section shown in Fig. 8.19, is of length L, and is subjected to an axial load through its shear center S. If the ends of the column are prevented from warping and twisting, determine the value of direct stress when failure occurs due to torsional buckling. See Ex. 1.1.

The torsion bending constant Γ is found using the method described in Section 27.2. The position of the shear center is given but is obvious by inspection. The swept area $2\Gamma A_{R,0}$ is determined as a function of s, and its distribution is shown in Fig. 8.20. The center of gravity of the "wire: is found by taking moments about the s axis.

Then,

$$2A'_R 5td = td\left(\frac{d^2}{2} + \frac{5d^2}{4} + \frac{3d^2}{2} + \frac{5d^2}{4} + \frac{d^2}{2}\right)$$

which gives

$$2A'_R = d^2$$

The torsion bending constant is then the "moment of inertia" of the "wire" and is

$$\Gamma = 2td\frac{1}{3}(d^2)^2 + \frac{td}{3}\left(\frac{d^2}{2}\right)^2 \times 2 + td\left(\frac{d^2}{2}\right)^2$$

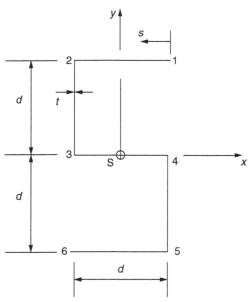

FIGURE 8.19 Section of Column of Example 8.5

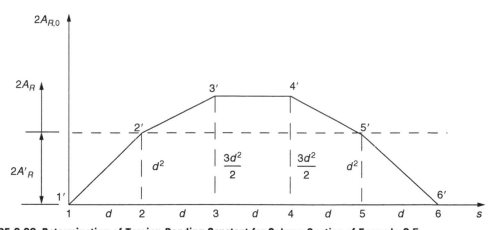

FIGURE 8.20 Determination of Torsion Bending Constant for Column Section of Example 8.5

from which

$$\Gamma = \frac{13}{12}td^5$$

Also, the torsion constant J is given by (see Section 3.4)

$$J = \sum \frac{st^3}{3} = \frac{5dt^3}{3}$$

The shear center of the section and the centroid of area coincide, so that the torsional buckling load is given by Eq. (8.76). Rewriting this equation,

$$\frac{d^4\theta}{dz^4} + \mu^2 \frac{d^2\theta}{dz^2} = 0 \tag{i}$$

where

$$\mu^2 = (\sigma I_0 - GJ)/E\Gamma, \quad (\sigma = P/A)$$

The solution of Eq. (i) is

$$\theta = A\cos\mu z + B\sin\mu z + Cz + D \tag{ii}$$

The boundary conditions are $\theta = 0$ when $z = 0$ and $z = L$, and since the warping is suppressed at the ends of the beam,

$$\frac{d\theta}{dz} = 0, \quad \text{when } z = 0 \ \text{ and } z = L \quad (\text{see Eq.}(18.19))$$

Putting $\theta = 0$ at $z = 0$ in Eq. (ii)

$$0 = A + D$$

or

$$A = -D$$

Also,

$$\frac{d\theta}{dz} = -\mu A\sin\mu z + \mu B\cos\mu z + C$$

and, since $(d\theta/dz) = 0$ at $z = 0$,

$$C = -\mu B$$

When $z = L$, $\theta = 0$, so that, from Eq. (ii),

$$0 = A\cos\mu L + B\sin\mu L + CL + D$$

which may be rewritten

$$0 = B(\sin\mu L - \mu L) + A(\cos\mu L - 1) \tag{iii}$$

Then, for $(d\theta/dz) = 0$ at $z = L$,

$$0 = \mu B\cos\mu L - \mu A\sin\mu L - \mu B$$

or

$$0 = B(\cos\mu L - 1) - A\sin\mu L \tag{iv}$$

Eliminating A from Eqs. (iii) and (iv),

$$0 = B[2(1 - \cos\mu L) - \mu L\sin\mu L] \tag{v}$$

Similarly, in terms of the constant C,

$$0 = -C[2(1 - \cos\mu L) - \mu L\sin\mu L] \tag{vi}$$

or

$$B = -C$$

But $B = -C/\mu$, so that to satisfy both equations, $B = C = 0$ and

$$\theta = A \cos \mu z - A = A(\cos \mu z - 1) \qquad \text{(vii)}$$

Since $\theta = 0$ at $z = l$,

$$\cos \mu L = 1$$

or

$$\mu L = 2n\pi$$

Therefore,

$$\mu^2 L^2 = 4n^2 \pi^2$$

or

$$\frac{\sigma I_0 - GJ}{E\Gamma} = \frac{4n^2 \pi^2}{L^2}$$

The lowest value of torsional buckling load corresponds to $n = 1$, so that, rearranging the preceding,

$$\sigma = \frac{1}{I_0}\left(GJ + \frac{4\pi^2 E\Gamma}{L^2}\right) \qquad \text{(viii)}$$

The polar second moment of area I_0 is given by

$$I_0 = I_{xx} + I_{yy} \quad \text{(see Megson}^2)$$

that is,

$$I_0 = 2\left(td\, d^2 + \frac{td^3}{3}\right) + \frac{3td^3}{12} + 2td\frac{d^2}{4}$$

which gives

$$I_0 = \frac{41}{12}td^3$$

Substituting for I_0, J, and Γ in Eq. (viii),

$$\sigma = \frac{4}{41d^2}\left(5Gt^2 + \frac{13\pi^2 Ed^4}{L^2}\right)$$

∎

References

[1] Timoshenko SP, Gere JM. Theory of Elastic Stability. 2nd ed. New York: McGraw-Hill; 1961.
[2] Megson THG. Structural and Stress Analysis. 2nd ed. Oxford: Elsevier; 2005.

PROBLEMS

P.8.1. The system shown in Fig. P.8.1 consists of two bars AB and BC, each of bending stiffness EI elastically hinged together at B by a spring of stiffness K (i.e., bending moment applied by spring $= K \times$ change in slope across B). Regarding A and C as simple pin joints, obtain an equation for the first buckling load of the system. What are the lowest buckling loads when (a) $K \to \infty$, (b) $EI \to \infty$. Note that B is free to move vertically.

Answer: $\mu K / \tan \mu l$.

P.8.2. A pin-ended column of length l and constant flexural stiffness EI is reinforced to give a flexural stiffness $4EI$ over its central half (see Fig. P.8.2). Considering symmetric modes of buckling only, obtain the equation whose roots yield the flexural buckling loads and solve for the lowest buckling load.

Answer: $\tan \mu l / 8 = 1/\sqrt{2}, \quad P = 24.2 EI / l^2$

P.8.3. A uniform column of length l and bending stiffness EI is built-in at one end and free at the other and has been designed so that its lowest flexural buckling load is P (see Fig. P.8.3). Subsequently it has to carry an increased load, and for this it is provided with a lateral spring at the free end. Determine the necessary spring stiffness k so that the buckling load becomes $4P$.

Answer: $k = 4P\mu / (\mu l - \tan \mu l)$.

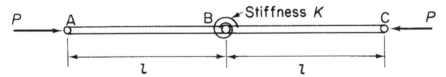

FIGURE P.8.1

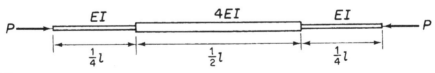

FIGURE P.8.2

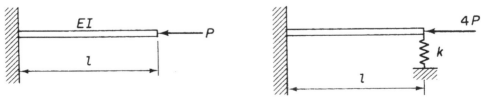

FIGURE P.8.3

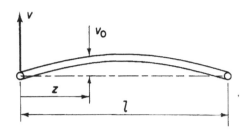

FIGURE P.8.4

P.8.4. A uniform, pin-ended column of length l and bending stiffness EI has an initial curvature such that the lateral displacement at any point between the column and the straight line joining its ends is given by

$$v_0 = a\frac{4z}{l^2}(l - z) \quad \text{(see Fig. P.8.4)}$$

Show that the maximum bending moment due to a compressive end load P is given by

$$M_{\max} = -\frac{8aP}{(\lambda l)^2}\left(\sec\frac{\lambda l}{2} - 1\right)$$

where

$$\lambda^2 = P/EI$$

P.8.5. The uniform pin-ended column shown in Fig. P.8.5 is bent at the center so that its eccentricity there is δ. If the two halves of the column are otherwise straight and have a flexural stiffness EI, find the value of the maximum bending moment when the column carries a compression load P.

Answer: $-P\dfrac{2\delta}{l}\sqrt{\dfrac{EI}{P}}\tan\sqrt{\dfrac{P}{EI}}\dfrac{l}{2}$.

P.8.6. A straight uniform column of length l and bending stiffness EI is subjected to uniform lateral loading w/unit length. The end attachments do not restrict rotation of the column ends. The longitudinal compressive force P has eccentricity e from the centroids of the end sections and is placed so as to oppose the bending effect of the lateral loading, as shown in Fig. P.8.6. The eccentricity e can be varied

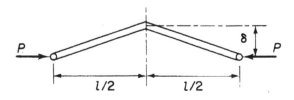

FIGURE P.8.5

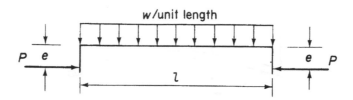

FIGURE P.8.6

and is to be adjusted to the value which, for given values of P and w, will result in the least maximum bending moment on the column. Show that

$$e = (w/P\mu^2)\tan^2\mu l/4$$

where

$$\mu^2 = P/EI$$

Deduce the end moment that gives the optimum condition when P tends to zero.

Answer: $wl^2/16$.

P.8.7. The relation between stress σ and strain ε in compression for a certain material is

$$10.5 \times 10^6\varepsilon = \sigma + 21{,}000\left(\frac{\sigma}{49{,}000}\right)^{16}$$

Assuming the tangent modulus equation to be valid for a uniform strut of this material, plot the graph of σ_b against l/r, where σ_b is the flexural buckling stress, l the equivalent pin-ended length, and r the least radius of gyration of the cross-section. Estimate the flexural buckling load for a tubular strut of this material, of 1.5 units outside diameter and 0.08 units wall thickness with effective length 20 units.

Answer: 14,454 force units

P.8.8. A rectangular portal frame ABCD is rigidly fixed to a foundation at A and D and is subjected to a compression load P applied at each end of the horizontal member BC (see Fig. P.8.8). If all the members have the same bending stiffness EI, show that the buckling loads for modes which are symmetrical about the vertical center line are given by the transcendental equation

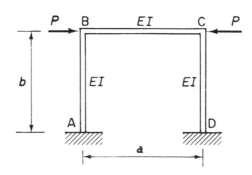

FIGURE P.8.8

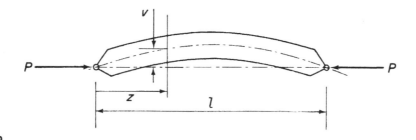

FIGURE P.8.9

$$\frac{\lambda a}{2} = -\frac{1}{2}\left(\frac{a}{b}\right)\tan\left(\frac{\lambda a}{2}\right)$$

where

$$\lambda^2 = P/EI$$

P.8.9. A compression member (Fig. P.8.9) is made of circular section tube, diameter d, thickness t. The member is not perfectly straight when unloaded, having a slightly bowed shape which may be represented by the expression

$$v = \delta\sin\left(\frac{\pi z}{l}\right)$$

Show that when the load P is applied, the maximum stress in the member can be expressed as

$$\sigma_{max} = \frac{P}{\pi dt}\left[1 + \frac{1}{1-\alpha}\frac{4\delta}{d}\right]$$

where

$$\alpha = P/P_e, \quad P_e = \pi^2 EI/l^2$$

Assume t is small compared with d, so that the following relationships are applicable:

Cross-sectional area of tube $= \pi dt$
Second moment of area of tube $= \pi d^3 t/8$

P.8.10. Figure P.8.10 illustrates an idealized representation of part of an aircraft control circuit. A uniform, straight bar of length a and flexural stiffness EI is built-in at the end A and hinged at B to a link BC, of length b, whose other end C is pinned, so that it is free to slide along the line ABC between smooth, rigid guides. A, B, and C are initially in a straight line and the system carries a compression force P, as shown. Assuming that the link BC has a sufficiently high flexural stiffness to prevent its buckling as a pin-ended strut, show, by setting up and solving the differential equation for flexure of AB, that buckling of the system, of the type illustrated in Fig. P.8.10, occurs when P has such a value that

$$\tan \lambda a = \lambda(a + b)$$

where

$$\lambda^2 = P/EI$$

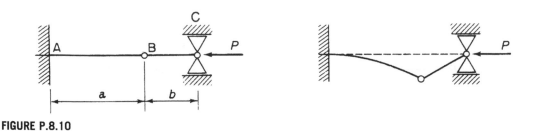

FIGURE P.8.10

P.8.11. A pin-ended column of length l has its central portion reinforced, the second moment of its area being I_2, while that of the end portions, each of length a, is I_1. Use the energy method to determine the critical load of the column, assuming that its centerline deflects into the parabola $v = kz(l - z)$ and taking the more accurate of the two expressions for the bending moment. In the case where $I_2 = 1.6I_1$ and $a = 0.2\,l$, find the percentage increase in strength due to the reinforcement, and compare it with the percentage increase in weight on the basis that the radius of gyration of the section is not altered.

Answer: $P_{CR} = 14.96EI_1/l^2, 52$ percent, 36 percent

P.8.11 MATLAB Use the MATLAB to repeat Problem P.8.11 for the following relations of I_1 and I_2.

	(i)	(ii)	(iii)	(iv)	(v)
I_2	$1.4\,I_1$	$1.5\,I_1$	$1.6\,I_1$	$1.7\,I_1$	$1.8\,I_1$

Answer: (i) $P_{CR} = 13.38EI_1/l^2, 36\%, 24\%$
(ii) $P_{CR} = 14.18EI_1/l^2, 44\%, 30\%$
(iii) $P_{CR} = 14.96EI_1/l^2, 52\%, 36\%$
(iv) $P_{CR} = 15.72EI_1/l^2, 59\%, 42\%$
(v) $P_{CR} = 16.47EI_1/l^2, 67\%, 48\%$

P.8.12. A tubular column of length l is tapered in wall thickness so that the area and the second moment of area of its cross-section decrease uniformly from A_1 and I_1 at its center to $0.2A_1$ and $0.2I_1$ at its ends. Assuming a deflected center-line of parabolic form and taking the more correct form for the bending moment, use the energy method to estimate its critical load when tested between pin-centers, in terms of the preceding data and Young's modulus E. Hence, show that the saving in weight by using such a column instead of one having the same radius of gyration and constant thickness is about 15 percent.

Answer: $7.01EI_1/l^2$

P.8.13. A uniform column (Fig. P.8.13), of length l and bending stiffness EI, is rigidly built-in at the end $z = 0$ and simply supported at the end $z = l$. The column is also attached to an elastic foundation of constant stiffness k/unit length. Representing the deflected shape of the column by a polynomial

$$v = \sum_{n=0}^{p} a_n \eta^n, \quad \text{where } \eta = z/l$$

determine the form of this function by choosing a minimum number of terms p such that all the kinematic (geometric) and static boundary conditions are satisfied, allowing for one arbitrary constant only.

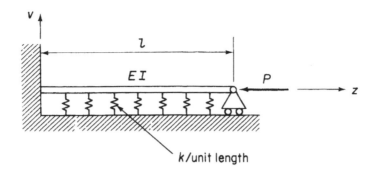

FIGURE P.8.13

Using the result thus obtained, find an approximation to the lowest flexural buckling load P_{CR} by the Rayleigh–Ritz method.

Answer: $P_{CR} = 21.05EI/l^2 + 0.09kl^2$

P.8.14. Figure P.8.14 shows the doubly symmetrical cross-section of a thin-walled column with rigidly fixed ends. Find an expression, in terms of the section dimensions and Poisson's ratio, for the column length for which the purely flexural and the purely torsional modes of instability occur at the same axial load. In which mode does failure occur if the length is less than the value found? The possibility of local instability is to be ignored.

Answer: $l = (2\pi b^2/t)\sqrt{(1+v)/255}$,   torsion

P.8.15. A column of length $2l$ with the doubly symmetric cross-section, shown in Fig. P.8.15, is compressed between the parallel platens of a testing machine which fully prevents twisting and warping of the ends. Using the data that follows, determine the average compressive stress at which the column first buckles in torsion:

$l = 500$ mm, $b = 25.0$ mm, $t - 2.5$ mm, $E = 70,000$ N/mm^2, $E/G = 2.6$

Answer: $\sigma_{CR} = 282$ N/mm^2.

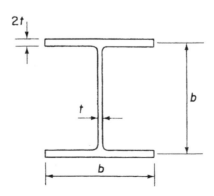

FIGURE P.8.14

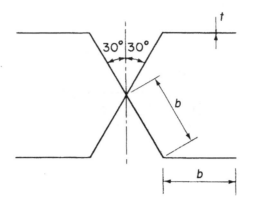

FIGURE P.8.15

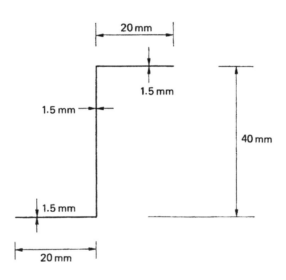

FIGURE P.8.16

P.8.16. A pin-ended column of length 1.0 m has the cross-section shown in Fig. P.8.16. If the ends of the column are free to warp, determine the lowest value of axial load which will causes the column to buckle and specify the mode. Take $E = 70,000$ N/mm^2 and $G = 25,000$ N/mm^2.

Answer: 5,527 N. The column buckles in bending about an axis in the plane of its web.

P.8.16. MATLAB Use MATLAB to repeat Problem P.8.16 for the following column lengths.

	(i)	(ii)	(iii)	(iv)	(v)	(vi)
L	.5 m	1 m	1.5 m	2 m	2.5 m	3 m

Answer: (i)   22108 N
           (ii) 5527 N
           (iii) 2456 N
           (iv) 1382 N
           (v)   884 N
           (vi) 614 N

The column buckles in bending about the y axis (plane of the web) for all selected values of L.

P.8.17. A pin-ended column of height 3.0 m has a circular cross-section of diameter 80 mm, wall thickness 2.0 mm, and is converted to an open section by a narrow longitudinal slit; the ends of the column are free to warp. Determine the values of axial load which would cause the column to buckle in (a) pure bending and (b) pure torsion. Hence, determine the value of the flexural–torsional buckling load. Take $E = 70,000$ N/mm^2 and $G = 22,000$ N/mm^2. *Note*: the position of the shear center of the column section may be found using the method described in Chapter 16.

Answer: (a) 3.09×10^4 N,   (b) 1.78×10^4 N, 1.19×10^4 N

Thin plates

We shall see in Chapter 11, when we examine the structural components of aircraft, that they consist mainly of thin plates stiffened by arrangements of ribs and stringers. Thin plates under relatively small compressive loads are prone to buckle and so must be stiffened to prevent this. The determination of buckling loads for thin plates in isolation is relatively straightforward, but when stiffened by ribs and stringers, the problem becomes complex and frequently relies on an empirical solution. In fact, the stiffeners may buckle before the plate and, depending on their geometry, may buckle as a column or suffer local buckling of, say, a flange.

In this chapter, we shall present the theory for the determination of buckling loads of flat plates and examine some of the different empirical approaches various researchers have suggested. In addition, we investigate the particular case of flat plates which, when reinforced by horizontal flanges and vertical stiffeners, form the spars of aircraft wing structures; these are known as *tension field beams*.

9.1 BUCKLING OF THIN PLATES

A thin plate may buckle in a variety of modes, depending upon its dimensions, the loading, and the method of support. Usually, however, buckling loads are much lower than those likely to cause failure in the material of the plate. The simplest form of buckling arises when compressive loads are applied to simply supported opposite edges and the unloaded edges are free, as shown in Fig. 9.1. A thin plate in this configuration behaves in exactly the same way as a pin-ended column, so that the critical load is that predicted by the Euler theory. Once this critical load is reached, the plate is incapable of supporting any further load. This is not the case, however, when the unloaded edges are supported against displacement out of the xy plane. Buckling, for such plates, takes the form of a bulging displacement of the central region of the plate while the parts adjacent to the supported edges remain straight. These parts enable the plate to resist higher loads; an important factor in aircraft design.

At this stage, we are not concerned with this postbuckling behavior but rather with the prediction of the critical load which causes the initial bulging of the central area of the plate. For the analysis, we may conveniently employ the method of total potential energy, since we already, in Chapter 7, derived expressions for strain and potential energy corresponding to various load and support configurations. In these expressions, we assumed that the displacement of the plate comprises bending deflections only and that these are small in comparison with the thickness of the plate. These restrictions therefore apply in the subsequent theory.

First, we consider the relatively simple case of the thin plate of Fig. 9.1, loaded as shown, but simply supported along all four edges. We saw in Chapter 7 that its true deflected shape may be represented by the infinite double trigonometrical series

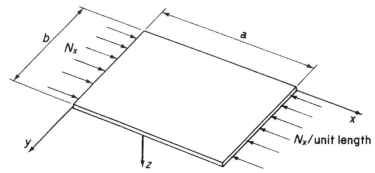

FIGURE 9.1 Buckling of a Thin Flat Plate

$$w = \sum_{m=1}^{\infty} \sum_{n=1}^{\infty} A_{mn} \sin\frac{m\pi x}{a} \sin\frac{n\pi y}{b}$$

Also, the total potential energy of the plate is, from Eqs. (7.37) and (7.45),

$$U + V = \frac{1}{2} \int_0^a \int_0^b \left[D\left\{ \left(\frac{\partial^2 w}{\partial x^2} + \frac{\partial^2 w}{\partial y^2}\right)^2 - 2(1-v)\left[\frac{\partial^2 w}{\partial x^2}\frac{\partial^2 w}{\partial y^2} - \left(\frac{\partial^2 w}{\partial x \partial y}\right)^2\right]\right\} - N_x\left(\frac{\partial w}{\partial x}\right)^2 \right] dx\, dy \tag{9.1}$$

The integration of Eq. (9.1) on substituting for w is similar to those integrations carried out in Chapter 7. Thus, by comparison with Eq. (7.47),

$$U + V = \frac{\pi^4 abD}{8} \sum_{m=1}^{\infty} \sum_{n=1}^{\infty} A_{mn}^2 \left(\frac{m^2}{a^2} + \frac{n^2}{b^2}\right) - \frac{\pi^2 b}{8a} N_x \sum_{m=1}^{\infty} \sum_{n=1}^{\infty} m^2 A_{mn}^2 \tag{9.2}$$

The total potential energy of the plate has a stationary value in the neutral equilibrium of its buckled state (i.e., $N_x = N_{x,CR}$). Therefore, differentiating Eq. (9.2) with respect to each unknown coefficient A_{mn}, we have

$$\frac{\partial(U+V)}{\partial A_{mn}} = \frac{\pi^4 abD}{4} A_{mn}\left(\frac{m^2}{a^2} + \frac{n^2}{b^2}\right)^2 - \frac{\pi^2 b}{4a} N_{x,CR} m^2 A_{mn} = 0$$

and, for a nontrivial solution,

$$N_{x,CR} = \pi^2 a^2 D \frac{1}{m^2} \left(\frac{m^2}{a^2} + \frac{n^2}{b^2}\right)^2 \tag{9.3}$$

Exactly the same result may have been deduced from Eq. (ii) of Example 7.3, where the displacement w becomes infinite for a negative (compressive) value of N_x equal to that of Eq. (9.3).

We observe, from Eq. (9.3), that each term in the infinite series for displacement corresponds, as in the case of a column, to a different value of critical load (note, the problem is an eigenvalue problem). The lowest value of critical load evolves from some critical combination of integers m and n, that is, the number of half-waves in the x and y directions, and the plate dimensions. Clearly $n = 1$ gives a minimum value, so that no matter what the values of m, a, and b, the plate buckles into a half sine wave in the y direction. Thus, we may write Eq. (9.3) as

$$N_{x,\text{CR}} = \pi^2 a^2 D \frac{1}{m^2} \left(\frac{m^2}{a^2} + \frac{1}{b^2} \right)^2$$

or

$$N_{x,\text{CR}} = \frac{k \pi^2 D}{b^2} \tag{9.4}$$

where the plate *buckling coefficient* k is given by the minimum value of

$$k = \left(\frac{mb}{a} + \frac{a}{mb} \right)^2 \tag{9.5}$$

for a given value of a/b. To determine the minimum value of k for a given value of a/b, we plot k as a function of a/b for different values of m, as shown by the dotted curves in Fig. 9.2. The minimum value of k is obtained from the lower envelope of the curves shown solid in the figure.

It can be seen that m varies with the ratio a/b and that k and the buckling load are a minimum when $k = 4$ at values of $a/b = 1, 2, 3, \ldots$. As a/b becomes large k approaches 4, so that long narrow plates tend to buckle into a series of squares.

The transition from one buckling mode to the next may be found by equating values of k for the m and $m + 1$ curves. Hence,

$$\frac{mb}{a} + \frac{a}{mb} = \frac{(m+1)b}{a} + \frac{a}{(m+1)b}$$

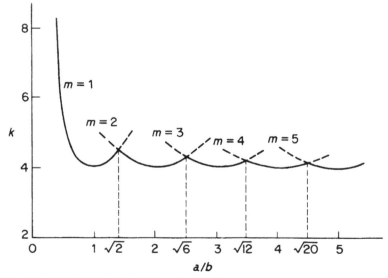

FIGURE 9.2 Buckling Coefficient k for Simply Supported Plates

giving

$$\frac{a}{b} = \sqrt{m(m+1)}$$

Substituting $m = 1$, we have $a/b = \sqrt{2} = 1.414$, and for $m = 2$, $a/b = \sqrt{6} = 2.45$, and so on.

For a given value of a/b, the critical stress, $\sigma_{CR} = N_{x,CR}/t$, is found from Eqs. (9.4) and (7.4); that is,

$$\sigma_{CR} = \frac{k\pi^2 E}{12(1-v^2)} \left(\frac{t}{b}\right)^2 \tag{9.6}$$

In general, the critical stress for a uniform rectangular plate, with various edge supports and loaded by constant or linearly varying in-plane direct forces (N_x, N_y) or constant shear forces (N_{xy}) along its edges, is given by Eq. (9.6). The value of k remains a function of a/b but depends also upon the type of loading and edge support. Solutions for such problems have been obtained by solving the appropriate differential equation or by using the approximate (Rayleigh–Ritz) energy method. Values of k for a variety of loading and support conditions are shown in Fig. 9.3. In Fig. 9.3(c), where k becomes the *shear buckling coefficient*, b is always the smaller dimension of the plate.

We see from Fig. 9.3 that k is very nearly constant for $a/b > 3$. This fact is particularly useful in aircraft structures where longitudinal stiffeners are used to divide the skin into narrow panels (having small values of b), thereby increasing the buckling stress of the skin.

9.2 INELASTIC BUCKLING OF PLATES

For plates having small values of b/t, the critical stress may exceed the elastic limit of the material of the plate. In such a situation, Eq. (9.6) is no longer applicable, since, as we saw in the case of columns, E becomes dependent on stress, as does Poisson's ratio v. These effects are usually included in a plasticity correction factor η, so that Eq. (9.6) becomes

$$\sigma_{CR} = \frac{\eta k\pi^2 E}{12(1-v^2)} \left(\frac{t}{b}\right)^2 \tag{9.7}$$

where E and v are elastic values of Young's modulus and Poisson's ratio. In the linearly elastic region, $\eta = 1$, which means that Eq. (9.7) may be applied at all stress levels. The derivation of a general expression for η is outside the scope of this book, but one[1] giving good agreement with experiment is

$$\eta = \frac{1-v_e^2}{1-v_p^2} \frac{E_s}{E} \left[\frac{1}{2} + \frac{1}{2}\left(\frac{1}{4} + \frac{3}{4}\frac{E_t}{E_s}\right)^{\frac{1}{2}}\right]$$

where E_t and E_s are the tangent modulus and secant modulus (stress/strain) of the plate in the inelastic region and v_e and v_p are Poisson's ratio in the elastic and inelastic ranges.

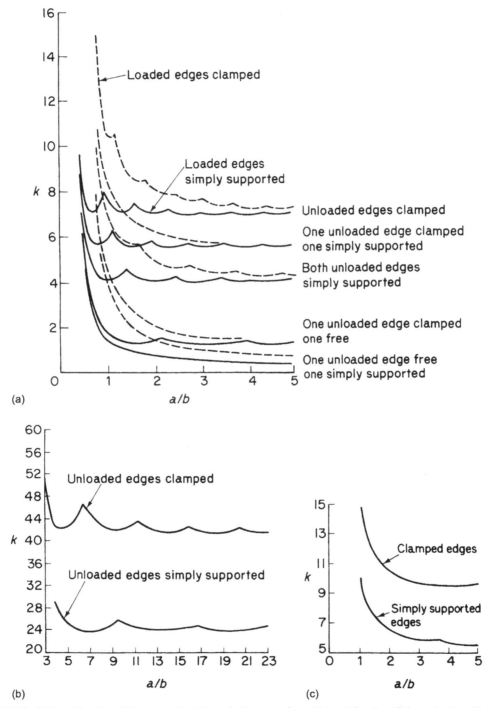

FIGURE 9.3 (a) Buckling Coefficients for Flat Plates in Compression; (b) Buckling Coefficients for Flat Plates in Bending; (c) Shear Buckling Coefficients for Flat Plates

9.3 EXPERIMENTAL DETERMINATION OF THE CRITICAL LOAD FOR A FLAT PLATE

In Section 8.3, we saw that the critical load for a column may be determined experimentally, without actually causing the column to buckle, by means of the Southwell plot. The critical load for an actual, rectangular, thin plate is found in a similar manner.

The displacement of an initially curved plate from the zero load position was found, in Section 7.5, to be

$$w_1 = \sum_{m=1}^{\infty} \sum_{n=1}^{\infty} B_{mn} \sin \frac{m\pi x}{a} \sin \frac{n\pi y}{b}$$

where

$$B_{mn} = \frac{A_{mn} N_x}{\dfrac{\pi^2 D}{a^2} \left(m + \dfrac{n^2 a^2}{mb^2} \right)^2 - N_x}$$

We see that the coefficients B_{mn} increase with an increase of compressive load intensity N_x. It follows that, when N_x approaches the critical value, $N_{x,\text{CR}}$, the term in the series corresponding to the buckled shape of the plate becomes the most significant. For a square plate, $n = 1$ and $m = 1$ give a minimum value of critical load, so that at the center of the plate

$$w_1 = \frac{A_{11} N_x}{N_{x,\text{CR}} - N_x}$$

or, rearranging,

$$w_1 = N_{x,\text{CR}} \frac{w_1}{N_x} - A_{11}$$

Thus, a graph of w_1 plotted against w_1/N_x has a slope, in the region of the critical load, equal to $N_{x,\text{CR}}$.

9.4 LOCAL INSTABILITY

We distinguished in the introductory remarks to Chapter 8 between primary and secondary (or local) instability. The latter form of buckling usually occurs in the flanges and webs of thin-walled columns having an effective slenderness ratio, $l_e/r < 20$. For $l_e/r > 80$, this type of column is susceptible to primary instability. In the intermediate range of l_e/r between 20 and 80, buckling occurs by a combination of both primary and secondary modes.

Thin-walled columns are encountered in aircraft structures in the shape of longitudinal stiffeners, which are normally fabricated by extrusion processes or by forming from a flat sheet. A variety of cross-sections are employed although each is usually composed of flat plate elements arranged to form angle, channel, Z, or "top hat" sections, as shown in Fig. 9.4. We see that the plate elements fall into two distinct categories: flanges, which have a free unloaded edge, and webs, which are supported by the adjacent plate elements on both unloaded edges.

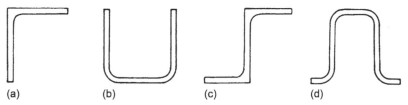

FIGURE 9.4 (a) Extruded Angle; (b) Formed Channel; (c) Extruded Z; (d) Formed "Top Hat"

In local instability, the flanges and webs buckle like plates, with a resulting change in the cross-section of the column. The wavelength of the buckle is of the order of the widths of the plate elements, and the corresponding critical stress is generally independent of the length of the column when the length is equal to or greater than three times the width of the largest plate element in the column cross-section.

Buckling occurs when the weakest plate element, usually a flange, reaches its critical stress, although in some cases all the elements reach their critical stresses simultaneously. When this occurs, the rotational restraint provided by adjacent elements to each other disappears and the elements behave as though they are simply supported along their common edges. These cases are the simplest to analyze and are found where the cross-section of the column is an equal-legged angle, T, cruciform, or a square tube of constant thickness. Values of local critical stress for columns possessing these types of section may be found using Eq. (9.7) and an appropriate value of k. For example, k for a cruciform section column is obtained from Fig. 9.3(a), for a plate which is simply supported on three sides with one edge free and has $a/b > 3$. Hence, $k = 0.43$; and if the section buckles elastically, then $\eta = 1$ and

$$\sigma_{CR} = 0.388E \left(\frac{t}{b}\right)^2, \quad (v = 0.3)$$

It must be appreciated that the calculation of local buckling stresses is generally complicated, with no particular method gaining universal acceptance, much of the information available being experimental. A detailed investigation of the topic is therefore beyond the scope of this book. Further information may be obtained from all the references listed at the end of this chapter.

9.5 INSTABILITY OF STIFFENED PANELS

It is clear from Eq. (9.7) that plates having large values of b/t buckle at low values of critical stress. An effective method of reducing this parameter is to introduce stiffeners along the length of the plate thereby dividing a wide sheet into a number of smaller and more stable plates. Alternatively, the sheet may be divided into a series of wide short columns by stiffeners attached across its width. In the former type of structure, the longitudinal stiffeners carry part of the compressive load, while in the latter, all the load is supported by the plate. Frequently, both methods of stiffening are combined to form a grid-stiffened structure.

Stiffeners in earlier types of stiffened panel possessed a relatively high degree of strength compared with the thin skin, resulting in the skin buckling at a much lower stress level than the stiffeners. Such panels may be analyzed by assuming that the stiffeners provide simply supported edge conditions to a series of flat plates.

A more efficient structure is obtained by adjusting the stiffener sections so that buckling occurs in both stiffeners and skin at about the same stress. This is achieved by a construction involving closely spaced stiffeners of comparable thickness to the skin. Since their critical stresses are nearly the same there is an appreciable interaction at buckling between skin and stiffeners so that the complete panel must be considered as a unit. However, caution must be exercised, since it is possible for the two simultaneous critical loads to interact and reduce the actual critical load of the structure[2] (see Example 8.4). Various modes of buckling are possible, including primary buckling, where the wavelength is of the order of the panel length, and local buckling, with wavelengths of the order of the width of the plate elements of the skin or stiffeners. A discussion of the various buckling modes of panels having Z-section stiffeners has been given by Argyris and Dunne.[3]

The prediction of critical stresses for panels with a large number of longitudinal stiffeners is difficult and relies heavily on approximate (energy) and semi-empirical methods. Bleich[4] and Timoshenko (see Ref. 1, Chapter 8) give energy solutions for plates with one and two longitudinal stiffeners and also consider plates having a large number of stiffeners. Gerard and Becker[5] summarize much of the work on stiffened plates, and a large amount of theoretical and empirical data is presented by Argyris and Dunne in the *Handbook of Aeronautics.*[3]

For detailed work on stiffened panels, reference should be made to as much as possible of the preceding work. The literature is extensive, however, so that here we present a relatively simple approach suggested by Gerard.[1] Figure 9.5 represents a panel of width w stiffened by longitudinal members which may be flats (as shown), Z, I, channel, or "top hat" sections. It is possible for the panel to behave as an Euler column, its cross-section being that shown in Fig. 9.5. If the equivalent length of the panel acting as a column is l_e, then the Euler critical stress is

$$\sigma_{CR,E} = \frac{\pi^2 E}{(l_e/r)^2}$$

as in Eq. (8.8). In addition to the column buckling mode, individual plate elements constituting the panel cross-section may buckle as long plates. The buckling stress is then given by Eq. (9.7); that is,

$$\sigma_{CR} = \frac{\eta k \pi^2 E}{12(1 - \nu^2)} \left(\frac{t}{b}\right)^2$$

where the values of k, t, and b depend upon the particular portion of the panel being investigated. For example, the portion of skin between stiffeners may buckle as a plate simply supported on all

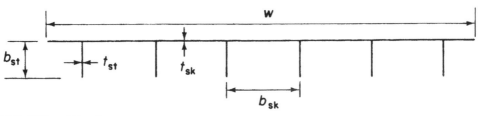

FIGURE 9.5 Stiffened Panel

four sides. Thus, for $a/b > 3$, $k = 4$ from Fig. 9.3(a), and, assuming that buckling takes place in the elastic range,

$$\sigma_{CR} = \frac{4\pi^2 E}{12(1 - v^2)} \left(\frac{t_{sk}}{b_{sk}}\right)^2$$

A further possibility is that the stiffeners may buckle as long plates simply supported on three sides with one edge free. Thus,

$$\sigma_{CR} = \frac{0.43\pi^2 E}{12(1 - v^2)} \left(\frac{t_{st}}{b_{st}}\right)^2$$

Clearly, the minimum value of these critical stresses is the critical stress for the panel taken as a whole.

The compressive load is applied to the panel over its complete cross-section. To relate this load to an applied compressive stress σ_A acting on each element of the cross-section, we divide the load per unit width, say N_x, by an equivalent skin thickness $\bar{t}$, hence,

$$\sigma_A = \frac{N_x}{\bar{t}}$$

where

$$\bar{t} = \frac{A_{st}}{b_{sk}} + t_{sk}$$

and A_{st} is the stiffener area.

The above remarks are concerned with the primary instability of stiffened panels. Values of local buckling stress have been determined by Boughan, Baab, and Gallaher for idealized web, Z, and T stiffened panels. The results are reproduced in Rivello[6] together with the assumed geometries.

Further types of instability found in stiffened panels occur where the stiffeners are riveted or spot welded to the skin. Such structures may be susceptible to *interrivet buckling*, in which the skin buckles between rivets with a wavelength equal to the rivet pitch, or *wrinkling*, where the stiffener forms an elastic line support for the skin. In the latter mode, the wavelength of the buckle is greater than the rivet pitch and separation of skin and stiffener does not occur. Methods of estimating the appropriate critical stresses are given in Rivello[6] and the *Handbook of Aeronautics*.[3]

9.6 FAILURE STRESS IN PLATES AND STIFFENED PANELS

The previous discussion on plates and stiffened panels investigated the prediction of buckling stresses. However, as we have seen, plates retain some of their capacity to carry load, even though a portion of the plate has buckled. In fact, the ultimate load is not reached until the stress in the majority of the plate exceeds the elastic limit. The theoretical calculation of the ultimate stress is difficult, since non-linearity results from both large deflections and the inelastic stress–strain relationship.

Gerard[1] proposes a semi-empirical solution for flat plates supported on all four edges. After elastic buckling occurs, theory and experiment indicate that the average compressive stress, $\overline{\sigma}_a$, in the plate and the unloaded edge stress, σ_e, are related by the following expression:

$$\frac{\overline{\sigma}_a}{\sigma_{CR}} = \alpha_1 \left(\frac{\sigma_e}{\sigma_{CR}} \right)^n \tag{9.8}$$

where

$$\sigma_{CR} = \frac{k\pi^2 E}{12(1 - v^2)} \left(\frac{t}{b} \right)^2$$

and α_1 is some unknown constant. Theoretical work by Stowell[7] and Mayers and Budiansky[8] shows that failure occurs when the stress along the unloaded edge is approximately equal to the compressive yield strength, σ_{cy}, of the material. Hence, substituting σ_{cy} for σ_e in Eq. (9.8) and rearranging gives

$$\frac{\overline{\sigma}_f}{\sigma_{cy}} = \alpha_1 \left(\frac{\sigma_{CR}}{\sigma_{cy}} \right)^{1-n} \tag{9.9}$$

where the average compressive stress in the plate has become the average stress at failure $\overline{\sigma}_f$. Substituting for σ_{CR} in Eq. (9.9) and putting

$$\frac{\alpha_1 \pi^{2(1-n)}}{[12(1 - v^2)]^{1-n}} = \alpha$$

yields

$$\frac{\overline{\sigma}_f}{\sigma_{cy}} = \alpha k^{1-n} \left[\frac{t}{b} \left(\frac{E}{\sigma_{cy}} \right)^{\frac{1}{2}} \right]^{2(1-n)} \tag{9.10}$$

or, in a simplified form,

$$\frac{\overline{\sigma}_f}{\sigma_{cy}} = \beta \left[\frac{t}{b} \left(\frac{E}{\sigma_{cy}} \right)^{\frac{1}{2}} \right]^m \tag{9.11}$$

where $\beta = \alpha k^{m/2}$. The constants β and m are determined by the best fit of Eq. (9.11) to test data.

Experiments on simply supported flat plates and square tubes of various aluminum and magnesium alloys and steel show that $\beta = 1.42$ and $m = 0.85$ fit the results within ± 10 percent up to the yield strength. Corresponding values for long clamped flat plates are $\beta = 1.80$, $m = 0.85$.

Gerard[9–12] extended this method to the prediction of local failure stresses for the plate elements of thin-walled columns. Equation (9.11) becomes

$$\frac{\overline{\sigma}_f}{\sigma_{cy}} = \beta_g \left[\left(\frac{gt^2}{A} \right) \left(\frac{E}{\sigma_{cy}} \right)^{\frac{1}{2}} \right]^m \tag{9.12}$$

where A is the cross-sectional area of the column, β_g and m are empirical constants, and g is the number of cuts required to reduce the cross-section to a series of flanged sections plus the number of flanges that would exist after the cuts are made. Examples of the determination of g are shown in Fig. 9.6.

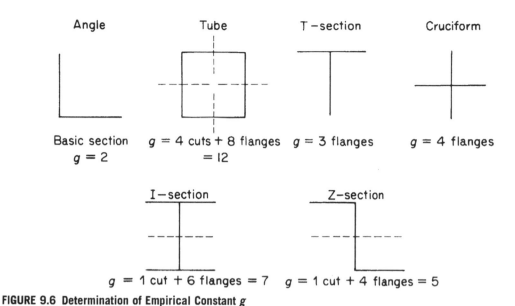

FIGURE 9.6 Determination of Empirical Constant g

The local failure stress in longitudinally stiffened panels was determined by Gerard[10,12] using a slightly modified form of Eqs. (9.11) and (9.12). Thus, for a section of the panel consisting of a stiffener and a width of skin equal to the stiffener spacing,

$$\frac{\overline{\sigma}_f}{\sigma_{cy}} = \beta_g \left[\frac{g t_{sk} t_{st}}{A} \left(\frac{E}{\overline{\sigma}_{cy}} \right)^{\frac{1}{2}} \right]^m \tag{9.13}$$

where t_{sk} and t_{st} are the skin and stiffener thicknesses, respectively. A weighted yield stress $\overline{\sigma}_{cy}$ is used for a panel in which the material of the skin and stiffener have different yield stresses, thus,

$$\overline{\sigma}_{cy} = \frac{\sigma_{cy} + \sigma_{cy,sk}[(\overline{t}/t_{st}) - 1]}{\overline{t}/t_{st}}$$

where $\overline{t}$ is the average or equivalent skin thickness previously defined. The parameter g is obtained in a similar manner to that for a thin-walled column, except that the number of cuts in the skin and the number of equivalent flanges of the skin are included. A cut to the left of a stiffener is not counted, since it is regarded as belonging to the stiffener to the left of that cut. The calculation of g for two types of skin/stiffener combination is illustrated in Fig. 9.7. Equation (9.13) is applicable to either monolithic or built-up panels when, in the latter case, interrivet buckling and wrinkling stresses are greater than the local failure stress.

The values of failure stress given by Eqs. (9.11), (9.12), and (9.13) are associated with local or secondary instability modes. Consequently, they apply when $l_e/r \leq 20$. In the intermediate range between the local and primary modes, failure occurs through a combination of both. At the moment, no theory satisfactorily predicts failure in this range, and we rely on test data and empirical methods. The NACA

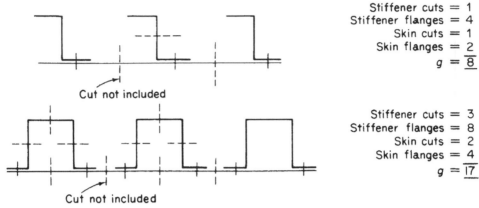

FIGURE 9.7 Determination of g for Two Types of Stiffener–Skin Combination

(now NASA) produced direct reading charts for the failure of "top hat," Z, and Y section stiffened panels; a bibliography of the results is given by Gerard.[10]

It must be remembered that research into methods of predicting the instability and postbuckling strength of the thin-walled types of structure associated with aircraft construction is a continuous process. Modern developments include the use of the computer-based finite element technique (see Chapter 6) and the study of the sensitivity of thin-walled structures to imperfections produced during fabrication; much useful information and an extensive bibliography is contained in Murray.[2]

9.7 TENSION FIELD BEAMS

The spars of aircraft wings usually comprise an upper and a lower flange connected by thin, stiffened webs. These webs are often of such a thickness that they buckle under shear stresses at a fraction of their ultimate load. The form of the buckle is shown in Fig. 9.8(a), where the web of the beam buckles under the action of internal diagonal compressive stresses produced by shear, leaving a wrinkled web capable of supporting diagonal tension only in a direction perpendicular to that of the buckle; the beam is then said to be a *complete tension field beam*.

9.7.1 Complete diagonal tension

The theory presented here is due to H. Wagner.

The beam shown in Fig. 9.8(a) has concentrated flange areas having a depth d between their centroids and vertical stiffeners spaced uniformly along the length of the beam. It is assumed that the flanges resist the internal bending moment at any section of the beam while the web, of thickness t, resists the vertical shear force. The effect of this assumption is to produce a uniform shear stress distribution through the depth of the web (see Section 19.3) at any section. Therefore, at a section of the beam where the shear force is S, the shear stress τ is given by

$$\tau = \frac{S}{td} \tag{9.14}$$

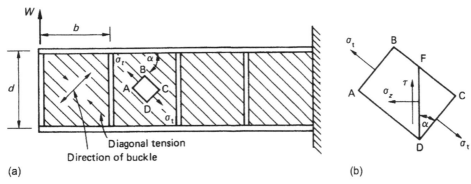

FIGURE 9.8 Diagonal Tension Field Beam

Consider now an element ABCD of the web in a panel of the beam, as shown in Fig. 9.8(a). The element is subjected to tensile stresses, σ_t, produced by the diagonal tension on the planes AB and CD; the angle of the diagonal tension is α. On a vertical plane FD in the element, the shear stress is τ and the direct stress σ_z. Now, considering the equilibrium of the element FCD (Fig. 9.8(b)) and resolving forces vertically, we have (see Section 1.6)

$$\sigma_t CDt \sin \alpha = \tau FDt$$

which gives

$$\sigma_t = \frac{\tau}{\sin \alpha \cos \alpha} = \frac{2\tau}{\sin 2\alpha} \tag{9.15}$$

or, substituting for τ from Eq. (9.14) and noting that, in this case, $S = W$ at all sections of the beam,

$$\sigma_t = \frac{2W}{td \sin 2\alpha} \tag{9.16}$$

Further, resolving forces horizontally for the element FCD,

$$\sigma_z FDt = \sigma_t CDt \cos \alpha$$

which gives

$$\sigma_z = \sigma_t \cos^2 \alpha$$

or, substituting for σ_t from Eq. (9.15),

$$\sigma_z = \frac{\tau}{\tan \alpha} \tag{9.17}$$

or, for this particular beam, from Eq. (9.14),

$$\sigma_z = \frac{W}{td \tan \alpha} \tag{9.18}$$

Since τ and σ_t are constant through the depth of the beam, it follows that σ_z is constant through the depth of the beam.

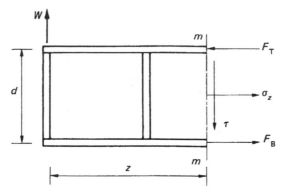

FIGURE 9.9 Determination of Flange Forces

The direct loads in the flanges are found by considering a length z of the beam, as shown in Fig. 9.9. On the plane mm, direct and shear stresses σ_z and τ are acting in the web, together with direct loads F_T and F_B in the top and bottom flanges, respectively. F_T and F_B are produced by a combination of the bending moment Wz at the section and the compressive action (σ_z) of the diagonal tension. Taking moments about the bottom flange,

$$Wz = F_T d - \frac{\sigma_z t d^2}{2}$$

Hence, substituting for σ_z from Eq. (9.18) and rearranging,

$$F_T = \frac{Wz}{d} + \frac{W}{2 \tan \alpha} \qquad (9.19)$$

Now, resolving forces horizontally,

$$F_B - F_T + \sigma_z t d = 0$$

which gives, on substituting for σ_z and F_T from Eqs. (9.18) and (9.19),

$$F_B = \frac{Wz}{d} - \frac{W}{2 \tan \alpha} \qquad (9.20)$$

The diagonal tension stress σ_t induces a direct stress σ_y on horizontal planes at any point in the web. Then, on a horizontal plane HC in the element ABCD of Fig. 9.8, there is a direct stress σ_y and a complementary shear stress τ, as shown in Fig. 9.10.

From a consideration of the vertical equilibrium of the element HDC we have

$$\sigma_y \text{HC} t = \sigma_t \text{CD} t \sin \alpha$$

which gives

$$\sigma_y = \sigma_t \sin^2 \alpha$$

Substituting for σ_t from Eq. (9.15),

$$\sigma_y = \tau \tan \alpha \qquad (9.21)$$

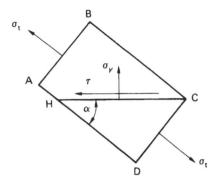

FIGURE 9.10 Stress System on a Horizontal Plane in the Beam Web

or, from Eq. (9.14), in which $S = W$,

$$\sigma_y = \frac{W}{td} \tan \alpha \tag{9.22}$$

The tensile stresses σ_y on horizontal planes in the web of the beam cause compression in the vertical stiffeners. Each stiffener may be assumed to support half of each adjacent panel in the beam, so that the compressive load P in a stiffener is given by

$$P = \sigma_y tb$$

which becomes, from Eq. (9.22),

$$P = \frac{Wb}{d} \tan \alpha \tag{9.23}$$

If the load P is sufficiently high, the stiffeners buckle. Tests indicate that they buckle as columns of equivalent length

$$\left. \begin{array}{ll} l_e = d/\sqrt{4 - 2b/d}, & \text{for } b \leq 1.5d \\ l_e = d, & \text{for } b \geq 1.5d \end{array} \right\} \tag{9.24}$$

In addition to causing compression in the stiffeners, the direct stress σ_y produces bending of the beam flanges between the stiffeners, as shown in Fig. 9.11. Each flange acts as a continuous beam carrying a uniformly distributed load of intensity $\sigma_y t$. The maximum bending moment in a continuous beam with ends fixed against rotation occurs at a support and is $wL^2/12$, in which w is the load intensity and L the beam span. In this case, therefore, the maximum bending moment M_{max} occurs at a stiffener and is given by

$$M_{max} = \frac{\sigma_y t b^2}{12}$$

or, substituting for σ_y from Eq. (9.22),

$$M_{max} = \frac{W b^2 \tan \alpha}{12d} \tag{9.25}$$

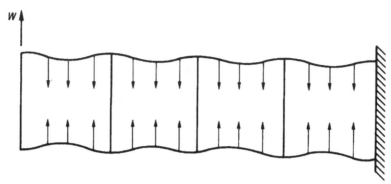

FIGURE 9.11 Bending of Flanges Due to Web Stress

Midway between the stiffeners this bending moment reduces to $Wb^2 \tan \alpha / 24d$.

The angle α adjusts itself such that the total strain energy of the beam is a minimum. If it is assumed that the flanges and stiffeners are rigid, then the strain energy comprises the shear strain energy of the web only and $\alpha = 45°$. In practice, both flanges and stiffeners deform, so that α is somewhat less than 45°, usually of the order of 40° and, in the type of beam common to aircraft structures, rarely below 38°. For beams having all components made of the same material, the condition of minimum strain energy leads to various equivalent expressions for α, one of which is

$$\tan^2 \alpha = \frac{\sigma_t + \sigma_F}{\sigma_t + \sigma_S} \tag{9.26}$$

in which σ_F and σ_S are the uniform direct *compressive* stresses induced by the diagonal tension in the flanges and stiffeners, respectively. Thus, from the second term on the right-hand side of either Eq. (9.19) or (9.20),

$$\sigma_F = \frac{W}{2A_F \tan \alpha} \tag{9.27}$$

in which A_F is the cross-sectional area of each flange. Also, from Eq. (9.23),

$$\sigma_S = \frac{Wb}{A_S d} \tan \alpha \tag{9.28}$$

where A_S is the cross-sectional area of a stiffener. Substitution of σ_t from Eq. (9.16) and σ_F and σ_S from Eqs. (9.27) and (9.28) into Eq. (9.26) produces an equation which may be solved for α. An alternative expression for α, again derived from a consideration of the total strain energy of the beam, is

$$\tan^4 \alpha = \frac{1 + td/2A_F}{1 + tb/A_S} \tag{9.29}$$

Example 9.1

The beam shown in Fig. 9.12 is assumed to have a complete tension field web. If the cross-sectional areas of the flanges and stiffeners are, respectively, 350 mm^2 and 300 mm^2 and the elastic section modulus of each flange is 750 mm^3, determine the maximum stress in a flange and also whether or not the stiffeners buckle. The thickness of the web is 2 mm and the second moment of area of a stiffener about an axis in the plane of the web is 2,000 mm^4; $E = 70,000$ N/mm^2. See Ex. 1.1.

From Eq. (9.29),

$$\tan^4 \alpha = \frac{1 + 2 \times 400/(2 \times 350)}{1 + 2 \times 300/300} = 0.7143$$

so that

$$\alpha = 42.6°$$

The maximum flange stress occurs in the top flange at the built-in end, where the bending moment on the beam is greatest and the stresses due to bending and diagonal tension are additive. Therefore, from Eq. (9.19),

$$F_T = \frac{5 \times 1200}{400} + \frac{5}{2 \tan 42.6°}$$

that is,

$$F_T = 17.7 \text{ kN}$$

Hence, the direct stress in the top flange produced by the externally applied bending moment and the diagonal tension is $17.7 \times 10^3/350 = 50.6$ N/mm^2. In addition to this uniform compressive stress, local bending of the type shown in Fig. 9.11 occurs. The local bending moment in the top flange at the built-in end is found using Eq. (9.25):

$$M_{\max} = \frac{5 \times 10^3 \times 300^2 \tan 42.6°}{12 \times 400} = 8.6 \times 10^4 \text{N mm}$$

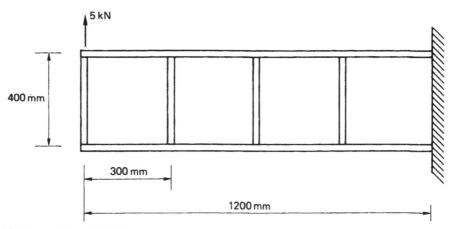

FIGURE 9.12 Beam of Example 9.1

The maximum compressive stress corresponding to this bending moment occurs at the lower extremity of the flange and is $8.6 \times 10^4/750 = 114.9$ N/mm^2. Thus, the maximum stress in the flange occurs on the inside of the top flange at the built-in end of the beam, is compressive, and is equal to $114.9 + 50.6 = 165.5$ N/mm^2.

The compressive load in a stiffener is obtained using Eq. (9.23):

$$P = \frac{5 \times 300 \tan 42.6°}{400} = 3.4 \text{ kN}$$

Since, in this case, $b < 1.5d$, the equivalent length of a stiffener as a column is given by the first of Eqs. (9.24):

$$l_e = 400/\sqrt{4 - 2 \times 300/400} = 253 \text{ mm}$$

From Eq. (8.7), the buckling load of a stiffener is then

$$P_{CR} = \frac{\pi^2 \times 70,000 \times 2,000}{253^2} = 22.0 \text{ kN}$$

Clearly, the stiffener does not buckle.

Example 9.1: MATLAB

Repeat Example 9.1 using MATLAB. See Ex. 1.1.

The maximum stress in a flange and the determination whether or not the stiffener buckles is obtained through the following MATLAB file:

```
% Declare any needed variables
t = 2;
d = 400;
b = 300;
A_F = 350;
A_S = 300;
L = 1200;
E = 70000;
S = 750;
I = 2000;
W = 5;

% Solve Eq. (9.29) for the angle of diagonal tension (alpha)
alpha = atan((((1+t*d/(2*A_F))/(1+t*b/A_S))^(1/4));

% The maximum flange stress and bending moment will occur in the top of the
% flange at the built-in end where the stresses due to bending and diagonal
% tension are additive.

% Evaluate Eq. (9.19) in the top of the flange at the built-in end
z = L;
F_T = W*z/d + 5/(2*tan(alpha)); % Eq. (9.19)
```

```
% Calculate the direct stress in the top flange produced by the externally
% applied bending moment
sig_d = F_T/A_F;

% Calculate the local bending moment in the top flange using Eq. (9.25)
M_max = W*b^2*tan(alpha)/(12*d);

% Calculate the corresponding maximum compressive stress due to M_max
sig_M = M_max/S;

% Calculate the combined maximum stress in a flange
sig_max = sig_d + sig_M;

% Calculate the compressive load in a stiffener using Eq. (9.23)
P = W*b*tan(alpha)/d;

% Substitute P into Eq. (9.24) to calculate the equivalent length of a
% stiffener as a column
if b < 1.5*d
l_e = d/sqrt(4 - 2*b/d);
else
l_e = d;
end

% Calculate the buckling load of a stiffener using Eq. (8.7)
P_CR = pi^2*E*I/(l_e^2);

% Check if the stiffener will buckle
if P_CR > P
 disp('The stiffener will not buckle')
else
 disp('The stiffener will buckle')
end

% Output the combined max stress to the Command Window
disp(['sig_max =' char(vpa(sig_max*1000,4)) 'N/mm^2'])
disp(['P_CR =' num2str(round(P_CR/1000)) 'kN'])
```

The Command Window output resulting from this MATLAB file is as follows:

```
The stiffener will not buckle
sig_max = 165.5 N/mm^2
P_CR = 22 kN
```

In Eqs. (9.28) and (9.29,) it is implicitly assumed that a stiffener is fully effective in resisting axial load. This is the case if the centroid of area of the stiffener lies in the plane of the beam web. Such a situation arises when the stiffener consists of two members symmetrically arranged on opposite sides of the web. In the case where the web is stiffened by a single member attached to one side, the compressive load P is offset from the stiffener axis, thereby producing bending in addition to axial load.

For a stiffener having its centroid a distance e from the center of the web, the combined bending and axial compressive stress, σ_c, at a distance e from the stiffener centroid, is

$$\sigma_c = \frac{P}{A_S} + \frac{Pe^2}{A_S r^2}$$

in which r is the radius of gyration of the stiffener cross-section about its neutral axis (note, second moment of area $I = Ar^2$). Then,

$$\sigma_c = \frac{P}{A_S}\left[1 + \left(\frac{e}{r}\right)^2\right]$$

or

$$\sigma_c = \frac{P}{A_{S_e}}$$

where

$$A_{S_e} = \frac{A_S}{1 + (e/r)^2} \tag{9.30}$$

and is termed the *effective stiffener area*.

9.7.2 Incomplete diagonal tension

In modern aircraft structures, beams having extremely thin webs are rare. They retain, after buckling, some of their ability to support loads, so that even near failure, they are in a state of stress somewhere between that of pure diagonal tension and the pre-buckling stress. Such a beam is described as an *incomplete diagonal tension field beam* and may be analyzed by semi-empirical theory as follows.

It is assumed that the nominal web shear $\tau\ (= S/td)$ may be divided into a "true shear" component τ_S and a diagonal tension component τ_{DT} by writing

$$\tau_{DT} = k\tau, \quad \tau_S = (1 - k)\tau \tag{9.31}$$

where k, the *diagonal tension factor*, is a measure of the degree to which the diagonal tension is developed. A completely unbuckled web has $k = 0$, whereas $k = 1$ for a web in complete diagonal tension. The value of k corresponding to a web having a critical shear stress τ_{CR} is given by the empirical expression

$$k = \tanh\left(0.5\log\frac{\tau}{\tau_{CR}}\right) \tag{9.32}$$

The ratio τ/τ_{CR} is known as the *loading ratio* or *buckling stress ratio*. The buckling stress τ_{CR} may be calculated from the formula

$$\tau_{CR,elastic} = k_{SS}E\left(\frac{t}{b}\right)^2\left[R_d + \frac{1}{2}(R_b - R_d)\left(\frac{b}{d}\right)^3\right] \tag{9.33}$$

where k_{ss} is the coefficient for a plate with simply supported edges and R_d and R_b are empirical restraint coefficients for the vertical and horizontal edges of the web panel, respectively. Graphs giving k_{ss}, R_d, and R_b are reproduced in Kuhn.[13]

The stress equations (9.27) and (9.28) are modified in the light of these assumptions and may be rewritten in terms of the applied shear stress τ as

$$\sigma_F = \frac{k\tau \cot \alpha}{(2A_F/td) + 0.5(1 - k)} \tag{9.34}$$

$$\sigma_S = \frac{k\tau \tan \alpha}{(A_S/tb) + 0.5(1 - k)} \tag{9.35}$$

Further, the web stress σ_t given by Eq. (9.15) becomes two direct stresses: σ_1 along the direction of α given by

$$\sigma_1 = \frac{2k\tau}{\sin 2\alpha} + \tau(1 - k) \sin 2\alpha \tag{9.36}$$

and σ_2 perpendicular to this direction given by

$$\sigma_2 = -\tau(1 - k) \sin 2\alpha \tag{9.37}$$

The secondary bending moment of Eq. (9.25) is multiplied by the factor k, while the effective lengths for the calculation of stiffener buckling loads become (see Eqs. (9.24))

$$\begin{aligned} l_e &= d_s/\sqrt{1 + k^2(3 - 2b/d_s)}, &&\text{for } b < 1.5d \\ l_e &= d_s, &&\text{for } b > 1.5d \end{aligned} \Biggr\}$$

where d_s is the actual stiffener depth, as opposed to the effective depth d of the web, taken between the web–flange connections, as shown in Fig. 9.13. We observe that Eqs. (9.34)–(9.37) are applicable to either incomplete or complete diagonal tension field beams, since, for the latter case, $k = 1$, giving the results of Eqs. (9.27), (9.28), and (9.15).

In some cases, beams taper along their lengths, in which case the flange loads are no longer horizontal but have vertical components which reduce the shear load carried by the web. Thus, in Fig. 9.14, where d is the depth of the beam at the section considered, we have, resolving forces vertically,

$$W - (F_T + F_B) \sin \beta - \sigma_t t(d \cos \alpha) \sin \alpha = 0 \tag{9.38}$$

For horizontal equilibrium,

$$(F_T - F_B) \cos \beta - \sigma_t td \cos^2 \alpha = 0 \tag{9.39}$$

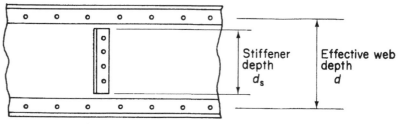

FIGURE 9.13 Calculation of Stiffener Buckling Load

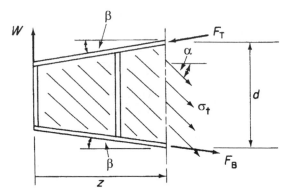

FIGURE 9.14 Effect of Taper on Diagonal Tension Field Beam Calculations

Taking moments about B,

$$Wz - F_\text{T}d \cos \beta + \frac{1}{2}\sigma_\text{t}td^2 \cos^2 \alpha = 0 \tag{9.40}$$

Solving Eqs. (9.38), (9.39), and (9.40) for σ_t, F_T, and F_B,

$$\sigma_\text{t} = \frac{2W}{td \sin 2\alpha}\left(1 - \frac{2z}{d}\tan \beta\right) \tag{9.41}$$

$$F_\text{T} = \frac{W}{d \cos \beta}\left[z + \frac{d \cot \alpha}{2}\left(1 - \frac{2z}{d}\tan \beta\right)\right] \tag{9.42}$$

$$F_\text{B} = \frac{W}{d \cos \beta}\left[z - \frac{d \cot \alpha}{2}\left(1 - \frac{2z}{d}\tan \beta\right)\right] \tag{9.43}$$

Equation (9.23) becomes

$$P = \frac{Wb}{d}\tan \alpha\left(1 - \frac{2z}{d}\tan \beta\right) \tag{9.44}$$

Also, the shear force S at any section of the beam is, from Fig. 9.14,

$$S = W - (F_\text{T} + F_\text{B})\sin \beta$$

or, substituting for F_T and F_B from Eqs. (9.42) and (9.43),

$$S = W\left(1 - \frac{2z}{d}\tan \beta\right) \tag{9.45}$$

9.7.3 Postbuckling behavior

Sections 9.7.1 and 9.7.2 are concerned with beams in which the thin webs buckle to form tension fields; the beam flanges are then regarded as being subjected to bending action, as in Fig. 9.11. It is possible, if the beam flanges are relatively light, for failure due to yielding to occur in the beam flanges after the web has buckled, so that plastic hinges form and a failure mechanism of the type shown

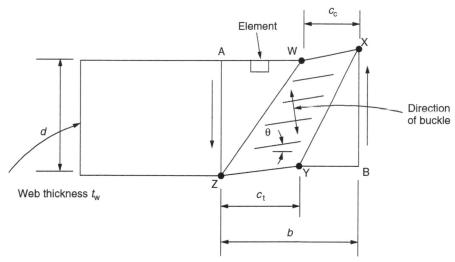

FIGURE 9.15 Collapse Mechanism of a Panel of a Tension Field Beam

in Fig. 9.15 exists. This postbuckling behavior was investigated by Evans, Porter, and Rockey,[14] who developed a design method for beams subjected to bending and shear. Their method of analysis is presented here.

Suppose that the panel AXBZ in Fig. 9.15 collapses due to a shear load S and a bending moment M; plastic hinges form at W, X, Y, and Z. In the initial stages of loading, the web remains perfectly flat until it reaches its critical stresses, that is, τ_{cr} in shear and σ_{crb} in bending. The values of these stresses may be found approximately from

$$\left(\frac{\sigma_{mb}}{\sigma_{crb}}\right)^2 + \left(\frac{\tau_m}{\tau_{cr}}\right)^2 = 1 \tag{9.46}$$

where σ_{crb} is the critical value of bending stress with $S = 0$, $M \neq 0$, and τ_{cr} is the critical value of shear stress when $S \neq 0$ and $M = 0$. Once the critical stress is reached, the web starts to buckle and cannot carry any increase in compressive stress, so that, as we have seen in Section 9.7.1, any additional load is carried by tension field action. It is assumed that the shear and bending stresses remain at their critical values τ_m and σ_{mb} and that there are *additional* stresses σ_t, which are inclined at an angle θ to the horizontal and which carry any increases in the applied load. At collapse, that is, at ultimate load conditions, the additional stress σ_t reaches its maximum value $\sigma_{t(max)}$ and the panel is in the collapsed state shown in Fig. 9.15.

Consider now the small rectangular element on the edge AW of the panel before collapse. The stresses acting on the element are shown in Fig. 9.16(a). The stresses on planes parallel to and perpendicular to the direction of the buckle may be found by considering the equilibrium of triangular elements within this rectangular element. Initially, we consider the triangular element CDE, which is subjected to the stress system shown in Fig. 9.16(b) and is in equilibrium under the action of the forces corresponding to these stresses. Note that the edge CE of the element is parallel to the direction of the buckle in the web.

For equilibrium of the element in a direction perpendicular to CE (see Section 1.6),

$$\sigma_\xi CE + \sigma_{mb}ED \cos \theta - \tau_m ED \sin \theta - \tau_m DC \cos \theta = 0$$

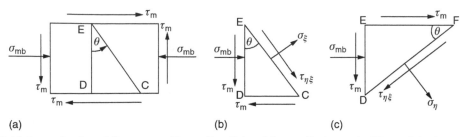

FIGURE 9.16 Determination of Stresses on Planes Parallel and Perpendicular to the Plane of the Buckle

Dividing through by CE and rearranging, we have

$$\sigma_\xi = -\sigma_{mb} \cos^2 \theta + \tau_m \sin 2\theta \tag{9.47}$$

Similarly, by considering the equilibrium of the element in the direction EC, we have

$$\tau_{\eta\xi} = -\frac{\sigma_{mb}}{2} \sin 2\theta - \tau_m \cos 2\theta \tag{9.48}$$

Further, the direct stress σ_η on the plane FD (Fig. 9.16(c)), which is perpendicular to the plane of the buckle, is found from the equilibrium of the element FED. Then,

$$\sigma_\eta FD + \sigma_{mb} ED \sin \theta + \tau_m EF \sin \theta + \tau_m DE \cos \theta = 0$$

Dividing through by FD and rearranging gives

$$\sigma_\eta = -\sigma_{mb} \sin^2 \theta - \tau_m \sin 2\theta \tag{9.49}$$

Note that the shear stress on this plane forms a complementary shear stress system with $\tau_{\eta\xi}$.

The failure condition is reached by adding $\sigma_{t(max)}$ to σ_ξ and using the von Mises theory of elastic failure[15]; that is,

$$\sigma_y^2 = \sigma_1^2 + \sigma_2^2 - \sigma_1 \sigma_2 + 3\tau^2 \tag{9.50}$$

where σ_y is the yield stress of the material, σ_1 and σ_2 are the direct stresses acting on two mutually perpendicular planes, and τ is the shear stress acting on the same two planes. Hence, when the yield stress in the web is σ_{yw}, failure occurs when

$$\sigma_{yw}^2 = (\sigma_\xi + \sigma_{t(max)})^2 + \sigma_\eta^2 - \sigma_\eta (\sigma_\xi + \sigma_{t(max)}) + 3\tau_{\eta\xi}^2 \tag{9.51}$$

Equations (9.47), (9.48), (9.49), and (9.51) may be solved for $\sigma_{t(max)}$, which is then given by

$$\sigma_{t(max)} = -\frac{1}{2}A + \frac{1}{2}\left[A^2 - 4\left(\sigma_{mb}^2 + 3\tau_m^2 - \sigma_{yw}^2\right)\right]^{\frac{1}{2}} \tag{9.52}$$

where

$$A = 3\tau_m \sin 2\theta + \sigma_{mb} \sin^2 \theta - 2\sigma_{mb} \cos^2 \theta \tag{9.53}$$

These equations have been derived for a point on the edge of the panel but are applicable to any point within its boundary. Therefore, the resultant force F_w, corresponding to the tension field in the web, may be calculated and its line of action determined.

If the average stresses in the compression and tension flanges are σ_{cf} and σ_{tf} and the yield stress of the flanges is σ_{yf}, the reduced plastic moments in the flanges are[16]

$$M'_{pc} = M_{pc}\left[1 - \left(\frac{\sigma_{cf}}{\sigma_{yf}}\right)^2\right] \quad \text{(compression flange)} \tag{9.54}$$

$$M'_{pt} = M_{pt}\left[1 - \left(\frac{\sigma_{tf}}{\sigma_{yf}}\right)\right] \quad \text{(tension flange)} \tag{9.55}$$

The position of each plastic hinge may be found by considering the equilibrium of a length of flange and employing the principle of virtual work. In Fig. 9.17, the length WX of the upper flange of the-beam is given a virtual displacement ϕ. The work done by the shear force at X is equal to the energy absorbed by the plastic hinges at X and W and the work done *against* the tension field stress $\sigma_{t(max)}$. Suppose the average value of the tension field stress is σ_{tc}, that is, the stress at the mid-point of WX.

Then,

$$S_x c_c \phi = 2M'_{pc}\phi + \sigma_{tc}t_w \ \sin^2\theta \frac{c_c^2}{2}\phi$$

The minimum value of S_x is obtained by differentiating with respect to c_c; that is,

$$\frac{dS_x}{dc_c} = -2\frac{M'_{pc}}{c_c^2} + \sigma_{tc}t_w \frac{\sin^2\theta}{2} = 0$$

which gives

$$c_c^2 = \frac{4M'_{pc}}{\sigma_{tc}t_w \ \sin^2\theta} \tag{9.56}$$

Similarly, in the tension flange,

$$c_t^2 = \frac{4M'_{pt}}{\sigma_{tt}t_w \ \sin^2\theta} \tag{9.57}$$

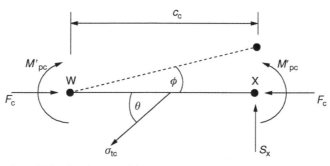

FIGURE 9.17 Determination of Plastic Hinge Position

Clearly, for the plastic hinges to occur within a flange, both c_c and c_t must be less than b. Therefore, from Eq. (9.56),

$$M'_{pc} < \frac{t_w b^2 \sin^2 \theta}{4} \sigma_{tc} \tag{9.58}$$

where σ_{tc} is found from Eqs. (9.52) and (9.53) at the mid-point of WX.

The average axial stress in the compression flange between W and X is obtained by considering the equilibrium of half of the length of WX (Fig. 9.18).

Then,

$$F_c = \sigma_{cf} A_{cf} + \sigma_{tc} t_w \frac{c_c}{2} \sin \theta \cos \theta + \tau_m t_w \frac{c_c}{2}$$

from which

$$\sigma_{cf} = \frac{F_c - \frac{1}{2}(\sigma_{tc} \sin \theta \cos \theta + \tau_m) t_w c_c}{A_{cf}} \tag{9.59}$$

where F_c is the force in the compression flange at W and A_{cf} is the cross-sectional area of the compression flange.

Similarly, for the tension flange,

$$\sigma_{tf} = \frac{F_t + \frac{1}{2}(\sigma_{tt} \sin \theta \cos \theta + \tau_m) t_w c_t}{A_{tf}} \tag{9.60}$$

The forces F_c and F_t are found by considering the equilibrium of the beam to the right of WY (Fig. 9.19). Then, resolving vertically and noting that $S_{cr} = \tau_m t_w d$,

$$S_{ult} = F_w \sin \theta + \tau_m t_w d + \sum W_n \tag{9.61}$$

Resolving horizontally and noting that $H_{cr} = \tau_m t_w (b - c_c - c_t)$,

$$F_c - F_t = F_w \cos \theta - \tau_m t_w (b - c_c - c_t) \tag{9.62}$$

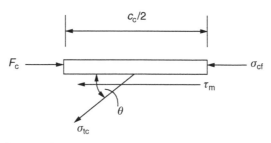

FIGURE 9.18 Determination of Flange Stress

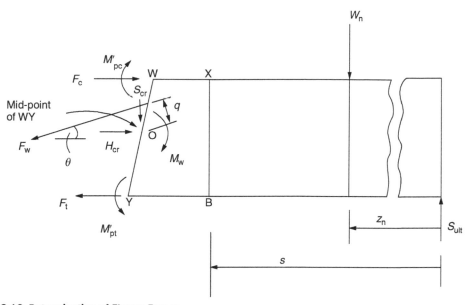

FIGURE 9.19 Determination of Flange Forces

Taking moments about O, we have

$$F_c + F_t = \frac{2}{d}\left[S_{ult}\left(s + \frac{b + c_c - c_t}{2}\right) + M'_{pt} - M'_{pc} + F_w q - M_w - \sum_n W_n z_n\right] \tag{9.63}$$

where W_1 to W_n are external loads applied to the beam to the right of WY and M_w is the bending moment in the web when it has buckled and become a tension field: that is,

$$M_w = \frac{\sigma_{mb} b d^2}{b} \tag{}$$

The flange forces are then

$$F_c = \frac{S_{ult}}{2d}(d\cot\theta + 2s + b + c_c - c_t) + \frac{1}{d}\left(M'_{pt} - M'_{pc} + F_w q - M_w - \sum_n W_n z_n\right)$$
$$-\frac{1}{2}\tau_m t_w(d\cot\theta + b - c_c - c_t) \tag{9.64}$$

$$F_t = \frac{S_{ult}}{2d}(d\cot\theta + 2s + b + c_c - c_t) + \frac{1}{d}\left(M'_{pt} - M'_{pc} - F_w q - M_w - \sum_n W_n z_n\right)$$
$$+\frac{1}{2}\tau_m t_w(d\cot\theta + b - c_c - c_t) \tag{9.65}$$

Evans, Porter, and Rockey adopted an iterative procedure for solving Eqs. (9.61)–(9.65), in which an initial value of θ was assumed and σ_{cf} and σ_{tf} were taken to be zero. Then, c_c and c_t were calculated and approximate values of F_c and F_t found giving better estimates for σ_{cf} and σ_{tf}. The procedure was repeated until the required accuracy was obtained.

References

[1] Gerard G. Introduction to structural stability theory. New York: McGraw-Hill; 1962.

[2] Murray NW. Introduction to the theory of thin-walled structures. Oxford: Oxford engineering science series; 1984.

[3] Handbook of aeronautics, No. 1, Structural principles and data. 4th ed. The royal aeronautical society; 1952.

[4] Bleich F. Buckling strength of metal structures. New York: McGraw-Hill; 1952.

[5] Gerard G, Becker H. Handbook of structural stability, Part I, Buckling of flat plates. NACA tech. note 3781, 1957.

[6] Rivello RM. Theory and analysis of flight structures. New York: McGraw-Hill; 1969.

[7] Stowell EZ. Compressive strength of flanges. NACA tech. note 1323, 1947.

[8] Mayers J, Budiansky B. Analysis of behaviour of simply supported flat plates compressed beyond the buckling load in the plastic range. NACA tech. note 3368, 1955.

[9] Gerard G, Becker H. Handbook of structural stability, Part IV, Failure of plates and composite elements. NACA Tech. Note 3784, 1957.

[10] Gerard G. Handbook of structural stability, Part V, Compressive strength of flat stiffened panels. NACA tech. note 3785, 1957.

[11] Gerard G, Becker H. Handbook of structural stability, Part VII, Strength of thin wing construction. NACA tech. note D-162, 1959.

[12] Gerard G. The crippling strength of compression elements. J Aeron Sci 1958;25(1):37–52.

[13] Kuhn P. Stresses in aircraft and shell structures. New York: McGraw-Hill; 1956.

[14] Evans HR, Porter DM, Rockey KC. The collapse behaviour of plate girders subjected to shear and bending. proc int assn bridge and struct eng P-18/78:1–20.

[15] Megson THG. Structural and stress analysis. 2nd ed. Oxford: Elsevier; 2005.

PROBLEMS

P.9.1 A thin square plate of side a and thickness t is simply supported along each edge and has a slight initial curvature giving an initial deflected shape:

$$w_0 = \delta \sin \frac{\pi x}{a} \sin \frac{\pi y}{a}$$

If the plate is subjected to a uniform compressive stress σ in the x direction (see Fig. P.9.1), find an expression for the *elastic* deflection w normal to the plate. Show also that the deflection at the mid-point of the plate can be presented in the form of a Southwell plot and illustrate your answer with a suitable sketch.

Answer: $\quad w = [\sigma t \delta / (4 \pi^2 D / a^2 - \sigma t)] \sin \frac{\pi x}{a} \sin \frac{\pi y}{a}$

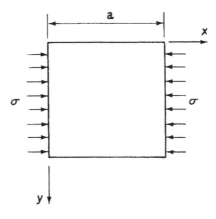

FIGURE P.9.1

P.9.2 A uniform flat plate of thickness t has a width b in the y direction and length l in the x direction (see Fig. P.9.2). The edges parallel to the x axis are clamped and those parallel to the y axis are simply supported. A uniform compressive stress σ is applied in the x direction along the edges parallel to the y axis. Using an energy method, find an approximate expression for the magnitude of the stress σ that causes the plate to buckle, assuming that the deflected shape of the plate is given by

$$w = a_{11} \sin\frac{m\pi x}{l} \sin^2\frac{\pi y}{b}$$

For this particular case, $l = 2b$, find the number of half waves m corresponding to the lowest critical stress, expressing the result to the nearest integer. Determine also the lowest critical stress.

Answer: $m = 3$, $\sigma_{CR} = [6E/(1 - v^2)](t/b)^2$

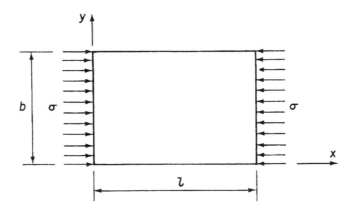

FIGURE P.9.2

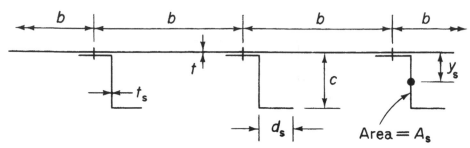

FIGURE P.9.3

P.9.3 A panel, comprising a flat sheet and uniformly spaced Z-section stringers, a part of whose cross-section is shown in Fig. P.9.3, is to be investigated for strength under uniform compressive loads in a structure in which it is to be stabilized by frames a distance l apart, l being appreciably greater than the spacing b.

    (a) State the modes of failure you would consider and how you would determine appropriate limiting stresses.

    (b) Describe a suitable test to verify your calculations, giving particulars of the specimen, the manner of support, and the measurements you would take. The last should enable you to verify the assumptions made as well as to obtain the load supported.

P.9.4 Part of a compression panel of internal construction is shown in Fig. P.9.4. The equivalent pin-center length of the panel is 500 mm. The material has a Young's modulus of 70,000 N/mm^2 and its elasticity may be taken as falling catastrophically when a compressive stress of 300 N/mm^2 is reached. Taking coefficients of 3.62 for buckling of a plate with simply supported sides and of 0.385 with one side simply supported and one free, determine (a) the load per millimetre width of panel when initial buckling may be expected and (b) the load per millimetre for ultimate failure. Treat the material as thin for calculating section constants and assume that, after initial buckling, the stress in the plate increases parabolically from its critical value in the center of sections.

    Answer: 613.8 N/mm, 844.7 N/mm

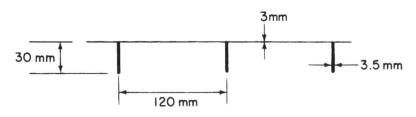

FIGURE P.9.4

P.9.4 MATLAB Use MATLAB to repeat part (a) of Problem P.9.4 for the following skin thicknesses (t_{sk}):

	(i)	(ii)	(iii)	(iv)	(v)
t_{sk}	2 mm	2.5 mm	3 mm	3.5 mm	4 mm

Answer: (i) $N_{x,CR} = 202.4$ N/mm
(ii) $N_{x,CR} = 371.25$ N/mm
(iii) $N_{x,CR} = 613.8$ N/mm
(iv) $N_{x,CR} = 650.1$ N/mm
(v) $N_{x,CR} = 677.1$ N/mm

P.9.5 A simply supported beam has a span of 2.4 m and carries a central concentrated load of 10 kN. Each of the flanges of the beam has a cross-sectional area of 300 mm^2 while that of the vertical web stiffeners is 280 mm^2. If the depth of the beam, measured between the centroids of area of the flanges, is 350 mm and the stiffeners are symmetrically arranged about the web and spaced at 300 mm intervals, determine the maximum axial load in a flange and the compressive load in a stiffener. It may be assumed that the beam web, of thickness 1.5 mm, is capable of resisting diagonal tension only.

Answer: 19.9 kN, 3.9 kN.

P.9.6 The spar of an aircraft is to be designed as an incomplete diagonal tension beam, the flanges being parallel. The stiffener spacing will be 250 mm, the effective depth of web will be 750 mm, and the depth between web-to-flange attachments is 725 mm. The spar is to carry an ultimate shear force of 100,000 N. The maximum permissible shear stress is 165 N/mm^2, but it is also required that the shear stress should not exceed 15 times the critical shear stress for the web panel. Assuming α to be 40° and using the following relationships
(a) Select the smallest suitable web thickness from the following range of standard thicknesses. (Take Young's Modulus E as 70,000 N/mm^2.)
0.7 mm, 0.9 mm, 1.2 mm, 1.6 mm
(b) Calculate the stiffener end load and the secondary bending moment in the flanges (assume stiffeners to be symmetrical about the web).
The shear stress buckling coefficient for the web may be calculated from the expression

$$K = 7.70\left[1 + 0.75(b/d)^2\right]$$

b and d having their usual significance. The relationship between the diagonal tension factor and buckling stress ratio is

τ/τ_{CR}	5	7	9	11	13	15
k	0.37	0.40	0.42	0.48	0.51	0.53

Note that α is the angle of diagonal tension measured from the spanwise axis of the beam, as in the usual notation.

Answer: 1.2 mm, $130A_S/(1 + 0.0113A_S)$, 238,910 N mm

P.9.6 MATLAB Use MATLAB to repeat Problem P.8.16 for the following stiffener spacings.

	(i)	(ii)	(iii)	(iv)
b	150 mm	250 mm	350 mm	450 mm

Answer:
    (i) 0.9 mm, $166A_S/(1 + 0.0247A_S)$, 83 910 N mm
    (ii) 1.2 mm, $130A_S/(1 + 0.0113A_S)$, 238 910 N mm
    (iii) 1.2 mm, $210A_S/(1 + 0.0101A_S)$, 605 317 N mm
    (iv) 1.6 mm, $106A_S/(1 + 0.00487A_S)$, 811 829 N mm

P.9.7 The main compressive wing structure of an aircraft consists of stringers, having the section shown in Fig. P.9.7(b), bonded to a thin skin (Fig. P.9.7(a)). Find suitable values for the stringer spacing b and rib spacing L if local instability, skin buckling, and panel strut instability all occur at the same stress. Note that, in Fig. P.9.7(a), only two of several stringers are shown for diagrammatic clarity. Also, the thin skin should be treated as a flat plate, since the curvature is small. For a flat plate simply supported along two edges, assume a buckling coefficient of 3.62. Take $E = 69,000$ N/mm^2.

Answer: $b = 56.5$ mm, $L = 700$ mm

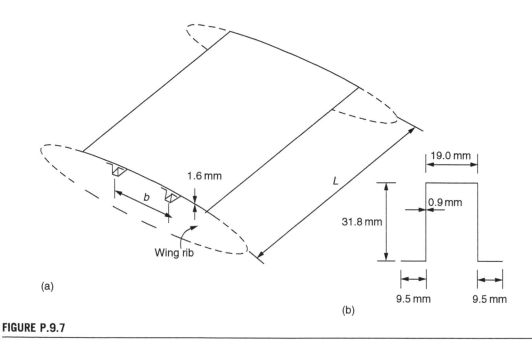

(a)

(b)

FIGURE P.9.7

Analysis of aircraft structures

Principles of stressed skin construction

B1

Materials

With the present chapter, we begin the purely aeronautical section of the book, where we consider structures peculiar to the field of aeronautical engineering. These structures are typified by arrangements of thin, load-bearing skins, frames, and stiffeners, fabricated from lightweight, high strength materials of which aluminum alloys are the most widely used examples.

As a preliminary to the analysis of the basic aircraft structural forms presented in subsequent chapters, we discuss the materials used in aircraft construction.

Several factors influence the selection of the structural material for an aircraft, but among these, strength allied to lightness is probably the most important. Other properties having varying, though sometimes critical significance include: stiffness; toughness; resistance to corrosion; fatigue and the effects of environmental heating; ease of fabrication; availability and consistency of supply; and, not least important, cost.

The main groups of materials used in aircraft construction have been wood, steel, aluminum alloys with, more recently, titanium alloys, and fiber-reinforced composites. In the field of engine design, titanium alloys are used in the early stages of a compressor, while nickel-based alloys or steels are used for the hotter later stages. As we are concerned primarily with the materials involved in the construction of the airframe, discussion of materials used in engine manufacture falls outside the scope of this book.

10.1 ALUMINUM ALLOYS

Pure aluminum is a relatively low strength, extremely flexible metal with virtually no structural applications. However, when alloyed with other metals, its properties are improved significantly. Three groups of aluminum alloy have been used in the aircraft industry for many years and still play a major role in aircraft construction. In the first of these, aluminum is alloyed with copper, magnesium, manganese, silicon, and iron and has a typical composition of 4 percent copper, 0.5 percent magnesium, 0.5 percent manganese, 0.3 percent silicon, and 0.2 percent iron with the remainder being aluminum. In the wrought, heat-treated, naturally aged condition, this alloy possesses a 0.1 percent proof stress not less than 230 N/mm^2, a tensile strength not less than 390 N/mm^2, and an elongation at fracture of 15 percent. Artificial ageing at a raised temperature of, for example, 170°C increases the proof stress to not less than 370 N/mm^2 and the tensile strength to not less than 460 N/mm^2 with an elongation of 8 percent.

The second group of alloys contain, in addition to the preceding, 1–2 percent of nickel, a higher content of magnesium, and possible variations in the amounts of copper, silicon, and iron. The most important property of these alloys is their retention of strength at high temperatures, which makes them particularly suitable for aero engine manufacture. A development of these alloys by Rolls-Royce and

High Duty Alloys Ltd. replaced some of the nickel by iron and reduced the copper content; these RR alloys, as they were called, were used for forgings and extrusions in aero engines and airframes.

The third group of alloys depends upon the inclusion of zinc and magnesium for their high strength and have a typical composition of 2.5 percent copper, 5 percent zinc, 3 percent magnesium, and up to 1 percent nickel with mechanical properties of 0.1 percent proof stress 510 N/mm^2, tensile strength 585 N/mm^2, and an elongation of 8 percent. In a modern development of this alloy, nickel has been eliminated and provision made for the addition of chromium and further amounts of manganese.

Alloys from each of these groups have been used extensively for airframes, skins, and other stressed components, the choice of alloy being influenced by factors such as strength (proof and ultimate stress), ductility, ease of manufacture (e.g., in extrusion and forging), resistance to corrosion and amenability to protective treatment, fatigue strength, freedom from liability to sudden cracking due to internal stresses, and resistance to fast crack propagation under load. Clearly, different types of aircraft have differing requirements. A military aircraft, for instance, having a relatively short life measured in hundreds of hours, does not call for the same degree of fatigue and corrosion resistance as a civil aircraft with a required life of 30,000 hours or more.

Unfortunately, as one particular property of aluminum alloys is improved, other desirable properties are sacrificed. For example, the extremely high static strength of the aluminum–zinc–magnesium alloys was accompanied for many years by a sudden liability to crack in an unloaded condition due to the retention of internal stresses in bars, forgings, and sheet after heat treatment. Although variations in composition eliminated this problem to a considerable extent, other deficiencies showed themselves. Early postwar passenger aircraft experienced large numbers of stress-corrosion failures of forgings and extrusions. The problem became so serious that, in 1953, it was decided to replace as many aluminum–zinc–manganese components as possible with the aluminum–4 percent copper Alloy L65 and to prohibit the use of forgings in zinc-bearing alloy in all future designs. However, improvements in the stress-corrosion resistance of the aluminum–zinc–magnesium alloys have resulted in recent years from British, American, and German research. Both British and American opinions agree on the benefits of including about 1 percent copper but disagree on the inclusion of chromium and manganese, while in Germany, the addition of silver has been found extremely beneficial. Improved control of casting techniques brought further improvements in resistance to stress corrosion. The development of aluminum–zinc–magnesium–copper alloys largely met the requirement for aluminum alloys possessing high strength, good fatigue crack growth resistance, and adequate toughness. Further development will concentrate on the production of materials possessing higher specific properties, bringing benefits in relation to weight saving rather than increasing strength and stiffness.

The first group of alloys possesses a lower static strength than the zinc-bearing alloys but are preferred for portions of the structure where fatigue considerations are of primary importance, such as the undersurfaces of wings, where tensile fatigue loads predominate. Experience has shown that the naturally aged version of these alloys has important advantages over the fully heat-treated forms in fatigue endurance and resistance to crack propagation. Furthermore, the inclusion of a higher percentage of magnesium was found, in America, to produce, in the naturally aged condition, mechanical properties between those of the normal naturally aged and artificially aged alloy. This alloy, designated 2024 (aluminum–copper alloys form the 2000 series) has the nominal composition: 4.5 percent copper, 1.5 percent magnesium, 0.6 percent manganese, with the remainder aluminum and appears to be a satisfactory compromise between the various important, but sometimes conflicting, mechanical properties.

Interest in aluminum–magnesium–silicon alloys has recently increased, although they have been in general use in the aerospace industry for decades. The reasons for this renewed interest are that they are potentially cheaper than aluminum–copper alloys and, being weldable, are capable of reducing manufacturing costs. In addition, variants, such as the ISO 6013 alloy, have improved property levels and, generally, possess a similar high fracture toughness and resistance to crack propagation as the 2000 series alloys.

Frequently, a particular form of an alloy is developed for a particular aircraft. An outstanding example of such a development is the use of Hiduminium RR58 as the basis for the main structural material, designated CM001, for Concorde. Hiduminium RR58 is a complex aluminum–copper–magnesium–nickel–iron alloy developed during the 1939–1945 war specifically for the manufacture of forged components in gas turbine aero engines. The chemical composition of the version used in Concorde, was decided on the basis of elevated temperature, creep, fatigue, and tensile testing programs,has the detailed specification of materials listed in Table 10.1. Generally, CM001 is found to possess better overall strength and fatigue characteristics over a wide range of temperatures than any of the other possible aluminum alloys.

Table 10.1 Materials of CM001

	%Cu	%Mg	%Si	%Fe	%Ni	%Ti	%Al
Minimum	2.25	1.35	0.18	0.90	1.0	—	Remainder
Maximum	2.70	1.65	0.25	1.20	1.30	0.20	

The latest aluminum alloys to find general use in the aerospace industry are the aluminum–lithium alloys. Of these, the aluminum–lithium–copper–manganese alloy, 8090, developed in the United Kingdom, is extensively used in the main fuselage structure of GKN Westland Helicopters' design EH101; it has also been qualified for Eurofighter 2000 (now named the Typhoon) but has yet to be embodied. In the United States, the aluminum–lithium–copper alloy, 2095, has been used in the fuselage frames of the F16 as a replacement for 2124, resulting in a fivefold increase in fatigue life and a reduction in weight. Aluminum–lithium alloys can be successfully welded, possess a high fracture toughness, and exhibit a high resistance to crack propagation.

10.2 STEEL

The use of steel for the manufacture of thin-walled, box-section spars in the 1930s has been superseded by the aluminum alloys described in Section 10.1. Clearly, its high specific gravity prevents steel's widespread use in aircraft construction, but it has retained some value as a material for castings for small components demanding high tensile strengths, high stiffness, and high resistance to wear. Such components include undercarriage pivot brackets, wing-root attachments, fasteners, and tracks.

Although the attainment of high and ultra-high tensile strengths presents no difficulty with steel, it is found that other properties are sacrificed and that it is difficult to manufacture into finished components. To overcome some of these difficulties, types of steel known as *maraging* steels were developed in 1961, from which carbon is either eliminated entirely or present only in very small amounts. Carbon, while producing the necessary hardening of conventional high-tensile steels, causes brittleness and distortion; the latter is not easily rectifiable as machining is difficult and cold forming impracticable. Welded fabrication is also almost impossible or very expensive. The hardening of maraging steels is achieved by the addition of other elements, such as nickel, cobalt, and molybdenum. A typical maraging steel has these elements present in the following proportions: nickel 17–19 percent, cobalt 8–9 percent, molybdenum 3–3.5 percent, with titanium 0.15–0.25 percent. The carbon content is a maximum of 0.03 percent, with traces of manganese, silicon, sulfur, phosphorus, aluminum, boron, calcium, and zirconium. Its 0.2 percent proof stress is nominally 1400 N/mm^2 and its modulus of elasticity 180,000 N/mm^2.

The main advantages of maraging steels over conventional low-alloy steels are: higher fracture toughness and notched strength, simpler heat treatment, much lower volume change and distortion during hardening, very much simpler to weld, easier to machine, and better resistance to stress corrosion and hydrogen embrittlement. On the other hand, the material cost of maraging steels is three or more times greater than the cost of conventional steels, although this may be more than offset by the increased cost of fabricating a complex component from the latter steel.

Maraging steels have been used in aircraft arrester hooks, rocket motor cases, helicopter undercarriages, gears, ejector seats, and various structural forgings.

In addition to the preceding, steel in its stainless form has found applications primarily in the construction of super- and hypersonic experimental and research aircraft, where temperature effects are considerable. Stainless steel formed the primary structural material in the Bristol 188, built to investigate kinetic heating effects, and also in the American rocket aircraft, the X-15, capable of speeds of the order of Mach 5–6.

10.3 TITANIUM

The use of titanium alloys increased significantly in the 1980s, particularly in the construction of combat aircraft as opposed to transport aircraft. This increase continued in the 1990s to the stage where, for combat aircraft, the percentage of titanium alloy as a fraction of structural weight is of the same order as that of aluminum alloy. Titanium alloys possess high specific properties, have a good fatigue strength/tensile strength ratio with a distinct fatigue limit, and some retain considerable strength at temperatures up to 400–500°C. Generally, there is also a good resistance to corrosion and corrosion fatigue although properties are adversely affected by exposure to temperature and stress in a salt environment. The latter poses particular problems in the engines of carrier-operated aircraft. Further disadvantages are a relatively high density, so that weight penalties are imposed if the alloy is extensively used, coupled with high primary and high fabrication costs, approximately seven times those of aluminum and steel.

In spite of this, titanium alloys were used in the airframe and engines of Concorde, while the Tornado wing carry-through box is fabricated from a weldable medium-strength titanium alloy. Titanium alloys are also used extensively in the F15 and F22 American fighter aircraft and are incorporated in the tail assembly of the Boeing 777 civil airliner. Other uses include forged components, such as flap and slat tracks and undercarriage parts.

New fabrication processes (e.g., superplastic forming combined with diffusion bonding) enable large and complex components to be produced, resulting in a reduction in production man-hours and weight. Typical savings are 30 percent in man-hours, 30 percent in weight, and 50 percent in cost compared with conventional riveted titanium structures. It is predicted that the number of titanium components fabricated in this way for aircraft will increase significantly and include items such as access doors, sheet for areas of hot gas impingement, and so forth.

10.4 PLASTICS

Plain plastic materials have specific gravities of approximately unity and are therefore considerably heavier than wood, although of comparable strength. On the other hand, their specific gravities are less than half those of the aluminum alloys, so that they find uses as windows or lightly stressed parts whose dimensions are established by handling requirements rather than strength. They are also particularly useful as electrical insulators and as energy-absorbing shields for delicate instrumentation and even structures where severe vibration, such as in a rocket or space shuttle launch, occurs.

10.5 GLASS

The majority of modern aircraft have cabins pressurized for flight at high altitudes. Windscreens and windows are therefore subjected to loads normal to their midplanes. Glass is frequently the material employed for this purpose, in the form of plain or laminated plate or heat-strengthened plate. The types of plate glass used in aircraft have a modulus of elasticity between 70,000 and 75,000 N/mm^2 with a modulus of rupture in bending of 45 N/mm^2. Heat-strengthened plate has a modulus of rupture of about four and a half times this figure.

10.6 COMPOSITE MATERIALS

Composite materials consist of strong fibers, such as glass or carbon, set in a matrix of plastic or epoxy resin, which is mechanically and chemically protective. The fibers may be continuous or discontinuous but possess a strength very much greater than that of the same bulk materials. For example, carbon fibers have a tensile strength of the order of 2,400 N/mm^2 and a modulus of elasticity of 400,000 N/mm^2.

A sheet of fiber-reinforced material is anisotropic; that is, its properties depend on the direction of the fibers. Generally, therefore, in structural form two or more sheets are sandwiched together to form a *lay-up* so that the fiber directions match those of the major loads.

In the early stages of the development of composite materials, glass fibers were used in a matrix of epoxy resin. This glass-reinforced plastic (GRP) was used for radomes and helicopter blades but found limited use in components of fixed wing aircraft, due to its low stiffness. In the 1960s, new fibrous reinforcements were introduced; Kevlar, for example, is an aramid material with the same strength as glass but stiffer. Kevlar composites are tough but poor in compression and difficult to machine, so they were used in secondary structures. Another composite, using boron fiber and developed in the United States, was the first to possess sufficient strength and stiffness for primary structures.

These composites have now been replaced by carbon-fiber-reinforced plastics (CFRP), which have similar properties to boron composites but are very much cheaper. Typically, CFRP has a modulus of the order of three times that of GRP, one and a half times that of a Kevlar composite, and twice that of aluminum alloy. Its strength is three times that of aluminum alloy, approximately the same as that of GRP, and slightly less than that of Kevlar composites. CFRP does, however, suffer from some disadvantages. It is a brittle material and therefore does not yield plastically in regions of high stress concentration. Its strength is reduced by impact damage, which may not be visible, and the epoxy resin matrices can absorb moisture over a long period, which reduces its matrix-dependent properties, such as its compressive strength; this effect increases with increase of temperature. Further, the properties of CFRP are subject to more random variation than those of metals. All these factors must be allowed for in design. On the other hand, the stiffness of CFRP is much less affected than its strength by such conditions and it is less prone to fatigue damage than metals. It is estimated that replacing 40 percent of an aluminum alloy structure by CFRP would result in a 12 percent saving in total structural weight.

CFRP is included in the wing, tailplane, and forward fuselage of the latest Harrier development; is used in the Tornado taileron; and has been used to construct a complete Jaguar wing and engine bay door for testing purposes. The use of CFRP in the fabrication of helicopter blades led to significant increases in their service life, where fatigue resistance rather than stiffness is of primary importance. Figure 10.1 shows the structural complexity of a Sea King helicopter rotor blade that incorporates CFRP, GRP, stainless steel, a honeycomb core, and foam filling. An additional advantage of the use of composites for helicopter rotor blades is that the molding techniques employed allow variations of cross-section along the span, resulting in substantial aerodynamic benefits. This approach is being employed in the fabrication of the main rotor blades of the GKN Westland Helicopters EH101.

A composite (fiberglass and aluminum) is used in the tail assembly of the Boeing 777, while the leading edge of the Airbus A310–300/A320 fin assembly is of conventional reinforced glass fiber

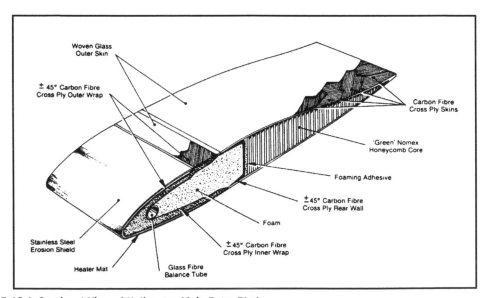

FIGURE 10.1 Sectional View of Helicopter Main Rotor Blade

(courtesy Royal Aeronautical Society, Aerospace magazine)

construction, reinforced at the nose to withstand bird strikes. A complete composite airframe was produced for the Beechcraft Starship turboprop executive aircraft, which, however, was not a commercial success due to its canard configuration, causing drag and weight penalties.

The development of composite materials is continuing with research into the removal of strength-reducing flaws and local imperfections from carbon fibers. Other matrices such as polyetheretherketone, which absorbs much less moisture than epoxy resin, has an indefinite shelf life, and performs well under impact, are being developed; fabrication, however, requires much higher temperatures. Metal matrix composites, such as graphite–aluminum and boron–aluminum, are lightweight and retain their strength at higher temperatures than aluminum alloys but are expensive to produce.

Generally, the use of composites in aircraft construction appears to have reached a plateau, particularly in civil subsonic aircraft, where the fraction of the structure comprising composites is approximately 15 percent. This is due largely to the greater cost of manufacturing composites than aluminum alloy structures, since composites require hand crafting of the materials and manual construction processes. These increased costs are particularly important in civil aircraft construction and are becoming increasingly important in military aircraft.

10.7 PROPERTIES OF MATERIALS

In Sections 10.1–10.6, we discussed the various materials used in aircraft construction and listed some of their properties. We now examine in more detail their behavior under load and also define different types of material.

Ductility

A material is said to be *ductile* if it is capable of withstanding large strains under load before fracture occurs. These large strains are accompanied by a visible change in cross-sectional dimensions and therefore give warning of impending failure. Materials in this category include mild steel, aluminum and some of its alloys, copper, and polymers.

Brittleness

A brittle material exhibits little deformation before fracture, the strain normally being below 5 percent. Brittle materials therefore may fail suddenly without visible warning. Included in this group are concrete, cast iron, high strength steel, timber, and ceramics.

Elastic materials

A material is said to be *elastic* if deformations disappear completely on removal of the load. All known engineering materials are, in addition, *linearly elastic* within certain limits of stress, so that strain, within these limits, is directly proportional to stress.

Plasticity

A material is perfectly *plastic* if no strain disappears after the removal of load. Ductile materials are *elastoplastic* and behave in an elastic manner until the *elastic limit* is reached, after which they behave plastically. When the stress is relieved the elastic component of the strain is recovered but the plastic strain remains as a *permanent set*.

Isotropic materials

In many materials, the elastic properties are the same in all directions at each point in the material, although they may vary from point to point; such a material is known as *isotropic*. An isotropic material having the same properties at all points is known as *homogeneous* (e.g., mild steel).

Anisotropic materials

Materials having varying elastic properties in different directions are known as *anisotropic*.

Orthotropic materials

Although a structural material may possess different elastic properties in different directions, this variation may be limited, as in the case of timber, which has just two values of Young's modulus, one in the direction of the grain and one perpendicular to the grain. A material whose elastic properties are limited to different values in three mutually perpendicular directions is known as *orthotropic*.

10.7.1 Testing of engineering materials

The properties of engineering materials are determined mainly by the mechanical testing of specimens machined to prescribed sizes and shapes. The testing may be static or dynamic in nature, depending on the particular property being investigated. Possibly the most common mechanical static tests are tensile and compressive tests which are carried out on a wide range of materials. Ferrous and nonferrous metals are subjected to both forms of test, while compression tests are usually carried out on many nonmetallic materials. Other static tests include bending, shear, and hardness tests, while the toughness of a material, in other words, its ability to withstand shock loads, is determined by impact tests.

Tensile tests

Tensile tests are normally carried out on metallic materials and, in addition, timber. Test pieces are machined from a batch of material, their dimensions being specified by Codes of Practice. They are commonly circular in cross-section, although flat test pieces having rectangular cross-sections are used when the batch of material is in the form of a plate. A typical test piece has the dimensions specified in Fig. 10.2. Usually, the diameter of a central portion of the test piece is fractionally less than that of the remainder to ensure that the test piece fractures between the gauge points.

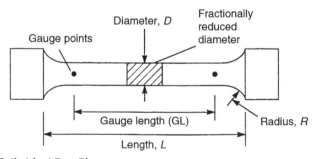

FIGURE 10.2 Standard Cylindrical Test Piece

Before the test begins, the mean diameter of the test piece is obtained by taking measurements at several sections using a micrometer screw gauge. Gauge points are punched at the required gauge length, the test piece is placed in the testing machine, and a suitable strain measuring device, usually an extensometer, is attached to the test piece at the gauge points, so that the extension is measured over the given gauge length. Increments of load are applied and the corresponding extensions recorded. This procedure continues until yield occurs, when the extensometer is removed as a precaution against the damage caused if the test piece fractures unexpectedly. Subsequent extensions are measured by dividers placed in the gauge points until, ultimately, the test piece fractures. The final gauge length and the diameter of the test piece in the region of the fracture are measured so that the percentage elongation and percentage reduction in area may be calculated. These two parameters give a measure of the ductility of the material.

A stress–strain curve is drawn (see Figs 10.9 and 10.13), the stress normally being calculated on the basis of the original cross-sectional area of the test piece, that is, a *nominal stress* as opposed to an *actual stress* (which is based on the actual area of cross-section).

For ductile materials, there is a marked difference in the latter stages of the test, as a considerable reduction in cross-sectional area occurs between yield and fracture. From the stress–strain curve, the ultimate stress, the yield stress, and Young's modulus, E, are obtained.

There are a number of variations on the basic tensile test just described. Some of these depend upon the amount of additional information required and some upon the choice of equipment. There is a wide range of strain measuring devices to choose from, extending from different makes of mechanical extensometer, such as Huggenberger, Lindley, and Cambridge, to the electrical resistance strain gauge. The last normally is used on flat test pieces, one on each face to eliminate the effects of possible bending. At the same time, a strain gauge could be attached in a direction perpendicular to the direction of loading, so that lateral strains are measured. The ratio lateral strain/longitudinal strain is Poisson's ratio, v.

Testing machines are usually driven hydraulically. More sophisticated versions employ load cells to record load and automatically plot load against extension or stress against strain on a pen recorder as the test proceeds, an advantage when investigating the distinctive behavior of mild steel at yield.

Compression tests

A compression test is similar in operation to a tensile test, with the obvious difference that the load transmitted to the test piece is compressive rather than tensile. This is achieved by placing the test piece between the platens of the testing machine and reversing the direction of loading. Test pieces are normally cylindrical and are limited in length to eliminate the possibility of failure being caused by instability. Again, contractions are measured over a given gauge length by a suitable strain measuring device.

Variations in test pieces occur when only the ultimate strength of the material in compression is required. For this purpose, concrete test pieces may take the form of cubes having edges approximately 10 cm long, while mild steel test pieces are still cylindrical in section but are of the order of 1 cm long.

Bending tests

Many structural members are subjected primarily to bending moments. Bending tests are therefore carried out on simple beams constructed from the different materials to determine their behavior under this type of load.

Two forms of loading are employed, the choice depending upon the type specified in Codes of Practice for the particular material. In the first, a simply supported beam is subjected to a "two-point"

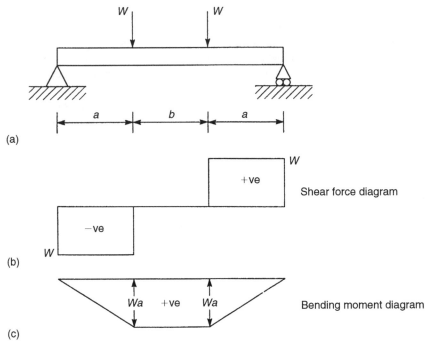

FIGURE 10.3 Bending Test on a Beam, "Two-Point" Load

loading system, as shown in Fig. 10.3(a). Two concentrated loads are applied symmetrically to the beam, producing zero shear force and constant bending moment in the central span of the beam (Figs. 10.3(b) and (c)). The condition of pure bending is therefore achieved in the central span.

The second form of loading system consists of a single concentrated load at midspan (Fig. 10.4(a)), which produces the shear force and bending moment diagrams shown in Figs. 10.4(b) and (c).

The loads may be applied manually by hanging weights on the beam or by a testing machine. Deflections are measured by a dial gauge placed underneath the beam. From the recorded results, a load–deflection diagram is plotted.

For most ductile materials, the test beams continue to deform without failure and fracture does not occur. Thus, plastic properties, for example, the ultimate strength in bending, cannot be determined for such materials. In the case of brittle materials, including cast iron, timber, and various plastics, failure does occur, so that plastic properties can be evaluated. For such materials, the ultimate strength in bending is defined by the *modulus of rupture*. This is taken to be the maximum direct stress in bending, $\sigma_{x,u}$, corresponding to the ultimate moment M_u, and is assumed to be related to M_u by the elastic relationship

$$\sigma_{x,u} = \frac{M_u}{I} y_{max}$$

Other bending tests are designed to measure the ductility of a material and involve the bending of a bar round a pin. The angle of bending at which the bar starts to crack is then taken as an indication of its ductility.

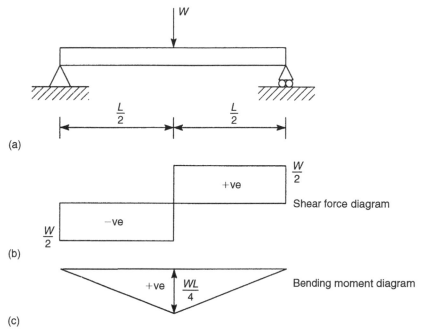

FIGURE 10.4 Bending Test on a Beam, Single Load

Shear tests

Two main types of shear test are used to determine the shear properties of materials. One type investigates the direct or transverse shear strength of a material and is used in connection with the shear strength of bolts, rivets, and beams. A typical arrangement is shown diagrammatically in Fig. 10.5, where the test piece is clamped to a block and the load is applied through the shear tool until failure occurs. In the arrangement shown the test piece is subjected to double shear, whereas if it is extended only partially across the gap in the block, it is subjected to single shear. In either case, the average shear strength is taken as the maximum load divided by the shear resisting area.

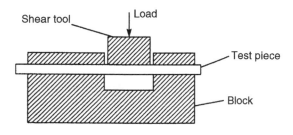

FIGURE 10.5 Shear Test

The other type of shear test is used to evaluate the basic shear properties of a material, such as the shear modulus, G, the shear stress at yield, and the ultimate shear stress. In the usual form of test, a solid circular-section test piece is placed in a torsion machine and twisted by controlled increments of torque. The corresponding angles of twist are recorded and torque–twist diagrams plotted, from which the shear properties of the material are obtained. The method is similar to that used to determine the tensile properties of a material from a tensile test and uses relationships derived in Chapter 3.

Hardness tests

The machinability of a material and its resistance to scratching or penetration are determined by its "hardness." There also appears to be a connection between the hardness of some materials and their tensile strength, so that hardness tests may be used to determine the properties of a finished structural member where tensile and other tests would be impracticable. Hardness tests are also used to investigate the effects of heat treatment, hardening and tempering, and cold forming. Two types of hardness test are in common use: *indentation tests* and *scratch and abrasion tests*.

Indentation tests may be subdivided into two classes: static and dynamic. Of the static tests, the *Brinell* is the most common. In this, a hardened steel ball is pressed into the material under test by a static load acting for a fixed period of time. The load in kilograms divided by the spherical area of the indentation in square millimeters is called the *Brinell hardness number* (BHN). In Fig. 10.6, if D is the diameter of the ball, F the load in kilograms, h the depth of the indentation, and d the diameter of the indentation, then

$$BHN = \frac{F}{\pi D h} = \frac{2F}{\pi D[D - \sqrt{D^2 - d^2}]}$$

In practice, the hardness number of a given material is found to vary with F and D, so that for uniformity the test is standardized. For steel and hard materials, $F = 3000$ kg and $D = 10$ mm, while for soft materials $F = 500$ kg and $D = 10$ mm; in addition, the load is usually applied for 15 s.

In the Brinell test, the dimensions of the indentation are measured by means of a microscope. To avoid this rather tedious procedure, direct reading machines have been devised, of which the *Rockwell* is typical. The indenting tool, again a hardened sphere, is first applied under a definite light load. This indenting tool is then replaced by a diamond cone with a rounded point, which is then applied under a specified indentation load. The difference between the depth of the indentation under the two loads is taken as a measure of the hardness of the material and is read directly from the scale.

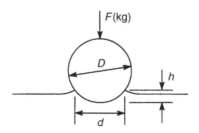

FIGURE 10.6 Brinell Hardness Test

A typical dynamic hardness test is performed by the *Shore Scleroscope*, which consists of a small hammer approximately 20 mm long and 6 mm in diameter fitted with a blunt, rounded, diamond point. The hammer is guided by a vertical glass tube and allowed to fall freely from a height of 25 cm onto the specimen, which it indents before rebounding. A certain proportion of the energy of the hammer is expended in forming the indentation, so that the height of the rebound, which depends upon the energy still possessed by the hammer, is taken as a measure of the hardness of the material.

A number of tests have been devised to measure the "scratch hardness" of materials. In one test, the smallest load in grams that, when applied to a diamond point, produces a scratch visible to the naked eye on a polished specimen of material is called its *hardness number*. In other tests, the magnitude of the load required to produce a definite width of scratch is taken as the measure of hardness. Abrasion tests, involving the shaking over a period of time of several specimens placed in a container, measure the resistance to wear of some materials. In some cases, there appears to be a connection between wear and hardness number, although the results show no level of consistency.

Impact tests

It has been found that certain materials, particularly heat-treated steels, are susceptible to failure under shock loading, whereas an ordinary tensile test on the same material shows no abnormality. Impact tests measure the ability of materials to withstand shock loads and provide an indication of their *toughness*. Two main tests are in use, the *Izod* and the *Charpy*.

Both tests rely on a striker or weight attached to a pendulum. The pendulum is released from a fixed height, the weight strikes a notched test piece, and the angle through which the pendulum then swings is a measure of the toughness of the material. The arrangement for the Izod test is shown diagrammatically in Fig. 10.7(a). The specimen and the method of mounting are shown in detail in Fig. 10.7(b). The Charpy test is similar in operation except that the test piece is supported in a different manner, as shown in the plan view in Fig. 10.8.

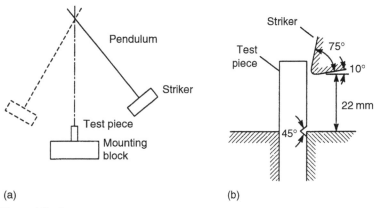

(a)   (b)

FIGURE 10.7 Izod Impact Test

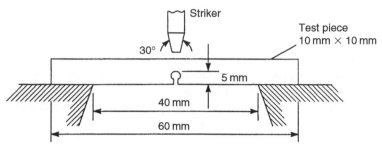

FIGURE 10.8 Charpy Impact Test

10.7.2 Stress–strain curves

We now examine in detail the properties of the different materials from the viewpoint of the results obtained from tensile and compression tests.

Low carbon steel (mild steel)

A nominal stress–strain curve for mild steel, a ductile material, is shown in Fig. 10.9. From 0 to a, the stress–strain curve is linear, the material in this range obeying Hooke's law. Beyond a, the *limit of proportionality*, stress is no longer proportional to strain and the stress–strain curve continues to b, the *elastic limit*, which is defined as the maximum stress that can be applied to a material without producing a permanent plastic deformation or *permanent set* when the load is removed. In other words, if the material is stressed beyond b and the load then removed, a residual strain exists at zero load. For many materials, it is impossible to detect a difference between the limit of proportionality and the elastic limit. From 0 to b, the material is said to be in the *elastic range*, while from b to fracture, the material is in the *plastic range*. The transition from the elastic to the plastic range may be explained by

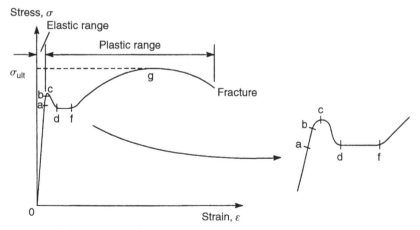

FIGURE 10.9 Stress–Strain Curve for Mild Steel

considering the arrangement of crystals in the material. As the load is applied, slipping occurs between the crystals aligned most closely to the direction of load. As the load is increased, more and more crystals slip with each equal load increment, until appreciable strain increments are produced and the plastic range is reached.

A further increase in stress from b results in the mild steel reaching its *upper yield point* at c, followed by a rapid fall in stress to its *lower yield point* at d. The existence of a lower yield point for mild steel is a peculiarity of the tensile test, wherein the movement of the ends of the test piece produced by the testing machine does not proceed as rapidly as its plastic deformation; the load therefore decreases, as does the stress. From d to f, the strain increases at a roughly constant value of stress until *strain hardening* again causes an increase in stress. This increase in stress continues, accompanied by a large increase in strain to g, the *ultimate stress*, σ_{ult}, of the material. At this point, the test piece begins, visibly, to "neck" as shown in Fig. 10.10. The material in the test piece in the region of the "neck" is almost perfectly plastic at this stage, and from this point onwards to fracture, there is a reduction in nominal stress.

For mild steel, yielding occurs at a stress of the order of 300 N/mm². At fracture, the strain (i.e., the elongation) is of the order of 30 percent. The gradient of the linear portion of the stress–strain curve gives a value for Young's modulus in the region of 200,000 N/mm².

The characteristics of the fracture are worthy of examination. In a cylindrical test piece, the two halves of the fractured test piece have ends that form a "cup and cone" (Fig. 10.11). The actual failure planes in this case are inclined at approximately 45° to the axis of loading and coincide with planes of maximum shear stress. Similarly, if a flat tensile specimen of mild steel is polished then stressed, a pattern of fine lines appears on the polished surface at yield. These lines, which were first discovered by Lüder in 1854, intersect approximately at right angles and are inclined at 45° to the axis of the specimen, thereby coinciding with planes of maximum shear stress. These forms of yielding and fracture suggest that the crystalline structure of the steel is relatively weak in shear with yielding taking the form of the sliding of one crystal plane over another rather than the tearing apart of two crystal planes.

The behavior of mild steel in compression is very similar to its behavior in tension, particularly in the elastic range. In the plastic range, it is not possible to obtain ultimate and fracture loads, since, due to compression, the area of cross-section increases as the load increases, producing a "barrelling"

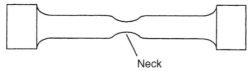

Neck

FIGURE 10.10 "Necking" of a Test Piece in the Plastic Range

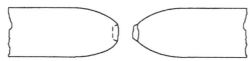

FIGURE 10.11 "Cup-and-Cone" Failure of a Mild Steel Test Piece

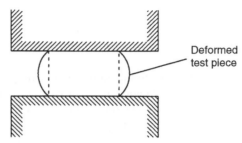

FIGURE 10.12 "Barrelling" of a Mild Steel Test Piece in Compression

effect, as shown in Fig. 10.12. This increase in cross-sectional area tends to decrease the true stress, thereby increasing the load resistance. Ultimately, a flat disc is produced. For design purposes, the ultimate stresses of mild steel in tension and compression are assumed to be the same.

Higher grades of steel have greater strengths than mild steel but are not as ductile. They also possess the same Young's modulus, so that the higher stresses are accompanied by higher strains.

Aluminum

Aluminum and some of its alloys are also ductile materials, although their stress–strain curves do not have the distinct yield stress of mild steel. A typical stress–strain curve is shown in Fig. 10.13. The points a and b again mark the limit of proportionality and elastic limit, respectively, but are difficult to determine experimentally. Instead, a *proof stress* is defined, which is the stress required to produce a given permanent strain on removal of the load. In Fig. 10.13, a line drawn parallel to the linear portion of the stress–strain curve from a strain of 0.001 (i.e., a strain of 0.1 percent) intersects the stress–strain curve at the 0.1 percent proof stress. For elastic design this, or the 0.2 percent proof stress, is taken as the working stress.

Beyond the limit of proportionality the material extends plastically, reaching its ultimate stress, σ_{ult}, at d before finally fracturing under a reduced nominal stress at f.

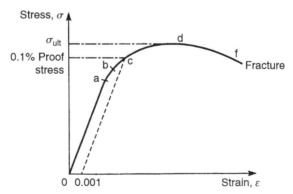

FIGURE 10.13 Stress–Strain Curve for Aluminum

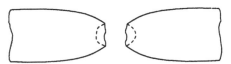

FIGURE 10.14 "Double-Cup" Failure of an Aluminum Alloy Test Piece

A feature of the fracture of aluminum alloy test pieces is the formation of a "double cup," as shown in Fig. 10.14, implying that failure was initiated in the central portion of the test piece while the outer surfaces remained intact. Again, considerable "necking" occurs.

In compression tests on aluminum and its ductile alloys, similar difficulties are encountered to those experienced with mild steel. The stress–strain curve is very similar in the elastic range to that obtained in a tensile test, but the ultimate strength in compression cannot be determined; in design, its value is assumed to coincide with that in tension.

Aluminum and its alloys can suffer a form of corrosion, particularly in the salt- laden atmosphere of coastal regions. The surface becomes pitted and covered by a white furry deposit. This can be prevented by an electrolytic process called *anodizing*, which covers the surface with an inert coating. Aluminum alloys also corrode if they are placed in direct contact with other metals, such as steel. To prevent this, plastic is inserted between the possible areas of contact.

Brittle materials

These include cast iron, high-strength steel, concrete, timber, ceramics, and glass. The plastic range for brittle materials extends to only small values of strain. A typical stress–strain curve for a brittle material under tension is shown in Fig. 10.15. Little or no yielding occurs and fracture takes place very shortly after the elastic limit is reached.

The fracture of a cylindrical test piece takes the form of a single failure plane approximately perpendicular to the direction of loading with no visible "necking" and an elongation of the order of 2–3 percent.

In compression, the stress–strain curve for a brittle material is very similar to that in tension, except that failure occurs at a much higher value of stress; for concrete, the ratio is of the order of 10:1.

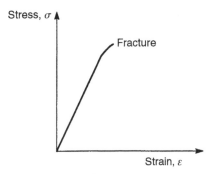

FIGURE 10.15 Stress–Strain Curve for a Brittle Material

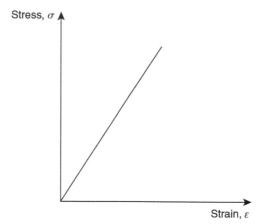

FIGURE 10.16 Stress–Strain Curve for a Fiber Composite

This is thought to be due to the presence of microscopic cracks in the material, giving rise to high stress concentrations, which are more likely to have a greater effect in reducing tensile strength than compressive strength.

Composites

Fiber composites have stress–strain characteristics indicating that they are brittle materials (Fig. 10.16). There is little or no plasticity, and the modulus of elasticity is less than that of steel and aluminum alloy. However, the fibers themselves can have much higher values of strength and modulus of elasticity than the composite. For example, carbon fibers have a tensile strength of the order 2,400 N/mm^2 and a modulus of elasticity of 400,000 N/mm^2.

Fiber composites are highly durable, require no maintenance, and can be used in hostile chemical and atmospheric environments; vinyls and epoxy resins provide the best resistance.

All the stress–strain curves described in the preceding discussion are those produced in tensile or compression tests in which the strain is applied at a negligible rate. A rapid strain application results in significant changes in the apparent properties of the materials, giving possible variations in yield stress of up to 100 percent.

10.7.3 Strain hardening

The stress–strain curve for a material is influenced by the *strain history*, or the loading and unloading of the material, within the plastic range. For example, in Fig. 10.17, a test piece is initially stressed in tension beyond the yield stress at a to a value at b. The material is then unloaded to c and reloaded to f, producing an increase in yield stress from the value at a to the value at d. Subsequent unloading to g and loading to j increases the yield stress still further to the value at h. This increase in strength resulting from the loading and unloading is known as *strain hardening*. It can be seen, from Fig. 10.17, that the stress–strain curve during the unloading and loading cycles form loops (the shaded areas in Fig. 10.17). These indicate that strain energy is lost during the cycle, the energy being dissipated in

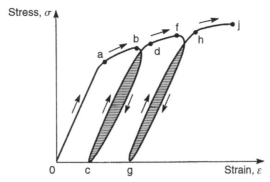

FIGURE 10.17 Strain Hardening of a Material

the form of heat produced by internal friction. This energy loss is known as *mechanical hysteresis* and the loops as *hysteresis loops*. Although the ultimate stress is increased by strain hardening, it is not influenced to the same extent as yield stress. The increase in strength produced by strain hardening is accompanied by decreases in toughness and ductility.

10.7.4 Creep and relaxation

We saw in Chapter 1 that a given load produces a calculable value of stress in a structural member and hence a corresponding value of strain once the full value of the load is transferred to the member. However, after this initial or 'instantaneous' stress and its corresponding value of strain have been attained, a great number of structural materials continue to deform slowly and progressively under load over a period of time. This behavior is known as *creep*. A typical creep curve is shown in Fig. 10.18.

Some materials, such as plastics and rubber, exhibit creep at room temperatures, but most structural materials require high temperatures or long-duration loading at moderate temperatures. In some "soft" metals, such as zinc and lead, creep occurs over a relatively short period of time, whereas materials such as concrete may be subject to creep over a period of years. Creep occurs in steel to a slight extent at normal temperatures but becomes very important at temperatures above 316°C.

Closely related to creep is *relaxation*. Whereas creep involves an increase in strain under constant stress, relaxation is the decrease in stress experienced over a period of time by a material subjected to a constant strain.

10.7.5 Fatigue

Structural members are frequently subjected to repetitive loading over a long period of time. For example, the members of a bridge structure suffer variations in loading possibly thousands of times a day as traffic moves over the bridge. In these circumstances, a structural member may fracture at a level of stress substantially below the ultimate stress for nonrepetitive static loads; this phenomenon is known as *fatigue*.

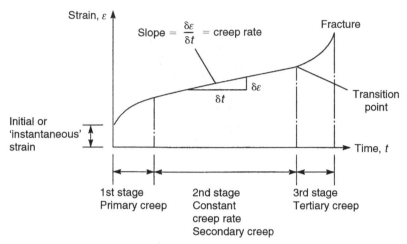

FIGURE 10.18 Typical Creep Curve

Fatigue cracks are most frequently initiated at sections in a structural member where changes in geometry, such as holes, notches, or sudden changes in section, cause *stress concentrations.* Designers seek to eliminate such areas by ensuring that rapid changes in section are as smooth as possible. At re-entrant corners for example, fillets are provided, as shown in Fig. 10.19.

Other factors which affect the failure of a material under repetitive loading are the type of loading (fatigue is primarily a problem with repeated tensile stresses, probably because microscopic cracks can propagate more easily under tension), temperature, the material, surface finish (machine marks are potential crack propagators), corrosion, and residual stresses produced by welding.

Frequently in structural members, an alternating stress, σ_{alt}, is superimposed on a static or mean stress, σ_{mean}, as illustrated in Fig. 10.20. The value of σ_{alt} is the most important factor in determining the number of cycles of load that produce failure. The stress σ_{alt} that can be withstood for a specified

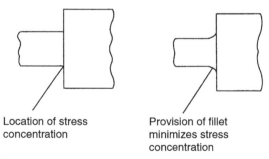

Location of stress
concentration

Provision of fillet
minimizes stress
concentration

FIGURE 10.19 Stress Concentration Location

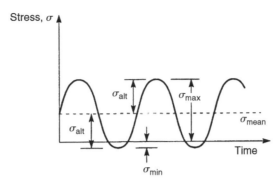

FIGURE 10.20 Alternating Stress in Fatigue Loading

number of cycles is called the *fatigue strength* of the material. Some materials, such as mild steel, possess a stress level that can be withstood for an indefinite number of cycles. This stress is known as the *endurance limit* of the material; no such limit has been found for aluminum and its alloys. Fatigue data are frequently presented in the form of an *S–n* curve or stress–endurance curve, as shown in Fig. 10.21.

In many practical situations, the amplitude of the alternating stress varies and is frequently random in nature. The *S–n* curve does not, therefore, apply directly, and an alternative means of predicting failure is required. *Miner's cumulative damage theory* suggests that failure occurs when

$$\frac{n_1}{N_1} + \frac{n_2}{N_2} + \ldots + \frac{n_r}{N_r} = 1 \tag{10.1}$$

where $n_1, n_2, \ldots, n_r$ are the number of applications of stresses σ_{alt}, σ_{mean} and $N_1, N_2, \ldots, N_r$ are the number of cycles to failure of stresses σ_{alt}, σ_{mean}.

We examine fatigue and its effect on aircraft design in much greater detail in Chapter 14.

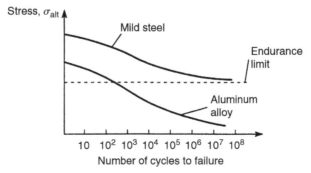

FIGURE 10.21 Stress–Endurance Curves

PROBLEMS

P.10.1 Describe a simple tensile test and show, with the aid of sketches, how measures of the ductility of the material of the specimen may be obtained. Sketch typical stress–strain curves for mild steel and an aluminum alloy showing their important features.

P.10.2 A bar of metal 25 mm in diameter is tested on a length of 250 mm. In tension, the results in Table P.10.2 were recorded. A torsion test gave the results also in Table P.10.2. Represent these results in graphical form and determine Young's modulus, E; the modulus of rigidity, G; Poisson's ratio, v; and the bulk modulus, K; for the metal.

Answer: $E \simeq 205,000 \text{ N/mm}^2$, $G \simeq 80,700 \text{ N/mm}^2$, $v \simeq 0.27$, $K \simeq 148,600 \text{ N/mm}^2$

P.10.2 MATLAB Use MATLAB to repeat Problem P.11.2 for the following bar diameter values:

	(i)	(ii)	(iii)	(iv)	(v)
d	24.8 mm	24.9 mm	25 mm	25.1 mm	25.2 mm

Assume that the slopes of the tension and torsion data provided in Table P.10.2 are 402.6 kN/mm and 12.38 kN m/rad, respectively.

Table P.10.2 Problem 11.2				
Tension:				
Load (kN)	10.4	31.2	52.0	72.8
Extension (mm)	0.036	0.089	0.140	0.191
Torsion:				
Torque (kN m)	0.051	0.152	0.253	0.354
Angle of twist (degrees)	0.24	0.71	1.175	1.642

Answer: (i) $E \approx 208,400 \text{ N/mm}^2$, $G \approx 83,300 \text{ N/mm}^2$, $v \approx 0.25$, $K \approx 138,900 \text{ N/mm}^2$
          (ii) $E \approx 206,700 \text{ N/mm}^2$, $G \approx 82,000 \text{ N/mm}^2$, $v \approx 0.26$, $K \approx 143,500 \text{ N/mm}^2$
          (iii) $E \approx 205,000 \text{ N/mm}^2$, $G \approx 80,700 \text{ N/mm}^2$, $v \approx 0.27$, $K \approx 148,600 \text{ N/mm}^2$
          (iv) $E \approx 203,400 \text{ N/mm}^2$, $G \approx 79,400 \text{ N/mm}^2$, $v \approx 0.28$, $K \approx 154,100 \text{ N/mm}^2$
          (v) $E \approx 201,800 \text{ N/mm}^2$, $G \approx 78,200 \text{ N/mm}^2$, $v \approx 0.29$, $K \approx 160,200 \text{ N/mm}^2$

P.10.3 The actual stress–strain curve for a particular material is given by $\sigma = C\varepsilon^n$, where C is a constant. Assuming that the material suffers no change in volume during plastic deformation, derive an expression for the nominal stress–strain curve and show that this has a maximum value when $\varepsilon = n/(1-n)$.

Answer: $\sigma_{nom} = C\varepsilon^n/(1+\varepsilon)$

P.10.4 A structural member is to be subjected to a series of cyclic loads that produce different levels of alternating stress, as shown in Table P.10.4. Determine whether or not a fatigue failure is probable.

Answer: Not probable $(n_1/N_1 + n_2/N_2 + \cdots = 0.39)$

Table P.10.4 Problem 10.4

Loading	Number of cycles	Number of cycles to failure
1	10^4	5×10^4
2	10^5	10^6
3	10^6	24×10^7
4	10^7	12×10^7

Structural components of aircraft

Aircraft are generally built up from the basic components of wings, fuselages, tail units, and control surfaces. There are variations in particular aircraft, for example, a delta wing aircraft would not necessarily possess a horizontal tail, although this is present in a canard configuration, such as that of the Eurofighter (Typhoon). Each component has one or more specific functions and must be designed to ensure that it can carry out these functions safely. In this chapter, we describe the various loads to which aircraft components are subjected, their function and fabrication, and the design of connections.

11.1 LOADS ON STRUCTURAL COMPONENTS

The structure of an aircraft is required to support two distinct classes of load: the first, termed *ground loads*, includes all loads encountered by the aircraft during movement or transportation on the ground, such as taxiing and landing loads, towing and hoisting loads; while the second, *air loads*, comprises loads imposed on the structure during flight by maneuvers and gusts. In addition, aircraft designed for a particular role encounter loads peculiar to their sphere of operation. Carrier-borne aircraft, for instance, are subjected to catapult take-off and arrested landing loads: most large civil and practically all military aircraft have pressurized cabins for high-altitude flying; amphibious aircraft must be capable of landing on water and aircraft designed to fly at high speed at low altitude, such as the Tornado, require a structure of above average strength to withstand the effects of flight in extremely turbulent air.

The two classes of loads may be further divided into *surface forces*, which act upon the surface of the structure, such as aerodynamic and hydrostatic pressure, and *body forces*, which act over the volume of the structure and are produced by gravitational and inertial effects. Calculation of the distribution of aerodynamic pressure over the various surfaces of an aircraft's structure is presented in numerous texts on aerodynamics and therefore is not attempted here. However, we discuss the types of load induced by these various effects and their action on the different structural components.

Basically, all air loads are the result of the pressure distribution over the surfaces of the skin produced by steady flight, maneuver, or gust conditions. Generally, these results cause direct loads, bending, shear, and torsion in all parts of the structure in addition to local, normal pressure loads imposed on the skin.

Conventional aircraft usually consist of fuselage, wings, and tailplane. The fuselage contains crew and payload, the latter being passengers, cargo, or weapons plus fuel, depending on the type of aircraft and its function, the wings provide the lift and the tailplane is the main contributor to directional control. In addition, ailerons, elevators, and the rudder enable the pilot to maneuver the aircraft and maintain its stability in flight, while wing flaps provide the necessary increase of lift for take-off and landing. Figure 11.1 shows typical aerodynamic force resultants experienced by an aircraft in steady flight.

Introduction to Aircraft Structural Analysis, Second Edition

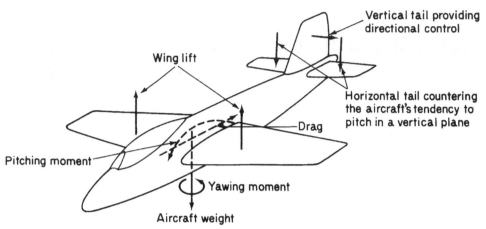

FIGURE 11.1 Principal Aerodynamic Forces on an Aircraft during Flight

The force on an aerodynamic surface (wing, vertical or horizontal tail) results from a differential pressure distribution caused by incidence, camber, or a combination of both. Such a pressure distribution, shown in Fig. 11.2(a), has vertical (lift) and horizontal (drag) resultants acting at a center of pressure (CP). (In practice, lift and drag are measured perpendicular and parallel to the flight path, respectively.) Clearly, the position of the CP changes as the pressure distribution varies with speed or wing incidence. However, there is, conveniently, a point in the aerofoil section about which the moment due to the lift and drag forces remains constant. We therefore replace the lift and drag forces acting at the CP by lift and drag forces acting at the aerodynamic center (AC) plus a constant moment M_0, as shown in Fig. 11.2(b). (Actually, at high Mach numbers, the position of the AC changes due to compressibility effects.)

While the chordwise pressure distribution fixes the position of the resultant aerodynamic load in the wing cross-section, the spanwise distribution locates its position in relation, say, to the wing root. A typical distribution for a wing–fuselage combination is shown in Fig. 11.3. Similar distributions occur on horizontal and vertical tail surfaces.

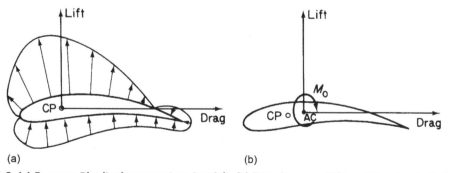

FIGURE 11.2 (a) Pressure Distribution around an Aerofoil; (b) Transference of Lift and Drag Loads to the AC

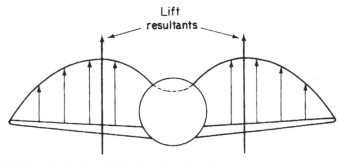

FIGURE 11.3 Typical Lift Distribution for a Wing–Fuselage Combination

We see therefore that wings, tailplane, and the fuselage are each subjected to direct, bending, shear, and torsional loads and must be designed to withstand critical combinations of these. Note that maneuvers and gusts do not introduce different loads but result only in changes of magnitude and position of the type of existing loads shown in Fig. 11.1. Over and above these basic in-flight loads, fuselages may be pressurized and thereby support hoop stresses; wings may carry weapons or extra fuel tanks with resulting additional aerodynamic and body forces contributing to the existing bending, shear, and torsion; and the thrust and weight of engines may affect either the fuselage or wings, depending on their relative positions.

Ground loads encountered in landing and taxiing subject the aircraft to concentrated shock loads through the undercarriage system. The majority of aircraft have their main undercarriage located in the wings, with a nosewheel or tail wheel in the vertical plane of symmetry. Clearly, the position of the main undercarriage should be such as to produce minimum loads on the wing structure compatible with the stability of the aircraft during ground maneuvers. This may be achieved by locating the undercarriage just forward of the flexural axis of the wing and as close to the wing root as possible. In this case, the shock landing load produces a given shear, minimum bending plus torsion, with the latter being reduced as far as practicable by offsetting the torque caused by the vertical load in the undercarriage leg by a torque in an opposite sense due to braking.

Other loads include engine thrust on the wings or fuselage, which acts in the plane of symmetry but may, in the case of engine failure, cause severe fuselage bending moments, as shown in Fig. 11.4; concentrated shock loads during a catapult launch; and hydrodynamic pressure on the fuselages or floats of seaplanes.

In Chapter 12, we examine in detail the calculation of ground and air loads for a variety of cases.

11.2 FUNCTION OF STRUCTURAL COMPONENTS

The basic functions of an aircraft's structure are to transmit and resist the applied loads and to provide an aerodynamic shape and to protect passengers, payload, systems, and the like from the environmental conditions encountered in flight. These requirements, in most aircraft, result in thin-shell structures where the outer surface or skin of the shell is usually supported by longitudinal stiffening members and transverse frames to enable it to resist bending, compressive, and torsional loads without buckling.

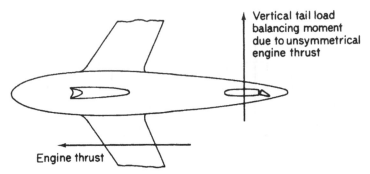

Engine thrust

Vertical tail load balancing moment due to unsymmetrical engine thrust

FIGURE 11.4 Fuselage and Wing Bending Caused by an Unsymmetrical Engine Load

Such structures are known as *semi-monocoque*, while thin shells that rely entirely on their skins for their capacity to resist loads are referred to as *monocoque*.

First, we consider wing sections, which, while performing the same function, can differ widely in their structural complexity, as can be seen by comparing Figs. 11.5 and 11.6. In Fig. 11.5, the wing of a small, light passenger aircraft, the De Havilland Canada Twin Otter, comprises a relatively simple arrangement of two spars, ribs, stringers, and skin, while the wing of the Harrier in Fig. 11.6 consists of numerous spars, ribs, and skin. However, no matter how complex the internal structural arrangement, the different components perform the same kind of function.

The shape of the cross-section is governed by aerodynamic considerations and clearly must be maintained for all combinations of load; this is one of the functions of the ribs. They also act with the skin in resisting the distributed aerodynamic pressure loads; they distribute concentrated loads (e.g., undercarriage and additional wing store loads) into the structure and redistribute stress around discontinuities, such as undercarriage wells, inspection panels, and fuel tanks, in the wing surface. Ribs increase the column buckling stress of the longitudinal stiffeners by providing end restraint and establishing their column length; in a similar manner, they increase the plate buckling stress of the skin panels. The dimensions of ribs are governed by their spanwise position in the wing and by the loads they are required to support. In the outer portions of the wing, where the cross-section may be relatively small if the wing is tapered and the loads are light, ribs act primarily as formers for the aerofoil shape. A light structure is sufficient for this purpose, whereas at sections closer to the wing root, where the ribs are required to absorb and transmit large concentrated applied loads, such as those from the undercarriage, engine thrust, and fuselage attachment point reactions, a much more rugged construction is necessary. Between these two extremes are ribs that support hinge reactions from ailerons, flaps, and other control surfaces, plus the many internal loads from fuel, armament, and systems installations.

The primary function of the wing skin is to form an impermeable surface for supporting the aerodynamic pressure distribution from which the lifting capability of the wing is derived. These aerodynamic forces are transmitted in turn to the ribs and stringers by the skin through plate and membrane action. Resistance to shear and torsional loads is supplied by shear stresses developed in the skin and spar webs, while axial and bending loads are reacted by the combined action of skin and stringers.

Although the thin skin is efficient for resisting shear and tensile loads, it buckles under comparatively low compressive loads. Rather than increase the skin thickness and suffer a consequent weight penalty, stringers are attached to the skin and ribs, thereby dividing the skin into small panels and

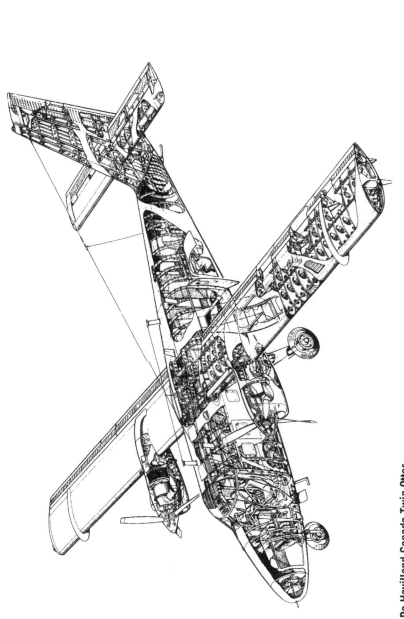

(courtesy of De Havilland Aircraft of Canada Ltd.)

FIGURE 11.5 De Havilland Canada Twin Otter

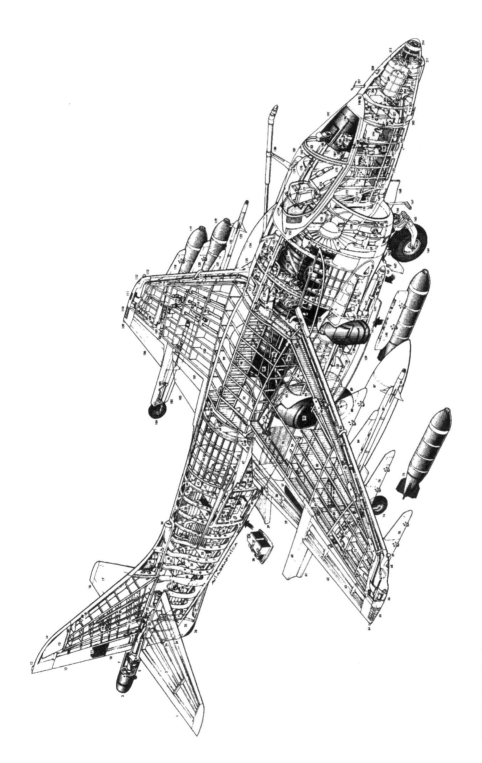

FIGURE 11.6 Harrier

1. Starboard all-moving tailplane;
2. Tailplane composite construction;
3. Tail radome;
4. Military equipment;
5. Tail pitch control air valve;
6. Yaw control air valves;
7. Tail "bullet" fairing;
8. Reaction control system air ducting;
9. Trim tab actuator;
10. Rudder trim tab;
11. Rudder composite construction;
12. Rudder;
13. Antenna;
14. Fin tip aerial fairing;
15. Upper broad band communications antenna;
16. Port tailplane;
17. Graphite epoxy tailplane skin;
18. Port side temperature probe;
19. MAD compensator;
20. Formation lighting strip;
21. Fin construction;
22. Fin attachment joint;
23. Tailplane pivot sealing plate;
24. Aerials;
25. Ventral fin;
26. Tail bumper;
27. Lower broad band communications antenna;
28. Tailplane hydraulic jack;
29. Heat exchanger air exhaust;
30. Aft fuselage frames;
31. Rudder hydraulic actuator;
32. Avionics equipment air conditioning plant;
33. Avionics equipment racks;
34. Heat exchanger ram air intake;
35. Electrical system circuit breaker panels, port and starboard;
36. Avionic equipment;
37. Chaff and flare dispensers;
38. Dispenser electronic control units;
39. Ventral airbrake;
40. Airbrake hydraulic jack;
41. Formation lighting strip;
42. Avionics bay access door, port and starboard;
43. Avionics equipment racks;
44. Fuselage frame and stringer construction;
45. Rear fuselage fuel tank;
46. Main undercarriage wheel bay;
47. Wing root fillet;
48. Wing spar–fuselage attachment joint;
49. Water filler cap;
50. Engine fire extinguisher bottle;
51. Anticollision light;
52. Water tank;
53. Flap hydraulic actuator;
54. Flap hinge fitting;
55. Nimonic fuselage heat shield;
56. Main undercarriage bay doors (closed after cycling of mainwheels);
57. Flap vane composite construction;
58. Flap composite construction;
59. Starboard slotted flap, lowered;
60. Outrigger wheel fairing;
61. Outrigger leg doors;
62. Starboard aileron;
63. Aileron composite construction;
64. Fuel jettison;
65. Formation lighting panel;
66. Roll control airvalve;
67. Wing tip fairing;
68. Starboard navigation light;
69. Radar warning aerial;
70. Outboard pylon;
71. Pylon attachment joint;
72. Graphite epoxy composite wing construction;
73. Aileron hydraulic actuator;
74. Starboard outrigger wheel;
75. BL755 600-lb. (272-kg) cluster bomb (CBU);
76. Intermediate pylon;
77. Reaction control air ducting;
78. Aileron control rod;
79. Outrigger hydraulic retraction jack;
80. Outrigger leg strut;
81. Leg pivot fixing;
82. Multispar wing construction;
83. Leading-edge wing fence;
84. Outrigger pylon;
85. Missile launch rail;
86. AIM-9L Sidewinder air-to-air missile;
87. External fuel tank, 300 U.S. gal (1 135 l);
88. Inboard pylon;
89. Aft retracting twin main wheels;
90. Inboard pylon attachment joint;
91. Rear (hot stream) swiveling exhaust nozzle;
92. Position of pressure refueling connection on port side;
93. Rear nozzle bearing;
94. Center fuselage flank fuel tank;
95. Hydraulic reservoir;
96. Nozzle bearing cooling air duct;
97. Engine exhaust divider duct;
98. Wing panel center rib;
99. Center section integral fuel tank;
100. Port wing integral fuel tank;
101. Flap vane;
102. Port slotted flap, lowered;
103. Outrigger wheel fairing;
104. Port outrigger wheel;
105. Torque scissor links;
106. Port aileron;
107. Aileron hydraulic actuator;
108. Aileron–air valve interconnection;
109. Fuel jettison;
110. Formation lighting panel;
111. Port roll control air valve;
112. Port navigation light;
113. Radar warning aerial;
114. Port wing reaction control air duct;
115. Fuel pumps;
116. Fuel system piping;
117. Port wing leading-edge fence;
118. Outboard pylon;
119. BL755 cluster bombs (maximum load, seven);
120. Intermediate pylon;
121. Port outrigger pylon;
122. Missile launch rail;
123. AIM-9L Sidewinder air-to-air missile;
124. Port leading-edge root extension (LERX);
125. Inboard pylon;
126. Hydraulic pumps;
127. APU intake;
128. Gas turbine starter–auxiliary power unit (APU);
129. Alternator cooling air exhaust;
130. APU exhaust;
131. Engine fuel control unit;
132. Engine bay venting ram air intake;
133. Rotary nozzle bearing;
134. Nozzle fairing construction;
135. Ammunition tank, 100 rounds;
136. Cartridge case collector box;
137. Ammunition feed chute;
138. Fuel vent;
139. Gun pack strake;
140. Fuselage centerline pylon;
141. Zero scarf forward (fan air) nozzle;
142. Ventral gun pack (two);
143. Aden 25-mm cannon;
144. Engine drain mast;
145. Hydraulic system ground connectors;
146. Forward fuselage flank fuel tank;
147. Engine electronic control units;
148. Engine accessory equipment gearbox;
149. Gearbox driven alternator;
150. Rolls-Royce Pegasus 11 Mk 105 vectored thrust turbofan;
151. Formation lighting strips;
152. Engine oil tank;
153. Bleed air spill duct;
154. Air conditioning intake scoops;
155. Cockpit air conditioning system heat exchanger;
156. Engine compressor–fan face;
157. Heat exchanger discharge to intake duct;
158. Nose undercarriage hydraulic retraction jack;
159. Intake blow-in doors;
160. Engine bay venting air scoop;
161. Cannon muzzle fairing;
162. Lift augmentation retractable cross-dam;
163. Cross-dam hydraulic jack;
164. Nosewheel;
165. Nosewheel forks;
166. Landing/taxiing lamp;
167. Retractable boarding step;
168. Nosewheel doors (closed after cycling of undercarriage);
169. Nosewheel door jack;
170. Boundary layer bleed air duct;
171. Nose undercarriage wheel bay;
172. Kick-in boarding steps;
173. Cockpit rear pressure bulkhead;
174. Starboard side console panel;
175. Martin-Baker Type 12 ejection seat;
176. Safety harness;
177. Ejection seat headrest;
178. Port engine air intake;
179. Probe hydraulic jack;
180. Retractable n-flight refueling probe (bolt-on pac-x);
181. Cockpit canopy cover;
182. Miniature detonating cord (MDC) canopy breaker;
183. Canopy frame;
184. Engine throttle and nozzle angle control levers;
185. Pilot's head-up display;
186. Instrument panel;
187. Moving map display;
188. Control column;
189. Central warning system panel;
190. Cockpit pressure floor;
191. Underfloor control runs;
192. Formation lighting strips;
193. Aileron trim actuator;
194. Rudder pedals;
195. Cockpit section composite construction;
196. Instrument panel shroud;
197. One-piece wraparound windscreen panel;
198. Ram air intake (cockpit fresh air);
199. Front pressure bulkhead;
200. Incidence vane;
201. Air data computer;
202. Pitot tube;
203. Lower IFF aerial;
204. Nose pitch control air valve;
205. Pitch trim control actuator;
206. Electrical system equipment;
207. Yaw vane;
208. Upper IFF aerial;
209. Avionic equipment;
210. ARBS heat exchanger;
211. MIRLS sensors;
212. Hughes Angle Rate Bombing System (ARBS);
213. Composite construction nose cone;
214. ARBS glazed aperture

(courtesy of Pilot Press Ltd.)

increasing the buckling and failing stresses. This stabilizing action of the stringers on the skin is, in fact, reciprocated to some extent although the effect normal to the surface of the skin is minimal. Stringers rely chiefly on rib attachments for preventing column action in this direction. We noted in the previous paragraph the combined action of stringers and skin in resisting axial and bending loads.

The role of spar webs in developing shear stresses to resist shear and torsional loads was mentioned previously; they perform a secondary but significant function in stabilizing, with the skin, the spar flanges, or caps, which are therefore capable of supporting large compressive loads from axial and bending effects. In turn, spar webs exert a stabilizing influence on the skin in a similar manner to the stringers.

While the majority of these remarks have been directed towards wing structures, they apply, as can be seen by referring to Figs 11.5 and 11.6, to all the aerodynamic surfaces, namely, wings, horizontal and vertical tails—except in the obvious cases of undercarriage loading—engine thrust, and so forth.

Fuselages, while of different shape to the aerodynamic surfaces, comprise members that perform functions similar to their counterparts in the wings and tailplane. However, there are differences in the generation of the various types of load. Aerodynamic forces on the fuselage skin are relatively low; on the other hand, the fuselage supports large concentrated loads, such as wing reactions, tailplane reactions, undercarriage reactions and it carries payloads of varying size and weight, which may cause large inertia forces. Furthermore, aircraft designed for high-altitude flight must withstand internal pressure. The shape of the fuselage cross-section is determined by operational requirements. For example, the most efficient sectional shape for a pressurized fuselage is circular or a combination of circular elements. Irrespective of shape, the basic fuselage structure is essentially a single-cell, thin-walled tube comprising skin, transverse frames, and stringers; transverse frames that extend completely across the fuselage are known as *bulkheads*. Three different types of fuselage are shown in Figs. 11.5–11.7. In Fig. 11.5, the fuselage is unpressurized, so that, in the passenger-carrying area, a more rectangular shape is employed to maximize space. The Harrier fuselage in Fig. 11.6 contains the engine, fuel tanks, and the like, so that its cross-sectional shape is, to some extent, predetermined, while in Fig. 11.7, the passenger-carrying fuselage of the British Aerospace 146 is pressurized and therefore circular in cross-section.

11.3 FABRICATION OF STRUCTURAL COMPONENTS

The introduction of all-metal, stressed skin aircraft resulted in methods and types of fabrication that remain in use to the present day. However, improvements in engine performance and advances in aerodynamics led to higher maximum lift, higher speeds, and therefore to higher wing loadings, so that improved techniques of fabrication are necessary, particularly in the construction of wings. The increase in wing loading from about 350 N/m^2 for 1917–1918 aircraft to around 4800 N/m^2 for modern aircraft, coupled with a drop in the structural percentage of the total weight from 30–40 to 22–25 percent, gives some indication of the improvements in materials and structural design.

For purposes of construction, aircraft are divided into a number of sub-assemblies. These are built in specially designed jigs, possibly in different parts of the factory or even different factories, before being forwarded to the final assembly shop. A typical breakdown into sub-assemblies of a medium-sized civil aircraft is shown in Fig. 11.8. Each sub-assembly relies on numerous minor assemblies such as spar webs, ribs, frames, and these, in turn, are supplied with individual components from the detail workshop.

Although the wings (and tail surfaces) of fixed wing aircraft generally consist of spars, ribs, skin, and stringers, methods of fabrication and assembly differ. The wing of the aircraft of Fig. 11.5 relies on fabrication techniques that have been employed for many years. In this form of construction, the spars

FIGURE 11.7 British Aerospace 146

(courtesy of British Aerospace)

comprise thin aluminum alloy webs and flanges, the latter being extruded or machined and bolted or riveted to the web. The ribs are formed in three parts from sheet metal by large presses and rubber dies and have flanges round their edges so that they can be riveted to the skin and spar webs; cut-outs around their edges allow the passage of spanwise stringers. Holes are cut in the ribs at positions of low stress for lightness and to accommodate control runs, fuel, and electrical systems.

Finally, the skin is riveted to the rib flanges and longitudinal stiffeners. Where the curvature of the skin is large, for example, at the leading edge, the aluminum alloy sheets are passed through "rolls" to preform them to the correct shape. A further, aerodynamic, requirement is that forward chordwise sections of the wing be as smooth as possible, to delay transition from laminar to turbulent flow. Therefore, countersunk rivets are used in these positions as opposed to dome-headed rivets nearer the trailing edge.

The wing is attached to the fuselage through reinforced fuselage frames, frequently by bolts. In some aircraft, the wing spars are continuous through the fuselage, depending on the demands of space.

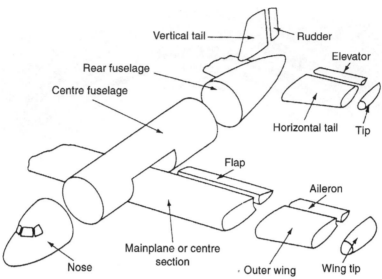

FIGURE 11.8 Typical Sub-assembly Breakdown

In a high wing aircraft (Fig. 11.5), deep spars passing through the fuselage would cause obstruction problems. In this case, a short third spar provides an additional attachment point. The ideal arrangement is obviously where continuity of the structure is maintained over the entire surface of the wing. In most practical cases, this is impossible since cut-outs in the wing surface are required for retracting undercarriages, bomb and gun bays, inspection panels, etc. The last are usually located on the undersurface of the wing and are fastened to stiffeners and rib flanges by screws, enabling them to resist direct and shear loads. Doors covering undercarriage wells and weapon bays are incapable of resisting wing stresses, so that provision must be made for transferring the loads from skin, flanges, and shear webs around the cut-out. This may be achieved by inserting strong bulkheads or increasing the spar flange areas, although, no matter the method employed, increased cost and weight result.

The different structural requirements of aircraft designed for differing operational roles lead to a variety of wing constructions. For instance, high-speed aircraft require relatively thin wing sections, which support high wing loadings. To withstand the correspondingly high surface pressures and to obtain sufficient strength, much thicker skins are necessary. Wing panels are therefore frequently machined integrally with stringers from solid slabs of material, as are the wing ribs. Figure 11.9 shows wing ribs for the European Airbus, in which web stiffeners, flanged lightness holes, and skin attachment lugs have been integrally machined from solid. This integral method of construction involves no new design principles and has the advantages of combining a high grade of surface finish, free from irregularities, with a more efficient use of material, since skin thicknesses are easily tapered to coincide with the spanwise decrease in bending stresses.

An alternative form of construction is the sandwich panel, which comprises a light honeycomb or corrugated metal core sandwiched between two outer skins of the stress-bearing sheet (see Fig. 11.10). The primary function of the core is to stabilize the outer skins, although it may be stress bearing as well. Sandwich panels are capable of developing high stresses, have smooth internal and external surfaces,

FIGURE 11.9 Wing Ribs for the European Airbus

(courtesy of British Aerospace)

and require small numbers of supporting rings or frames. They also possess a high resistance to fatigue from jet efflux. The uses of this method of construction include lightweight "planks" for cabin furniture, monolithic fairing shells generally having plastic facing skins, and the stiffening of flying control surfaces. Thus, for example, the ailerons and rudder of the British Aerospace Jaguar are fabricated from aluminum honeycomb, while fiberglass- and aluminum-faced honeycomb are used extensively in the wings and tail surfaces of the Boeing 747. Some problems, mainly disbonding and internal corrosion, have been encountered in service.

The general principles relating to wing construction are applicable to fuselages, with the exception that integral construction is not used in fuselages for obvious reasons. Figures 11.5, 11.6, and 11.7 show that the same basic method of construction is employed in aircraft having widely differing roles. Generally, the fuselage frames that support large concentrated floor loads or loads from wing or tailplane attachment points are heavier than lightly loaded frames and require stiffening, with additional provision for transmitting the concentrated load into the frame and the skin.

With the frames in position in the fuselage jig, stringers, passing through cut-outs, are riveted to the frame flanges. Before the skin is riveted to the frames and stringers, other subsidiary frames, such as door and window frames, are riveted or bolted in position. The areas of the fuselage in the regions of these cut-outs are reinforced by additional stringers, portions of frame, and increased skin thickness, to react to the high shear flows and direct stresses developed.

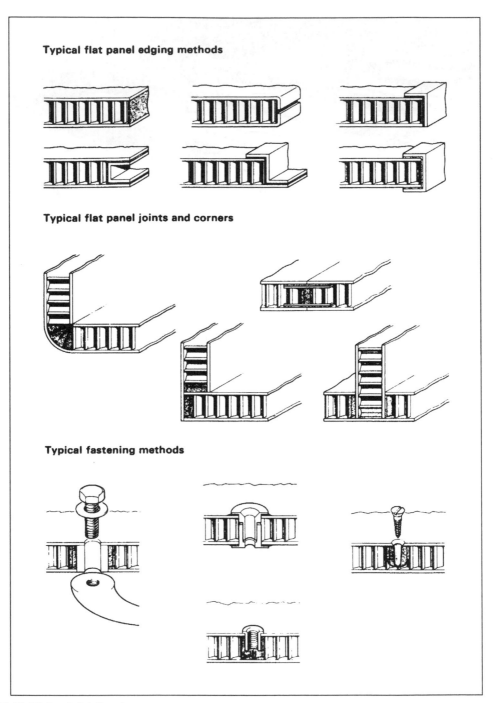

FIGURE 11.10 Sandwich Panels

(courtesy of Ciba-Geigy Plastics)

On completion, the various subassemblies are brought together for final assembly. Fuselage sections are usually bolted together through flanges around their peripheries, while wings and the tailplane are attached to pick-up points on the relevant fuselage frames. Wing spars on low wing civil aircraft usually pass completely through the fuselage, simplifying wing design and the method of attachment. On smaller, military aircraft, engine installations frequently prevent this, so that wing spars are attached directly to and terminate at the fuselage frame. Clearly, at these positions, frame–stringer–skin structures require reinforcement.

11.4 CONNECTIONS

The fabrication of aircraft components generally involves the joining of one part of the component to another. For example, fuselage skins are connected to stringers and frames, while wing skins are connected to stringers and wing ribs unless, as in some military aircraft with high wing loadings, the stringers are machined integrally with the wing skin (see Section 11.3). With the advent of all-metal, that is, aluminum alloy, construction, riveted joints became the main form of connection with some welding, although aluminum alloys are difficult to weld, and in the modern era, some glued joints, which use epoxy resin. In this section, we concentrate on the still predominant method of connection, riveting.

In general, riveted joints are stressed in complex ways, and an accurate analysis is very often difficult to achieve because of the discontinuities in the region of the joint. Fairly crude assumptions as to joint behavior are made, but, when combined with experience, safe designs are produced.

11.4.1 Simple lap joint

Figure 11.11 shows two plates of thickness t connected by a single line of rivets; this type of joint, termed a *lap joint*, is one of the simplest used in construction.

Suppose that the plates carry edge loads of P/unit width, that the rivets are of diameter d and are spaced at a distance b apart, and that the distance from the line of rivets to the edge of each plate is a.

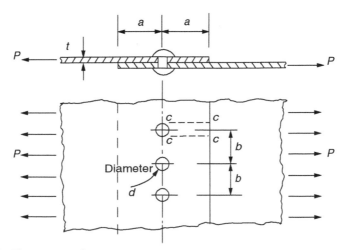

FIGURE 11.11 Simple Riveted Lap Joint

Four possible modes of failure must be considered: rivet shear, bearing pressure, plate failure in tension, and shear failure in a plate.

Rivet shear

The rivets may fail by shear across their diameter at the interface of the plates. Then, if the maximum shear stress the rivets will withstand is τ_1, failure occurs when

$$Pb = \tau_1 \left(\frac{\pi d^2}{4}\right)$$

which gives

$$P = \frac{\pi d^2 \tau_1}{4b} \tag{11.1}$$

Bearing pressure

Either the rivet or plate may fail due to bearing pressure. Suppose that p_b is this pressure then failure occurs when

$$\frac{Pb}{td} = p_b$$

so that

$$P = \frac{p_b td}{b} \tag{11.2}$$

Plate failure in tension

The area of plate in tension along the line of rivets is reduced due to the presence of rivet holes. Therefore, if the ultimate tensile stress in the plate is σ_{ult}, failure occurs when

$$\frac{Pb}{t(b-d)} = \sigma_{\text{ult}}$$

from which

$$P = \frac{\sigma_{\text{ult}} t(b-d)}{b} \tag{11.3}$$

Shear failure in a plate

Shearing of the plates may occur on the planes cc, resulting in the rivets being dragged out of the plate. If the maximum shear stress at failure of the material of the plates is τ_2, then a failure of this type occurs when

$$Pb = 2at\,\tau_2$$

which gives

$$P = \frac{2at\,\tau_2}{b} \tag{11.4}$$

Example 11.1

A joint in a fuselage skin is constructed by riveting the abutting skins between two straps, as shown in Fig. 11.12. The fuselage skins are 2.5 mm thick and the straps are each 1.2 mm thick; the rivets have a diameter of 4 mm. If the tensile stress in the fuselage skin must not exceed 125 N/mm^2 and the shear stress in the rivets is limited to 120 N/mm^2, determine the maximum allowable rivet spacing such that the joint is equally strong in shear and tension.

A tensile failure in the plate occurs on the reduced plate cross-section along the rivet lines. This area is given by

$$A_p = (b - 4) \times 2.5 \, \text{mm}^2$$

The failure load/unit width P_f is then given by

$$P_f b = (b - 4) \times 2.5 \times 125 \tag{i}$$

The area of cross-section of each rivet is

$$A_r = \frac{\pi \times 4^2}{4} = 12.6 \, \text{mm}^2$$

Since each rivet is in double shear (i.e., two failure shear planes), the area of cross-section in shear is

$$2 \times 12.6 = 25.2 \, \text{mm}^2$$

Then, the failure load/unit width in shear is given by

$$P_f b = 25.2 \times 120 \tag{ii}$$

For failure to occur simultaneously in shear and tension, that is, equating Eqs. (i) and (ii),

$$25.2 \times 120 = (b - 4) \times 2.5 \times 12.5$$

from which

$$b = 13.7 \, \text{mm}$$

Say, a rivet spacing of 13 mm.

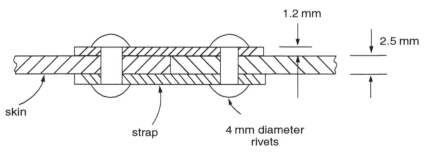

FIGURE 11.12 Joint of Example 11.1

11.4.2 Joint efficiency

The efficiency of a joint or connection is measured by comparing the actual failure load with that which would apply if there were no rivet holes in the plate. Then, for the joint shown in Fig. 11.11, the joint efficiency η is given by

$$\eta = \frac{\sigma_{ult} \, t(b-d)/b}{\sigma_{ult} \, t} = \frac{b-d}{b} \tag{11.5}$$

11.4.3 Group-riveted joints

Rivets may be grouped on each side of a joint such that the efficiency of the joint is a maximum. Suppose that two plates are connected as shown in Fig. 11.13 and that six rivets are required on each side. If it is assumed that each rivet is equally loaded, then the single rivet on the line aa takes one-sixth of the total load. The two rivets on the line bb then share two-sixths of the load, while the three rivets on the line cc share three-sixths of the load. On the line bb the area of cross-section of the plate is reduced by two rivet holes and that on the line cc by three rivet holes, so that, relatively, the joint is as strong at these sections as at aa. Therefore, a more efficient joint is obtained than if the rivets were arranged in, say, two parallel rows of three.

11.4.4 Eccentrically loaded riveted joints

The bracketed connection shown in Fig. 11.14 carries a load P offset from the centroid of the rivet group. The rivet group is then subjected to a shear load P through its centroid and a moment or torque Pe about its centroid.

It is assumed that the shear load P is distributed equally among the rivets, causing a shear force in each rivet parallel to the line of action of P. The moment Pe is assumed to produce a shear force S in each rivet, where S acts in a direction perpendicular to the line joining a particular rivet to the centroid

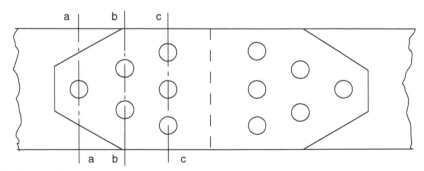

FIGURE 11.13 A Group-Riveted Joint

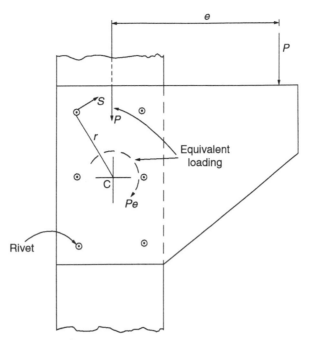

FIGURE 11.14 Eccentrically Loaded Joint

of the rivet group. Furthermore, the value of S is assumed to be proportional to the distance of the rivet from the centroid of the rivet group. Then,

$$Pe = \sum Sr$$

If $S = kr$, where k is a constant for all rivets, then

$$Pe = k\sum r^2$$

from which

$$k = Pe/\sum r^2$$

and

$$S = \frac{Pe}{\sum r^2} r \tag{11.6}$$

The resultant force on a rivet is then the vector sum of the forces due to P and Pe.

Example 11.2

The bracket shown in Fig. 11.15 carries an offset load of 5 kN. Determine the resultant shear forces in the rivets A and B.

The vertical shear force on each rivet is $5/6 = 0.83$ kN. The moment (Pe) on the rivet group is $5 \times 75 = 375$ kNmm. The distance of rivet A (and B, G, and H) from the centroid C of the rivet group is given by

$$r(20^2 + 25^2)^{1/2} = (1025)^{1/2} = 32.02\,\text{mm}$$

The distance of D (and F) from C is 20 mm. Therefore,

$$\sum r^2 = 2 \times 400 + 4 \times 1025 = 4900$$

From Eq. (11.6), the shear forces on rivets A and B due to the moment are

$$S = \frac{375}{4900} \times 32.02 = 2.45 \text{ kN}$$

On rivet A, the force system due to P and Pe is that shown in Fig. 11.16(a), while that on B is shown in Fig. 11.16(b).

The resultant forces may then be calculated using the rules of vector addition or determined graphically using the parallelogram of forces[1].

The design of riveted connections is carried out in the actual design of the rear fuselage of a single-engine trainer/semi-aerobatic aircraft in the Appendix.

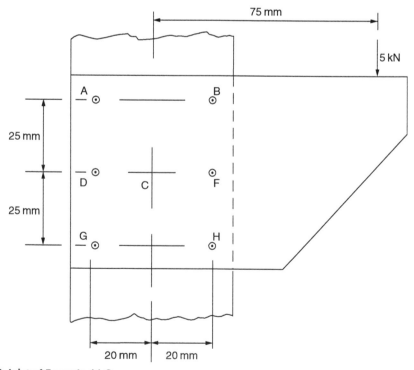

FIGURE 11.15 Joint of Example 11.2

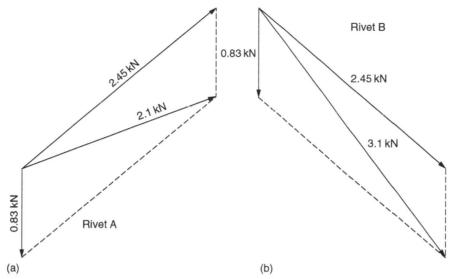

FIGURE 11.16 Force Diagrams for Rivets of Example 11.2

11.4.5 Use of adhesives

In addition to riveted connections, adhesives have and are being used in aircraft construction, although, generally, they are employed in areas of low stress, since their application is still a matter of research. Of these adhesives, epoxy resins are the most frequently used, since they have the advantages over, say, polyester resins, of good adhesive properties, low shrinkage during cure so that residual stresses are reduced, good mechanical properties, and thermal stability. The modulus and ultimate strength of epoxy resin are, typically, 5000 and 100 N/mm^2. Epoxy resins are now found extensively as the matrix component in fibrous composites.

Reference

[1] Megson THG. Structural and stress analysis. 2nd ed. Oxford: Elsevier; 2005.

PROBLEMS

P.11.1. Examine possible uses of new materials in future aircraft manufacture.

P.11.2. Describe the main features of a stressed skin structure. Discuss the structural functions of the various components, with particular reference either to the fuselage or to the wing of a medium-sized transport aircraft.

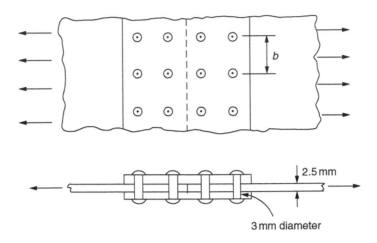

FIGURE P.11.3

P.11.3. The double riveted butt joint shown in Fig. P.11.3 connects two plates, which are each 2.5 mm thick, the rivets have a diameter of 3 mm. If the failure strength of the rivets in shear is 370 N/mm^2 and the ultimate tensile strength of the plate is 465 N/mm^2, determine the necessary rivet pitch if the joint is to be designed so that failure due to shear in the rivets and failure due to tension in the plate occur simultaneously. Calculate also the joint efficiency.

Answer: The rivet pitch is 12 mm, joint efficiency is 75 percent.

P.11.4. The rivet group shown in Fig. P.11.4 connects two narrow lengths of plate one of which carries a 15 kN load positioned as shown. If the ultimate shear strength of a rivet is 350 N/mm^2 and its failure strength in compression is 600 N/mm^2, determine the minimum allowable values of rivet diameter and plate thickness.

Answer: The rivet diameter is 4.2 mm, plate thickness is 1.93 mm.

P.11.4. MATLAB Use MATLAB to repeat Problem P.11.4 for load values (P) from 13 kN to 17 kN in increments of 0.5 kN.

Answer:
(i) For $P = 13$ kN, rivet diameter is 3.91 mm, plate thickness is 1.8 mm.
(ii) For $P = 13.5$ kN, rivet diameter is 3.99 mm, plate thickness is 1.83 mm.
(iii) For $P = 14$ kN, rivet diameter is 4.06 mm, plate thickness is 1.86 mm.
(iv) For $P = 14.5$ kN, rivet diameter is 4.13 mm, plate thickness is 1.9 mm.
(v) For $P = 15$ kN, rivet diameter is 4.2 mm, plate thickness is 1.93 mm.
(vi) For $P = 15.5$ kN, rivet diameter is 4.27 mm, plate thickness is 1.96 mm.
(vii) For $P = 16$ kN, rivet diameter is 4.34 mm, plate thickness is 1.99 mm.
(viii) For $P = 16.5$ kN, rivet diameter is 4.41 mm, plate thickness is 2.02 mm.
(ix) For $P = 17$ kN, rivet diameter is 4.48 mm, plate thickness is 2.05 mm.

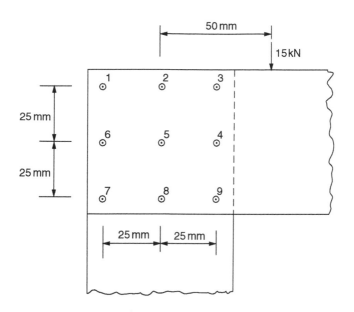

FIGURE P.11.4

Airworthiness and airframe loads

B2

Airworthiness

The airworthiness of an aircraft is concerned with the standards of safety incorporated in all aspects of its construction. These range from structural strength to the provision of certain safeguards in the event of crash landings and include design requirements relating to aerodynamics, performance, and electrical and hydraulic systems. The selection of minimum standards of safety is largely the concern of "national and international" airworthiness authorities, who prepare handbooks of official requirements. The handbooks include operational requirements, minimum safety requirements, recommended practices, design data, and so forth.

In this chapter, we concentrate on the structural aspects of airworthiness that depend chiefly on the strength and stiffness of the aircraft. Stiffness problems may be conveniently grouped under the heading *aeroelasticity* and are discussed in Section B6. Strength problems arise, as we have seen, from ground and air loads, and their magnitudes depend on the selection of maneuvering and other conditions applicable to the operational requirements of a particular aircraft.

12.1 FACTORS OF SAFETY-FLIGHT ENVELOPE

The control of weight in aircraft design is of extreme importance. Increases in weight require stronger structures to support them, which in turn lead to further increases in weight, and so on. Excesses of structural weight mean lesser amounts of payload, thereby affecting the economic viability of the aircraft. The aircraft designer is therefore constantly seeking to pare his aircraft's weight to the minimum compatible with safety. However, to ensure general minimum standards of strength and safety, airworthiness regulations lay down several factors that the primary structure of the aircraft must satisfy. These are the *limit load*, which is the maximum load that the aircraft is expected to experience in normal operation; the *proof load*, which is the product of the limit load and the *proof factor* (1.0–1.25); and the *ultimate load*, which is the product of the limit load and the *ultimate factor* (usually 1.5). The aircraft's structure must withstand the proof load without detrimental distortion and should not fail until the ultimate load has been achieved. The proof and ultimate factors may be regarded as factors of safety and provide for various contingencies and uncertainties, which are discussed in greater detail in Section 12.2.

The basic strength and flight performance limits for a particular aircraft are selected by the airworthiness authorities and are contained in the *flight envelope* or *V-n* diagram shown in Fig. 12.1. The curves OA and OF correspond to the stalled condition of the aircraft and are obtained from the well-known aerodynamic relationship

$$\text{Lift} = nW = \frac{1}{2}\rho V^2 SC_{L,\max}$$

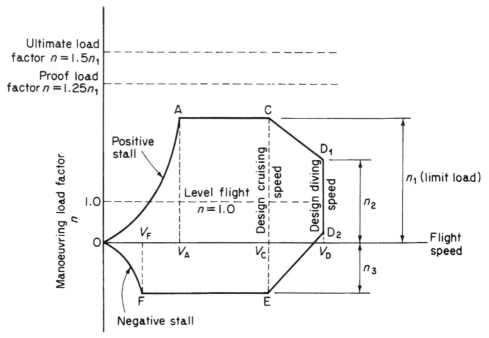

FIGURE 12.1 Flight Envelope

Therefore, for speeds below V_A (positive wing incidence) and V_F (negative incidence), the maximum loads that can be applied to the aircraft are governed by $C_{L,max}$. As the speed increases, it is possible to apply the positive and negative limit loads, corresponding to n_1 and n_3, without stalling the aircraft, so that AC and FE represent maximum operational load factors for the aircraft. Above the design cruising speed V_C, the cut-off lines CD_1 and D_2E relieve the design cases to be covered, since it is not expected that the limit loads are applied at maximum speed. Values of n_1, n_2, and n_3 are specified by the airworthiness authorities for particular aircraft; typical load factors are shown in Table 12.1.

A particular flight envelope is applicable to one altitude only, since $C_{L,max}$ is generally reduced with an increase in altitude, and the speed of sound decreases with altitude, thereby reducing the critical Mach number and hence the design diving speed V_D. Flight envelopes are therefore drawn for a range of altitudes from sea level to the operational ceiling of the aircraft.

Table 12.1 Typical load factors

Load factor n	Category		
	Normal	Semi-aerobatic	Aerobatic
n_1	$2.1 + 24,000/(W + 10,000)$	4.5	6.0
n_2	$0.75 n_1$ but $n_2 \not< 2.0$	3.1	4.5
n_3	1.0	1.8	3.0

12.2 **LOAD FACTOR DETERMINATION**

Several problems require solution before values for the various load factors in the flight envelope can be determined. The limit load, for example, may be produced by a specified maneuver or by an encounter with a particularly severe gust (gust cases and the associated gust envelope are discussed in Section 14.4). Clearly some knowledge of possible gust conditions is required to determine the limiting case. Furthermore, the fixing of the proof and ultimate factors also depends upon the degree of uncertainty in design, variations in structural strength, structural deterioration, and the like. We now investigate some of these problems to see their comparative influence on load factor values.

12.2.1 **Limit load**

An aircraft is subjected to a variety of loads during its operational life, the main classes of which are maneuver loads, gust loads, undercarriage loads, cabin pressure loads, buffeting, and induced vibrations. Of these, maneuver, undercarriage, and cabin pressure loads are determined with reasonable simplicity, since maneuver loads are controlled design cases, undercarriages are designed for given maximum descent rates, and cabin pressures are specified. The remaining loads depend to a large extent on the atmospheric conditions encountered during flight. Estimates of the magnitudes of such loads are possible therefore only if in-flight data on these loads are available. It obviously requires a great number of hours of flying, if the experimental data are to include possible extremes of atmospheric conditions. In practice, the amount of data required to establish the probable period of flight time before an aircraft encounters, say, a gust load of a given severity, is a great deal more than that available. It therefore becomes a problem in statistics to extrapolate the available data and calculate the probability of an aircraft being subjected to its proof or ultimate load during its operational life. The aim would be for a zero or negligible rate of occurrence of its ultimate load and an extremely low rate of occurrence of its proof load. Having decided on an ultimate load, the limit load may be fixed as defined in Section 12.1, although the value of the ultimate factor includes, as already noted, allowances for uncertainties in design, variation in structural strength, and structural deterioration.

12.2.2 **Uncertainties in design and structural deterioration**

Neither of these presents serious problems in modern aircraft construction and therefore do not require large factors of safety to minimize their effects. Modern methods of aircraft structural analysis are refined and, in any case, tests to determine actual failure loads are carried out on representative full-scale components to verify design estimates. The problem of structural deterioration due to corrosion and wear may be largely eliminated by close inspection during service and the application of suitable protective treatments.

12.2.3 **Variation in structural strength**

To minimize the effect of the variation in structural strength between two apparently identical components, strict controls are employed in the manufacture of materials and in the fabrication of the structure. Material control involves the observance of strict limits in chemical composition and close supervision of manufacturing methods such as machining, heat treatment, and rolling. In addition, the inspection of samples by visual, radiographic, and other means and the carrying out of strength tests on specimens enable below-limit batches to be isolated and rejected. Thus, if a sample of a batch of material falls below a specified minimum strength, then the batch is rejected. This means of course that an actual structure always comprises materials with properties equal to or better than those assumed for design purposes, an added but unallowed for "bonus" in considering factors of safety.

Similar precautions are applied to assembled structures with regard to dimension tolerances, quality of assembly, welding, and so forth. Again, visual and other inspection methods are employed and, in certain cases, strength tests are carried out on sample structures.

12.2.4 Fatigue

Although adequate precautions are taken to ensure that an aircraft's structure possesses sufficient strength to withstand the most severe expected gust or maneuver load, the problem of fatigue remains. Practically all components of the aircraft's structure are subjected to fluctuating loads, which occur a great many times during the life of the aircraft. It has been known for many years that materials fail under fluctuating loads at much lower values of stress than their normal static failure stress. A graph of failure stress against number of repetitions of this stress has the typical form shown in Fig. 12.2. For some materials, such as mild steel, the curve (usually known as an *S–N* curve or diagram) is asymptotic to a certain minimum value, which means that the material has an actual infinite-life stress. Curves for other materials, for example, aluminum and its alloys, do not always appear to have asymptotic values so that these materials may not possess an infinite-life stress. We discuss the implications of this a little later.

Prior to the mid-1940s, little attention had been paid to fatigue considerations in the design of aircraft structures. It was felt that sufficient static strength would eliminate the possibility of fatigue

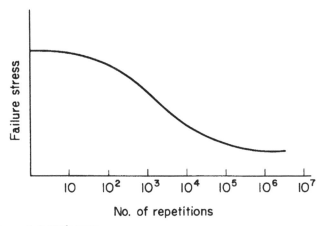

FIGURE 12.2 Typical Form of *S–N* Diagram

failure. However, evidence began to accumulate that several aircraft crashes had been caused by fatigue failure. The seriousness of the situation was highlighted in the early 1950s by catastrophic fatigue failures of two Comet airliners. These were caused by the once-per-flight cabin pressurization cycle that produced circumferential and longitudinal stresses in the fuselage skin. Although these stresses were well below the allowable stresses for single cycle loading, stress concentrations occurred at the corners of the windows and around rivets, which raised local stresses considerably above the general stress level. Repeated cycles of pressurization produced fatigue cracks that propagated disastrously, causing an explosion of the fuselage at high altitude.

Several factors contributed to the emergence of fatigue as a major factor in design. For example, aircraft speeds and sizes increased, calling for higher wing and other loadings. Consequently, the effect of turbulence was magnified and the magnitudes of the fluctuating loads became larger. In civil aviation, airliners had a greater utilization and a longer operational life. The new zinc-rich alloys, used for their high static strength properties, did not show a proportional improvement in fatigue strength, exhibited high crack propagation rates, and were extremely notch sensitive.

Even though the causes of fatigue were reasonably clear at that time, its elimination as a threat to aircraft safety was a different matter. The fatigue problem has two major facets: the prediction of the fatigue strength of a structure and a knowledge of the loads causing fatigue. Information was lacking on both counts. The Royal Aircraft Establishment (RAE) and the aircraft industry therefore embarked on an extensive test programme to determine the behavior of complete components, joints, and other detail parts under fluctuating loads. These included fatigue testing by the RAE of some 50 Meteor 4 tailplanes at a range of temperatures, plus research, also by the RAE, into the fatigue behavior of joints and connections. Further work was undertaken by some universities and by the industry itself into the effects of stress concentrations.

In conjunction with their fatigue strength testing, the RAE initiated research to develop a suitable instrument for counting and recording gust loads over long periods of time. Such an instrument was developed by J. Taylor in 1950 and was designed so that the response fell off rapidly above 10 Hz. Crossings of g thresholds from 0.2 to 1.8 g at 0.1 g intervals were recorded (note that steady level flight is 1 g flight) during experimental flying at the RAE on three different aircraft over 28,000 km, and the best techniques for extracting information from the data established. Civil airlines cooperated by carrying the instruments on their regular air services for a number of years. Eight types of aircraft were equipped, so that by 1961, records had been obtained for regions including Europe, the Atlantic, Africa, India, and the Far East, representing 19,000 hours and 8 million km of flying.

Atmospheric turbulence and the cabin pressurization cycle are only two of the many fluctuating loads that cause fatigue damage in aircraft. On the ground, the wing is supported on the undercarriage and experiences tensile stresses in its upper surfaces and compressive stresses in its lower surfaces. In flight, these stresses are reversed, as aerodynamic lift supports the wing. Also, the impact of landing and ground maneuvering on imperfect surfaces cause stress fluctuations, while, during landing and take-off, flaps are lowered and raised, producing additional load cycles in the flap support structure. Engine pylons are subjected to fatigue loading from thrust variations in take-off and landing and also from inertia loads produced by lateral gusts on the complete aircraft.

A more detailed investigation of fatigue and its associated problems is presented in Chapter 14 while a fuller discussion of airworthiness as applied to civil jet aircraft is presented in Jenkinson, Simpkin, and Rhodes[1].

Reference

[1] Jenkinson LR, Simpkin P, Rhodes D. Civil jet aircraft design. London: Arnold; 1999.

PROBLEMS

P.12.1 A radar dome weighing 300 kg is positioned on the top of the fuselage of a surveillance aircraft. If the dome is attached to the fuselage frames by four bolts, each having an ultimate shear strength of 5000 N and the aircraft is subjected to a maximum acceleration of 3 g in a vertical climb, determine (a) the limit load per bolt, (b) the ultimate load per bolt, (c) the ultimate margin of safety.

Answer: (a) 2943 N, (b) 4415 N, (c) 0.133

P.12.2 The relative positions of the center of gravity and the centers of pressure of the wing and tailplane of an aircraft are shown in Fig. P.12.2; the total weight of the aircraft is 667.5 N. If the fuselage and its contents weigh 26.3 kN/m and the weight of the tailplane is 8.9 kN, calculate the ultimate shear force in the fuselage at the section AA for a maneuver load factor of 3 g including gravity. Assume a factor of safety of 1.5.

Answer: 212.2 kN

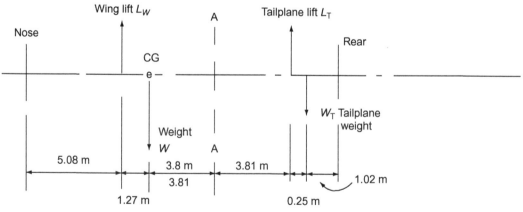

FIGURE P.12.2

Airframe loads

In Chapter 11, we discussed in general terms the types of load to which aircraft are subjected during their operational life. We shall now examine in more detail the loads produced by various maneuvers and the manner in which they are calculated.

13.1 AIRCRAFT INERTIA LOADS

The maximum loads on the components of an aircraft's structure generally occur when the aircraft is undergoing some form of acceleration or deceleration, such as in landings, take-offs, and maneuvers within the flight and gust envelopes. Therefore, before a structural component can be designed, the inertia loads corresponding to these accelerations and decelerations must be calculated. For these purposes, we shall suppose that an aircraft is a rigid body and represent it by a rigid mass, m, as shown in Fig. 13.1. We also, at this stage, consider motion in the plane of the mass which would correspond to pitching of the aircraft without roll or yaw. We further suppose that the center of gravity (CG) of the mass has coordinates $\bar{x}, \bar{y}$, referred to x and y axes having an arbitrary origin O; the mass is rotating about an axis through O perpendicular to the xy plane with a constant angular velocity ω.

The acceleration of any point, a distance r from O, is $\omega^2 r$ and is directed toward O. Thus, the inertia force acting on the element, δm, is $\omega^2 r \delta m$ in a direction opposite to the acceleration, as shown in Fig. 13.1. The components of this inertia force, parallel to the x and y axes, are $\omega^2 r \delta m \cos \theta$ and $\omega^2 r \delta m \sin\theta$, respectively, or, in terms of x and y, $\omega^2 x \delta m$ and $\omega^2 y \delta m$. The resultant inertia forces, F_x and F_y, are given by

$$F_x = \int \omega^2 x \, dm = \omega^2 \int x \, dm$$
$$F_y = \int \omega^2 y \, dm = \omega^2 \int y \, dm$$

in which we note that the angular velocity ω is constant and may therefore be taken outside the integral sign. In these expressions, $\int x \, dm$ and $\int y \, dm$ are the moments of the mass, m, about the y and x axes, respectively, so that

$$F_x = \omega^2 \bar{x} m \tag{13.1}$$

and

$$F_y = \omega^2 \bar{y} m \tag{13.2}$$

If the CG lies on the x axis, $\bar{y} = 0$ and $F_y = 0$. Similarly, if the CG lies on the y axis, $F_x = 0$. Clearly, if O coincides with the CG, $\bar{x} = \bar{y} = 0$ and $F_x = F_y = 0$.

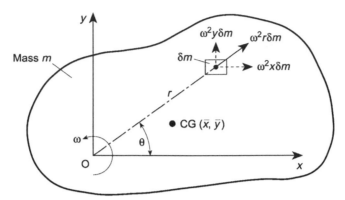

FIGURE 13.1 Inertia Forces on a Rigid Mass Having a Constant Angular Velocity

Suppose now that the rigid body is subjected to an angular acceleration (or deceleration) α in addition to the constant angular velocity, ω, as shown in Fig. 13.2. An additional inertia force, $\alpha r \delta m$, acts on the element δm in a direction perpendicular to r and in the opposite sense to the angular acceleration. This inertia force has components $\alpha r \delta m \cos\theta$ and $\alpha r \delta m \sin\theta$, that is, $\alpha x \delta m$ and $\alpha y \delta m$, in the y and x directions, respectively. The resultant inertia forces, F_x and F_y, are then given by

$$F_x = \int \alpha y \, dm = \alpha \int y \, dm$$

and

$$F_y = -\int \alpha x \, dm = -\alpha \int x \, dm$$

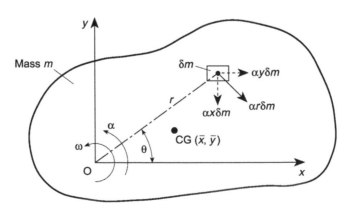

FIGURE 13.2 Inertia Forces on a Rigid Mass Subjected to an Angular Acceleration

for $\underline{\alpha}$ in the direction shown. Then, as before,

$$F_x = \alpha \bar{y} m \qquad (13.3)$$

and

$$F_y = \alpha \bar{x} m \qquad (13.4)$$

Also, if the CG lies on the x axis, $\bar{y} = 0$ and $F_x = 0$. Similarly, if the CG lies on the y axis, $\bar{x} = 0$ and $F_y = 0$.

The torque about the axis of rotation produced by the inertia force corresponding to the angular acceleration on the element δm is given by

$$\delta T_O = \alpha r^2 \delta m$$

Thus, for the complete mass,

$$T_O = \int \alpha r^2 \, dm = \alpha \int r^2 \, dm$$

The integral term in this expression is the moment of inertia, I_O, of the mass about the axis of rotation. Thus,

$$T_O = \alpha I_O \qquad (13.5)$$

Equation (13.5) may be rewritten in terms of I_{CG}, the moment of inertia of the mass about an axis perpendicular to the plane of the mass through the CG. Hence, using the parallel axes theorem,

$$I_O = m(\bar{r})^2 + I_{CG}$$

where $\bar{r}$ is the distance between O and the CG. Then,

$$I_O = m[(\bar{x})^2 + (\bar{y})^2] + I_{CG}$$

and

$$T_O = m[(\bar{x})^2 + (\bar{y})^2]\alpha + I_{CG}\alpha \qquad (13.6)$$

Example 13.1

An aircraft having a total weight of 45 kN lands on the deck of an aircraft carrier and is brought to rest by means of a cable engaged by an arrester hook, as shown in Fig. 13.3. If the deceleration induced by the cable is 3g, determine the tension, T, in the cable, the load on an undercarriage strut, and the shear and axial loads in the fuselage at the section AA; the weight of the aircraft aft of AA is 4.5 kN. Calculate also the length of deck covered by the aircraft before it is brought to rest if the touch-down speed is 25 m/s. See Ex. 1.1.

The aircraft is subjected to a horizontal inertia force ma where m is the mass of the aircraft and a its deceleration. Thus, resolving forces horizontally

$$T \cos 10° - ma = 0$$

that is,

$$T \cos 10° - \frac{45}{g} 3g = 0$$

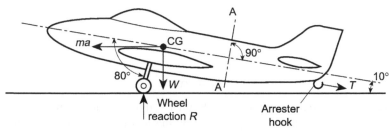

FIGURE 13.3 Forces on the Aircraft of Example 13.1

which gives

$$T = 137.1 \text{ kN}$$

Now, resolving forces vertically,

$$R - W - T\sin10° = 0$$

that is,

$$R = 45 + 137.1\sin10° = 68.8 \text{ kN}$$

Assuming two undercarriage struts, the load in each strut is $(R/2)/\cos20° = 36.6$ kN.

Let N and S be the axial and shear loads at the section AA, as shown in Fig. 13.4. The inertia load acting at the CG of the fuselage aft of AA is m_1a, where m_1 is the mass of the fuselage aft of AA. Then,

$$m_1a = \frac{4.5}{g}3\,g = 13.5 \text{ kN}$$

Resolving forces parallel to the axis of the fuselage,

$$N - T + m_1a\cos10° - 4.5\sin10° = 0$$

that is,

$$N - 137.1 + 13.5\cos10° - 4.5\sin10° = 0$$

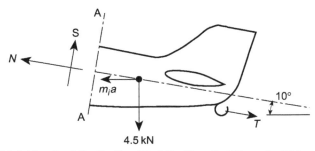

FIGURE 13.4 Shear and Axial Loads at the Section AA of the Aircraft of Example 13.1

from which

$$N = 124.6 \text{ kN}$$

Now, resolving forces perpendicular to the axis of the fuselage,

$$S - m_1 a \sin 10° - 4.5 \cos 10° = 0$$

that is,

$$S - 13.5 \sin 10° - 4.5 \cos 10° = 0$$

so that

$$S = 6.8 \text{ kN}$$

Note that, in addition to the axial load and shear load at the section AA, there will also be a bending moment.
    Finally, from elementary dynamics,

$$v^2 = v_0^2 + 2as$$

where v_0 is the touch-down speed, v the final speed (= 0), and s the length of deck covered. Then,

$$v_0^2 = -2as$$

that is,

$$25^2 = -2(-3 \times 9.81)s$$

which gives

$$s = 10.6 \text{ m}$$

Example 13.2

An aircraft having a weight of 250 kN and a tricycle undercarriage lands at a vertical velocity of 3.7 m/s, such that the vertical and horizontal reactions on the main wheels are 1,200 kN and 400 kN, respectively; at this instant, the nosewheel is 1.0 m from the ground, as shown in Fig. 13.5. If the moment of inertia of the aircraft about its CG is 5.65×10^8 Ns^2 mm, determine the inertia forces on the aircraft, the time taken for its vertical velocity to become zero, and its angular velocity at this instant. See Ex. 1.1.

    The horizontal and vertical inertia forces ma_x and ma_y act at the CG, as shown in Fig. 13.5; m is the mass of the aircraft; and a_x and a_y its accelerations in the horizontal and vertical directions, respectively. Then, resolving forces horizontally,

$$ma_x - 400 = 0$$

from which

$$ma_x = 400 \text{ kN}$$

Now, resolving forces vertically,

$$ma_y + 250 - 1,200 = 0$$

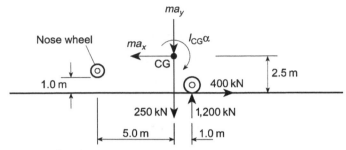

FIGURE 13.5 Geometry of the Aircraft of Example 13.2

which gives

$$ma_y = 950 \text{ kN}$$

Then,

$$a_y = \frac{950}{m} = \frac{950}{250/g} = 3.8 \, g \tag{i}$$

Now, taking moments about the CG,

$$I_{CG}\alpha - 1,200 \times 1.0 - 400 \times 2.5 = 0 \tag{ii}$$

from which

$$I_{CG}\alpha = 2,200 \text{ m kN}$$

Note that the units of $I_{CG}\alpha$ are mkNrad but radians are non-dimensional, so are not included but implicit.

Hence,

$$\alpha = \frac{I_{CG}\alpha}{I_{CG}} = \frac{2,200 \times 10^6}{5.65 \times 10^8} = 3.9 \text{ rad/s}^2 \tag{iii}$$

From Eq. (i), the aircraft has a vertical deceleration of 3.8 g from an initial vertical velocity of 3.7 m/s. Therefore, from elementary dynamics, the time, t, taken for the vertical velocity to become zero is given by

$$v = v_0 + a_y t \tag{iv}$$

in which $v = 0$ and $v_0 = 3.7$ m/s. Hence,

$$0 = 3.7 - 3.8 \times 9.81t$$

from which

$$t = 0.099 \text{ s}$$

In a similar manner to Eq. (iv), the angular velocity of the aircraft after 0.099 s is given by

$$\omega = \omega_0 + \alpha t$$

in which $\omega_0 = 0$ and $\alpha = 3.9$ rad/s^2. Hence,

$$\omega = 3.9 \times 0.099$$

that is,

$$\omega = 0.39 \text{ rad/s}$$

Example 13.3

The aircraft shown in Fig. 13.6 has a moment of inertia of 5,460 kNs^{2}m and an all-up weight of 430 kN. During landing the aerodynamic lift is equal to 90 percent of its all-up weight and it is subjected to the ground loads shown. Determine (a) the limit load factor at its center of gravity, (b) the limit pitching acceleration.

(a) Wing lift $= 0.9 \times 430 = 387$ kN. Resolving vertically, the total vertical load on the aircraft $= 387 + 1,112.5 - 430 = 1,069.5$ kN. Therefore the limit load factor $= (1,069.5/430) + 1 = 3.5$

(b) Taking moments about the CG,

$$3.81L_W + 222.5 \times 2.54 + 1,112.5 \times 2.54 = I_0\alpha = 5460\alpha$$

from which

$$\alpha = 0.89 \text{ rad/s}^2$$

13.2 SYMMETRIC MANEUVER LOADS

We shall now consider the calculation of aircraft loads corresponding to the flight conditions specified by flight envelopes. In fact, an infinite number of flight conditions lie within the boundary of the flight envelope, although, structurally, those represented by the boundary are the most severe. Furthermore, it

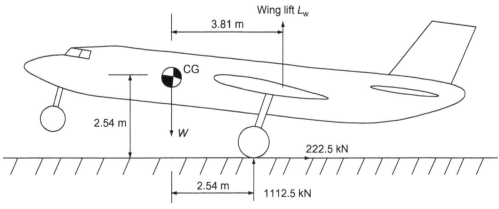

FIGURE 13.6 Aircraft of Example 13.3

is usually found that the corners A, C, D_1, D_2, E, and F (see Fig. 13.1) are more critical than points on the boundary between the corners, so that, in practice, only the six conditions corresponding to these corner points need be investigated for each flight envelope.

In symmetric maneuvers, we consider the motion of the aircraft initiated by movement of the control surfaces in the plane of symmetry. Examples of such maneuvers are loops, straight pullouts, and bunts; and the calculations involve the determination of lift, drag, and tailplane loads at given flight speeds and altitudes. The effects of atmospheric turbulence and gusts are discussed in Section 13.4.

13.2.1 Level flight

Although steady level flight is not a maneuver in the strict sense of the word, it is a useful condition to investigate initially, since it establishes points of load application and gives some idea of the equilibrium of an aircraft in the longitudinal plane. The loads acting on an aircraft in steady flight are shown in Fig. 13.7, with the following notation:

L is the lift acting at the aerodynamic center of the wing.
D is the aircraft drag.
M_0 is the aerodynamic pitching moment of the aircraft *less* its horizontal tail.
P is the horizontal tail load acting at the aerodynamic center of the tail, usually taken to be at approximately one-third of the tailplane chord.
W is the aircraft weight acting at its CG.
T is the engine thrust, assumed here to act parallel to the direction of flight in order to simplify calculation.

The loads are in static equilibrium, since the aircraft is in a steady, unaccelerated, level flight condition. Thus, for vertical equilibrium,

$$L + P - W = 0 \tag{13.7}$$

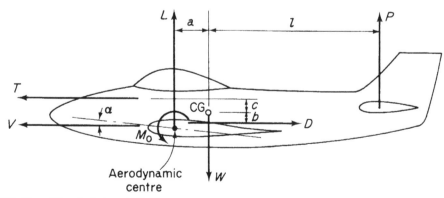

FIGURE 13.7 Aircraft Loads in Level Flight

for horizontal equilibrium,

$$T - D = 0 \tag{13.8}$$

and taking moments about the aircraft's CG in the plane of symmetry,

$$La - Db - Tc - M_0 - Pl = 0 \tag{13.9}$$

For a given aircraft weight, speed, and altitude, Eqs. (13.7)–(13.9) may be solved for the unknown lift, drag, and tail loads. However, other parameters in these equations, such as M_0, depend upon the wing incidence α, which in turn is a function of the required wing lift, so that, in practice, a method of successive approximation is found to be the most convenient means of solution.

As a first approximation, we assume that the tail load P is small compared with the wing lift L, so that, from Eq. (13.7), $L \approx W$. From aerodynamic theory with the usual notation,

$$L = \frac{1}{2}\rho V^2 S C_L$$

Hence,

$$\frac{1}{2}\rho V^2 S C_L \approx W \tag{13.10}$$

Equation (13.10) gives the approximate lift coefficient C_L and thus (from C_L–α curves established by wind tunnel tests), the wing incidence α. The drag load D follows (knowing V and α) and hence we obtain the required engine thrust T from Eq. (13.8). Also M_0, a, b, c, and l may be calculated (again, since V and α are known) and Eq. (13.9) solved for P. As a second approximation, this value of P is substituted in Eq. (13.7) to obtain a more accurate value for L and the procedure is repeated. Usually, three approximations are sufficient to produce reasonably accurate values.

In most cases P, D, and T are small compared with the lift and aircraft weight. Therefore, from Eq. (13.7), $L \approx W$ and substitution in Eq. (13.9) gives, neglecting D and T,

$$P \approx W\frac{a}{l} - \frac{M_0}{l} \tag{13.11}$$

We see, from Eq. (13.11), that, if a is large, then P most likely will be positive. In other words, the tail load acts upward when the CG of the aircraft is far aft. When a is small or negative, that is, a forward CG, then P probably will be negative and act downward.

13.2.2 General case of a symmetric maneuver

In a rapid pull-out from a dive, a downward load is applied to the tailplane, causing the aircraft to pitch nose upward. The downward load is achieved by a backward movement of the control column, thereby applying negative incidence to the elevators, or horizontal tail if the latter is all-moving. If the maneuver is carried out rapidly, the forward speed of the aircraft remains practically constant, so that increases in lift and drag result from the increase in wing incidence only. Since the lift is now greater than that required to balance the aircraft weight, the aircraft experiences an upward acceleration normal to its flight path. This normal acceleration combined with the aircraft's speed in the dive results in the curved flight path shown in Fig. 13.8. As the drag load builds up with an increase of incidence, the

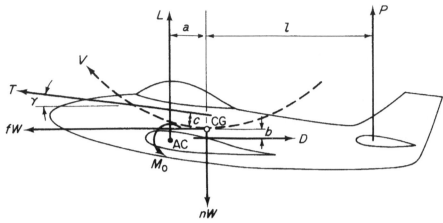

FIGURE 13.8 Aircraft Loads in a Pull-out from a Dive

forward speed of the aircraft falls, since the thrust is assumed to remain constant during the maneuver. It is usual, as we observed in the discussion of the flight envelope, to describe the maneuvers of an aircraft in terms of a maneuvering load factor n. For steady level flight $n = 1$, giving 1 g flight, although in fact the acceleration is zero. What is implied in this method of description is that the inertia force on the aircraft in the level flight condition is 1.0 times its weight. It follows that the vertical inertia force on an aircraft carrying out an ng maneuver is nW. We therefore replace the dynamic conditions of the accelerated motion by an equivalent set of static conditions, in which the applied loads are in equilibrium with the inertia forces. Thus, in Fig. 13.8, n is the maneuver load factor, while f is a similar factor giving the horizontal inertia force. Note that the actual normal acceleration in this particular case is $(n - 1)$ g.

For vertical equilibrium of the aircraft, we have, referring to Fig. 13.8, where the aircraft is shown at the lowest point of the pullout,

$$L + P + T \sin\gamma - nW = 0 \tag{13.12}$$

For horizontal equilibrium,

$$T \cos\gamma + fW - D = 0 \tag{13.13}$$

and, for pitching moment equilibrium about the aircraft's CG,

$$La - Db - Tc - M_0 - Pl = 0 \tag{13.14}$$

Equation (13.14) contains no terms representing the effect of pitching acceleration of the aircraft; this is assumed to be negligible at this stage.

Again, the method of successive approximation is found to be most convenient for the solution of Eqs. (13.12)–(13.14). There is, however, a difference to the procedure described for the steady level flight case. The engine thrust T is no longer directly related to the drag D, as the latter changes during the maneuver. Generally, the thrust is regarded as remaining constant and equal to the value appropriate to conditions before the maneuver began.

Example 13.4

The curves C_D, α, and $C_{M,CG}$ for a light aircraft are shown in Fig. 13.9(a). The aircraft weight is 8,000 N, its wing area 14.5 m², and its mean chord 1.35 m. Determine the lift, drag, tail load, and forward inertia force for a symmetric maneuver corresponding to $n = 4.5$ and a speed of 60 m/s. Assume that engine-off conditions apply and that the air density is 1.223 kg/m³. Figure 13.9(b) shows the relevant aircraft dimensions. See Ex. 1.1.

As a first approximation, we neglect the tail load P. Therefore, from Eq. (13.12), since $T = 0$, we have

$$L \approx nW \tag{i}$$

Hence,

$$C_L = \frac{L}{\frac{1}{2}\rho V^2 S} \approx \frac{4.5 \times 8000}{\frac{1}{2} \times 1.223 \times 60^2 \times 14.5} = 1.113$$

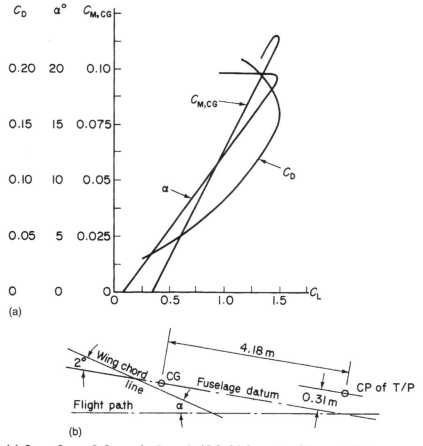

(a)

(b)

FIGURE 13.9 (a) C_D, α, $C_{M,CG}$ –C_L Curves for Example 13.3; (b) Geometry of Example 13.4

From Fig. 13.9(a), $\alpha = 13.75°$ and $C_{M,CG} = 0.075$. The tail arm l, from Fig. 13.9(b), is

$$l = 4.18\cos(\alpha - 2) + 0.31\sin(\alpha - 2) \qquad \text{(ii)}$$

Substituting this value of α gives $l = 4.123$ m. In Eq. (13.14), the terms $La - Db - M_0$ are equivalent to the aircraft pitching moment M_{CG} about its CG. Equation (13.14) may therefore be written

$$M_{CG} - Pl = 0$$

or

$$Pl = \frac{1}{2}\rho V^2 ScC_{M,CG} \qquad \text{(ii)}$$

where c = wing mean chord. Substituting P from Eq. (iii) into Eq. (13.12), we have

$$L + \frac{\frac{1}{2}\rho V^2 ScC_{M,CG}}{l} = nW$$

or dividing through by $\frac{1}{2}\rho V^2 S$,

$$C_L + \frac{c}{l}C_{M,CG} = \frac{nW}{\frac{1}{2}\rho V^2 S} \qquad \text{(iv)}$$

We now obtain a more accurate value for C_L from Eq. (iv),

$$C_L = 1.113 - \frac{1.35}{4.123} \times 0.075 = 1.088$$

giving $\alpha = 13.3°$ and $C_{M,CG} = 0.073$.

Substituting this value of α into Eq. (ii) gives a second approximation for l, namely, $l = 4.161$ m.

Equation (iv) now gives a third approximation for C_L, that is, $C_L = 1.099$. Since all three calculated values of C_L are extremely close, further approximations will not give values of C_L very much different to them. Therefore, we shall take $C_L = 1.099$. From Fig. 13.9(a), $C_D = 0.0875$.

The values of lift, tail load, drag, and forward inertia force then follow:

$$\text{Lift } L = \frac{1}{2}\rho V^2 SC_L = \frac{1}{2} \times 1.223 \times 60^2 \times 14.5 \times 1.099 = 35,000 \text{ N}$$

$$\text{Tail load } P = nW - L = 4.5 \times 8,000 - 35,000 = 1,000 \text{ N}$$

$$\text{Drag } D = \frac{1}{2}\rho V^2 SC_D = \frac{1}{2} \times 1.223 \times 60^2 \times 14.5 \times 0.0875 = 2,790 \text{ N}$$

$$\text{Forward inertia force } fW = D(\text{from Eq.}(14.13)) = 2,790\text{N}$$

13.3 NORMAL ACCELERATIONS ASSOCIATED WITH VARIOUS TYPES OF MANEUVER

In Section 13.2, we determined aircraft loads corresponding to a given maneuver load factor n. Clearly, it is necessary to relate this load factor to given types of maneuver. Two cases arise: the first involves a steady pull-out from a dive and the second, a correctly banked turn. Although the latter is not a symmetric maneuver in the strict sense of the word, it gives rise to normal accelerations in the plane of symmetry and is therefore included.

13.3.1 **Steady pull-out**

Let us suppose that the aircraft has just begun its pull-out from a dive, so that it is describing a curved flight path but is not yet at its lowest point. The loads acting on the aircraft at this stage of the maneuver are shown in Fig. 13.10, where R is the radius of curvature of the flight path. In this case, the lift vector must equilibrate the normal (to the flight path) component of the aircraft weight and provide the force producing the centripetal acceleration V^2/R of the aircraft towards the center of curvature of the flight path. Thus,

$$L = \frac{WV^2}{gR} + W\cos\theta$$

or, since $L = nW$ (see Section 13.2),

$$n = \frac{V^2}{gR} + \cos\theta \tag{13.15}$$

At the lowest point of the pullout, $\theta = 0$, and

$$n = \frac{V^2}{gR} + 1 \tag{13.16}$$

We see, from either Eq. (13.15) or (13.16), that the smaller the radius of the flight path, that is, the more severe the pull-out, the greater is the value of n. It is quite possible therefore for a severe pull-out to overstress the aircraft by subjecting it to loads that lie outside the flight envelope and which may even exceed the proof or ultimate loads. In practice, the control surface movement may be limited by stops incorporated in the control circuit. These stops usually operate only above a

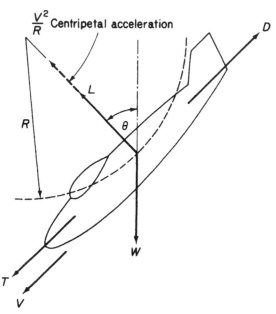

FIGURE 13.10 Aircraft Loads and Acceleration during a Steady Pull-out

certain speed, giving the aircraft adequate maneuverability at lower speeds. For hydraulically operated controls, "artificial feel" is built in to the system, whereby the stick force increases progressively as the speed increases, a necessary precaution in this type of system, since the pilot is merely opening and closing valves in the control circuit and therefore receives no direct physical indication of control surface forces.

Alternatively, at low speeds, a severe pull-out or pull-up may stall the aircraft. Again, safety precautions are usually incorporated in the form of stall warning devices, since, for modern high-speed aircraft, a stall can be disastrous, particularly at low altitude.

Example 13.5

A semi-aerobatic aircraft has reached its design diving speed of 185 m/s in a dive inclined at 45° to the horizontal ground. If the maximum maneuver load factor for the aircraft is 5.5, determine the height at which the pull-out from the dive must begin for straight and level flight to be achieved at a height of 500 m. See Ex. 1.1.

Assuming the design diving speed is maintained throughout the pull-out, from Eq.(13.16),

$$5.5 = \left(185^2/gR\right) + 1$$

so that

$$R = 775.3 \text{ m}$$

Referring to Fig. 13.11,

$$a = R - R\cos45°$$

Substituting for R from the preceding,

$$a = 227 \text{ m}$$

Therefore,

$$h = 727 \text{ m}$$

13.3.2 Correctly banked turn

In this maneuver, the aircraft flies in a horizontal turn with no sideslip at constant speed. If the radius of the turn is R and the angle of bank ϕ, then the forces acting on the aircraft are those shown in Fig. 13.12. The horizontal component of the lift vector in this case provides the force necessary to produce the centripetal acceleration of the aircraft toward the center of the turn. Then,

$$L \sin\phi = \frac{WV^2}{gR} \tag{13.17}$$

and, for vertical equilibrium,

$$L \cos\phi = W \tag{13.18}$$

or

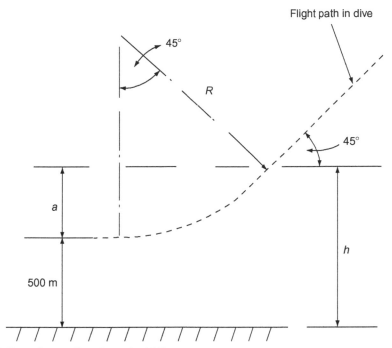

FIGURE 13.11 Pull-out from a dive, Example 13.5

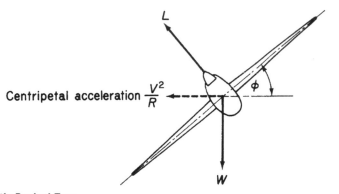

FIGURE 13.12 Correctly Banked Turn

$$L = W \sec\phi \qquad (13.19)$$

From Eq. (13.19), we see that the load factor n in the turn is given by

$$n = \sec\phi \qquad (13.20)$$

Also, dividing Eq. (13.17) by Eq. (13.18),

$$\tan\phi = \frac{V^2}{gR} \tag{13.21}$$

Examination of Eq. (13.21) reveals that the tighter the turn, the greater is the angle of bank required to maintain horizontal flight. Furthermore, we see, from Eq. (13.20), that an increase in bank angle results in an increased load factor. Aerodynamic theory shows that, for a limiting value of n, the minimum time taken to turn through a given angle at a given value of engine thrust occurs when the lift coefficient C_L is a maximum; that is, with the aircraft on the point of stalling.

Example 13.6

The wing of a military aircraft has a maximum lift coefficient of 1.25 and an area of 16 m²; the maximum maneuver load factor is 6.0. If the weight of the aircraft is 50 kN, determine the angle of bank required at a speed of 180 m/s. Calculate also the radius of turn. Take $\rho = 1.223$ kg/m³. See Ex. 1.1.

The maximum available lift is given by

$$L_{max} = (1/2)\rho V^2 S C_{L,max}$$

that is,

$$L_{max} = (1/2) \times 1.223 \times 180^2 \times 16 \times 1.25 = 396.3 \text{ kN}$$

Then,

$$n = (396.3/50) = 7.9$$

which means that the aircraft exceeds the maximum maneuver load at the point of maximum lift. The maneuver load factor is therefore the critical criterion.

From Eq. (13.20),

$$\sec\phi = 6.0$$

which gives

$$\phi = 80.4°$$

Then, from Eq.(13.21),

$$\tan 80.4° = \left(180^2/9.81R\right)$$

So that

$$R = 558.6 \text{ m}$$

13.4 GUST LOADS

In Section 13.2, we considered aircraft loads resulting from prescribed maneuvers in the longitudinal plane of symmetry. Other types of in-flight load are caused by air turbulence. The movements of the air in turbulence are generally known as *gusts* and produce changes in wing incidence, thereby subjecting the aircraft to sudden or gradual increases or decreases in lift from which normal accelerations result.

These may be critical for large, high-speed aircraft and may possibly cause higher loads than control initiated maneuvers.

At the present time two approaches are employed in gust analysis. One method, which has been in use for a considerable number of years, determines the aircraft response and loads due to a single or "discrete" gust of a given profile. This profile is defined as a distribution of vertical gust velocity over a given finite length or given period of time. Examples of these profiles are shown in Fig. 13.13.

Early airworthiness requirements specified an instantaneous application of gust velocity u, resulting in the "sharp-edged" gust of Fig. 13.13(a). Calculations of normal acceleration and aircraft response were based on the assumptions that the aircraft's flight is undisturbed while the aircraft passes from still air into the moving air of the gust and during the time taken for the gust loads to build up; that the aerodynamic forces on the aircraft are determined by the instantaneous incidence of the particular lifting surface; and finally that the aircraft's structure is rigid. The second assumption here relating the aerodynamic force on a lifting surface to its instantaneous incidence neglects the fact that, in a disturbance such as a gust, there is a gradual growth of circulation and hence of lift to a steady state value (Wagner effect). This in general leads to an overestimation of the upward acceleration of an aircraft and therefore of gust loads.

The sharp-edged gust was replaced when it was realized that the gust velocity built up to a maximum over a period of time. Airworthiness requirements were modified, on the assumption that the gust velocity increased linearly to a maximum value over a specified gust gradient distance H; hence, the "graded" gust of Fig. 13.13(b). In the United Kingdom, H is taken as 30.5 m. Since, as far as the aircraft is concerned, the gust velocity builds up to a maximum over a period of time, it is no longer allowable to ignore the change of flight path as the aircraft enters the gust. By the time the gust has attained its maximum value, the aircraft has developed a vertical component of velocity and, in addition, may be pitching, depending on its longitudinal stability characteristics. The effect of the former is to reduce the

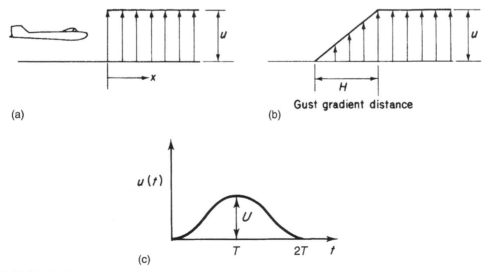

(a)

(b)

Gust gradient distance

(c)

FIGURE 13.13 (a) Sharp-Edged Gust; (b) Graded Gust; (c) 1 – cosine Gust

severity of the gust, while the latter may either increase or decrease the loads involved. To evaluate the corresponding gust loads, the designer may either calculate the complete motion of the aircraft during the disturbance, and hence obtain the gust loads, or replace the graded gust by an equivalent sharp-edged gust producing approximately the same effect. We shall discuss the latter procedure in greater detail later.

The calculation of the complete response of the aircraft to a graded gust may be obtained from its response to a sharp-edged or "step" gust, by treating the former as comprising a large number of small steps and superimposing the responses to each of these. Such a process is known as *convolution* or *Duhamel integration*. This treatment is desirable for large or unorthodox aircraft, where aeroelastic (structural flexibility) effects on gust loads may be appreciable or unknown. In such cases, the assumption of a rigid aircraft may lead to an underestimation of gust loads. The equations of motion are therefore modified to allow for aeroelastic in addition to aerodynamic effects. For small and medium-sized aircraft having orthodox aerodynamic features, the equivalent sharp-edged gust procedure is satisfactory.

While the graded or "ramp" gust is used as a basis for gust load calculations, other shapes of gust profile are in current use. Typical of these is the "1 – cosine" gust of Fig. 13.13(c), where the gust velocity u is given by $u(t) = (U/2)[1 - \cos(\pi t/T)]$. Again the aircraft response is determined by superimposing the responses to each of a large number of small steps.

Although the discrete gust approach still finds widespread use in the calculation of gust loads, alternative methods based on *power spectral* analysis are being investigated. The advantage of the power spectral technique lies in its freedom from arbitrary assumptions of gust shapes and sizes. It is assumed that gust velocity is a random variable that may be regarded for analysis as consisting of a large number of sinusoidal components whose amplitudes vary with frequency. The *power spectrum* of such a function is then defined as the distribution of energy over the frequency range. This may then be related to gust velocity. Establishing appropriate amplitude and frequency distributions for a particular random gust profile requires a large amount of experimental data. The collection of such data was referred to in Section 13.2.

Calculations of the complete response of an aircraft and detailed assessments of the discrete gust and power spectral methods of analysis are outside the scope of this book. More information may be found in references [1–4] at the end of the chapter. Our present analysis is confined to the discrete gust approach, in which we consider the sharp-edged gust and the equivalent sharp-edged gust derived from the graded gust.

13.4.1 Sharp-edged gust

The simplifying assumptions introduced in the determination of gust loads resulting from the sharp-edged gust have been discussed in the earlier part of this section. In Fig. 13.14, the aircraft is flying at a speed V with wing incidence α_0 in still air. After entering the gust of upward velocity u, the incidence increases by an amount $\tan^{-1} u/V$, or since u is usually small compared with V, u/V. This is accompanied by an increase in aircraft speed from V to $(V^2 + u^2)^{1/2}$, but again this increase is neglected since u is small. The increase in wing lift ΔL is then given by

$$\Delta L = \frac{1}{2}\rho V^2 S \frac{\partial C_L}{\partial \alpha} \frac{u}{V} = \frac{1}{2}\rho V S \frac{\partial C_L}{\partial \alpha} u \qquad (13.22)$$

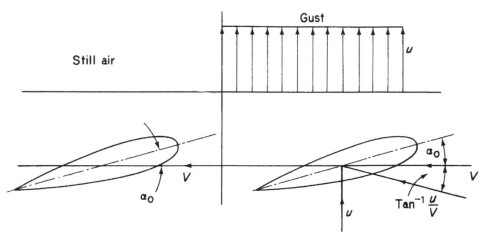

FIGURE 13.14 Increase in Wing Incidence Due to a Sharp-Edged Gust

where $\partial C_L/\partial\alpha$ is the wing lift–curve slope. Neglecting the change of lift on the tailplane as a first approximation, the gust load factor Δn produced by this change of lift is

$$\Delta n = \frac{\frac{1}{2}\rho VS(\partial C_L/\partial\alpha)u}{W} \tag{13.23}$$

where W is the aircraft weight. Expressing Eq. (13.23) in terms of the wing loading, $w = W/S$, we have

$$\Delta n = \frac{\frac{1}{2}\rho V(\partial C_L/\partial\alpha)u}{w} \tag{13.24}$$

This increment in gust load factor is additional to the steady level flight value $n = 1$. Therefore, as a result of the gust, the total gust load factor is

$$n = 1 + \frac{\frac{1}{2}\rho V(\partial C_L/\partial\alpha)u}{w} \tag{13.25}$$

Similarly, for a downgust,

$$n = 1 - \frac{\frac{1}{2}\rho V(\partial C_L/\partial\alpha)u}{w} \tag{13.26}$$

If flight conditions are expressed in terms of equivalent sea-level conditions, then V becomes the equivalent airspeed (EAS), V_E, u becomes u_E and the air density ρ is replaced by the sea-level value ρ_0. Equations (13.25) and (13.26) are written

$$n = 1 + \frac{\frac{1}{2}\rho_0 V_E(\partial C_L/\partial\alpha)u_E}{w} \tag{13.27}$$

and

$$n = 1 - \frac{\frac{1}{2}\rho_0 V_E(\partial C_L/\partial\alpha)u_E}{w} \tag{13.28}$$

We observe from Eqs. (13.25)–(13.28) that the gust load factor is directly proportional to aircraft speed but inversely proportional to wing loading. It follows that high-speed aircraft with low or moderate wing loadings are most likely to be affected by gust loads.

The contribution to normal acceleration of the change in tail load produced by the gust may be calculated using the same assumptions as before. However, the change in tailplane incidence is not equal to the change in wing incidence, due to downwash effects at the tail. Thus, if ΔP is the increase (or decrease) in tailplane load, then

$$\Delta P = \frac{1}{2} \rho_0 V_E^2 S_T \Delta C_{L,T} \tag{13.29}$$

where S_T is the tailplane area and $\Delta C_{L,T}$ the increment of tailplane lift coefficient given by

$$\Delta C_{L,T} = \frac{\partial C_{LT,}}{\partial \alpha} \frac{u_E}{V_E} \tag{13.30}$$

in which $\partial C_{L,T}/\partial \alpha$ is the rate of change of tailplane lift coefficient with wing incidence. From aerodynamic theory,

$$\frac{\partial C_{L,T}}{\partial \alpha} = \frac{\partial C_{L,T}}{\partial \alpha_T} \left(1 - \frac{\partial \varepsilon}{\partial \alpha} \right)$$

where $\partial C_{L,T}/\partial \alpha_T$ is the rate of change of $C_{L,T}$ with tailplane incidence and $\partial \varepsilon/\partial \alpha$ the rate of change of downwash angle with wing incidence. Substituting for $\Delta C_{L,T}$ from Eq. (13.30) into Eq. (13.29), we have

$$\Delta P = \frac{1}{2} \rho_0 V_E S_T \frac{\partial C_{L,T}}{\partial \alpha} u_E \tag{13.31}$$

For positive increments of wing lift and tailplane load,

$$\Delta n W = \Delta L + \Delta P$$

or, from Eqs. (13.27) and (13.31),

$$\Delta n = \frac{\frac{1}{2} \rho_0 V_E (\partial C_L/\partial \alpha) u_E}{w} \left(1 + \frac{S_T}{S} \frac{\partial C_{L,T}/\partial \alpha}{\partial C_L/\partial \alpha} \right) \tag{13.32}$$

13.4.2 The graded gust

The graded gust of Fig. 13.13(b) may be converted to an equivalent sharp-edged gust by multiplying the maximum velocity in the gust by a *gust alleviation factor, F*. Equation (13.27) then becomes

$$n = 1 + \frac{\frac{1}{2} \rho_0 V_E (\partial C_L/\partial \alpha) F u_E}{w} \tag{13.33}$$

Similar modifications are carried out on Eqs. (13.25), (13.26), (13.28), and (13.32). The gust alleviation factor allows for some of the dynamic properties of the aircraft, including unsteady lift, and has been calculated taking into account the heaving motion (i.e., the up and down motion with zero rate of pitch) of the aircraft only[5].

Horizontal gusts cause lateral loads on the vertical tail or fin. Their magnitudes may be calculated in an identical manner to the preceding, except that areas and values of lift–curve slope are referred to the vertical tail. Also, the gust alleviation factor in the graded gust case becomes F_1 and includes allowances for the aerodynamic yawing moment produced by the gust and the yawing inertia of the aircraft.

13.4.3 Gust envelope

Airworthiness requirements usually specify that gust loads shall be calculated at certain combinations of gust and flight speed. The equations for gust load factor in the above analysis show that n is proportional to aircraft speed for a given gust velocity. Therefore, we plot a gust envelope similar to the flight envelope of Fig. 13.1, as shown in Fig. 13.15. The gust speeds $\pm U_1$, $\pm U_2$, and $\pm U_3$ are high, medium, and low velocity gusts, respectively. Cut-offs occur at points where the lines corresponding to each gust velocity meet specific aircraft speeds. For example, A and F denote speeds at which a gust of velocity $\pm U_1$ would stall the wing.

The lift coefficient–incidence curve is, as we noted in connection with the flight envelope, affected by compressibility and therefore altitude, so that a series of gust envelopes should be drawn for different altitudes. An additional variable in the equations for gust load factor is the wing loading w. Further gust envelopes should therefore be drawn to represent different conditions of aircraft loading.

Typical values of U_1, U_2, and U_3 are 20 m/s, 15.25 m/s, and 7.5 m/s. It can be seen from the gust envelope that the maximum gust load factor occurs at the cruising speed V_C. If this value of n exceeds that for the corresponding flight envelope case, that is n_1, then the gust case is the most critical in the cruise.

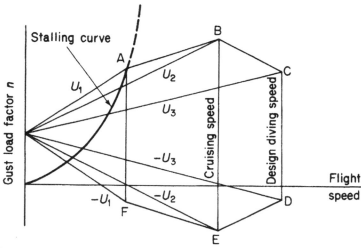

FIGURE 13.15 Typical Gust Envelope

Example 13.7

A civil, non-aerobatic aircraft has a wing loading of 2,400 N/m^2 and $\partial C_L / \partial \alpha = 5.0$/rad. If $n_1 = 2.5$ and $F = 0.715$, calculate the cruising speed for the gust case to be critical. See Ex. 1.1.

From Eq. (13.33),

$$n = 1 + \frac{\frac{1}{2} \times 1.223 \, V_C \times 5.0 \times 0.715 \times 15.25}{2,400}$$

giving $n = 1 + 0.0139V_C$, where the cruising speed V_C is expressed as an EAS. For the gust case to be critical,

$$1 + 0.0139V_C > 2.5$$

or

$$V_C > 108 \text{ m/s}$$

Thus, for civil aircraft of this type having cruising speeds in excess of 108 m/s, the gust case is the most critical. This, in fact, applies to most modern civil airliners.

Although the same combination of V and n in the flight and gust envelopes produce the same total lift on an aircraft, the individual wing and tailplane loads are different, as shown previously (see the derivation of Eq. (13.33)). This situation can be important for aircraft such as the Airbus, which has a large tailplane and a CG forward of the aerodynamic center. In the flight envelope case, the tail load is downward, whereas in the gust case, it is upward; clearly, there will be a significant difference in wing load.

The transference of maneuver and gust loads into bending, shear, and torsional loads on wings, fuselage, and tailplanes has been discussed in Section 12.1. Further loads arise from aileron application, in undercarriages during landing, on engine mountings, and during crash landings. Analysis and discussion of these may be found in the *Handbook of Aeronautics*[6].

Example 13.8

An aircraft of all-up weight 150,000 N has wings of area 60 m^2 and mean chord 2.6 m. The drag coefficient for the complete aircraft is given by $C_D = 0.02 + 0.04C_L^2$ and the lift–curve slope of the wings is 4.5. The tailplane has an area of 10 m^2 and a lift–curve slope of 2.2, allowing for the effects of downwash. The pitching moment coefficient about the aerodynamic center (of the complete aircraft less tailplane) based on wing area is –0.03. Geometric data are given in Fig.13.16.

During a steady glide with zero thrust at 245 m/s EAS in which $C_L = 0.09$, the aircraft meets an upgust of equivalent sharp-edged speed 5 m/s. Calculate the tail load, the gust load factor, and the forward inertia force. Take $\rho_0 = 1.223$ kg/m^3. See Ex. 1.1.

As a first approximation, suppose that the wing lift is equal to the aircraft weight. Then,

$$L = (1/2)\rho V^2 S (\partial C_L / \partial \alpha) \, \alpha_w = 150{,}000 \text{ N}$$

so that

$$\alpha_w = 150{,}000 / \left[(1/2) \times 1.223 \times 245^2 \times 60 \times 4.6 \right] = 0.015 \text{ rad} = 0.87°$$

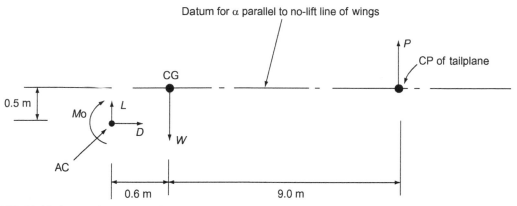

FIGURE 13.16 Geometric Data for Aircraft of Example 13.8

Also,

$$C_D = 0.02 + 0.04 \times 0.09^2 = 0.02$$

Taking moments about the CG and noting that $\cos 0.87° = 1$ approximately,

$$0.6L - 0.5D + M_o = 9.0P$$

that is,

$$0.6(150{,}000 - P) - 0.5 \times (1/2)\,\rho V^2 S C_D + (1/2)\rho V^2\,cC_{M,o} = 9.0P$$

Substituting the appropriate values gives

$$P = -10{,}813 \text{ N}$$

Then,

$$L = W - P = 150{,}000 + 10{,}813 = 160{,}813 \text{ N}$$

The change, ΔP, in the tailplane load due to the gust is, from Eqs. (13.29) and (13.30),

$$\Delta P = (1/2)\rho V^2 S_T(\partial L_{L,T}/\partial\alpha)(u_E/V_E)$$

so that

$$\Delta P = (1/2) \times 1.223 \times 245^2 \times 10 \times 2.2 \times (5/245)$$

that is,

$$\Delta P = 16{,}480 \text{ N}$$

The total load on the tail is then $16{,}480 - 10{,}813 = 5{,}667$ N (upwards).

The increase in wing lift, ΔL, due to the gust is given by

$$\Delta L = (1/2)\rho V^2 S(\partial C_L/\partial\alpha)(u_E/V_E)$$

Substituting the appropriate values gives

$$\Delta L = 202{,}254 \text{ N}$$

Then,

$$n = 1 + [(202{,}254 + 16{,}480)/150{,}000)] = 2.46$$

The forward inertia force, FW, is given by

$$FW = D = (1/2)\rho V^2 S C_D = (1/2) \times 1.223 \times 245^2 \times 60 \times 0.02$$

that is,

$$FW = 44{,}046 \text{ N}$$

References

[1] Zbrozek JK. Atmospheric gusts—present state of the art and further research. J Roy Aero Soc January 1965.
[2] Cox RA. A comparative study of aircraft gust analysis procedures. J Roy Aero Soc October 1970.
[3] Bisplinghoff RL, Ashley H, Halfman RL. Aeroelasticity. Cambridge, MA: Addison-Wesley; 1955.
[4] Babister AW. Aircraft stability and control. London: Pergamon Press; 1961.
[5] Zbrozek JK. Gust alleviation factor, R. and M. No. 2970, May 1953.
[6] Handbook of Aeronautics, No. 1, Structural principles and data. 4th ed. The Royal aeronautical society; 1952.

PROBLEMS

P.13.1 The aircraft shown in Fig. P.13.1(a) weighs 135 kN and has landed such that, at the instant of impact, the ground reaction on each main undercarriage wheel is 200 kN and its vertical velocity is 3.5 m/s. If each undercarriage wheel weighs 2.25 kN and is attached to an oleo strut, as shown in Fig. P.8.1(b), calculate the axial load and bending moment in the strut; the strut may be assumed to be vertical. Determine also the shortening of the strut when the vertical velocity of the aircraft is zero. Finally, calculate the shear force and bending moment in the wing at the section AA if the wing, outboard of this section, weighs 6.6 kN and has its CG 3.05 m from AA.

Answer: 193.3 kN, 29.0 kN m (clockwise); 0.32 m; 19.5 kN, 59.6 kN m (counterclockwise)

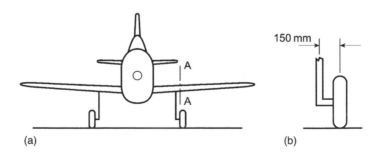

(a)   (b)

FIGURE P.13.1

P.13.1 MATLAB Use MATLAB to repeat Problem P.13.1 for the following aircraft weight values:

	(i)	(ii)	(iii)	(iv)	(v)
W	125 kN	130 kN	135 kN	140 kN	145 kN

Answer: (i) 192.8 kN, 28.9 kN m (clockwise); 0.28 m; 21.1 kN, 64.4 kN m (counterclockwise)
(ii) 193.1 kN, 29 kN m (clockwise); 0.3 m; 20.3 kN, 62 kN m (counterclockwise)
(iii) 193.3 kN, 29 kN m (clockwise); 0.32 m; 19.5 kN, 59.6 kN m (counterclockwise)
(iv) 193.6 kN, 29 kN m (clockwise); 0.34 m; 18.9 kN, 57.6 kN m (counterclockwise)
(v) 193.8 kN, 29.1 kN m (clockwise); 0.35 m; 18.2 kN, 55.6 kN m (counterclockwise)

P.13.2 Determine, for the aircraft of Example 13.2, the vertical velocity of the nosewheel when it hits the ground.

Answer: 3.1 m/s

P.13.3 Figure P.13.3 shows the plan view of an aircraft with wing-tip drop tanks, each of which, when full, weighs 15,575 N and has a radius of gyration in the yaw plane of 0.6 m about its own center of gravity. The aircraft is in steady level flight at 260 m/s and has a total weight of 169,000 N with tip

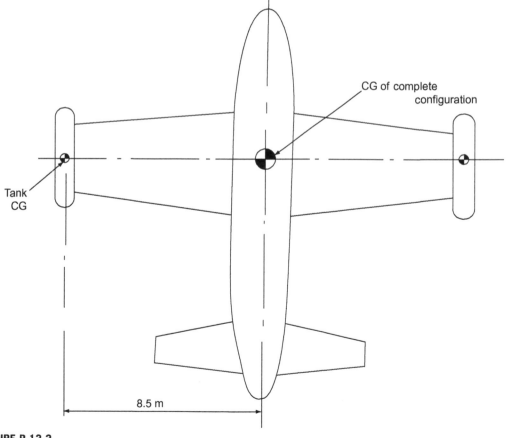

FIGURE P.13.3

tanks full. The radius of gyration in yaw of the aircraft without its tip tanks is 2.5 m and the drag coefficient of the complete configuration is 0.02 (based on a wing area of 60 m^2) of which 8 percent is due to each tip tank. If, at the instant of release, only one tank is dropped, calculate the total fore-and-aft loading on the mounting of the remaining tip tank. Take $\rho = 1.223$ kg/m^3.

Answer: 2,312 N

P.13.4 Figure P.13.4 shows the idealized plan view of a four-engine jet aircraft. The weight distribution of the wing plus fuel is assumed to be uniform and to act along the solid lines shown. Similarly, the fuselage plus payload weight distribution is also assumed to be uniform and to act along the fuselage center line. The weights of the engines and tail unit are assumed to be concentrated at the positions shown. If each engine delivers 66,750 N of thrust in a certain flight condition determine the instantaneous yawing acceleration if two engines on one side instantaneously lose thrust.

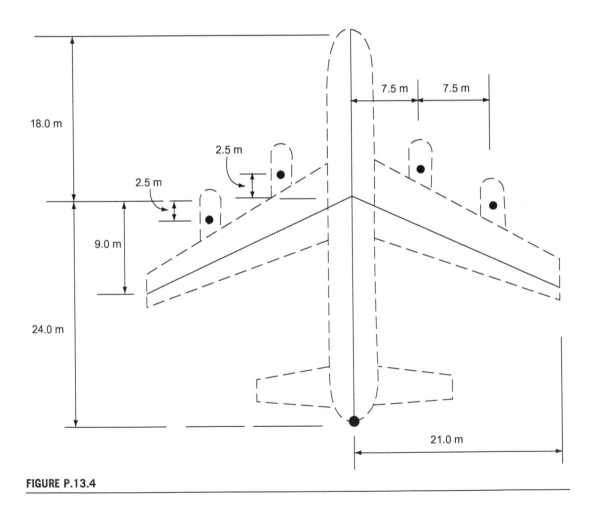

FIGURE P.13.4

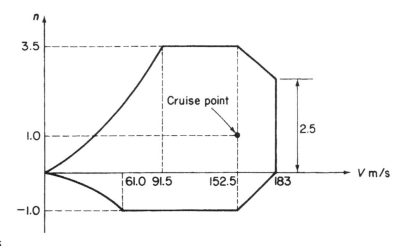

FIGURE P.13.5

The fraction of the all-up weight, W, contributed by the major components of the aircraft are
Wings + fuel: 0.564, Fuselage + payload: 0.310,
Each engine: 0.027, Tail unit: 0.018

Answer: $17{,}182/W \text{ rad/s}^2$

P.13.5 Figure P.13.5 shows the flight envelope at sea-level for an aircraft of wing span 27.5 m, average wing chord 3.05 m and total weight 196,000 N. The aerodynamic center is 0.915 m forward of the CG and the center of lift for the tail unit is 16.7 m aft of the CG. The pitching moment coefficient is

$$C_{M,0} = -0.0638 \text{ (nose-up positive)}$$

both $C_{M,0}$ and the position of the aerodynamic center are specified for the complete aircraft less tail unit. For steady cruising flight at sea level, the fuselage bending moment at the CG is 600,000 Nm. Calculate the maximum value of this bending moment for the given flight envelope. For this purpose, it may be assumed that the aerodynamic loadings on the fuselage itself can be neglected, that is, the only loads on the fuselage structure aft of the CG are those due to the tail lift and the inertia of the fuselage.

Answer: $1{,}549{,}500 \text{ Nm at } n = 3.5, V = 152.5\text{m/s}$

P.13.6 An aircraft weighing 238,000 N has wings 88.5 m² in area, for which $C_D = 0.0075 + 0.045C_L^2$. The extra-to-wing drag coefficient based on wing area is 0.0128 and the pitching moment coefficient for all parts excluding the tailplane about an axis through the CG is given by $C_M \cdot c = (0.427C_L - 0.061)$ m. The radius from the CG to the line of action of the tail lift may be taken as constant at 12.2 m. The moment of inertia of the aircraft for pitching is 204,000 kg m². During a pull-out from a dive with zero thrust at 215 m/s EAS when the flight path is at 40° to the horizontal with a radius of curvature of 1525 m, the angular velocity of pitch is checked by applying a retardation of 0.25 rad/s². Calculate the maneuver load factor both at the CG and at the tailplane CP, the forward inertia coefficient, and the tail lift.

Answer: $n = 3.78(\text{CG}), n = 5.19 \text{ at TP}, f = -0.370, P = 18{,}925 \text{ N}$

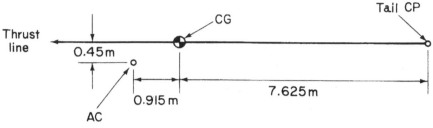

FIGURE P.13.7

P.13.7 An aircraft flies at sea level in a correctly banked turn of radius 610 m at a speed of 168 m/s. Figure P.13.7 shows the relative positions of the CG, aerodynamic center of the complete aircraft less tailplane, and the tailplane center of pressure for the aircraft at zero lift incidence. Calculate the tail load necessary for equilibrium in the turn. The necessary data are given in the usual notation as follows:

$$\text{Weight } W = 133,500 \text{ N}, \qquad dC_L/d\alpha = 4.5/\text{rad},$$
$$\text{Wing area } S = 46.5 \text{ m}^2, \qquad C_D = 0.01 + 0.05C_L^2,$$
$$\text{Wing mean chord } \bar{c} = 3.0 \text{ m}, \qquad C_{M,0} = -0.03$$

Answer: 73,160 N

P.13.8 The aircraft for which the stalling speed V_s in level flight is 46.5 m/s has a maximum allowable maneuver load factor n_1 of 4.0. In assessing gyroscopic effects on the engine mounting, the following two cases are to be considered:
(a) Pull-out at maximum permissible rate from a dive in symmetric flight, the angle of the flight path to the horizontal being limited to 60° for this aircraft.
(b) Steady, correctly banked turn at the maximum permissible rate in horizontal flight.
Find the corresponding maximum angular velocities in yaw and pitch.

Answer: (a) Pitch, 0.37 rad/s, (b) Pitch, 0.41 rad/s, Yaw, 0.103 rad/s

P.13.9 A tail-first supersonic airliner, whose essential geometry is shown in Fig. P.13.9, flies at 610 m/s true airspeed at an altitude of 18,300 m. Assuming that thrust and drag forces act in the same straight line, calculate the tail lift in steady straight and level flight.

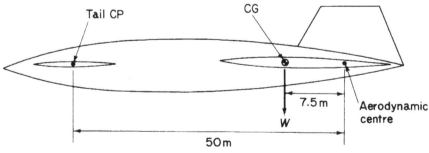

FIGURE P.13.9

If, at the same altitude, the aircraft encounters a sharp-edged vertical upgust of 18 m/s true airspeed, calculate the changes in the lift and tail load and also the resultant load factor n.

The relevant data in the usual notation are as follows

$$
\begin{aligned}
\text{Wing: } S &= 280\text{m}^2, & \partial C_L/\partial\alpha &= 1.5, \\
\text{Tail: } S_T &= 28\text{m}^2, & \partial C_{L,T}/\partial\alpha &= 2.0, \\
\text{Weight } W &= 1{,}600{,}000 \text{ N}, \\
C_{M,0} &= -0.01, \\
\text{Mean chord } \bar{c} &= 22.8\text{m}
\end{aligned}
$$

At 18,300 m,

$$\rho = 0.116 \text{ kg/m}^3$$

Answer: $P = 267{,}852$ N, $\Delta P = 36{,}257$ N, $\Delta L = 271{,}931$ N, $n = 1.19$

P.13.10 An aircraft of all-up weight 145,000 N has wings of area 50 m^2 and mean chord 2.5 m. For the whole aircraft, $C_D = 0.021 + 0.041\ C_L{}^2$, for the wings $dC_L/d\alpha = 4.8$, for the tailplane of area 9.0 m^2, $dC_{L,T}/d\alpha = 2.2$ allowing for the effects of downwash, and the pitching moment coefficient about the aerodynamic center (of complete aircraft less tailplane) based on wing area is $C_{M,0} = -0.032$. Geometric data are given in Fig. P.13.10. During a steady glide with zero thrust at 250 m/s EAS in which $C_L = 0.08$, the aircraft meets a downgust of equivalent sharp-edged speed 6 m/s. Calculate the tail load, the gust load factor and the forward inertia force, $\rho_0 = 1.223$ kg/m^3.

Answer: $P = -28{,}902$ N(down), $n = -0.64$, forward inertia force $= 40{,}703$ N

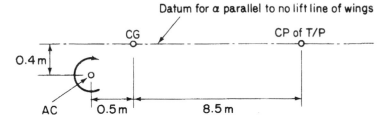

FIGURE P.13.10

Fatigue

14

Fatigue was discussed briefly in Section 11.7, when we examined the properties of materials, and also in Section 13.2, as part of the chapter on airworthiness. We now look at fatigue in greater detail and consider factors affecting the life of an aircraft, including safe life and fail safe structures, designing against fatigue, the fatigue strength of components, the prediction of aircraft fatigue life, and crack propagation.

Fatigue is defined as the progressive deterioration of the strength of a material or structural component during service, such that failure can occur at much lower stress levels than the ultimate stress level. As we have seen, fatigue is a dynamic phenomenon that initiates small (micro) cracks in the material or component and causes them to grow into large (macro) cracks; these, if not detected, can result in catastrophic failure.

Fatigue damage can be produced in a variety of ways. *Cyclic fatigue* is caused by repeated fluctuating loads. *Corrosion fatigue* is fatigue accelerated by surface corrosion of the material penetrating inward so that the material strength deteriorates. Small-scale rubbing movements and abrasion of adjacent parts cause *fretting fatigue*, while *thermal fatigue* is produced by stress fluctuations induced by thermal expansions and contractions; the last does not include the effect on material strength of heat. Finally, high-frequency stress fluctuations, due to vibrations excited by jet or propeller noise, cause *sonic* or *acoustic fatigue*.

Clearly an aircraft's structure must be designed so that fatigue does not become a problem. For aircraft in general, the requirements that the strength of an aircraft throughout its operational life shall be such as to ensure that the possibility of a disastrous fatigue failure shall be extremely remote (i.e., the probability of failure is less than 10^{-7}) under the action of the repeated loads of variable magnitude expected in service. Also, it is required that the principal parts of the primary structure of the aircraft be subjected to a detailed analysis and to load tests that demonstrate a *safe life* or that the parts of the primary structure have *fail-safe* characteristics. These requirements do not apply to light aircraft, provided that zinc-rich aluminum alloys are not used in their construction and that wing stress levels are kept low, that is, provided that a 3.05 m/s upgust causes no greater stress than 14 N/mm^2.

14.1 SAFE LIFE AND FAIL-SAFE STRUCTURES

The danger of a catastrophic fatigue failure in the structure of an aircraft may be eliminated completely or may become extremely remote if the structure is designed to have a safe life or to be fail safe. In the former approach, the structure is designed to have a minimum life during which it is known that no catastrophic damage will occur. At the end of this life, the structure must be replaced even though there may be no detectable signs of fatigue. If a structural component is not economically replaceable when its safe life has been reached, the complete structure must be written off. Alternatively, it is possible for easily replaceable components, such as undercarriage legs and mechanisms, to have a safe life less than that of the complete aircraft, since it would probably be more economical to use, say, two lightweight undercarriage systems during the life of the aircraft rather than carry a heavier undercarriage which has the same safe life as the aircraft.

The fail-safe approach relies on the fact that the failure of a member in a redundant structure does not necessarily lead to the collapse of the complete structure, provided the remaining members are able to carry the load shed by the failed member and can withstand further repeated loads until the presence of the failed member is discovered. Such a structure is called a *fail-safe structure* or a *damage tolerant structure*.

Generally, it is more economical to design some parts of the structure to be fail-safe rather than to have a long safe life, since such components can be lighter. When failure is detected, either through a routine inspection or by some malfunction, such as fuel leakage from a wing crack, the particular aircraft may be taken out of service and repaired. However, the structure must be designed and the inspection intervals arranged such that a failure, for example, a crack, too small to be noticed at one inspection must not increase to a catastrophic size before the next. The determination of crack propagation rates is discussed later.

Some components must be designed to have a safe life; these include landing gear, major wing joints, wing–fuselage joints, and hinges on all-moving tailplanes or on variable geometry wings. Components which may be designed to be fail safe include wing skins stiffened by stringers and fuselage skins stiffened by frames and stringers; the stringers and frames prevent skin cracks spreading disastrously for a sufficient period of time for them to be discovered at a routine inspection.

14.2 DESIGNING AGAINST FATIGUE

Various precautions may be taken to ensure that an aircraft has an adequate fatigue life. We saw, in Chapter 10, that the early aluminum–zinc alloys possessed high ultimate and proof stresses but were susceptible to early failure under fatigue loading; choice of materials is therefore important. The naturally aged aluminum–copper alloys possess good fatigue resistance but with lower static strengths. Modern research is concentrating on alloys which combine high strength with high fatigue resistance.

Attention to detail design is equally important. Stress concentrations can arise at sharp corners and abrupt changes in section. Fillets should therefore be provided at re-entrant corners, and cutouts, such as windows and access panels, should be reinforced. In machined panels, the material thickness should be increased around bolt holes, while holes in primary bolted joints should be reamed to improve surface finish; surface scratches and machine marks are sources of fatigue crack initiation. Joggles in highly stressed members should be avoided, while asymmetry can cause additional stresses due to bending.

In addition to sound structural and detail design, an estimation of the number, frequency, and magnitude of the fluctuating loads an aircraft encounters is necessary. The *fatigue load spectrum* begins when the aircraft taxis to its take-off position. During taxiing, the aircraft may be maneuvering over uneven ground with a full payload, so that wing stresses, for example, are greater than in the static case. Also, during take-off and climb and descent and landing, the aircraft is subjected to the greatest load fluctuations. The undercarriage is retracted and lowered; flaps are raised and lowered; there is the impact on landing; the aircraft has to carry out maneuvers; and, finally, the aircraft, as we shall see, experiences a greater number of gusts than during the cruise.

The loads corresponding to these various phases must be calculated before the associated stresses can be obtained. For example, during take-off, wing bending stresses and shear stresses due to shear and torsion are based on the total weight of the aircraft, including full fuel tanks, and maximum payload all factored by 1.2 to allow for a bump during each take-off on a hard runway or by 1.5 for a take-off from grass. The loads produced during level flight and symmetric maneuvers are calculated using the methods described in Section 14.2. From these values, distributions of shear force, bending moment, and torque may be found in, say, the wing by integrating the lift distribution. Loads due to gusts are calculated using the methods described in Section 14.4. Thus, due to a single equivalent sharp-edged gust, the load factor is given by either Eq, (14.25) or (14.26).

Although it is a relatively simple matter to determine the number of load fluctuations during a ground–air–ground cycle caused by standard operations, such as raising and lowering flaps or retracting and lowering the undercarriage, it is more difficult to estimate the number and magnitude of gusts an aircraft will encounter. For example, there is a greater number of gusts at low altitude (during take-off, climb, and descent) than at high altitude (during cruise). Terrain (sea, flat land, mountains) also affects the number and magnitude of gusts, as does weather. The use of radar enables aircraft to avoid cumulus, where gusts are prevalent, but has little effect at low altitude in the climb and descent, where clouds cannot easily be avoided. The ESDU (Engineering Sciences Data Unit) produced gust data based on information collected by gust recorders carried by aircraft. These show, in graphical form (l_{10} versus h curves, h is altitude), the average distance flown at various altitudes for a gust having a velocity greater than ± 3.05 m/s to be encountered. In addition, *gust frequency curves* give the number of gusts of a given velocity per 1000 gusts of velocity 3.05 m/s. Combining both sets of data enables the *gust exceedance* to be calculated, that is, the number of gust cycles having a velocity greater than or equal to a given velocity encountered per kilometre of flight.

Since an aircraft is subjected to the greatest number of load fluctuations during taxi–take-off–climb and descent–standoff–landing while little damage is caused during cruise, the fatigue life of an aircraft does not depend on the number of flying hours but on the number of flights. However, the operational requirements of aircraft differ from class to class. The Airbus is required to have a life free from fatigue cracks of 24,000 flights or 30,000 hours, while its economic repair life is 48,000 flights or 60,000 hours; its landing gear, however, is designed for a safe life of 32,000 flights, after which it must be replaced. On the other hand, the BAe 146, with a greater number of shorter flights per day than the Airbus, has a specified crack-free life of 40,000 flights and an economic repair life of 80,000 flights. Although these figures are operational requirements, the nature of fatigue is such that it is unlikely that all of a given type of aircraft satisfies them. Of the total number of Airbus aircraft, at least 90 percent achieve these values and 50 percent are better; clearly, frequent inspections are necessary during an aircraft's life.

14.3 FATIGUE STRENGTH OF COMPONENTS

In Section 13.2.4, we discussed the effect of stress level on the number of cycles to failure of a material such as mild steel. As the stress level is decreased, the number of cycles to failure increases, resulting in a fatigue endurance curve (the *S–N* curve) of the type shown in Fig. 13.2. Such a curve corresponds to the average value of N at each stress amplitude, since the given stress has a wide range of values of N; even under carefully controlled conditions, the ratio of maximum N to minimum N may be as high as 10:1. Two other curves may therefore be drawn, as shown in Fig. 14.1, enveloping all or nearly all the

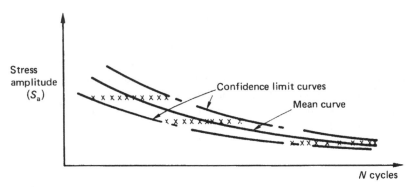

FIGURE 14.1 *S–N* Diagram

experimental results; these curves are known as the *confidence limits*. If 99.9 percent of all the results lie between the curves, that is, only 1 in 1000 falls outside, they represent the 99.9 percent confidence limits. If 99.99999 percent of results lie between the curves only 1 in 10^7 results falls outside them and they represent the 99.99999 percent confidence limits.

The results from tests on a number of specimens may be represented as a histogram, in which the number of specimens failing within certain ranges R of N is plotted against N. Then, if N_{av} is the average value of N at a given stress amplitude, the probability of failure occurring at N cycles is given by

$$p(N) = \frac{1}{\sigma\sqrt{2\pi}} \exp\left[-\frac{1}{2}\left(\frac{N - N_{av}}{\sigma}\right)^2\right] \tag{14.1}$$

in which σ is the standard deviation of the whole population of N values. The derivation of Eq. (14.1) depends on the histogram approaching the profile of a continuous function close to the *normal distribution*, which it does as the interval N_{av}/R becomes smaller and the number of tests increases. The *cumulative probability*, which gives the probability that a particular specimen fails at or below N cycles, is defined as

$$P(N) = \int_{-\infty}^{N} p(N)\, dN \tag{14.2}$$

The probability that a specimen endures more than N cycles is then $1 - P(N)$. The normal distribution allows negative values of N, which is clearly impossible in a fatigue testing situation. Other distributions, *extreme value distributions*, are more realistic and allow the existence of minimum fatigue endurances and fatigue limits.

The damaging portion of a fluctuating load cycle occurs when the stress is tensile; this causes cracks to open and grow. Therefore, if a steady tensile stress is superimposed on a cyclic stress, the maximum tensile stress during the cycle increases and the number of cycles to failure decreases. Conversely, if the steady stress is compressive, the maximum tensile stress decreases and the number of cycles to failure increases. An approximate method of assessing the effect of a steady mean value of stress is provided by a Goodman diagram, shown in Fig. 14.2. This shows the cyclic stress amplitudes which can be superimposed upon different mean stress levels to give a constant fatigue life. In Fig. 14.2, S_a is the allowable

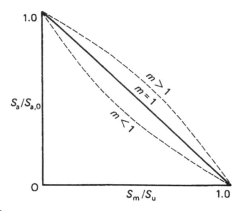

FIGURE 14.2 Goodman Diagram

stress amplitude, $S_{a,0}$ is the stress amplitude required to produce fatigue failure at N cycles with zero mean stress, S_m is the mean stress, and S_u the ultimate tensile stress. If $S_m = S_u$, any cyclic stress causes failure, while if $S_m = 0$, the allowable stress amplitude is $S_{a,0}$. The equation of the straight line portion of the diagram is

$$\frac{S_a}{S_{a,0}} = \left(1 - \frac{S_m}{S_u}\right) \tag{14.3}$$

Experimental evidence suggests a nonlinear relationship for particular materials. Equation (14.3) then becomes

$$\frac{S_a}{S_{a,0}} = \left[1 - \left(\frac{S_m}{S_u}\right)^m\right] \tag{14.4}$$

in which m lies between 0.6 and 2. The relationship for the case of $m = 2$ is known as the *Gerber parabola*.

Example 14.1

The allowable stress amplitude for a low-carbon steel is $\pm225\text{N/mm}^2$ and its ultimate tensile stress is 750 N/mm^2. The steel is subjected to a repeated cycle of stress in which the minimum stress is zero. Calculate the safe range of stress based on the Goodman and Berger predictions. See Ex. 1.1.

For the Goodman prediction, Eq. (14.3) applies in which $S_{a,0} = 450$ N/mm^2 and $S_m = S_a/2$. Then,

$$S_a = 450[1 - (S_a/2 \times 750)]$$

from which

$$S_a = 346 \text{ N/mm}^2$$

For the Gerber prediction, $m = 2$ in Eq. (14.4). Then,

$$S_a = 450[1 - (S_a/2 \times 750)^2]$$

which simplifies to

$$S_a^2 + 5000S_a - 2.25 \times 10^6 = 0$$

Solving gives

$$S_a = 415 \, \text{N/mm}^2$$

Example 14.2

If the steel in Example 14.1 is subjected to an alternating cycle of tensile stress about a mean stress of 180 N/mm², calculate the safe range of stress based on the Goodman prediction and also the maximum and minimum stress values. See Ex. 1.1.

From Eq. (14.3),

$$S_a = 450[1 - (180/750)]$$

that is,

$$S_a = 342 \, \text{N/mm}^2$$

Now

$$S_a = \sigma_{max} - \sigma_{min} \tag{i}$$

and

$$S_m = (\sigma_{max} + \sigma_{min})/2 \tag{ii}$$

Adding Eqs. (i) and (ii),

$$S_a + 2S_m = 2\sigma_{max}$$

that is,

$$342 + 2 \times 180 = 2\sigma_{max}$$

so that

$$\sigma_{max} = 351 \, \text{N/mm}^2$$

Then, from Eq. (i),

$$\sigma_{min} = 351 - 342 = 9 \, \text{N/mm}^2$$

In practical situations, fatigue is not caused by a large number of identical stress cycles but by many stress amplitude cycles. The prediction of the number of cycles to failure therefore becomes complex. Miner and Palmgren proposed a *linear cumulative damage law*, as follows. If N cycles of stress amplitude S_a cause fatigue failure then one cycle produces $1/N$ of the total damage to cause failure. Therefore, if r different cycles are applied, in which a stress amplitude S_j $(j = 1, 2, \ldots, r)$ would cause failure in N_j cycles, the number of cycles n_j required to cause total fatigue failure is given by

$$\sum_{j=1}^{r} \frac{n_j}{N_j} = 1 \tag{14.5}$$

Example 14.3

In a fatigue test, a steel specimen is subjected to a reversed cyclic loading in a continuous sequence of four stages as follows:

200 cycles at ± 150 N/mm^2
250 cycles at ± 125 N/mm^2
400 cycles at ± 120 N/mm^2
550 cycles at ± 100 N/mm^2

If the loading is applied at the rate of 80 cycles/hour and the fatigue lives at these stress levels are $10^4, 10^5, 1.5 \times 10^5$, and 2×10^5, respectively, calculate the life of the specimen. See Ex. 1.1.

Suppose that the specimen fails after P sequences of the four stages. Then, from Eq. (14.5),

$$P[(200/10^4) + (250/10^5) + (400/1.5 \times 10^5) + (550/2 \times 10^5)] = 1$$

which gives

$$P = 35.8$$

The total number of cycles in the four stages is 1400, so that, for a loading rate of 80 cycles/hour, the total number of hours to fracture is given by

$$\text{Total hours to fracture} = 35.8 \times 1400/80 = 626.5$$

Although S–N curves may be readily obtained for different materials by testing a large number of small specimens (*coupon tests*), it is not practical to adopt the same approach for aircraft components, since these are expensive to manufacture and the test program too expensive to run for long periods of time. However, such a program was initiated in the early 1950s to test the wings and tailplanes of Meteor and Mustang fighters. These were subjected to constant amplitude loading until failure, with different specimens being tested at different load levels. Stresses were measured at points where fatigue was expected (and actually occurred) and S–N curves plotted for the complete structure. The curves had the usual appearance and at low stress levels had such large endurances that fatigue did not occur; thus, a fatigue limit existed. It was found that the average S–N curve could be approximated to by the equation

$$S_a = 10.3(1 + 1000/\sqrt{N}) \tag{14.6}$$

in which the mean stress was 90 N/mm^2. In general terms, Eq. (14.6) may be written as

$$S_a = S_\infty(1 + C/\sqrt{N}) \tag{14.7}$$

in which S_∞ is the fatigue limit and C is a constant. Thus, $S_a \rightarrow S_\infty$ as $N \rightarrow \infty$. Equation (14.7) may be rearranged to give the endurance directly:

$$N = C^2 \left(\frac{S_\infty}{S_a - S_\infty} \right)^2 \tag{14.8}$$

which shows clearly that, as $S_a \rightarrow S_\infty$, $N \rightarrow \infty$.

Example 14.4

Determine the number of cycles to fracture for a specimen of low-carbon steel whose fatigue limit is 60 N/mm^2 and which is subjected to a stress amplitude of ± 100 N/mm^2. See Ex. 1.1.

Assume that C in Eq. (14.8) is 1000, then

$$N = 1000^2 [60/(200 - 60)]^2$$

which gives

$$N = 1.8 \times 10^5 \text{cycles}$$

It has been found experimentally that N is inversely proportional to the mean stress, as the latter varies in the region of 90 N/mm^2 while C is virtually constant. This suggests a method of determining a "standard" endurance curve (corresponding to a mean stress level of 90 N/mm^2) from tests carried out on a few specimens at other mean stress levels. Suppose S_m is the mean stress level, not 90 N/mm^2, in tests carried out on a few specimens at an alternating stress level $S_{a,m}$, where failure occurs at a mean number of cycles N_m. Then, assuming that the S–N curve has the same form as Eq. (14.7),

$$S_{a,m} = S_{\infty,m}(1 + C/\sqrt{N_m}) \tag{14.9}$$

in which $C = 1000$ and $S_{\infty,m}$ is the fatigue limit stress corresponding to the mean stress S_m. Rearranging Eq. (14.9), we have

$$S_{\infty,m} = S_{a,m}/(1 + C/\sqrt{N_m}) \tag{14.10}$$

The number of cycles to failure at a mean stress of 90 N/mm^2 would have been, from the above,

$$N' = \frac{S_m}{90} N_m \tag{14.11}$$

The corresponding fatigue limit stress would then have been, from a comparison with Eq. (14.10),

$$S'_{\infty,m} = S_{a,m}/(1 + C/\sqrt{N'}) \tag{14.12}$$

The standard endurance curve for the component at a mean stress of 90 N/mm^2 is, from Eq. (14.7),

$$S_a = S'_{\infty,m}/(1 + C/\sqrt{N}) \tag{14.13}$$

Substituting in Eq. (14.13) for $S'_{\infty,m}$ from Eq. (14.12), we have

$$S_a = \frac{S_{a,m}}{(1 + C/\sqrt{N'})}(1 + C/\sqrt{N}) \tag{14.14}$$

in which N' is given by Eq. (14.11).

Equation (14.14) is based on a few test results so that a "safe" fatigue strength is usually taken to be three standard deviations below the mean fatigue strength. Hence, we introduce *a scatter factor* K_n (>1) to allow for this; Eq. (14.14) then becomes

$$S_a = \frac{S_{a,m}}{K_n(1 + C/\sqrt{N'})}(1 + C/\sqrt{N}) \tag{14.15}$$

K_n varies with the number of test results available and, for a coefficient of variation of 0.1, $K_n = 1.45$ for 6 specimens, $K_n = 1.445$ for 10 specimens, $K_n = 1.44$ for 20 specimens, and for 100 specimens or more, $K_n = 1.43$. For typical S–N curves, a scatter factor of 1.43 is equivalent to a life factor of 3 to 4.

Example 14.5

A fatigue test is carried out on 20 specimens at a mean stress level of 120 N/mm^2, where the alternating stress is ±150 N/mm^2. If the average number of cycles at fracture is 10^5, determine the equation of the S-N curve. Assume that the constant C is equal to 1000 and determine also the number of cycles to fracture for a stress amplitude of 200 N/mm^2. See Ex. 1.1.

From Eq. (14.11),

$$N' = (120/90) \times 10^5 = 1.33 \times 10^5$$

Therefore, from Eq. (14.15) and since $K_n = 1.44$ for 20 specimens,

$$S_a = \{300/[1.44(1 + 1000/(1.33 \times 10^5)^{1/2})]\}[1 + (1000/(N)^{1/2}]$$

so that

$$S_a = 55.7[1 + 1000/(N)^{1/2}) \tag{i}$$

From Eq. (i),

$$200 = 55.7[1 + 1000/(N)^{1/2}]$$

which gives

$$N = 1.5 \times 10^5 \text{cycles}$$

14.4 PREDICTION OF AIRCRAFT FATIGUE LIFE

We see that an aircraft suffers fatigue damage during all phases of the ground–air–ground cycle. The various contributions to this damage may be calculated separately and hence the safe life of the aircraft in terms of the number of flights calculated.

In the ground–air–ground cycle, the maximum vertical acceleration during takeoff is 1.2 g for a take-off from a runway or 1.5 g for a take-off from grass. It is assumed that these accelerations occur at zero lift and therefore produce compressive (negative) stresses, $-S_{TO}$, in critical components such as the undersurface of wings. The maximum positive stress for the same component occurs in level flight (at 1g) and is $+S_{1g}$. The ground–air–ground cycle produces, on the undersurface of the wing, a fluctuating stress $S_{GAG} = (S_{1g} + S_{TO})/2$ about a mean stress $S_{GAG(mean)} = (S_{1g} - S_{TO})/2$. Suppose that tests show that for this stress cycle and mean stress, failure occurs after N_G cycles. For a life factor of 3, the safe life is $N_G/3$, so that the damage done during one cycle is $3/N_G$. This damage is multiplied by a factor of 1.5 to allow for the variability of loading between different aircraft of the same type, so that the damage per flight D_{GAG} from the ground–air–ground cycle is given by

$$D_{GAG} = 4.5/N_G \qquad (14.16)$$

Fatigue damage is also caused by gusts encountered in flight, particularly during the climb and descent. Suppose that a gust of velocity u_e causes a stress S_u about a mean stress corresponding to level flight and suppose also that the number of stress cycles of this magnitude required to cause failure is $N(S_u)$; the damage caused by one cycle is then $1/N(S_u)$. Therefore, from the Palmgren–Miner hypothesis, when sufficient gusts of this and all other magnitudes together with the effects of all other load cycles produce a cumulative damage of 1.0, fatigue failure occurs. It is therefore necessary to know the number and magnitude of gusts likely to be encountered in flight.

Gust data have been accumulated over a number of years from accelerometer records from aircraft flying over different routes and terrains, at different heights, and at different seasons. The ESDU data sheets[1] present the data in two forms, as we previously noted. First, l_{10} against altitude curves show the distance that must be flown at a given altitude in order that a gust (positive or negative) having a velocity ≥ 3.05 m/s be encountered. It follows that $1/l_{10}$ is the number of gusts encountered in unit distance (1 km) at a particular height. Second, gust frequency distribution curves, $r(u_e)$ against u_e, give the number of gusts of velocity u_e for every 1000 gusts of velocity 3.05 m/s.

From these two curves, the gust exceedance $E(u_e)$ is obtained; $E(u_e)$ is the number of times a gust of a given magnitude (u_e) will be equaled or exceeded in 1 km of flight. Thus, from the preceding,

Number of gusts ≥ 3.05 m/s per km $= 1/l_{10}$
Number of gusts equal to u_e per 1000 gusts equal to 3.05 m/s $= r(u_e)$

Hence,

Number of gusts equal to u_e per single gust equal to 3.05 m/s $= r(u_e)/1000$

It follows that the gust exceedance $E(u_e)$ is given by

$$E(u_e) = \frac{r(u_e)}{1000 l_{10}} \qquad (14.17)$$

in which l_{10} is dependent on height. A good approximation for the curve of $r(u_e)$ against u_e in the region $u_e = 3.05$ m/s is

$$r(u_e) = 3.23 \times 10^5 u_e^{-5.26} \qquad (14.18)$$

Consider now the typical gust exceedance curve shown in Fig. 14.3. In 1 km of flight, there are likely to be $E(u_e)$ gusts exceeding u_e m/s and $E(u_e) - \delta E(u_e)$ gusts exceeding $u_e + \delta u_e$ m/s. Thus, there will be

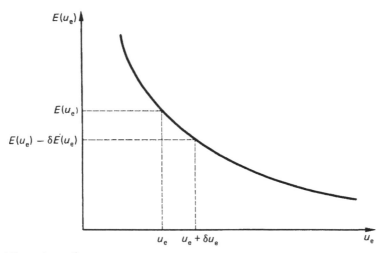

FIGURE 14.3 Gust Exceedance Curve

$\delta E(u_e)$ *fewer* gusts exceeding $u_e + \delta u_e$ m/s than u_e m/s and the increment in gust speed δu_e corresponds to a number $-\delta E(u_e)$ of gusts at a gust speed close to u_e. Half of these gusts will be positive (upgusts) and half negative (downgusts), so that if it is assumed that each upgust is followed by a downgust of equal magnitude, the number of complete gust cycles will be $-\delta E(u_e)/2$. Suppose that each cycle produces a stress $S(u_e)$ and that the number of these cycles required to produce failure is $N(S_{u,e})$. The damage caused by one cycle is then $1/N(S_{u,e})$ and over the gust velocity interval δu_e the total damage δD is given by

$$\delta D = -\frac{\delta E(u_e)}{2N(S_{u,e})} = -\frac{dE(u_e)}{du_e}\frac{\delta u_e}{2N(S_{u,e})} \tag{14.19}$$

Integrating Eq. (14.19) over the whole range of gusts likely to be encountered, we obtain the total damage D_g per kilometre of flight:

$$D_g = -\int_0^\infty \frac{1}{2N(S_{u,e})}\frac{dE(u_e)}{du_e}\,du_e \tag{14.20}$$

Further, if the average block length journey of an aircraft is R_{av}, the average gust damage per flight is $D_g R_{av}$. Also, some aircraft in a fleet experience more gusts than others, since the distribution of gusts is random. Therefore, if, for example, it is found that one particular aircraft encounters 50 percent more gusts than the average, its gust fatigue damage is $1.5\,D_g$/km.

The gust damage predicted by Eq. (14.20) is obtained by integrating over a complete gust velocity range from zero to infinity. Clearly, there will be a gust velocity below which no fatigue damage occurs, since the cyclic stress produced is below the fatigue limit stress of the particular component. Equation (14.20) is therefore rewritten

$$D_g = -\int_{u_f}^\infty \frac{1}{2N(S_{u,e})}\frac{dE(u_e)}{du_e}\,du_e \tag{14.21}$$

in which u_f is the gust velocity required to produce the fatigue limit stress.

We noted previously that more gusts are encountered during climb and descent than during cruise. Altitude therefore affects the amount of fatigue damage caused by gusts and its effects may be determined as follows. Substituting for the gust exceedance $E(u_e)$ in Eq. (14.21) from Eq. (14.17), we obtain

$$D_g = -\frac{1}{1000 l_{10}} \int_{u_f}^{\infty} \frac{1}{2N(S_{u,e})} \frac{dr(u_e)}{du_e} \, du_e$$

or

$$D_g = \frac{1}{l_{10}} d_g \text{ per kilometre} \tag{14.22}$$

in which l_{10} is a function of height h and

$$d_g = -\frac{1}{1000} \int_{u_f}^{\infty} \frac{1}{2N(S_{u,e})} \frac{dr(u_e)}{du_e} \, du_e$$

Suppose that the aircraft is climbing at a speed V with a rate of climb (ROC). The time taken for the aircraft to climb from a height h to a height $h + \delta h$ is $\delta h/\text{ROC}$, during which time it travels a distance $V\delta h/\text{ROC}$. Hence, from Eq. (14.22), the fatigue damage experienced by the aircraft in climbing through a height δh is

$$\frac{1}{l_{10}} d_g \frac{V}{\text{ROC}} \delta h$$

The total damage produced during a climb from sea level to an altitude H at a constant speed V and ROC is

$$D_{g,\text{climb}} = d_g \frac{V}{\text{ROC}} \int_0^H \frac{dh}{l_{10}} \tag{14.23}$$

Plotting $1/l_{10}$ against h from ESDU data sheets for aircraft having cloud warning radar and integrating gives

$$\int_0^{3000} \frac{dh}{l_{10}} = 303, \quad \int_{3000}^{6000} \frac{dh}{l_{10}} = 14, \quad \int_{6000}^{9000} \frac{dh}{l_{10}} = 3.4$$

From this, $\int_0^{9000} dh/l_{10} = 320.4$, from which it can be seen that approximately 95 percent of the total damage in the climb occurs in the first 3000 m.

An additional factor influencing the amount of gust damage is forward speed. For example, the change in wing stress produced by a gust may be represented by

$$S_{u,e} = k_1 u_e V_e \quad \text{(see Eq. (14.24))} \tag{14.24}$$

in which the forward speed of the aircraft is in equivalent airspeed (EAS). From Eq. (14.24), we see that the gust velocity u_f required to produce the fatigue limit stress S_∞ is

$$u_f = S_\infty/k_1 V_e \tag{14.25}$$

The gust damage per kilometre at different forward speeds V_e is then found using Eq. (14.21) with the appropriate value of u_f as the lower limit of integration. The integral may be evaluated by using the known approximate forms of $N(S_{u,e})$ and $E(u_e)$ from Eqs. (14.15) and (14.17). From Eq. (14.15),

$$S_a = S_{u,e} = \frac{S'_{\infty,m}}{K_n}\left[1 + C/\sqrt{N(S_{u,e})}\right]$$

from which

$$N(S_{u,e}) = \left(\frac{C}{K_n}\right)^2 \left(\frac{S'_{\infty,m}}{S_{u,e} - S'_{\infty,m}}\right)^2$$

where $S_{u,e} = k_1 V_e u_e$ and $S'_{\infty,m} = k_1 V_e u_f$. Also, Eq. (14.17) is

$$E(u_e) = \frac{r(u_e)}{1000l_{10}}$$

or, substituting for $r(u_e)$ from Eq. (14.18),

$$E(u_e) = \frac{3.23 \times 10^5 u_e^{-5.26}}{1000l_{10}}$$

Equation (14.21) then becomes

$$D_g = -\int_{u_f}^{\infty} \frac{1}{2}\left(\frac{K_n}{C}\right)^2 \left(\frac{S_{u,e} - S'_{\infty,m}}{S'_{\infty,m}}\right)^2 \left(\frac{-3.23 \times 5.26 \times 10^5 u_e^{-5.26}}{1000l_{10}}\right) du_e$$

Substituting for $S_{u,e}$ and $S'_{\infty,m}$, we have

$$D_g = \frac{16.99 \times 10^2}{2l_{10}}\left(\frac{K_n}{C}\right)^2 \int_{u_f}^{\infty} \left(\frac{u_e - u_f}{u_f}\right) u_e^{-6.26} \, du_e$$

or

$$D_g = \frac{16.99 \times 10^2}{2l_{10}}\left(\frac{K_n}{C}\right)^2 \int_{u_f}^{\infty} \left(\frac{u_e^{-4.26}}{u_f^2} - \frac{2u_e^{-5.26}}{u_f} + u_e^{-6.26}\right) du_e$$

from which

$$D_g = \frac{46.55}{2l_{10}}\left(\frac{K_n}{C}\right)^2 u_f^{-5.26}$$

or, in terms of the aircraft speed V_e,

$$D_g = \frac{46.55}{2l_{10}}\left(\frac{K_n}{C}\right)^2 \left(\frac{k_1 V_e}{S'_{\infty,m}}\right)^{5.26} \text{ per kilometre} \tag{14.26}$$

It can be seen from Eq. (14.26) that gust damage increases in proportion to $V_e^{5.26}$, so that increasing forward speed has a dramatic effect on gust damage.

The total fatigue damage suffered by an aircraft per flight is the sum of the damage caused by the ground–air–ground cycle, the damage produced by gusts, and the damage due to other causes, such as pilot-induced maneuvers, ground turning and braking, and landing and take-off load fluctuations. The damage produced by these other causes can be determined from load exceedance data. Thus, if this extra damage per flight is D_{extra}, the total fractional fatigue damage per flight is

$$D_{total} = D_{GAG} + D_g R_{av} + D_{extra}$$

or

$$D_{total} = 4.5/N_G + D_g R_{av} + D_{extra} \tag{14.27}$$

and the life of the aircraft in terms of flights is

$$N_{flight} = 1/D_{total} \tag{14.28}$$

Example 14.6

A passenger aircraft cruises at 250 m/s EAS and its average block length journey is 1,500 km. The fatigue limit stress for critical components of the aircraft is ±230 N/mm^2 based on a mean stress of 90 N/mm^2. Assuming that the damage caused during the ground–air–ground cycle is 10 percent of the total damage per flight, calculate the life of the aircraft in terms of the number of flights. Take $k_1 = 2.97$. See Ex. 1.1.

The damage per kilometre due to gusts is given by Eq. (14.26), where we assume that $K_n = 1.43$ and $C = 1000$ (see Section 14.3). From gust data $l_{10} = 5,000$ m. Substituting the appropriate values in Eq.(14.26),

$$D_g = (46.55/2 \times 5000)(1.43/1000)^2(2.97 \times 250/460)^{5.26}$$

so that

$$D_g = 1.18 \times 10^{-7}/km$$

Now,

$$D_{total} = D_g R_{av} + 0.1 D_{total}$$

Then,

$$0.9 D_{total} = 1.18 \times 10^{-7} \times 1500 = 1.77 \times 10^{-4}$$

and

$$D_{total} = 1.97 \times 10^{-4}$$

Therefore, the life of the aircraft in terms of the number of flights is

$$\text{No. of flights} = 1/1.97 \times 10^{-4} = 5,076, \text{ say, } 5,000$$

14.5 **CRACK PROPAGATION**

We have seen that the concept of fail-safe structures in aircraft construction relies on a damaged structure being able to retain enough of its load-carrying capacity to prevent catastrophic failure, at least until the damage is detected. It is therefore essential that the designer be able to predict how and at what rate a fatigue crack grows. The ESDU data sheets provide a useful introduction to the study of crack propagation; some of the results are presented here.

The analysis of stresses close to a crack tip using elastic stress concentration factors breaks down, since the assumption that the crack tip radius approaches zero results in the stress concentration factor tending to infinity. Instead, linear elastic fracture mechanics analyses the stress field around the crack tip and identifies features of the field common to all cracked elastic bodies.

14.5.1 **Stress concentration factor**

There are three basic modes of crack growth are shown in Fig. 14.4. Generally, the stress field in the region of the crack tip is described by a two-dimensional model, which may be used as an approximation for many practical three-dimensional loading cases. Thus, the stress system at a distance r ($r \leq a$) from the tip of a crack of length $2a$, shown in Fig. 14.5, can be expressed in the form[2]

$$S_r, \; S_\theta, \; S_{r,\theta} = \frac{K}{(2\pi r)^{\frac{1}{2}}} f(\theta) \tag{14.29}$$

in which $f(\theta)$ is a different function for each of the three stresses and K is the *stress intensity factor*; K is a function of the nature and magnitude of the applied stress levels and also of the crack size. The terms $(2\pi r)^{1/2}$ and $f(\theta)$ map the stress field in the vicinity of the crack and are the same for all cracks under external loads that cause crack openings of the same type.

Equation (14.29) applies to all modes of crack opening, with K having different values, depending on the geometry of the structure, the nature of the applied loads, and the type of crack.

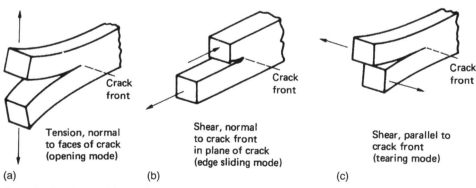

(a) Tension, normal to faces of crack (opening mode)

(b) Shear, normal to crack front in plane of crack (edge sliding mode)

(c) Shear, parallel to crack front (tearing mode)

FIGURE 14.4 Basic Modes of Crack Growth

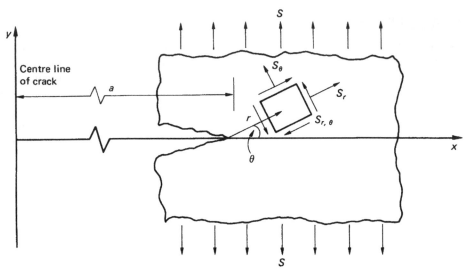

FIGURE 14.5 Stress Field in the Vicinity of a Crack

Experimental data show that crack growth and residual strength data are better correlated using K than any other parameter. K may be expressed as a function of the nominal applied stress S and the crack length in the form

$$K = S(\pi a)^{\frac{1}{2}}\alpha \tag{14.30}$$

in which α is a non-dimensional coefficient usually expressed as the ratio of crack length to any convenient local dimension in the plane of the component; for a crack in an infinite plate under an applied uniform stress level S remote from the crack, $\alpha = 1.0$. Alternatively, in cases where opposing loads P are applied at points close to the plane of the crack,

$$K = \frac{P\alpha}{(\pi a)^{\frac{1}{2}}} \tag{14.31}$$

in which P is the load/unit thickness. Equations (14.30) and (14.31) may be rewritten as

$$K = K_0\alpha \tag{14.32}$$

where K_0 is a reference value of the stress intensity factor, which depends upon the loading. For the simple case of a remotely loaded plate in tension,

$$K_0 = S(\pi\alpha)^{\frac{1}{2}} \tag{14.33}$$

and Eqs. (14.32) and (14.30) are identical, so that for a given ratio of crack length to plate width, α is the same in both formulations. In more complex cases, for example, the in-plane bending of a plate of width $2b$ and having a central crack of length $2a$,

$$K_0 = \frac{3Ma}{4b^3}(\pi a)^{\frac{1}{2}} \tag{14.34}$$

in which M is the bending moment per unit thickness. Comparing Eqs. (14.34) and (14.30), we see that $S = 3\,Ma/4b^3$, which is the value of direct stress given by basic bending theory at a point a distance $\pm a/2$ from the central axis. However, if S is specified as the bending stress in the outer fibers of the plate, that is, at $\pm b$, then $S = 3\,M/2b^2$; clearly, the different specifications of S require different values of α. On the other hand, the final value of K must be independent of the form of presentation used. Use of Eqs. (14.30)–(14.32) depends on the form of the solution for K_0 and care must be taken to ensure that the formula used and the way in which the nominal stress is defined are compatible with those used in the derivation of α.

There are a number of methods available for determining the values of K and α. In one method, the solution for a component subjected to more than one type of loading is obtained from available standard solutions using superposition or, if the geometry is not covered, two or more standard solutions may be compounded.[1] Alternatively, a finite element analysis may be used.

The coefficient α in Eq. (14.30) has, as we have noted, different values depending on the plate and crack geometries. The following are values of α for some of the more common cases.

(i) A semi-infinite plate having an edge crack of length a; $\alpha = 1.12$.
(ii) An infinite plate having an embedded circular crack or a semi-circular surface crack, each of radius a, lying in a plane normal to the applied stress; $\alpha = 0.64$.
(iii) An infinite plate having an embedded elliptical crack of axes $2a$ and $2b$ or a semi-elliptical crack of width $2b$ in which the depth a is less than half the plate thickness, each lying in a plane normal to the applied stress; $\alpha = 1.12\Phi$ in which Φ varies with the ratio b/a as follows:

b/a	0	0.2	0.4	0.6	0.8
Φ	1.0	1.05	1.15	1.28	1.42

For $b/a = 1$, the situation is identical to case (ii).
(iv) A plate of finite width w having a central crack of length $2a$, where $a \le 0.3w$; $\alpha = [\sec(a\pi/w)]^{1/2}$.
(v) For a plate of finite width w having two symmetrical edge cracks each of depth $2a$, Eq. (14.30) becomes

$$K = S\left[w\tan(\pi a/w) + (0.1w)\sin(2\pi a/w)\right]^{1/2}$$

From Eq. (14.29), it can be seen that the stress intensity at a point ahead of a crack can be expressed in terms of the parameter K. Failure occurs when K reaches a critical value K_c. This is known as the *fracture toughness* of the material and has units $MN/m^{3/2}$ or $N/mm^{3/2}$.

Example 14.7

An infinite plate has a fracture toughness of 3,300 $N/mm^{3/2}$. If the plate contains an embedded circular crack of 3 mm radius, calculate the maximum allowable stress that could be applied around the boundary of the plate. See Ex. 1.1.

In this case, Eq. (14.30) applies with $\alpha = 0.64$. Then,

$$S = 3,300/[(\pi \times 3)^{1/2} \times 0.64]$$

So that

$$S = 1,680\ N/mm^2$$

Example 14.8

If the steel plate of Example 14.7 develops an elliptical crack of length 6 mm and width 2.4 mm calculate the allowable stress that could be applied around the boundary of the plate. See Ex. 1.1.

In this case, $b/a = 1.2/3 = 0.4$. Then, $\alpha = 1.12 \times 1.15 = 1.164$ and, from Eq. (14.30),

$$S = 3,300/[(\pi \times 3)^{1/2} \times 1.164]$$

that is,

$$S = 931.5 \text{ N/mm}^2$$

Example 14.9

Suppose that the plate of Example 14.7 has a finite width of 50 mm and develops a central crack of length 6 mm. What is the allowable stress that could be applied around the boundary of the plate. See Ex. 1.1.

For this case,

$$\alpha = [\sec(\pi a/w)]^{1/2}$$

that is,

$$\alpha = [\sec(\pi \times 3/50)]^{1/2} = 1.018$$

so that

$$S = 3,300/[(\pi \times 3)^{1/2} \times 1.018]$$

which gives

$$S = 1,056 \text{ N/mm}^2$$

14.5.2 Crack tip plasticity

In certain circumstances, it may be necessary to account for the effect of plastic flow in the vicinity of the crack tip. This may be allowed for by estimating the size of the plastic zone and adding this to the actual crack length to form an effective crack length $2a_1$. Thus, if r_p is the radius of the plastic zone, $a_1 = a + r_p$ and Eq. (14.30) becomes

$$K_p = S(\pi a_1)^{\frac{1}{2}}\alpha_1 \qquad (14.35)$$

in which K_p is the stress intensity factor corrected for plasticity and α_1 corresponds to a_1. Thus, for $r_p/t > 0.5$, that is, a condition of plane stress,

$$r_p = \frac{1}{2\pi}\left(\frac{K}{f_y}\right)^2 \quad \text{or} \quad r_p = \frac{a}{2}\left(\frac{S}{f_y}\right)^2 \alpha^2 \quad \text{(Ref. 3)} \qquad (14.36)$$

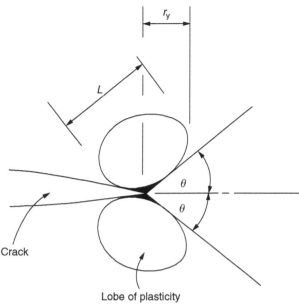

FIGURE 14.6 Plane Strain Plasticity

in which f_y is the yield proof stress of the material. For $r_p/t < 0.02$, a condition of plane strain,

$$r_p = \frac{1}{6\pi} \left(\frac{K}{f_y}\right)^2 \tag{14.37}$$

For intermediate conditions the correction should be such as to produce a conservative solution.

Dugdale[4] showed that the fracture toughness parameter K_c is highly dependent on plate thickness. In general, since the toughness of a material decreases with decreasing plasticity, it follows that the true fracture toughness is that corresponding to a plane strain condition. This lower limiting value is particularly important to consider in high-strength alloys, since these are prone to brittle failure. In addition, the assumption that the plastic zone is circular is not representative in plane strain conditions. Rice and Johnson[5] showed that, for a small amount of plane strain yielding, the plastic zone extends as two lobes (Fig. 14.6), each inclined at an angle θ to the axis of the crack, where $\theta = 70°$ and the greatest extent L and forward penetration (r_y for $\theta = 0$) of plasticity are given by

$$L = 0.155(K/f_y)^2$$
$$r_y = 0.04(K/f_y)^2$$

14.5.3 Crack propagation rates

Having obtained values of the stress intensity factor and the coefficient α, fatigue crack propagation rates may be estimated. From these, the life of a structure containing cracks or crack-like defects may be determined; alternatively, the loading condition may be modified or inspection periods arranged so that the crack is detected before failure.

Under constant amplitude loading, the rate of crack propagation may be represented graphically by curves described in general terms by the law[6]

$$\frac{da}{dN} = f(R, \Delta K) \tag{14.38}$$

in which ΔK is the stress intensity factor range and $R = S_{min}/S_{max}$. If Eq. (14.30) is used,

$$\Delta K = (S_{max} - S_{min})(\pi a)^{\frac{1}{2}}\alpha \tag{14.39}$$

Equation (14.39) may be corrected for plasticity under cyclic loading and becomes

$$\Delta K_p = (S_{max} - S_{min})(\pi a_1)^{\frac{1}{2}}\alpha_1 \tag{14.40}$$

in which $a_1 = a + r_p$, where, for plane stress,[7]

$$r_p = \frac{1}{8\pi}\left(\frac{\Delta K}{f_y}\right)^2$$

The curves represented by Eq. (14.38) may be divided into three regions. The first corresponds to a very slow crack growth rate ($<10^{-8}$ m/cycle), where the curves approach a threshold value of stress intensity factor ΔK^{th} corresponding to 4×10^{-11} m/cycle, that is, no crack growth. In the second region (10^{-8}–10^{-6} m/cycle), much of the crack life takes place and, for small ranges of ΔK, Eq. (14.38) may be represented by[8]

$$\frac{da}{dN} = C(\Delta K)^n \tag{14.41}$$

in which C and n depend on the material properties; over small ranges of da/dN and ΔK, C and n remain approximately constant. The third region corresponds to crack growth rates $>10^{-6}$ m/cycle, where instability and final failure occur.

An attempt has been made to describe the complete set of curves by the relationship[9]

$$\frac{da}{dN} = \frac{C(\Delta K)^n}{(1 - R)K_c - \Delta K} \tag{14.42}$$

in which K_c is the fracture toughness of the material obtained from toughness tests. Integration of Eq. (14.41) or (14.42) analytically or graphically gives an estimate of the crack growth rate for failure of the structure, that is, the number of cycles required for a crack to grow from an initial size to an unacceptable length, or the crack growth rate for failure, whichever is the design criterion. Thus, for example, integration of Eq. (14.41) gives, for an infinite width plate for which $\alpha = 1.0$,

$$[N]_{N_i}^{N_f} = \frac{1}{C[(S_{max} - S_{min})\pi^{\frac{1}{2}}]^n}\left[\frac{a^{(1-n/2)}}{1 - n/2}\right]_{a_i}^{a_f} \tag{14.43}$$

for $n > 2$. An analytical integration may be carried out only if n is an integer and α is in the form of a polynomial; otherwise, graphical or numerical techniques must be employed.

Substituting the limits in Eq. (14.43) and taking $N_i = 0$, the number of cycles to failure is given by

$$N_f = \frac{2}{C(n - 2)[(S_{max} - S_{min})\pi^{1/2}]^n}\left[\frac{1}{a_i^{(n-2)/2}} - \frac{1}{a_f^{(n-2)/2}}\right] \tag{14.44}$$

Example 14.10

An infinite plate contains a crack having an initial length of 0.2 mm and is subjected to a cyclic repeated stress range of 175 N/mm^2. If the fracture toughness of the plate is 1708 N/mm$^{3/2}$ and the rate of crack growth is $40 \times 10^{-15} (\Delta K)^4$ mm/cycle, determine the number of cycles to failure. See Ex. 1.1.

The crack length at failure is given by Eq. (14.30), in which $\alpha = 1$, $K = 1{,}708$ N/mm$^{3/2}$, and $S = 175$ N/mm^2; that is,

$$a_f = \frac{1{,}708^2}{\pi \times 175^2} = 30.3 \text{ mm}$$

Also, it can be seen from Eq. (14.41) that $C = 40 \times 10^{-15}$ and $n = 4$. Then, substituting the relevant parameters in Eq. (14.44) gives

$$N_f = 1/(40 \times 10^{-15}[175 \times \pi^{1/2}]^4)(1/0.1 - 1/30.3)$$

From which

$$N_f = 26{,}919 \text{ cycles}$$

Example 14.10 MATLAB

Use MATLAB to repeat Example 14.10 for the following initial crack lengths:

	(i)	(ii)	(iii)	(iv)
a_i	0.1 m	0.2 m	0.3 m	0.4 m

The number of cycles to failure for each case is obtained through the following MATLAB file:

```
% Declare any needed variables
a_i = [0.1 0.2 0.3 0.4];
alpha = 1;
K = 1708;
S = 175;
C = 40e-15;
n = 4;

% Calculate a_f using Eq. (14.30)
a_f = K^2/(pi*S^2*alpha^2);

% Calculate the number of cycles using Eq. (14.44)
N_f = 2/(C*(n-2)*(S*sqrt(pi))^n)*(1./a_i.^((n-2)/2) - 1/a_f^((n-2)/2));

% Output the results for each case to the Command Window
for i=1:1:length(a_i)
    disp(['For a_i =' num2str(a_i(i)) 'm: N_f =' num2str(round(N_f(i))) 'cycles'])
    disp(' ')
end
```

The Command Window outputs resulting from this MATLAB file are as follows:

```
For a_i = 0.1 m: N_f = 26919 cycles
For a_i = 0.2 m: N_f = 13415 cycles
For a_i = 0.3 m: N_f = 8914 cycles
For a_i = 0.4 m: N_f = 6663 cycles
```

References

[1] ESDU data sheets. Fatigue. No. 80036.
[2] Knott JF. Fundamentals of fracture mechanics. London: Butterworths; 1973.
[3] McClintock FA, Irwin GR. Plasticity aspects of fracture mechanics. In: Fracture toughness testing and its applications. Philadelphia: American Society for Testing Materials; ASTM STP 381, April 1965.
[4] Dugdale DS. J Mech Phys Solids 1960;8.
[5] Rice JR, Johnson MA. Inelastic behaviour of solids. New York: McGraw Hill; 1970.
[6] Paris PC, Erdogan F. A critical analysis of crack propagation laws. Trans Am Soc Mech Engrs, Series D, December 1963;85:(4).
[7] Rice JR. Mechanics of crack tip deformation and extension by fatigue. In: Fatigue crack propagation. Philadelphia: American Society for Testing Materials; ASTM STP 415, June 1967.
[8] Paris PC. The fracture mechanics approach to fatigue. In: Fatigue—an interdisciplinary approach. Syracuse, NY: Syracuse University Press; 1964.
[9] Forman RG. Numerical analysis of crack propagation in cyclic-loaded structures. Trans Am Soc Mech Engrs, Series D, September 1967;89(3).

Further reading

Freudenthal AM. Fatigue in aircraft structures. New York: Academic Press; 1956.

PROBLEMS

P.14.1. A material has a fatigue limit of ±230 N/mm^2 and an ultimate tensile strength of 870 N/mm^2. If the safe range of stress is determined by the Goodman prediction, calculate its value.

Answer: 363 N/mm^2

P.14.2. A more accurate estimate for the safe range of stress for the material of P.14.1 is given by the Gerber prediction. Calculate its value.

Answer: 432 N/mm^2

P.14.3. A steel component is subjected to a reversed cyclic loading of 100 cycles/day over a period of time in which ±160 N/mm^2 is applied for 200 cycles, ±140 N/mm^2 is applied for 200 cycles and ±100 N/mm^2 is applied for 600 cycles. If the fatigue life of the material at each of these stress levels is 10^4, 10^5, and 2×10^5 cycles, respectively, estimate the life of the component using Miner's law.

Answer: 400 days

P.14.3. MATLAB Use MATLAB to repeat Problem P.14.3 for the following application cycles of the 160 N/mm^2 load:

(i)	(ii)	(iii)	(iv)	(v)	(vi)
100 cycles	200 cycles	300 cycles	400 cycles	500 cycles	600 cycles

All other loads and cycles are as provided in P.14.3.

Answer: (i) 600 days
(ii) 400 days
(iii) 314 days
(iv) 276 days
(v) 236 days
(vi) 215 days

P.14.4. An aircraft's cruise speed is increased from 200 m/s to 220 m/s. Determine the percentage increase in gust damage this would cause.

Answer: 65 percent

P.14.5. The average block length journey of an executive jet airliner is 1,000 km and its cruise speed is 240 m/s. If the damage during the ground–air–ground cycle may be assumed to be 10 percent of the total damage during a complete flight, determine the percentage increase in the life of the aircraft when the cruising speed is reduced to 235 m/s.

Answer: 12 percent

P.14.6. A small executive jet cruises at 220 m/s EAS and has an average block journey distance of 1,200 km. The most critical portion of the aircraft's structure is the undersurface of the wing, which has a fatigue limit stress of $\pm 200 \text{ N/mm}^2$ based on a mean stress of 90 N/mm^2. If the ground–air–ground cycle and other minor loads produce 8 percent of the total damage, estimate the number of hours of the life of the aircraft. Take $k_1 = 3.0$ and $l_{10} = 6000$ m.

Answer: 11,400 hours

P.14.7. An infinite steel plate has a fracture toughness of $3,320 \text{ N/mm}^{3/2}$ and contains a 4 mm long crack. Calculate the maximum allowable design stress that could be applied around the boundary of the plate.

Answer: $1,324 \text{ N/mm}^2$

P.14.8. A semi-infinite plate has an edge crack of length 0.4 mm. If the plate is subjected to a cyclic repeated stress loading of 180 N/mm^2, its fracture toughness is $1,800 \text{ N/mm}^{3/2}$, and the rate of crack growth is $30 \times 10^{-15} (\Delta K)^4$ mm/cycle determine the crack length at failure and the number of cycles to failure.

Answer: 25.4 mm, 7,916 cycles

P.14.8 MATLAB Use MATLAB to repeat Problem P.14.8 for the following cyclic repeated stress loadings:

X	(i)	(ii)	(iii)	(iv)	(v)	(vi)
S	150 N/mm^2	160 N/mm^2	170 N/mm^2	180 N/mm^2	190 N/mm^2	200 N/mm^2

Answer: (i) 36.5 mm, 16,496 cycles
(ii) 32.1 mm, 12,723 cycles
(iii) 28.4 mm, 9,967 cycles
(iv) 25.4 mm, 7,916 cycles
(v) 22.8 mm, 6,365 cycles
(vi) 20.6 mm, 5,174 cycles

P.14.9. A steel plate 50 mm wide is 5 mm thick and carries an in-plane bending moment. If the plate develops an elliptical crack of length 6 mm and width 2.4 mm, calculate the maximum bending moment the plate can withstand if the fracture toughness of the steel is 3,500 N/mm$^{3/2}$.

Answer: 4,300 Nmm

Bending, shear and torsion of thin-walled beams

B3

Bending of open and closed, thin-walled beams

In Chapter 11, we discussed the various types of structural component found in aircraft construction and the various loads they support. We saw that an aircraft is basically an assembly of stiffened shell structures, ranging from the single-cell closed section fuselage to multicellular wings and tail surfaces, each subjected to bending, shear, torsional, and axial loads. Other, smaller portions of the structure consist of thin-walled channel, T, Z, "top-hat" or I sections, which are used to stiffen the thin skins of the cellular components and provide support for internal loads from floors, engine mountings, etc. Structural members such as these are known as *open section* beams, while the cellular components are termed *closed section* beams; clearly, both types of beam are subjected to axial, bending, shear, and torsional loads.

In this chapter, we investigate the stresses and displacements in thin-walled open and single-cell closed section beams produced by bending loads.

In Chapter 1, we saw that an axial load applied to a member produces a uniform direct stress across the cross-section of the member. A different situation arises when the applied loads cause a beam to bend, which, if the loads are vertical, takes up a sagging ($\smile$) or hogging ($\frown$) shape. This means that, for loads which cause a beam to sag, the upper surface of the beam must be shorter than the lower surface, as the upper surface becomes concave and the lower one convex; the reverse is true for loads which cause hogging. The strains in the upper regions of the beam, therefore, are different from those in the lower regions, and since we established that stress is directly proportional to strain (Eq. (1.40)), it follows that the stress varies through the depth of the beam.

The truth of this can be demonstrated by a simple experiment. Take a reasonably long rectangular rubber eraser and draw three or four lines on its longer faces as shown in Fig. 15.1(a); the reason for this becomes clear a little later. Now hold the eraser between the thumb and forefinger at each end and apply pressure as shown by the direction of the arrows in Fig. 15.1(b). The eraser bends into the shape shown and the lines on the side of the eraser *remain straight* but are now further apart at the top than at the bottom.

Since, in Fig. 15.1(b), the upper fibers are stretched and the lower fibers compressed, there are fibers somewhere in between which are neither stretched nor compressed; the plane containing these fibers is called the *neutral plane*.

Now, rotate the eraser so that its shorter sides are vertical and apply the same pressure with your fingers. The eraser again bends but now requires much less effort. It follows that the geometry and orientation of a beam section must affect its *bending stiffness*. This is more readily demonstrated with a plastic ruler. When flat, it requires hardly any effort to bend it but when held with its width vertical it becomes almost impossible to bend.

Introduction to Aircraft Structural Analysis, Second Edition

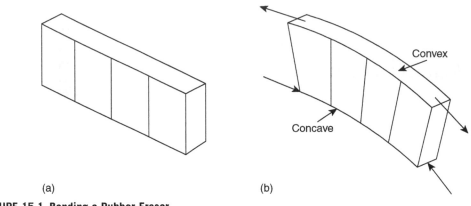

(a)   (b)

FIGURE 15.1 Bending a Rubber Eraser

15.1 SYMMETRICAL BENDING

Although symmetrical bending is a special case of the bending of beams of arbitrary cross-section, we shall investigate the former first, so that the more complex, general case may be more easily understood.

Symmetrical bending arises in beams which have either singly or doubly symmetrical cross-sections; examples of both types are shown in Fig. 15.2.

Suppose that a length of beam, of rectangular cross-section, say, is subjected to a pure, sagging bending moment, M, applied in a vertical plane. We define this later as a negative bending moment. The length of beam bends into the shape shown in Fig. 15.3(a), in which the upper surface is concave and the lower convex. It can be seen that the upper longitudinal fibers of the beam are compressed while the lower fibers are stretched. It follows that, as in the case of the eraser, between these two extremes are fibers that remain unchanged in length.

The direct stress therefore varies through the depth of the beam, from compression in the upper fibers to tension in the lower. Clearly, the direct stress is zero for the fibers that do not change in length; we called the plane containing these fibers the *neutral plane*. The line of intersection of the neutral plane and any cross-section of the beam is termed the *neutral axis* (Fig. 15.3(b)).

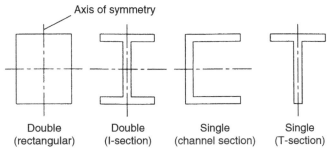

Double   Double   Single   Single
(rectangular)   (I-section)   (channel section)   (T-section)

FIGURE 15.2 Symmetrical Section Beams

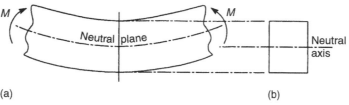

FIGURE 15.3 Beam Subjected to a Pure Sagging Bending Moment

The problem, therefore, is to determine the variation of direct stress through the depth of the beam, determine the values of the stresses, and subsequently find the corresponding beam deflection.

15.1.1 Assumptions

The primary assumption made in determining the direct stress distribution produced by pure bending is that plane cross-sections of the beam remain plane and normal to the longitudinal fibers of the beam after bending. Again, we see this from the lines on the side of the eraser. We also assume that the material of the beam is linearly elastic, that is, it obeys Hooke's law, and that the material of the beam is homogeneous.

15.1.2 Direct stress distribution

Consider a length of beam (Fig. 15.4(a)) that is subjected to a pure, sagging bending moment, M, applied in a vertical plane; the beam cross-section has a vertical axis of symmetry, as shown in Fig. 15.4(b). The bending moment causes the length of beam to bend in a manner similar to that shown in Fig. 15.3(a), so that a neutral plane exists, which is, as yet, unknown distances y_1 and y_2 from the top and bottom of the beam, respectively. Coordinates of all points in the beam are referred to axes $Oxyz$ in which the origin O lies in the neutral plane of the beam. We now investigate the behavior of an elemental length, δz, of the beam formed by parallel sections MIN and PGQ (Fig. 15.4(a)) and also

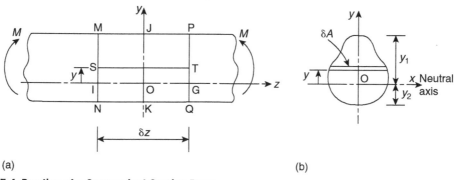

FIGURE 15.4 Bending of a Symmetrical Section Beam

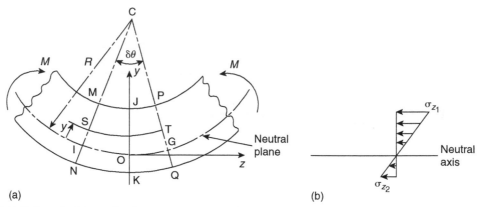

FIGURE 15.5 Length of Beam Subjected to a Pure Bending Moment

the fiber ST of cross-sectional area δA a distance y above the neutral plane. Clearly, before bending takes place $MP = IG = ST = NQ = \delta z$.

The bending moment M causes the length of beam to bend about a *center of curvature* C as shown in Fig. 15.5(a). Since the element is small in length and a pure moment is applied, we can take the curved shape of the beam to be circular with a *radius of curvature R* measured to the neutral plane. This is a useful reference point, since, as we saw, strains and stresses are zero in the neutral plane.

The previously parallel plane sections MIN and PGQ remain plane, as we demonstrated, but are now inclined at an angle $\delta\theta$ to each other. The length MP is now shorter than δz, as is ST, while NQ is longer; IG, being in the neutral plane, is still of length δz. Since the fiber ST has changed in length, it has suffered a strain ε_z which is given by

$$\varepsilon_z = \frac{\text{change in length}}{\text{original length}}$$

Then,

$$\varepsilon_z = \frac{(R - y)\delta\theta - \delta z}{\delta z}$$

that is,

$$\varepsilon_z = \frac{(R - y)\delta\theta - R\delta\theta}{R\delta\theta}$$

so that

$$\varepsilon_z = -\frac{y}{R} \tag{15.1}$$

The negative sign in Eq. (15.1) indicates that fibers in the region where y is positive shorten when the bending moment is negative. Then, from Eq. (1.40), the direct stress σ_z in the fiber ST is given by

$$\sigma_z = -E\frac{y}{R} \tag{15.2}$$

The direct or normal force on the cross-section of the fiber ST is $\sigma_z \delta A$. However, since the direct stress in the beam section is due to a pure bending moment, in other words, there is no axial load, the resultant normal force on the complete cross-section of the beam must be zero. Then,

$$\int_A \sigma_z \, dA = 0 \tag{15.3}$$

where A is the area of the beam cross-section.

Substituting for σ_z in Eq. (15.3) from (15.2) gives

$$-\frac{E}{R} \int_A y \, dA = 0 \tag{15.4}$$

in which both E and R are constants for a beam of a given material subjected to a given bending moment. Therefore,

$$\int_A y \, dA = 0 \tag{15.5}$$

Equation (15.5) states that the first moment of the area of the cross-section of the beam with respect to the neutral axis, that is, the x axis, is equal to zero. Therefore, we see that *the neutral axis passes through the centroid of area of the cross-section*. Since the y axis in this case is also an axis of symmetry, it must also pass through the centroid of the cross-section. Hence, the origin, O, of the coordinate axes coincides with the centroid of area of the cross-section.

Equation (15.2) shows that, for a sagging (i.e., negative) bending moment, the direct stress in the beam section is negative (i.e., compressive) when y is positive and positive (i.e., tensile) when y is negative.

Consider now the elemental strip δA in Fig. 15.4(b); this is, in fact, the cross-section of the fiber ST. The strip is above the neutral axis, so that there will be a *compressive* force acting on its cross-section of $\sigma_z \delta A$, which is *numerically* equal to $(Ey/R)\delta A$ from Eq. (15.2). Note that this force acts on all sections along the length of ST. At S, this force exerts a clockwise moment $(Ey/R)y\delta A$ about the neutral axis, while at T the force exerts an identical counterclockwise moment about the neutral axis. Considering either end of ST, we see that the moment resultant about the neutral axis of the stresses on all such fibers must be *equivalent* to the applied negative moment M; that is,

$$M = -\int_A E\frac{y^2}{R} \, dA$$

or

$$M = -\frac{E}{R} \int_A y^2 \, dA \tag{15.6}$$

The term $\int_A y^2 dA$ is known as the *second moment of area* of the cross-section of the beam about the neutral axis and is given the symbol I. Rewriting Eq. (15.6), we have

$$M = -\frac{EI}{R} \tag{15.7}$$

or, combining this expression with Eq. (15.2),

$$\frac{M}{I} = -\frac{E}{R} = \frac{\sigma_z}{y} \tag{15.8}$$

From Eq. (15.8), we see that

$$\sigma_z = \frac{My}{I} \tag{15.9}$$

The direct stress, σ_z, at any point in the cross-section of a beam is therefore directly proportional to the distance of the point from the neutral axis and so varies linearly through the depth of the beam as shown, for the section JK, in Fig. 15.5(b). Clearly, for a positive bending moment σ_z is positive, that is, tensile, when y is positive and compressive (i.e., negative) when y is negative. Thus, in Fig. 15.5(b),

$$\sigma_{z,1} = \frac{My_1}{I} \text{ (compression)}, \quad \sigma_{z,2} = \frac{My_2}{I} \text{ (tension)} \tag{15.10}$$

Furthermore, we see from Eq. (15.7) that the curvature, $1/R$, of the beam is given by

$$\frac{1}{R} = \frac{M}{EI} \tag{15.11}$$

and is therefore directly proportional to the applied bending moment and inversely proportional to the product EI, which is known as the *flexural rigidity* of the beam.

Example 15.1

The cross-section of a beam has the dimensions shown in Fig. 15.6(a). If the beam is subjected to a negative bending moment of 100 kN m applied in a vertical plane, determine the distribution of direct stress through the depth of the section.

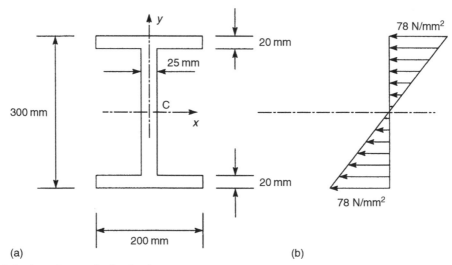

FIGURE 15.6 Direct Stress Distribution in the Beam of Example 15.1

The cross-section of the beam is doubly symmetrical, so that the centroid, C, of the section, and therefore the origin of axes, coincides with the mid-point of the web. Furthermore, the bending moment is applied to the beam section on a vertical plane, so that the x axis becomes the neutral axis of the beam section; we therefore need to calculate the second moment of area, I_{xx}, about this axis.

$$I_{xx} = \frac{200 \times 300^3}{12} - \frac{175 \times 260^3}{12} = 193.7 \times 10^6 \text{mm}^4 \qquad \text{(see Section 15.4)}$$

From Eq. (15.9), the distribution of direct stress, σ_z, is given by

$$\sigma_z = -\frac{100 \times 10^6}{193.7 \times 10^6} y = -0.52y \tag{i}$$

The direct stress, therefore, varies linearly through the depth of the section from a value

$$-0.52 \times (+150) = -78 \text{ N/mm}^2 \text{ (compression)}$$

at the top of the beam to

$$-0.52 \times (-150) = +78 \text{ N/mm}^2 \text{ (tension)}$$

at the bottom, as shown in Fig. 15.6(b).

Example 15.2

Now determine the distribution of direct stress in the beam of Example 15.1 if the bending moment is applied on a horizontal plane and in a clockwise sense about Cy when viewed in the direction yC. See Ex. 1.1.

In this case, the beam bends about the vertical y axis, which therefore becomes the neutral axis of the section. Thus, Eq. (15.9) becomes

$$\sigma_z = \frac{M}{I_{yy}} x \tag{i}$$

where I_{yy} is the second moment of area of the beam section about the y axis. Again, from Section 15.4,

$$I_{yy} = 2 \times \frac{20 \times 200^3}{12} + \frac{260 \times 25^3}{12} = 27.0 \times 10^6 \text{ mm}^4$$

Hence, substituting for M and I_{yy} in Eq. (i),

$$\sigma_z = \frac{100 \times 10^6}{27.0 \times 10^6} x = 3.7x$$

We have not specified a sign convention for bending moments applied on a horizontal plane. However, a physical appreciation of the problem shows that the left-hand edges of the beam are in compression while the right-hand edges are in tension. Again, the distribution is linear and varies from $3.7 \times (-100) = -370$ N/mm^2 (compression) at the left-hand edges of each flange to $3.7 \times (+100) = +370$ N/mm^2 (tension) at the right-hand edges.

We note that the maximum stresses in this example are very much greater than those in Example 15.1. This is because the bulk of the material in the beam section is concentrated in the region of the neutral axis, where the stresses are low. The use of an I section in this manner would therefore be structurally inefficient.

Example 15.3

The beam section of Example 15.1 is subjected to a bending moment of 100 kNm applied in a plane parallel to the longitudinal axis of the beam but inclined at 30° to the left of vertical. The sense of the bending moment is clockwise when viewed from the left-hand edge of the beam section. Determine the distribution of direct stress. See Ex. 1.1.

The bending moment is first resolved into two components, M_x in a vertical plane and M_y in a horizontal plane. Equation (15.9) may then be written in two forms:

$$\sigma_z = \frac{M_x}{I_{xx}}y, \quad \sigma_z = \frac{M_y}{I_{yy}}x \tag{i}$$

The separate distributions can then be determined and superimposed. A more direct method is to combine the two equations (i) to give the total direct stress at any point (x, y) in the section. Thus

$$\sigma_z = \frac{M_x}{I_{xx}}y + \frac{M_y}{I_{yy}}x \tag{ii}$$

Now,

$$\left.\begin{array}{l} M_x = 100 \, \cos 30° = 86.6 \text{ kNm} \\ M_y = 100 \, \sin 30° = 50.0 \text{ kNm} \end{array}\right\} \tag{iii}$$

M_x is, in this case, a positive bending moment producing tension in the upper half of the beam, where y is positive. Also, M_y produces tension in the left-hand half of the beam, where x is negative; we therefore call M_y a *negative bending moment*. Substituting the values of M_x and M_y from Eq. (iii) but with the appropriate sign in Eq. (ii), together with the values of I_{xx} and I_{yy} from Examples 15.1 and 15.2, we obtain

$$\sigma_z = \frac{86.6 \times 10^6}{193.7 \times 10^6}y - \frac{50.0 \times 10^6}{27.0 \times 10^6}x \tag{iv}$$

or

$$\sigma_z = 0.45y - 1.85x \tag{v}$$

Equation (v) gives the value of direct stress at any point in the cross-section of the beam and may also be used to determine the distribution over any desired portion. Thus, on the upper edge of the top flange, $y = +150$ mm, $100 \text{ mm} \geq x \geq -100$ mm, so that the direct stress varies linearly with x. At the top left-hand corner of the top flange,

$$\sigma_z = 0.45 \times (+150) - 1.85 \times (-100) = +252.5 \text{ N/mm}^2 \text{ (tension)}$$

At the top right-hand corner,

$$\sigma_z = 0.45 \times (+150) - 1.85 \times (+100) = -117.5 \text{ N/mm}^2 \text{ (compression)}$$

The distributions of direct stress over the outer edge of each flange and along the vertical axis of symmetry are shown in Fig. 15.7. Note that the neutral axis of the beam section does not, in this case, coincide with either the x or y axis, although it still passes through the centroid of the section. Its inclination, α, to the x axis, say, can be found by setting $\sigma_z = 0$ in Eq. (v). Then,

$$0 = 0.45y - 1.85x$$

or

$$\frac{y}{x} = \frac{1.85}{0.45} = 4.11 = \tan \alpha$$

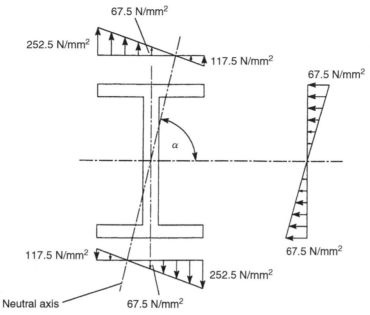

FIGURE 15.7 Direct Stress Distribution in the Beam of Example 15.3

which gives

$$\alpha = 76.3°$$

Note that α may be found in general terms from Eq. (ii) by again setting $\sigma_z = 0$. Hence,

$$\frac{y}{x} = -\frac{M_y I_{xx}}{M_x I_{yy}} = \tan \alpha \tag{vi}$$

or

$$\tan \alpha = \frac{M_y I_{xx}}{M_x I_{yy}}$$

since y is positive and x is positive for a positive value of α. We define α in a slightly different way in Section 15.2.4 for beams of unsymmetrical section.

15.1.3 Anticlastic bending

In the rectangular beam section shown in Fig. 15.8(a), the direct stress distribution due to a negative bending moment applied in a vertical plane varies from compression in the upper half of the beam to tension in the lower half (Fig. 15.8(b)). However, due to the Poisson effect, the compressive stress produces a lateral elongation of the upper fibers of the beam section, while the tensile stress produces a

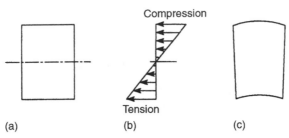

FIGURE 15.8 Anticlastic Bending of a Beam Section

lateral contraction of the lower. The section does not therefore remain rectangular but distorts as shown in Fig. 15.8(c); the effect is known as *anticlastic bending*.

Anticlastic bending is of interest in the analysis of thin-walled box beams in which the cross-sections are maintained by stiffening ribs. The prevention of anticlastic distortion induces local variations in stress distributions in the webs and covers of the box beam and also in the stiffening ribs.

15.2 UNSYMMETRICAL BENDING

We have shown that the value of direct stress at a point in the cross-section of a beam subjected to bending depends on the position of the point, the applied loading, and the geometric properties of the cross-section. It follows that it is of no consequence whether or not the cross-section is open or closed. We therefore derive the theory for a beam of arbitrary cross-section and discuss its application to thin-walled open and closed section beams subjected to bending moments.

The assumptions are identical to those made for symmetrical bending and are listed in Section 15.1.1. However, before we derive an expression for the direct stress distribution in a beam subjected to bending, we establish sign conventions for moments, forces, and displacements; investigate the effect of choice of section on the positive directions of these parameters; and discuss the determination of the components of a bending moment applied in any longitudinal plane.

15.2.1 Sign conventions and notation

Forces, moments, and displacements are referred to an arbitrary system of axes $Oxyz$, of which Oz is parallel to the longitudinal axis of the beam and Oxy are axes in the plane of the cross-section. We assign the symbols M, S, P, T, and w to bending moment, shear force, axial or direct load, torque, and distributed load intensity, respectively, with suffixes where appropriate to indicate sense or direction. Thus, M_x is a bending moment about the x axis, S_x is a shear force in the x direction, and so on. Figure 15.9 shows positive directions and senses for these loads and moments applied externally to a beam and also the positive directions of the components of displacement u, v, and w of any point in the beam cross-section parallel to the x, y, and z axes, respectively. A further condition defining the signs of the bending moments M_x and M_y is that they are positive when they induce tension in the positive xy quadrant of the beam cross-section.

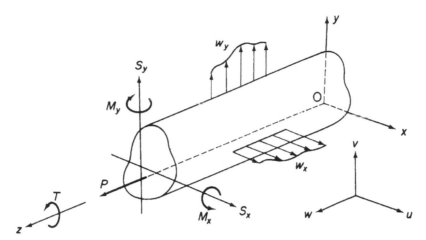

FIGURE 15.9 Notation and Sign Convention for Forces, Moments, and Displacements

If we refer internal forces and moments to that face of a section which is seen when viewed in the direction zO, then, as shown in Fig. 15.10, positive internal forces and moments are in the same direction and sense as the externally applied loads, whereas on the opposite face, they form an opposing system. The former system, which we use, has the advantage that direct and shear loads are always positive in the positive directions of the appropriate axes, whether they are internal loads or not. It must be realized, though, that internal stress resultants then become equivalent to externally applied forces and moments and are not in equilibrium with them.

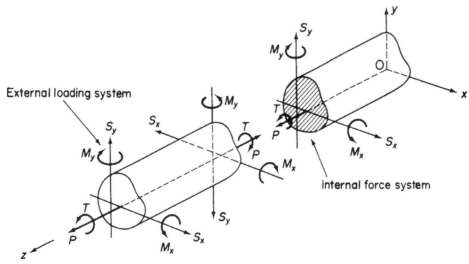

FIGURE 15.10 Internal Force System

15.2.2 **Resolution of bending moments**

A bending moment M applied in any longitudinal plane parallel to the z axis may be resolved into components M_x and M_y by the normal rules of vectors. However, a visual appreciation of the situation is often helpful. Referring to Fig. 15.11, we see that a bending moment M in a plane at an angle θ to Ox may have components of differing sign depending on the size of θ. In both cases, for the sense of M shown,

$$M_x = M \sin \theta$$

$$M_y = M \cos \theta$$

which give, for $\theta < \pi/2$, M_x and M_y positive (Fig. 15.11(a)) and for $\theta > \pi/2$, M_x positive and M_y negative (Fig. 15.11(b)).

15.2.3 **Direct stress distribution due to bending**

Consider a beam having the arbitrary cross-section shown in Fig. 15.12(a). The beam supports bending moments M_x and M_y and bends about some axis in its cross-section, which is therefore an axis of zero stress or a neutral axis (NA). Let us suppose that the origin of axes coincides with the centroid C of the cross-section and that the neutral axis is a distance p from C. The direct stress σ_z on an element of area δA at a point (x,y) and a distance ξ from the neutral axis is, from the third of Eq. (1.42),

$$\sigma_z = E\varepsilon_z \tag{15.12}$$

If the beam is bent to a radius of curvature ρ about the neutral axis at this particular section, then, since plane sections are assumed to remain plane after bending and by a comparison with symmetrical bending theory,

$$\varepsilon_z = \frac{\xi}{\rho}$$

Substituting for ε_z in Eq. (15.12), we have

$$\sigma_z = \frac{E\xi}{\rho} \tag{15.13}$$

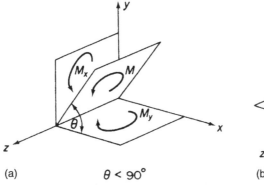
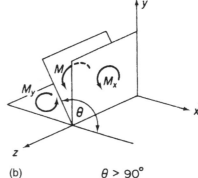

(a) $\theta < 90°$ (b) $\theta > 90°$

FIGURE 15.11 Resolution of Bending Moments

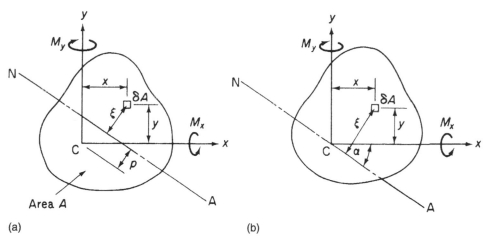

FIGURE 15.12 Determination of Neutral Axis Position and Direct Stress Due to Bending

The beam supports pure bending moments so that the resultant normal load on any section must be zero. Hence,

$$\int_A \sigma_z \, dA = 0$$

Therefore, replacing σ_z in this equation from Eq. (15.13) and cancelling the constant E/ρ gives

$$\int_A \xi \, dA = 0$$

that is, the first moment of area of the cross-section of the beam about the neutral axis is zero. It follows that the neutral axis passes through the centroid of the cross-section as shown in Fig. 15.12(b) which is the result we obtained for the case of symmetrical bending.

Suppose that the inclination of the neutral axis to Cx is α (measured clockwise from Cx), then

$$\xi = x \sin \alpha + y \cos \alpha \tag{15.14}$$

and from Eq. (15.13)

$$\sigma_z = \frac{E}{\rho}(x \sin \alpha + y \cos \alpha) \tag{15.15}$$

The moment resultants of the internal direct stress distribution have the same sense as the applied moments M_x and M_y. Therefore

$$M_x = \int_A \sigma_z y \, dA, \quad M_y = \int_A \sigma_z x \, dA \tag{15.16}$$

Substituting for σ_z from Eq. (15.15) in (15.16) and defining the second moments of area of the section about the axes Cx, Cy as

$$I_{xx} = \int_A y^2 \, dA, \quad I_{yy} = \int_A x^2 \, dA, \quad I_{xy} = \int_A xy \, dA$$

gives

$$M_x = \frac{E \sin \alpha}{\rho} I_{xy} + \frac{E \cos \alpha}{\rho} I_{xx}, \quad M_y = \frac{E \sin \alpha}{\rho} I_{yy} + \frac{E \cos \alpha}{\rho} I_{xy}$$

or, in matrix form

$$\left\{ \begin{array}{c} M_x \\ M_y \end{array} \right\} = \frac{E}{\rho} \left[\begin{array}{cc} I_{xy} & I_{xx} \\ I_{yy} & I_{xy} \end{array} \right] \left\{ \begin{array}{c} \sin \alpha \\ \cos \alpha \end{array} \right\}$$

from which

$$\frac{E}{\rho} \left\{ \begin{array}{c} \sin \alpha \\ \cos \alpha \end{array} \right\} = \left[\begin{array}{cc} I_{xy} & I_{xx} \\ I_{yy} & I_{xy} \end{array} \right]^{-1} \left\{ \begin{array}{c} M_x \\ M_y \end{array} \right\}$$

that is,

$$\frac{E}{\rho} \left\{ \begin{array}{c} \sin \alpha \\ \cos \alpha \end{array} \right\} = \frac{1}{I_{xx}I_{yy} - I_{xy}^2} \left[\begin{array}{cc} -I_{xy} & I_{xx} \\ I_{yy} & -I_{xy} \end{array} \right] \left\{ \begin{array}{c} M_x \\ M_y \end{array} \right\}$$

so that, from Eq. (15.15),

$$\sigma_z = \left(\frac{M_y I_{xx} - M_x I_{xy}}{I_{xx}I_{yy} - I_{xy}^2} \right) x + \left(\frac{M_x I_{yy} - M_y I_{xy}}{I_{xx}I_{yy} - I_{xy}^2} \right) y \tag{15.17}$$

Alternatively, Eq. (15.17) may be rearranged in the form

$$\sigma_z = \frac{M_x(I_{yy}y - I_{xy}x)}{I_{xx}I_{yy} - I_{xy}^2} + \frac{M_y(I_{xx}x - I_{xy}y)}{I_{xx}I_{yy} - I_{xy}^2} \tag{15.18}$$

From Eq. (15.18), it can be seen that, if, say, $M_y = 0$, the moment M_x produces a stress which varies with both x and y; similarly for M_y if $M_x = 0$. In the case where the beam cross-section has *either* (or both) Cx or Cy as an axis of symmetry, the product second moment of area I_{xy} is zero and Cxy are *principal axes*. Equation (15.18) then reduces to

$$\sigma_z = \frac{M_x}{I_{xx}} y + \frac{M_y}{I_{yy}} x \tag{15.19}$$

Further, if either M_y or M_x is zero, then

$$\sigma_z = \frac{M_x}{I_{xx}} y \quad \text{or} \quad \sigma_z = \frac{M_y}{I_{yy}} x \tag{15.20}$$

Equations (15.19) and (15.20) are those derived for the bending of beams having at least a singly symmetrical cross-section (see Section 15.1). It may also be noted that. in Eq. (15.20). $\sigma_z = 0$ when, for the first equation, $y = 0$ and, for the second equation, when $x = 0$. Therefore, in symmetrical bending theory, the x axis becomes the neutral axis when $M_y = 0$ and the y axis becomes the neutral axis when $M_x = 0$. We see that the position of the neutral axis depends on the form of the applied loading as well as the geometrical properties of the cross-section.

There exists, in any unsymmetrical cross-section, a centroidal set of axes for which the product second moment of area is zero[1]. These axes are then principal axes and the direct stress distribution referred to these axes takes the simplified form of Eq. (15.19) or (15.20). It therefore appears that the amount of computation can be reduced if these axes are used. This is not the case, however, unless the principal axes are obvious from inspection, since the calculation of the position of the principal axes, the principal sectional properties, and the coordinates of points at which the stresses are to be determined consumes a greater amount of time than direct use of Eqs. (15.17) or (15.18) for an arbitrary but convenient set of centroidal axes.

15.2.4 Position of the neutral axis

The neutral axis always passes through the centroid of area of a beam's cross-section, but its inclination α (see Fig. 15.12(b)) to the x axis depends on the form of the applied loading and the geometrical properties of the beam's cross-section.

At all points on the neutral axis, the direct stress is zero. Therefore, from Eq. (15.17),

$$0 = \left(\frac{M_y I_{xx} - M_x I_{xy}}{I_{xx} I_{yy} - I_{xy}^2}\right) x_{\text{NA}} + \left(\frac{M_x I_{yy} - M_y I_{xy}}{I_{xx} I_{yy} - I_{xy}^2}\right) y_{\text{NA}}$$

where x_{NA} and y_{NA} are the coordinates of any point on the neutral axis. Hence,

$$\frac{y_{\text{NA}}}{x_{\text{NA}}} = -\frac{M_y I_{xx} - M_x I_{xy}}{M_x I_{yy} - M_y I_{xy}}$$

or, referring to Fig. 15.12(b) and noting that, when α is positive, x_{NA} and y_{NA} are of opposite sign,

$$\tan \alpha = \frac{M_y I_{xx} - M_x I_{xy}}{M_x I_{yy} - M_y I_{xy}} \tag{15.21}$$

Example 15.4

A beam having the cross-section shown in Fig. 15.13 is subjected to a bending moment of 1,500 Nm in a vertical plane. Calculate the maximum direct stress due to bending stating the point at which it acts. See Ex. 1.1.

The position of the centroid of the section may be found by taking moments of areas about some convenient point. Thus,

$$(120 \times 8 + 80 \times 8)\bar{y} = 120 \times 8 \times 4 + 80 \times 8 \times 48$$

giving

$$\bar{y} = 21.6 \text{ mm}$$

and

$$(120 \times 8 + 80 \times 8)\bar{x} = 80 \times 8 \times 4 + 120 \times 8 \times 24$$

giving

$$\bar{x} = 16 \text{ mm}$$

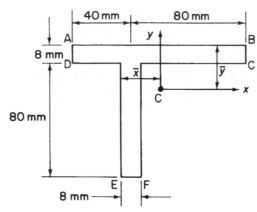

FIGURE 15.13 Cross-Section of Beam in Example 15.4

The next step is to calculate the section properties referred to axes Cxy (see Section 15.4):

$$I_{xx} = \frac{120 \times (8)^3}{12} + 120 \times 8 \times (17.6)^2 + \frac{8 \times (80)^3}{12} + 80 \times 8 \times (26.4)^2$$

$$= 1.09 \times 10^6 \text{ mm}^4$$

$$I_{yy} = \frac{8 \times (120)^3}{12} + 120 \times 8 \times (8)^2 + \frac{80 \times (8)^3}{12} + 80 \times 8 \times (12)^2$$

$$= 1.31 \times 10^6 \text{ mm}^4$$

$$I_{xy} = 120 \times 8 \times 8 \times 17.6 + 80 \times 8 \times (-12) \times (-26.4)$$
$$= 0.34 \times 10^6 \text{ mm}^4$$

Since $M_x = 1,500$ Nm and $M_y = 0$, we have, from Eq. (15.18),

$$\sigma_z = 1.5y - 0.39x \qquad \text{(i)}$$

in which the units are N and mm.

By inspection of Eq. (i), we see that σ_z is a maximum at F, where $x = -8$ mm, $y = -66.4$ mm. Thus,

$$\sigma_{z,\max} = -96 \text{ N/mm}^2 \text{ (compressive)}$$

In some cases the maximum value cannot be obtained by inspection, so that values of σ_z at several points must be calculated.

15.2.5 Load intensity, shear force, and bending moment relationships, the general case

Consider an element of length δz of a beam of unsymmetrical cross-section subjected to shear forces, bending moments, and a distributed load of varying intensity, all in the yz plane, as shown in Fig. 15.14. The forces and moments are positive in accordance with the sign convention previously adopted. Over

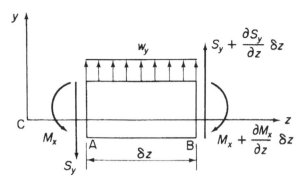

FIGURE 15.14 Equilibrium of a Beam Element Supporting a General Force System in the *yz* Plane

the length of the element, we assume that the intensity of the distributed load is constant. Therefore, for equilibrium of the element in the y direction,

$$\left(S_y + \frac{\partial S_y}{\partial z}\delta z\right) + w_y\delta z - S_y = 0$$

from which

$$w_y = -\frac{\partial S_y}{\partial z}$$

Taking moments about A, we have

$$\left(M_x + \frac{\partial M_x}{\partial z}\delta z\right) - \left(S_y + \frac{\partial S_y}{\partial z}\delta z\right)\delta z - w_y\frac{(\delta z)^2}{2} - M_x = 0$$

or, when second-order terms are neglected,

$$S_y = \frac{\partial M_x}{\partial z}$$

We may combine these results into a single expression:

$$-w_y = \frac{\partial S_y}{\partial z} = \frac{\partial^2 M_x}{\partial z^2} \tag{15.22}$$

Similarly, for loads in the xz plane,

$$-w_x = \frac{\partial S_x}{\partial z} = \frac{\partial^2 M_y}{\partial z^2} \tag{15.23}$$

15.3 DEFLECTIONS DUE TO BENDING

We noted that a beam bends about its neutral axis whose inclination relative to arbitrary centroidal axes is determined from Eq. (15.21). Suppose that, at some section of an unsymmetrical beam, the deflection normal to the neutral axis (and therefore an absolute deflection) is ζ, as shown in Fig. 15.15. In other

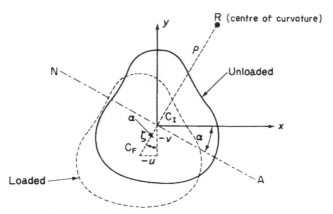

FIGURE 15.15 Determination of Beam Deflection Due to Bending

words, the centroid C is displaced from its initial position C_I through an amount ζ to its final position C_F. Suppose also that the center of curvature R of the beam at this particular section is on the opposite side of the neutral axis to the direction of the displacement ζ and that the radius of curvature is ρ. For this position of the center of curvature and from the usual approximate expression for curvature, we have

$$\frac{1}{\rho} = \frac{d^2\zeta}{dz^2} \tag{15.24}$$

The components u and v of ζ are in the negative directions of the x and y axes, respectively, so that

$$u = -\zeta \sin \alpha, \quad v = -\zeta \cos \alpha \tag{15.25}$$

Differentiating Eqs. (15.25) twice with respect to z and substituting for ζ from Eq. (15.24), we obtain

$$\frac{\sin \alpha}{\rho} = -\frac{d^2u}{dz^2}, \quad \frac{\cos \alpha}{\rho} = -\frac{d^2v}{dz^2} \tag{15.26}$$

In the derivation of Eq. (15.17), we see that

$$\frac{1}{\rho} \begin{Bmatrix} \sin \alpha \\ \cos \alpha \end{Bmatrix} = \frac{1}{E\left(I_{xx}I_{yy} - I_{xy}^2\right)} \begin{bmatrix} -I_{xy} & I_{xx} \\ I_{yy} & -I_{xy} \end{bmatrix} \begin{Bmatrix} M_x \\ M_y \end{Bmatrix} \tag{15.27}$$

Substituting in Eqs. (15.27) for $\sin\alpha/\rho$ and $\cos\alpha/\rho$ from Eqs (15.26) and writing $u'' = d^2u/dz^2$, $v'' = d^2v/dz^2$, we have

$$\begin{Bmatrix} u'' \\ v'' \end{Bmatrix} = \frac{-1}{E\left(I_{xx}I_{yy} - I_{xy}^2\right)} \begin{bmatrix} -I_{xy} & I_{xx} \\ I_{yy} & -I_{xy} \end{bmatrix} \begin{Bmatrix} M_x \\ M_y \end{Bmatrix} \tag{15.28}$$

It is instructive to rearrange Eq. (15.28) as follows:

$$\begin{Bmatrix} M_x \\ M_y \end{Bmatrix} = -E \begin{bmatrix} I_{xy} & I_{xx} \\ I_{yy} & I_{xy} \end{bmatrix} \begin{Bmatrix} u'' \\ v'' \end{Bmatrix} \quad \text{(see derivation of Eq. (15.17))} \tag{15.29}$$

that is,

$$\left. \begin{array}{l} M_x = -EI_{xy}u'' - EI_{xx}v'' \\ M_y = -EI_{yy}u'' - EI_{xy}v'' \end{array} \right\} \tag{15.30}$$

The first of Eqs. (15.30) shows that M_x produces curvatures, that is, deflections, on both the xz and yz planes even though $M_y = 0$; similarly for M_y when $M_x = 0$. Thus, for example, an unsymmetrical beam deflects both vertically and horizontally, even though the loading is entirely in a vertical plane. Similarly, vertical and horizontal components of deflection in an unsymmetrical beam are produced by horizontal loads.

For a beam having either Cx or Cy (or both) as an axis of symmetry, $I_{xy} = 0$ and Eqs. (15.28) reduce to

$$u'' = -\frac{M_y}{EI_{yy}}, \quad v'' = -\frac{M_x}{EI_{xx}} \tag{15.31}$$

Example 15.5

Determine the deflection curve and the deflection of the free end of the cantilever shown in Fig. 15.16(a); the flexural rigidity of the cantilever is EI and its section is doubly symmetrical. See Ex. 1.1.

The load W causes the cantilever to deflect such that its neutral plane takes up the curved shape shown Fig. 15.16(b); the deflection at any section Z is then v, while that at its free end is v_{tip}. The axis system is chosen so that the origin coincides with the built-in end, where the deflection is clearly zero.

The bending moment, M, at the section Z is, from Fig. 15.16(a),

$$M = W(L - z) \tag{i}$$

Substituting for M in the second of Eq. (15.31),

$$v'' = -\frac{W}{EI}(L - z)$$

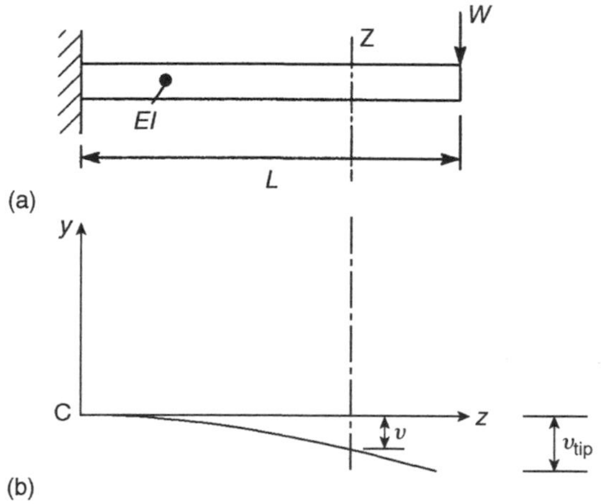

FIGURE 15.16 Deflection of a Cantilever Beam Carrying a Concentrated Load at Its Free End (Example 15.5)

or, in more convenient form,

$$EIv'' = -W(L - z) \tag{ii}$$

Integrating Eq. (ii) with respect to z gives

$$EIv' = -W\left(Lz - \frac{z^2}{2}\right) + C_1$$

where C_1 is a constant of integration, which is obtained from the boundary condition that $v' = 0$ at the built-in end, where $z = 0$. Hence, $C_1 = 0$ and

$$EIv' = -W\left(Lz - \frac{z^2}{2}\right) \tag{iii}$$

Integrating Eq. (iii), we obtain

$$EIv = -W\left(\frac{Lz^2}{2} - \frac{z^3}{6}\right) + C_2$$

in which C_2 is again a constant of integration. At the built-in end, $v = 0$ when $z = 0$, so that $C_2 = 0$. Hence, the equation of the deflection curve of the cantilever is

$$v = -\frac{W}{6EI}(3Lz^2 - z^3) \tag{iv}$$

The deflection, v_{tip}, at the free end is obtained by setting $z = L$ in Eq. (iv). Then,

$$v_{\text{tip}} = -\frac{WL^3}{3EI} \tag{v}$$

and is clearly negative and downward.

Example 15.6

Determine the deflection curve and the deflection of the free end of the cantilever shown in Fig. 15.17(a). The cantilever has a doubly symmetrical cross-section. See Ex. 1.1.

The bending moment, M, at any section Z is given by

$$M = \frac{w}{2}(L - z)^2 \tag{i}$$

Substituting for M in the second of Eqs. (15.31) and rearranging, we have

$$EIv'' = -\frac{w}{2}(L - z)^2 = -\frac{w}{2}(L^2 - 2Lz + z^2) \tag{ii}$$

Integration of Eq. (ii) yields

$$EIv' = -\frac{w}{2}\left(L^2 z - Lz^2 + \frac{z^3}{3}\right) + C_1$$

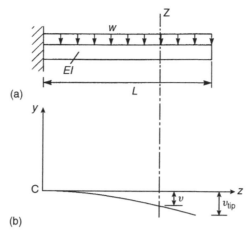

(a)

(b)

FIGURE 15.17 Deflection of a Cantilever Beam Carrying a Uniformly Distributed Load

When $z=0$ at the built-in end, $v'=0$, so that $C_1=0$ and

$$EIv' = -\frac{w}{2}\left(L^2z - Lz^2 + \frac{z^3}{3}\right) \tag{iii}$$

Integrating Eq. (iii), we have

$$EIv = -\frac{w}{2}\left(L^2\frac{z^2}{2} - \frac{Lz^3}{3} + \frac{z^4}{12}\right) + C_2$$

and, since $v=0$ when $x=0$, $C_2=0$. The deflection curve of the beam therefore has the equation

$$v = -\frac{w}{24EI}\left(6L^2z^2 - 4Lz^3 + z^4\right) \tag{iv}$$

and the deflection at the free end, where $x=L$, is

$$v_{tip} = -\frac{wL^4}{8EI} \tag{v}$$

which is again negative and downward.

Example 15.7

Determine the deflection curve and the mid-span deflection of the simply supported beam shown in Fig. 15.18(a); the beam has a doubly symmetrical cross-section. See Ex. 1.1.

The support reactions are each $wL/2$ and the bending moment, M, at any section Z, a distance z from the left-hand support is

$$M = -\frac{wL}{2}z + \frac{wz^2}{2} \tag{i}$$

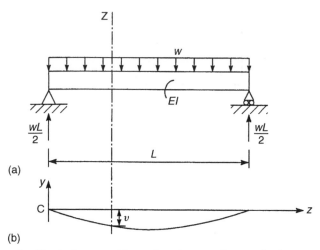

FIGURE 15.18 Deflection of a Simply Supported Beam Carrying a Uniformly Distributed Load (Example 15.7)

Substituting for M in the second of Eqs. (15.31), we obtain

$$EIv'' = \frac{w}{2}(Lz - z^2) \tag{ii}$$

Integrating, we have

$$EIv' = \frac{w}{2}\left(\frac{Lz^2}{2} - \frac{z^3}{3}\right) + C_1$$

From symmetry, it is clear that, at the mid-span section, the gradient $v' = 0$. Hence,

$$0 = \frac{w}{2}\left(\frac{L^3}{8} - \frac{L^3}{24}\right) + C_1$$

which gives

$$C_1 = -\frac{wL^3}{24}$$

Therefore,

$$EIv' = \frac{w}{24}(6Lz^2 - 4z^3 - L^3) \tag{iii}$$

Integrating again gives

$$EIv = \frac{w}{24}(2Lz^3 - z^4 - L^3z) + C_2$$

Since $v = 0$ when $z = 0$ (or since $v = 0$ when $z = L$), it follows that $C_2 = 0$ and the deflected shape of the beam has the equation

$$v = \frac{w}{24EI}(2Lz^3 - z^4 - L^3z) \tag{iv}$$

The maximum deflection occurs at mid-span, where $z = L/2$ and is

$$v_{\text{mid-span}} = -\frac{5wL^4}{384EI} \tag{v}$$

So far, the constants of integration were determined immediately they arose. However, in some cases, a relevant boundary condition, say, a value of gradient, is not obtainable. The method is then to carry the unknown constant through the succeeding integration and use known values of deflection at two sections of the beam. Thus, in the previous example, Eq. (ii) is integrated twice to obtain

$$EIv = \frac{w}{2}\left(\frac{Lz^3}{6} - \frac{z^4}{12}\right) + C_1 z + C_2$$

The relevant boundary conditions are $v = 0$ at $z = 0$ and $z = L$. The first of these gives $C_2 = 0$, while from the second, we have $C_1 = -wL^3/24$. Thus, the equation of the deflected shape of the beam is

$$v = \frac{w}{24EI}\left(2Lz^3 - z^4 - L^3 z\right)$$

as before.

Example 15.8

Figure 15.19(a) shows a simply supported beam carrying a concentrated load W at mid-span. Determine the deflection curve of the beam and the maximum deflection if the beam section is doubly symmetrical. See Ex. 1.1.

The support reactions are each $W/2$ and the bending moment M at a section Z a distance $z\ (\leq L/2)$ from the left-hand support is

$$M = -\frac{W}{2}z \tag{i}$$

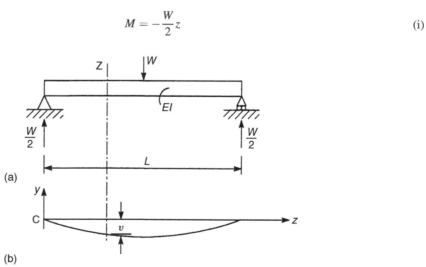

FIGURE 15.19 Deflection of a Simply Supported Beam Carrying a Concentrated Load at Mid-span (Example 15.8)

From the second of Eqs. (15.31), we have

$$EIv'' = \frac{W}{2}z \qquad \text{(ii)}$$

Integrating we obtain

$$EIv' = \frac{W}{2}\frac{z^2}{2} + C_1$$

From symmetry, the slope of the beam is zero at mid-span, where $z = L/2$. Thus, $C_1 = -WL^2/16$ and

$$EIv' = \frac{W}{16}(4z^2 - L^2) \qquad \text{(iii)}$$

Integrating Eq. (iii), we have

$$EIv = \frac{W}{16}\left(\frac{4z^3}{3} - L^2z\right) + C_2$$

and, when $z = 0$, $v = 0$, so that $C_2 = 0$. The equation of the deflection curve is, therefore,

$$v = \frac{W}{48EI}(4z^3 - 3L^2z) \qquad \text{(iv)}$$

The maximum deflection occurs at mid-span and is

$$v_{\text{mid-span}} = -\frac{WL^3}{48EI} \qquad \text{(v)}$$

Note that, in this problem, we could not use the boundary condition that $v = 0$ at $z = L$ to determine C_2, since Eq. (i) applies only for $0 \le z \le L/2$; it follows that Eqs. (iii) and (iv) for slope and deflection apply only for $0 \le z \le L/2$, although the deflection curve is clearly symmetrical about mid-span.

Examples 15.5–15.8 are frequently regarded as 'standard' cases of beam deflection.

15.3.1 Singularity functions

The double integration method used in Examples 15.5–15.8 becomes extremely lengthy when even relatively small complications, such as the lack of symmetry due to an offset load, are introduced. For example, the addition of a second concentrated load on a simply supported beam results in a total of six equations for slope and deflection, producing six arbitrary constants. Clearly, the computation involved in determining these constants is tedious, even though a simply supported beam carrying two concentrated loads is a comparatively simple practical case. An alternative approach is to introduce so-called *singularity* or *half-range* functions. Such functions were first applied to beam deflection problems by Macauley in 1919 and hence the method is frequently known as *Macauley's method*.

We now introduce a quantity $[z - a]$ and define it to be zero if $(z - a) < 0$, that is, $z < a$, and to be simply $(z - a)$ if $z > a$. The quantity $[z - a]$ is known as a *singularity* or *half-range function* and is defined to have a value only when the argument is positive, in which case the square brackets behave in an identical manner to ordinary parentheses.

Example 15.9

Determine the position and magnitude of the maximum upward and downward deflections of the beam shown in Fig. 15.20. See Ex. 1.1.

A consideration of the overall equilibrium of the beam gives the support reactions; thus,

$$R_A = \frac{3}{4}W \text{ (upward)}, \quad R_F = \frac{3}{4}W \text{ (downward)}$$

Using the method of singularity functions and taking the origin of axes at the left-hand support, we write an expression for the bending moment, M, at any section Z between D and F, *the region of the beam furthest from the origin*:

$$M = -R_A z + W[z - a] + W[z - 2a] - 2W[z - 3a] \tag{i}$$

Substituting for M in the second of Eqs. (15.31), we have

$$EIv'' = \frac{3}{4}Wz - W[z - a] - W[z - 2a] + 2W[z - 3a] \tag{ii}$$

Integrating Eq. (ii) and retaining the square brackets, we obtain

$$EIv' = \frac{3}{8}Wz^2 - \frac{W}{2}[z - a]^2 - \frac{W}{2}[z - 2a]^2 + W[z - 3a]^2 + C_1 \tag{iii}$$

and

$$EIv = \frac{1}{8}Wz^3 - \frac{W}{6}[z - a]^3 - \frac{W}{6}[z - 2a]^3 + \frac{W}{3}[z - 3a]^3 + C_1 z + C_2 \tag{iv}$$

in which C_1 and C_2 are arbitrary constants. When $z = 0$ (at A), $v = 0$ and hence $C_2 = 0$. Note that the second, third, and fourth terms on the right-hand side of Eq. (iv) disappear for $z < a$. Also $v = 0$ at $z = 4a$ (F), so that, from Eq. (iv), we have

$$0 = \frac{W}{8}64a^3 - \frac{W}{6}27a^3 - \frac{W}{6}8a^3 + \frac{W}{3}a^3 + 4aC_1$$

which gives

$$C_1 = -\frac{5}{8}Wa^2$$

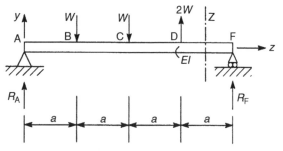

FIGURE 15.20 Macauley's Method for the Deflection of a Simply Supported Beam (Example 15.9)

Equations (iii) and (iv) now become

$$EIv' = \frac{3}{8}Wz^2 - \frac{W}{2}[z-a]^2 - \frac{W}{2}[z-2a]^2 + W[z-3a]^2 - \frac{5}{8}Wa^2 \tag{v}$$

and

$$EIv = \frac{1}{8}Wz^3 - \frac{W}{6}[z-a]^3 - \frac{W}{6}[z-2a]^3 + \frac{W}{3}[z-3a]^3 - \frac{5}{8}Wa^2z \tag{vi}$$

respectively.

To determine the maximum upward and downward deflections, we need to know in which bays $v' = 0$ and thereby which terms in Eq. (v) disappear when the exact positions are being located. One method is to select a bay and determine the sign of the slope of the beam at the extremities of the bay. A change of sign indicates that the slope is zero within the bay.

By inspection of Fig. 15.20, it seems likely that the maximum downward deflection occurs in BC. At B, using Eq. (v),

$$EIv' = \frac{3}{8}Wa^2 - \frac{5}{8}Wa^2$$

which is clearly negative. At C,

$$EIv' = \frac{3}{8}W4a^2 - \frac{W}{2}a^2 - \frac{5}{8}Wa^2$$

which is positive. Therefore, the maximum downward deflection does occur in BC and its exact position is located by equating v' to zero for any section in BC. Thus, from Eq. (v),

$$0 = \frac{3}{8}Wz^2 - \frac{W}{2}[z-a]^2 - \frac{5}{8}Wa^2$$

or, simplifying,

$$0 = z^2 - 8az + 9a^2 \tag{vii}$$

Solution of Eq. (vii) gives

$$z = 1.35a$$

so that the maximum downward deflection is, from Eq. (vi),

$$EIv = \frac{1}{8}W(1.35a)^3 - \frac{W}{6}(0.35a)^3 - \frac{5}{8}Wa^2(1.35a)$$

that is,

$$v_{max}(\text{downward}) = -\frac{0.54Wa^3}{EI}$$

In a similar manner it can be shown that the maximum upward deflection lies between D and F at $z = 3.42a$ and that its magnitude is

$$v_{max}(\text{upward}) = \frac{0.04Wa^3}{EI}$$

An alternative method of determining the position of maximum deflection is to select a possible bay, set $V = 0$ for that bay and solve the resulting equation in z. If the solution gives a value of z that lies within the bay, then the selection is correct; otherwise, the procedure must be repeated for a second and possibly a third and a fourth bay. This method is quicker than the former if the correct bay is selected initially; if not, the equation corresponding to each selected bay must be completely solved, a procedure clearly longer than determining the sign of the slope at the extremities of the bay.

Example 15.10

Determine the position and magnitude of the maximum deflection in the beam of Fig. 15.21. See Ex. 1.1.

Following the method of Example 15.9, we determine the support reactions and find the bending moment, M, at any section Z in the bay furthest from the origin of the axes.

Then,

$$M = -R_A z + w\frac{L}{4}\left[z - \frac{5L}{8}\right] \tag{i}$$

Examining Eq. (i), we see that the singularity function $[z - 5L/8]$ does not become zero until $z \leq 5L/8$, although Eq. (i) is valid only for $z \geq 3L/4$. To obviate this difficulty, we extend the distributed load to the support D while simultaneously restoring the status quo by applying an upward distributed load of the same intensity and length as the additional load (Fig. 15.22).

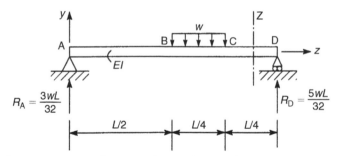

FIGURE 15.21 Deflection of a Beam Carrying a Part Span Uniformly Distributed Load (Example 15.10)

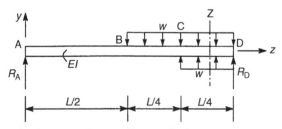

FIGURE 15.22 Method of Solution for a Part Span Uniformly Distributed Load

At the section Z, a distance z from A, the bending moment is now given by

$$M = -R_A z + \frac{w}{2}\left[z - \frac{L}{2}\right]^2 - \frac{w}{2}\left[z - \frac{3L}{4}\right]^2 \tag{ii}$$

Equation (ii) is valid for all sections of the beam if the singularity functions are discarded as they become zero. Substituting Eq. (ii) into the second of Eqs. (15.31), we obtain

$$EIv'' = \frac{3}{32}wLz - \frac{w}{2}\left[z - \frac{L}{2}\right]^2 + \frac{w}{2}\left[z - \frac{3L}{4}\right]^2 \tag{iii}$$

Integrating Eq. (iii) gives

$$EIv' = \frac{3}{64}wLz^2 - \frac{w}{6}\left[z - \frac{L}{2}\right]^3 + \frac{w}{6}\left[z - \frac{3L}{4}\right]^3 + C_1 \tag{iv}$$

$$EIv = \frac{wLz^3}{64} - \frac{w}{24}\left[z - \frac{L}{2}\right]^4 + \frac{w}{24}\left[z - \frac{3L}{4}\right]^4 + C_1 z + C_2 \tag{v}$$

where C_1 and C_2 are arbitrary constants. The required boundary conditions are $v=0$ when $z=0$ and $z=L$. From the first of these, we obtain $C_2=0$, while the second gives

$$0 = \frac{wL^4}{64} - \frac{w}{24}\left(\frac{L}{2}\right)^4 + \frac{w}{24}\left(\frac{L}{4}\right)^4 + C_1 L$$

from which

$$C_1 = -\frac{27wL^3}{2,048}$$

Equations (iv) and (v) then become

$$EIv' = \frac{3}{64}wLz^2 - \frac{w}{6}\left[z - \frac{L}{2}\right]^3 + \frac{w}{6}\left[z - \frac{3L}{4}\right]^3 - \frac{27wL^3}{2,048} \tag{vi}$$

and

$$EIv = \frac{wLz^3}{64} - \frac{w}{24}\left[z - \frac{L}{2}\right]^4 + \frac{w}{24}\left[z - \frac{3L}{4}\right]^4 - \frac{27wL^3}{2,048}z \tag{vii}$$

In this problem, the maximum deflection clearly occurs in the region BC of the beam. Therefore, equating the slope to zero for BC, we have

$$0 = \frac{3}{64}wLz^2 - \frac{w}{6}\left[z - \frac{L}{2}\right]^3 - \frac{27wL^3}{2,048}$$

which simplifies to

$$z^3 - 1.78Lz^2 + 0.75zL^2 - 0.046L^3 = 0 \tag{viii}$$

Solving Eq. (viii) by trial and error, we see that the slope is zero at $z \simeq 0.6\,L$. Hence, from Eq. (vii), the maximum deflection is

$$v_{max} = -\frac{4.53 \times 10^{-3}wL^4}{EI}$$

Example 15.11

Determine the deflected shape of the beam shown in Fig. 15.23. See Ex. 1.1.

In this problem, an external moment M_0 is applied to the beam at B. The support reactions are found in the normal way and are

$$R_A = -\frac{M_0}{L} \text{ (downward)}, \quad R_C = \frac{M_0}{L} \text{ (upward)}$$

The bending moment at any section Z between B and C is then given by

$$M = -R_A z - M_0 \tag{i}$$

Equation (i) is valid only for the region BC and clearly does not contain a singularity function which would cause M_0 to vanish for $z \le b$. We overcome this difficulty by writing

$$M = -R_A z - M_0 [z - b]^0 \qquad (\text{Note} : [z - b]^0 = 1) \tag{ii}$$

Equation (ii) has the same value as Eq. (i) but is now applicable to all sections of the beam, since $[z - b]^0$ disappears when $z \le b$. Substituting for M from Eq. (ii) in the second of Eqs. (15.31), we obtain

$$EIv'' = R_A z + M_0 [z - b]^0 \tag{iii}$$

Integration of Eq. (iii) yields

$$EIv' = R_A \frac{z^2}{2} + M_0[z - b] + C_1 \tag{vi}$$

and

$$EIv = R_A \frac{z^3}{6} + \frac{M_0}{2}[z - b]^2 + C_1 z + C_2 \tag{v}$$

where C_1 and C_2 are arbitrary constants. The boundary conditions are $v = 0$ when $z = 0$ and $z = L$. From the first of these, we have $C_2 = 0$, while the second gives

$$0 = -\frac{M_0}{L} \frac{L^3}{6} + \frac{M_0}{2}[L - b]^2 + C_1 L$$

from which

$$C_1 = -\frac{M_0}{6L}(2L^2 - 6Lb + 3b^2) \tag{vi}$$

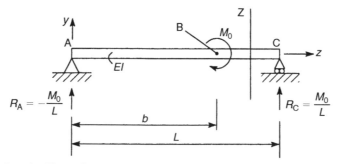

FIGURE 15.23 Deflection of a Simply Supported Beam Carrying a Point Moment (Example 15.11)

The equation of the deflection curve of the beam is then

$$v = \frac{M_0}{6EIL}\{z^3 + 3L[z-b]^2 - (2L^2 - 6Lb + 3b^2)z\} \tag{vii}$$

Example 15.11 MATLAB

Use MATLAB to repeat Example 15.11. See Ex. 1.1.

　The beam deflection curve equation is obtained through the following MATLAB file:

```
% Declare any needed variables
syms M_0 L b EI z a o C_1 C_2

% Define the support reactions shown in Fig. 15.23
R_A=-M_0/L;
R_C=M_0/L;

% Define the equation for the bending moment at any section Z between B and C
M=[-R_A*z -M_0*(a)^o]; % For z<=b: a=0; For z>b: a=z-b

% Substitute M into the second of Eq. (15.32)
v_zz=-M/EI;

% Integrate v_zz to get v_z and v
v_z=[int(v_zz(1),z)+C_1/EI int(v_zz(2),a)];
v=[int(v_z(1),z)+C_2/EI int(v_z(2),a)];
v=sum(subs(v,o,0));

% Use boundary conditions to determine C_1 and C_2
  % BC #1: v=0 when z=0
  c_2=solve(subs(subs(v*EI,a,0),z,0),C_2);
  v=subs(v,C_2,c_2);
  % BC #1: v=0 when z=0
  c_1=solve(subs(subs(v*EI,a,z-b),z,L),C_1);
  v=simplify(subs(v,C_1,c_1));

% Output the resulting deflection equation to the Command Window
disp('The equation for the deflection curve of the beam is:')
disp(['v=' char(v)])
disp('Where: a=0 for z<=b, and a=z-b for z>b')
```

The Command Window outputs resulting from this MATLAB file are as follows. The equation for the deflection curve of the beam is

```
v=-(M_0*(2*L^2*z - 3*L*a^2 - 6*L*b*z+3*b^2*z+z^3))/(6*EI*L)
Where: a=0 for z<=b, and a=z-b for z>b
```

Example 15.12

Determine the horizontal and vertical components of the tip deflection of the cantilever shown in Fig. 15.24. The second moments of area of its unsymmetrical section are I_{xx}, I_{yy}, and I_{xy}. See Ex. 1.1.

From Eqs. (15.28),

$$u'' = \frac{M_x I_{xy} - M_y I_{xx}}{E(I_{xx}I_{yy} - I_{xy}^2)} \tag{i}$$

In this case, $M_x = W(L - z)$, $M_y = 0$, so that Eq. (i) simplifies to

$$u'' = \frac{W I_{xy}}{E(I_{xx}I_{yy} - I_{xy}^2)}(L - z) \tag{ii}$$

Integrating Eq. (ii) with respect to z,

$$u' = \frac{W I_{xy}}{E(I_{xx}I_{yy} - I_{xy}^2)}\left(Lz - \frac{z^2}{2} + A\right) \tag{iii}$$

and

$$u = \frac{W I_{xy}}{E(I_{xx}I_{yy} - I_{xy}^2)}\left(L\frac{z^2}{2} - \frac{z^3}{6} + Az + B\right) \tag{iv}$$

in which u' denotes du/dz and the constants of integration A and B are found from the boundary conditions, namely, $u' = 0$ and $u = 0$ when $z = 0$. From the first of these and Eq. (iii), $A = 0$, while from the second and Eq. (iv), $B = 0$. Hence, the deflected shape of the beam in the xz plane is given by

$$u = \frac{W I_{xy}}{E(I_{xx}I_{yy} - I_{xy}^2)}\left(L\frac{z^2}{2} - \frac{z^3}{6}\right) \tag{v}$$

At the free end of the cantilever ($z = L$), the horizontal component of deflection is

$$u_{\text{f.e.}} = \frac{W I_{xy} L^3}{3E(I_{xy}I_{yy} - I_{xy}^2)} \tag{vi}$$

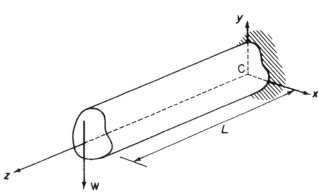

FIGURE 15.24 Determination of the Deflection of a Cantilever

Similarly, the vertical component of the deflection at the free end of the cantilever is

$$v_{\text{f.e.}} = \frac{-WI_{yy}L^3}{3E(I_{xx}I_{yy} - I_{xy}^2)}$$

(vii)

The actual deflection $\delta_{\text{f.e.}}$ at the free end is then given by

$$\delta_{\text{f.e.}} = (u_{\text{f.e.}}^2 + v_{\text{f.e.}}^2)^{\frac{1}{2}}$$

at an angle of $\tan^{-1} u_{\text{f.e.}}/v_{\text{f.e.}}$ to the vertical.

Note that, if either Cx or Cy were an axis of symmetry, $I_{xy} = 0$ and Eqs. (vi) and (vii) reduce to

$$u_{\text{f.e.}} = 0, \quad v_{\text{f.e.}} = \frac{-WL^3}{3EI_{xx}}$$

the well-known results for the bending of a cantilever having a symmetrical cross-section and carrying a concentrated vertical load at its free end (see Example 15.5).

15.4 CALCULATION OF SECTION PROPERTIES

It will be helpful at this stage to discuss the calculation of the various section properties required in the analysis of beams subjected to bending. Initially, however, two useful theorems are quoted.

15.4.1 Parallel axes theorem

Consider the beam section shown in Fig. 15.25 and suppose that the second moment of area, I_C, about an axis through its centroid C is known. The second moment of area, I_N, about a parallel axis, NN, a distance b from the centroidal axis is then given by

$$I_N = I_C + Ab^2$$

(15.32)

15.4.2 Theorem of perpendicular axes

In Fig. 15.26, the second moments of area, I_{xx} and I_{yy}, of the section about Ox and Oy are known. The second moment of area about an axis through O perpendicular to the plane of the section (i.e., a *polar second moment of area*) is

$$I_o = I_{xx} + I_{yy}$$

(15.33)

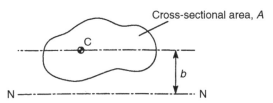

FIGURE 15.25 Parallel Axes Theorem

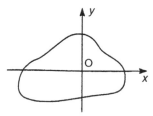

FIGURE 15.26 Theorem of Perpendicular Axes

15.4.3 Second moments of area of standard sections

Many sections may be regarded as comprising a number of rectangular shapes. The problem of determining the properties of such sections is simplified if the second moments of area of the rectangular components are known and use is made of the parallel axes theorem. Thus, for the rectangular section of Fig. 15.27,

$$I_{xx} = \int_A y^2 \, dA = \int_{-d/2}^{d/2} by^2 \, dy = b\left[\frac{y^3}{3}\right]_{-d/2}^{d/2}$$

which gives

$$I_{xx} = \frac{bd^3}{12} \tag{15.34}$$

Similarly,

$$I_{yy} = \frac{db^3}{12} \tag{15.35}$$

Frequently, it is useful to know the second moment of area of a rectangular section about an axis which coincides with one of its edges. Therefore, in Fig. 15.27 and using the parallel axes theorem,

$$I_N = \frac{bd^3}{12} + bd\left(-\frac{d}{2}\right)^2 = \frac{bd^3}{3} \tag{15.36}$$

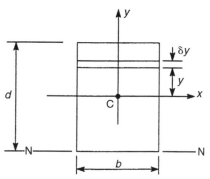

FIGURE 15.27 Second Moments of Area of a Rectangular Section

Example 15.13

Determine the second moments of area I_{xx} and I_{yy} of the I section shown in Fig. 15.28. See Ex. 1.1.

Using Eq. (15.34),

$$I_{xx} = \frac{bd^3}{12} - \frac{(b - t_w)d_w^3}{12}$$

Alternatively, using the parallel axes theorem in conjunction with Eq. (15.34),

$$I_{xy} = 2\left[\frac{bt_f^3}{12} + bt_f\left(\frac{d_w + t_f}{2}\right)^2\right] + \frac{t_w d_w^3}{12}$$

The equivalence of these two expressions for I_{xx} is most easily demonstrated by a numerical example.

Also, from Eq. (15.35),

$$I_{yy} = 2\frac{t_f b^3}{12} + \frac{d_w t_w^3}{12}$$

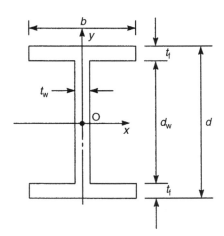

FIGURE 15.28 Second Moments of Area of an I Section

It is also useful to determine the second moment of area, about a diameter, of a circular section. In Fig. 15.29, where the x and y axes pass through the centroid of the section,

$$I_{xx} = \int_A y^2 \, dA = \int_{-d/2}^{d/2} 2\left(\frac{d}{2}\cos\theta\right)y^2 \, dy \tag{15.37}$$

Integration of Eq. (i) is simplified if an angular variable, θ, is used. Thus,

$$I_{xx} = \int_{-\pi/2}^{\pi/2} d\cos\theta\left(\frac{d}{2}\sin\theta\right)^2\frac{d}{2}\cos\theta \, d\theta$$

that is,

$$I_{xx} = \frac{d^4}{8}\int_{-\pi/2}^{\pi/2}\cos^2\theta\,\sin^2\theta \, d\theta$$

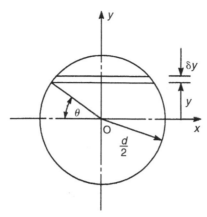

FIGURE 15.29 Second Moments of Area of a Circular Section

which gives

$$I_{xx} = \frac{\pi d^4}{64} \tag{15.38}$$

Clearly, from symmetry,

$$I_{yy} = \frac{\pi d^4}{64} \tag{15.39}$$

Using the theorem of perpendicular axes, the polar second moment of area, I_0, is given by

$$I_0 = I_{xx} + I_{yy} = \frac{\pi d^4}{32} \tag{15.40}$$

15.4.4 Product second moment of area

The product second moment of area, I_{xy}, of a beam section with respect to x and y axes is defined by

$$I_{xy} = \int_A xy \, dA \tag{15.41}$$

Thus, each element of area in the cross-section is multiplied by the product of its coordinates and the integration is taken over the complete area. Although second moments of area are always positive, since elements of area are multiplied by the square of one of their coordinates, it is possible for I_{xy} to be negative if the section lies predominantly in the second and fourth quadrants of the axes system. Such a situation would arise in the case of the Z section of Fig. 15.30(a), where the product second moment of area of each flange is clearly negative.

A special case arises when one (or both) of the coordinate axes is an axis of symmetry, so that for any element of area, δA, having the product of its coordinates positive, there is an identical element

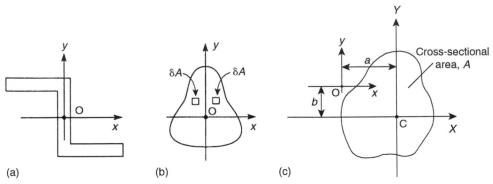

(a)   (b)   (c)

FIGURE 15.30 Product Second Moment of Area

for which the product of its coordinates is negative (Fig. 15.30 ((b))). Summation (i.e., integration) over the entire section of the product second moment of area of all such pairs of elements results in a zero value for I_{xy}.

We have shown previously that the parallel axes theorem may be used to calculate second moments of area of beam sections comprising geometrically simple components. The theorem can be extended to the calculation of product second moments of area. Let us suppose that we wish to calculate the product second moment of area, I_{xy}, of the section shown in Fig. 15.30(c) about axes xy when I_{XY} about its own, say, centroidal, axes system CXY is known. From Eq. (15.41),

$$I_{xy} = \int_A xy \, dA$$

or

$$I_{xy} = \int_A (X - a)(Y - b) dA$$

which, on expanding, gives

$$I_{xy} = \int_A XY \, dA - b \int_A X dA - a \int_A Y \, dA + ab \int_A dA$$

If X and Y are centroidal axes, then $\int_A X \, dA = \int_A Y \, dA = 0$. Hence,

$$I_{xy} = I_{XY} + abA \tag{15.42}$$

It can be seen from Eq. (15.42) that, if either CX or CY is an axis of symmetry, that is. $I_{XY}=0$, then

$$I_{xy} = abA \tag{15.43}$$

Therefore, for a section component having an axis of symmetry that is parallel to either of the section reference axes, the product second moment of area is the product of the coordinates of its centroid multiplied by its area.

15.4.5 **Approximations for thin-walled sections**

We may exploit the thin-walled nature of aircraft structures to make simplifying assumptions in the determination of stresses and deflections produced by bending. Thus, the thickness t of thin-walled sections is assumed to be small compared with their cross-sectional dimensions, so that stresses may be regarded as constant across the thickness. Furthermore, we neglect squares and higher powers of t in the computation of sectional properties and take the section to be represented by the mid-line of its wall. As an illustration of the procedure we shall consider the channel section of Fig. 15.31(a). The section is singly symmetric about the x axis so that $I_{xy}=0$. The second moment of area I_{xx} is then given by

$$I_{xx} = 2 \left[\frac{(b + t/2)t^3}{12} + \left(b + \frac{t}{2}\right) th^2 \right] + t \frac{[2(h - t/2)]^3}{12}$$

Expanding the cubed term, we have

$$I_{xx} = 2 \left[\frac{(b + t/2)t^3}{12} + \left(b + \frac{t}{2}\right) th^2 \right] + \frac{t}{12} \left[(2)^3 \left(h^3 - 3h^2 \frac{t}{2} + 3h \frac{t^2}{4} - \frac{t^3}{8}\right) \right]$$

which reduces, after powers of t^2 and upward are ignored, to

$$I_{xx} = 2bth^2 + t \frac{(2h)^3}{12}$$

The second moment of area of the section about Cy is obtained in a similar manner.

We see therefore that, for the purpose of calculating section properties, we may regard the section as represented by a single line, as shown in Fig. 15.31(b).

Thin-walled sections frequently have inclined or curved walls, which complicate the calculation of section properties. Consider the inclined thin section of Fig. 15.32. Its second moment of area about a horizontal axis through its centroid is given by

$$I_{xx} = 2 \int_0^{a/2} ty^2 \, \mathrm{d}s = 2 \int_\circ^{a/2} t(s \sin \beta)^2 \, \mathrm{d}s$$

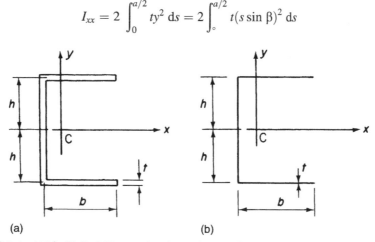

(a)   (b)

FIGURE 15.31 (a) Actual Thin-Walled Channel Section; (b) Approximate Representation of Section

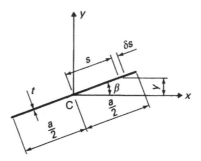

FIGURE 15.32 Second Moments of Area of an Inclined Thin Section

from which

$$I_{xx} = \frac{a^3 t \, s \sin^2 \beta}{12}$$

Similarly,

$$I_{yy} = \frac{a^3 t \, \cos^2 \beta}{12}$$

The product second moment of area is

$$I_{xy} = 2 \int_0^{a/2} txy \, ds$$

$$= 2 \int_0^{a/2} t(s \cos \beta)(s \sin \beta) ds$$

which gives

$$I_{xy} = \frac{a^3 t \sin 2\beta}{24}$$

We note here that these expressions are approximate, in that their derivation neglects powers of t^2 and upward by ignoring the second moments of area of the element δs about axes through its own centroid.

Properties of thin-walled curved sections are found in a similar manner. Thus, I_{xx} for the semi-circular section of Fig. 15.33, is

$$I_{xx} = \int_0^{\pi r} ty^2 \, ds$$

FIGURE 15.33 Second Moment of Area of a Semicircular Section

Expressing y and s in terms of a single variable θ simplifies the integration; hence,

$$I_{xx} = \int_0^\pi t(r\cos\theta)^2 r\,d\theta$$

from which

$$I_{xx} = \frac{\pi r^3 t}{2}$$

Example 15.14

Determine the direct stress distribution in the thin-walled Z section shown in Fig. 15.34, produced by a positive bending moment M_x. See Ex. 1.1.

The section is antisymmetrical, with its centroid at the mid-point of the vertical web. Therefore, the direct stress distribution is given by either of Eq. (15.17) or (15.18), in which $M_y = 0$. From Eq. (15.18),

$$\sigma_z = \frac{M_x(I_{yy}y - I_{xy}x)}{I_{xx}I_{yy} - I_{xy}^2} \tag{i}$$

The section properties are calculated as follows:

$$I_{xx} = 2\frac{ht}{2}\left(\frac{h}{2}\right)^2 + \frac{th^3}{12} = \frac{h^3 t}{3}$$

$$I_{yy} = 2\frac{t}{3}\left(\frac{h}{2}\right)^3 = \frac{h^3 t}{12}$$

$$I_{xy} = \frac{ht}{2}\left(\frac{h}{4}\right)\left(\frac{h}{2}\right) + \frac{ht}{2}\left(-\frac{h}{4}\right)\left(-\frac{h}{2}\right) = \frac{h^3 t}{8}$$

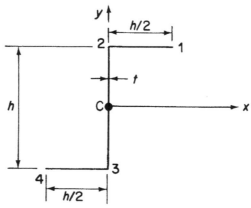

FIGURE 15.34 *Z* Section Beam of Example 15.14

Substituting these values in Eq. (i),

$$\sigma_z = \frac{M_x}{h^3 t}(6.86y - 10.30x)$$ (ii)

On the top flange, $y = h/2$, $0 \leq x \leq h/2$ and the distribution of direct stress is given by

$$\sigma_z = \frac{M_x}{h^3 t}(3.43h - 10.30x)$$

which is linear. Hence,

$$\sigma_{z,1} = -\frac{1.72M_x}{h^3 t} \quad \text{(compressive)}$$

$$\sigma_{z,2} = +\frac{3.43M_x}{h^3 t} \quad \text{(tensile)}$$

In the web, $-h/2 \leq y \leq h/2$ and $x = 0$. Again, the distribution is of linear form and is given by the equation

$$\sigma_z = \frac{M_x}{h^3 t}6.86y$$

from which

$$\sigma_{z,2} = +\frac{3.43M_x}{h^3 t} \quad \text{(tensile)}$$

and

$$\sigma_{z,3} = -\frac{3.43M_x}{h^3 t} \quad \text{(compressive)}$$

The distribution in the lower flange may be deduced from antisymmetry; the complete distribution is as shown in Fig. 15.35.

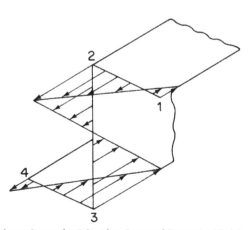

FIGURE 15.35 Distribution of Direct Stress in *Z* Section Beam of Example 15.14

15.5 APPLICABILITY OF BENDING THEORY

The expressions for direct stress and displacement derived in the preceding theory are based on the assumptions that the beam is of uniform, homogeneous cross-section and that plane sections remain plane after bending. The latter assumption is strictly true only if the bending moments M_x and M_y are constant along the beam. Variation of bending moment implies the presence of shear loads, and the effect of these is to deform the beam section into a shallow, inverted s (see Section 2.6). However, shear stresses in beams whose cross-sectional dimensions are small in relation to their lengths are comparatively low so that the basic theory of bending may be used with reasonable accuracy.

In thin-walled sections, shear stresses produced by shear loads are not small and must be calculated, although the direct stresses may still be obtained from the basic theory of bending so long as axial constraint stresses are absent; this effect is discussed in Chapters 26 and 27. Deflections in thin-walled structures are assumed to result primarily from bending strains; the contribution of shear strains may be calculated separately if required.

15.6 TEMPERATURE EFFECTS

In Section 1.15.1, we considered the effect of temperature change on stress–strain relationships, while in Section 5.11, we examined the effect of a simple temperature gradient on a cantilever beam of rectangular cross-section using an energy approach. However, as we saw, beam sections in aircraft structures are generally thin walled and do not necessarily have axes of symmetry. We now investigate how the effects of temperature on such sections may be determined.

We saw that the strain produced by a temperature change ΔT is given by

$$\varepsilon = \alpha \Delta T \qquad \text{(see Eq.(1.55))}$$

It follows from Eq. (1.40) that the direct stress on an element of cross-sectional area δA is

$$\sigma = E\alpha\Delta T\delta A \tag{15.44}$$

Consider now the beam section shown in Fig. 15.36 and suppose that a temperature variation ΔT is applied to the complete cross-section, that is, ΔT is a function of both x and y.

The total normal force due to the temperature change on the beam cross-section is then given by

$$N_T = \iint_A E\alpha\Delta T \, dA \tag{15.45}$$

Further, the moments about the x and y axes are

$$M_{xT} = \iint_A E\alpha\Delta Ty \, dA \tag{15.46}$$

and

$$M_{yT} = \iint_A E\alpha\Delta Tx \, dA \tag{15.47}$$

respectively.

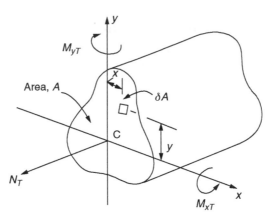

FIGURE 15.36 Beam Section Subjected to a Temperature Rise

We have noted that beam sections in aircraft structures are generally thin-walled, so that Eqs. (15.41)–(15.43) may be more easily integrated for such sections by dividing them into thin rectangular components, as we did when calculating section properties. We then use the Riemann integration technique, in which we calculate the contribution of each component to the normal force and moments and sum them to determine each result. Equations (15.45)–(15.47) then become

$$N_T = \Sigma E\alpha\Delta T A_i \tag{15.48}$$

$$M_{xT} = \Sigma E\alpha\Delta T \bar{y}_i A_i \tag{15.49}$$

$$M_{yT} = \Sigma E\alpha\Delta T \bar{x}_i A_i \tag{15.50}$$

in which A_i is the cross-sectional area of a component and $\bar{x}_i$ and $\bar{y}_i$ are the coordinates of its centroid.

Example 15.15

The beam section shown in Fig. 15.37 is subjected to a temperature rise of $2T_0$ in its upper flange, a temperature rise of T_0 in its web, and no temperature change in its lower flange. Determine the normal force on the beam section and the moments about the centroidal x and y axes. The beam section has a Young's modulus E and the coefficient of linear expansion of the material of the beam is α. See Ex. 1.1.

From Eq. (15.48),

$$N_T = E\alpha(2T_0 at + T_0\, 2at) = 4E\alpha at T_0$$

From Eq. (15.49),

$$M_{xT} = E\alpha[2T_0 at(a) + T_0\, 2at(0)] = 2E\alpha a^2 t T_0$$

and from Eq. (15.50),

$$M_{yT} = E\alpha[2T_0 at(-a/2) + T_0\, 2at(0)] = -E\alpha a^2 t T_0$$

Note that M_{yT} is negative, which means that the upper flange tends to rotate out of the paper about the web, which agrees with a temperature rise for this part of the section. The stresses corresponding to these stress resultants are calculated in the normal way and are added to those produced by any applied loads.

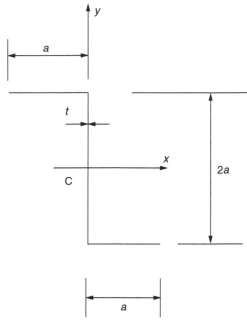

FIGURE 15.37 Beam Section of Example 15.15

In some cases, the temperature change is not conveniently constant in the components of a beam section and must then be expressed as a function of x and y. Consider the thin-walled beam section shown in Fig. 15.38 and suppose that a temperature change $\Delta T(x, y)$ is applied.

The direct stress on an element δs in the wall of the section is then, from Eq. (15.44),

$$\sigma = E\alpha\Delta T(x,y)t\delta s$$

Equations (15.45)–(15.47) then become

$$N_T = \int_A E\alpha\Delta T(x,y)t \, ds \tag{15.51}$$

$$M_{xT} = \int_A E\alpha\Delta T(x,y)ty \, ds \tag{15.52}$$

$$M_{yT} = \int_A E\alpha\Delta T(x,y)tx \, ds \tag{15.53}$$

Example 15.16

If, in the beam section of Example 15.15, the temperature change in the upper flange is $2T_0$ but in the web varies linearly from $2T_0$ at its junction with the upper flange to zero at its junction with the lower flange, determine the values of the stress resultants; the temperature change in the lower flange remains zero.

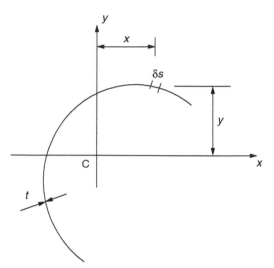

FIGURE 15.38 Thin-walled Beam Section Subjected to a Varying Temperature Change

The temperature change at any point in the web is given by

$$T_w = 2T_0(a+y)/2a = \frac{T_0}{a}(a+y)$$

Then, from Eqs. (15.48) and (15.51),

$$N_T = E\alpha 2T_0 at + \int_{-a}^{a} E\alpha \frac{T_0}{a}(a+y)t \, ds$$

that is,

$$N_T = E\alpha T_0 \left\{ 2at + \frac{1}{a} \left[ay + \frac{y^2}{2} \right]_{-a}^{a} \right\}$$

which gives

$$N_T = 4E\alpha T_0 at$$

Note that, in this case, the answer is identical to that in Example 15.15, which is to be expected, since the average temperature change in the web is $(2\,T_0+0)/2 = T_0$, which is equal to the constant temperature change in the web in Example 15.15.

From Eqs. (15.49) and (15.52),

$$M_{xT} = E\alpha 2T_0 at(a) + \int_{-a}^{a} E\alpha \frac{T_0}{a}(a+y)yt \, ds$$

that is,

$$M_{xT} = E\alpha T_0 \left\{ 2a^2 t + \frac{1}{a} \left[\frac{ay^2}{2} + \frac{y^3}{3} \right]_{-a}^{a} \right\}$$

from which

$$M_{xT} = \frac{8E\alpha a^2 t T_0}{3}$$

Alternatively, the average temperature change T_0 in the web may be considered to act at the centroid of the temperature change distribution. Then,

$$M_{xT} = E\alpha 2T_0 at(a) + E\alpha T_0 2at \left(\frac{a}{3}\right)$$

that is,

$$M_{xT} = \frac{8E\alpha a^2 t T_0}{3}$$

as before. The contribution of the temperature change in the web to M_{yT} remains zero, since the section centroid is in the web; the value of M_{yT} is therefore $-E\alpha a^2 t T_0$ as in Example 15.14.

Reference

[1] Megson THG. Structures and stress analysis. 2nd ed. Oxford: Elsevier; 2005.

PROBLEMS

P.15.1 Figure P.15.1 shows the section of an angle purlin. A bending moment of 3,000 Nm is applied to the purlin in a plane at an angle of 30° to the vertical y axis. If the sense of the bending moment is such that both its components M_x and M_y produce tension in the positive xy quadrant, calculate the maximum direct stress in the purlin, stating clearly the point at which it acts.

Answer: $\sigma_{z,max} = -63.3 \text{ N/mm}^2$ at C

P.15.2 A thin-walled, cantilever beam of unsymmetrical cross-section supports shear loads at its free end as shown in Fig. P.15.2. Calculate the value of direct stress at the extremity of the lower flange (point A) at a section halfway along the beam if the position of the shear loads is such that no twisting of the beam occurs.

Answer: 194.5 N/mm^2 (tension)

P.15.2 MATLAB If the 400 and 800 N shear loads in Fig. P.15.2 are labeled S_x and S_y, respectively, use MATLAB to repeat Problem P.15.2 for the following (S_x,S_y) combinations, assuming positive values are in the direction of the load arrows shown:

	(i)	(ii)	(iii)	(iv)	(v)	(vi)	(vii)	(viii)	(ix)
S_y(N)	400	500	600	700	800	900	1000	1100	1200
S_x(N)	600	600	600	400	400	−400	−400	−600	−600

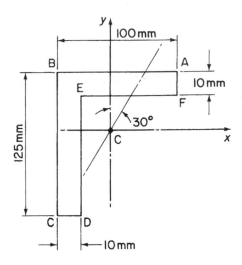

FIGURE P.15.1

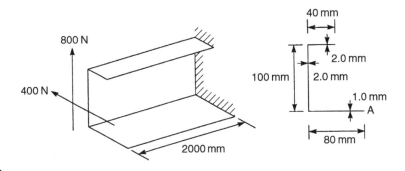

FIGURE P.15.2

Answer: (i)   269.9 N/mm^2 (tension)
         (ii)   272.6 N/mm^2 (tension)
      (iii)   275.3 N/mm^2 (tension)
      (iv)   191.8 N/mm^2 (tension)
       (v)   194.5 N/mm^2 (tension)
      (vi)   148.1 N/mm^2 (compression)
    (vii)   145.3 N/mm^2 (compression)
   (viii)   228.9 N/mm^2 (compression)
     (ix)   226.2 N/mm^2 (compression)

P.15.3 A beam, simply supported at each end, has a thin-walled cross-section, as shown in Fig. P.15.3. If a uniformly distributed loading of intensity w/unit length acts on the beam in the plane of the lower, horizontal flange, calculate the maximum direct stress due to bending of the beam and show

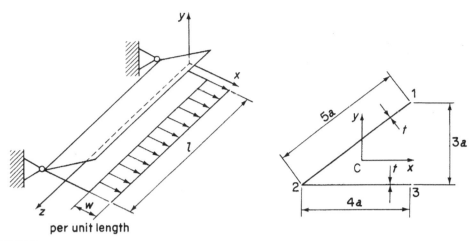

FIGURE P.15.3

diagrammatically the distribution of the stress at the section where the maximum occurs. The thickness t is to be taken as small in comparison with the other cross-sectional dimensions in calculating the section properties I_{xx}, I_{yy} and I_{xy}.

Answer: $\sigma_{z,max} = \sigma_{z,3} = 13wl^2/384a^2t$, $\sigma_{z,1} = wl^2/96a^2t$, $\sigma_{z,2} = -wl^2/48a^2t$

P.15.4 A thin-walled cantilever with walls of constant thickness t has the cross-section shown in Fig. P.15.4. It is loaded by a vertical force W at the tip and a horizontal force $2W$ at the mid-section, both forces acting through the shear center. Determine and sketch the distribution of direct stress, according to the basic theory of bending, along the length of the beam for the points 1 and 2 of the cross-section. The wall thickness t can be taken as very small in comparison with d in calculating the sectional properties I_{xx}, I_{xy}, and so forth.

Answer: $\sigma_{z,1}$ (mid-point) $= -0.05\ Wl/td^2$, $\sigma_{z,1}$ (built-in end) $= -1.85\ Wl/td^2$,
$\sigma_{z,2}$ (mid-point) $= -0.63\ Wl/td^2$, $\sigma_{z,2}$ (built-in end) $= 0.1\ Wl/td^2$

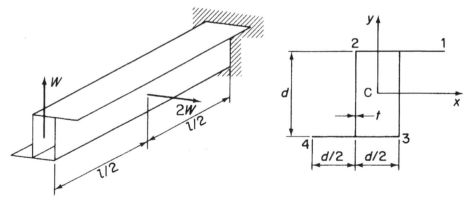

FIGURE P.15.4

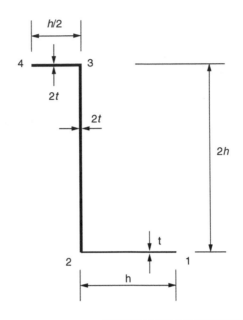

FIGURE P.15.5

P. 15.5 A thin-walled beam has the cross-section shown in Fig. P.15.5. If the beam is subjected to a bending moment M_x in the plane of the web 23, calculate and sketch the distribution of direct stress in the beam cross-section.

Answer: At 1, $0.92M_x/th^2$; at 2, $-0.65M_x/th^2$; at 3, $0.65M_x/th^2$; at 4, $-0.135M_x/th^2$

P.15.6 The thin-walled beam section shown in Fig. P.15.6 is subjected to a bending moment M_x applied in a negative sense. Find the position of the neutral axis and the maximum direct stress in the section.

Answer: NA inclined at 40.9° to Cx. $\pm0.74\ M_x/ta^2$ at 1 and 2, respectively

P.15.6 MATLAB If the angle from horizontal of the web in the section shown in Fig. P.15.6 is labeled θ, use MATLAB to repeat Problem P.15.6 for the following values of θ:

	(i)	(ii)	(iii)	(iv)	(v)	(vi)
θ	30°	40°	50°	60°	70°	80°

Answer:
(i) NA inclined at 23.4° to Cx. $\pm1.03M_x/a^2t$ at 1 and 2, respectively
(ii) NA inclined at 30.5° to Cx. $\pm0.86M_x/a^2t$ at 1 and 2, respectively
(iii) NA inclined at 36.6° to Cx. $\pm0.78M_x/a^2t$ at 1 and 2, respectively
(iv) NA inclined at 40.9° to Cx. $\pm0.74M_x/a^2t$ at 1 and 2, respectively
(v) NA inclined at 41.2° to Cx. $\pm0.70M_x/a^2t$ at 1 and 2, respectively
(vi) NA inclined at 31.4° to Cx. $\pm0.59M_x/a^2t$ at 1 and 2, respectively

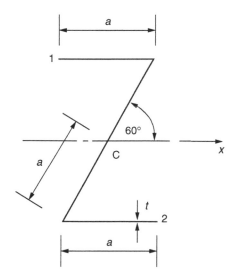

FIGURE P.15.6

P.15.7 A thin-walled cantilever has a constant cross-section of uniform thickness with the dimensions shown in Fig. P.15.7. It is subjected to a system of point loads acting in the planes of the walls of the section in the directions shown. Calculate the direct stresses according to the basic theory of bending at the points 1, 2, and 3 of the cross-section at the built-in end and halfway along the beam. Illustrate your answer by means of a suitable sketch. The thickness is to be taken as small in comparison with the other cross-sectional dimensions in calculating the section properties I_{xx}, I_{xy}, and so on.

Answer: At built-in end, $\sigma_{z,1} = -11.4\ \text{N/mm}^2$, $\sigma_{z,2} = -18.9\ \text{N/mm}^2$, $\sigma_{z,3} = 39.1\ \text{N/mm}^2$
Halfway, $\sigma_{z,1} = -20.3\ \text{N/mm}^2$, $\sigma_{z,2} = -1.1\ \text{N/mm}^2$, $\sigma_{z,3} = 15.4\text{N/mm}^2$.

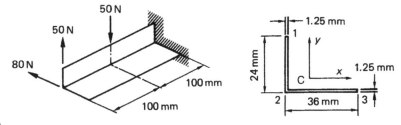

FIGURE P.15.7

P.15.8 A uniform thin-walled beam has the open cross-section shown in Fig. P.15.8. The wall thickness t is constant. Calculate the position of the neutral axis and the maximum direct stress for a bending moment $M_x = 3.5$ Nm applied about the horizontal axis Cx. Take $r = 5$ mm, $t = 0.64$ mm.

Answer: $\alpha = 51.9°$, $\sigma_{z,\text{max}} = 101\ \text{N/mm}^2$.

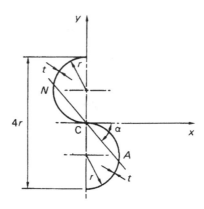

FIGURE P.15.8

P.15.9 A uniform beam is simply supported over a span of 6 m. It carries a trapezoidally distributed load with intensity varying from 30 kN/m at the left-hand support to 90 kN/m at the right-hand support. Find the equation of the deflection curve and hence the deflection at the mid-span point. The second moment of area of the cross-section of the beam is 120×10^6 mm^4 and Young's modulus $E = 206{,}000$ N/mm^2.

Answer: 41 mm (downward)

P.15.10 A cantilever of length L and having a flexural rigidity EI carries a distributed load that varies in intensity from w/unit length at the built-in end to zero at the free end. Find the deflection of the free end.

Answer: $wL^4/30EI$ (downward)

P.15.11 Determine the position and magnitude of the maximum deflection of the simply supported beam shown in Fig. P.15.11 in terms of its flexural rigidity EI.

Answer: $38.8/EI$ m downwards at 2.9 m from left-hand support.

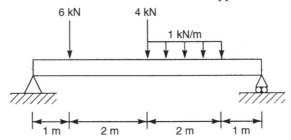

FIGURE P.15.11

P.15.12 Determine the equation of the deflection curve of the beam shown in Fig. P.15.12. The flexural rigidity of the beam is EI.

Answer: $v = -\dfrac{1}{EI}\left(\dfrac{125}{6}z^3 - 50[z-1]^2 + \dfrac{50}{12}[z-2]^4 - \dfrac{50}{12}[z-4]^4 - \dfrac{525}{6}[z-4]^3 + 237.5z\right)$

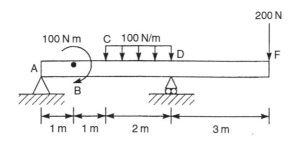

FIGURE P.15.12

P.15.13 A uniform thin-walled beam ABD of open cross-section (Fig. P.15.13) is simply supported at points B and D with its web vertical. It carries a downward vertical force W at the end A in the plane of the web. Derive expressions for the vertical and horizontal components of the deflection of the beam midway between the supports B and D. The wall thickness t and Young's modulus E are constant throughout.

Answer: $u = 0.186Wl^3/Ea^3t$, $v = 0.177Wl^3/Ea^3t$.

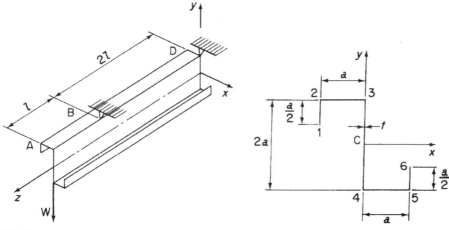

FIGURE P.15.13

P.15.14 A uniform cantilever of arbitrary cross-section and length l has section properties I_{xx}, I_{yy}, and I_{xy} with respect to the centroidal axes shown in Fig. P.15.14. It is loaded in the vertical (yz) plane with a uniformly distributed load of intensity w/unit length. The tip of the beam is hinged to a horizontal link which constrains it to move in the vertical direction only (provided that the actual deflections are small). Assuming that the link is rigid and that there are no twisting effects, calculate

(a) The force in the link;
(b) The deflection of the tip of the beam.

Answer: (a) $3wlI_{xy}/8I_{xx}$; (b) $wl^4/8EI_{xx}$.

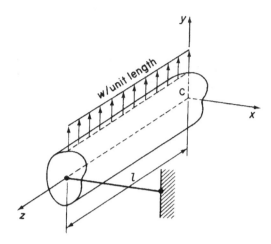

FIGURE P.15.14

P.15.15 A uniform beam of arbitrary, unsymmetrical cross-section and length $2l$ is built-in at one end and simply supported in the vertical direction at a point halfway along its length. This support, however, allows the beam to deflect freely in the horizontal x direction (Fig. P.15.15). For a vertical load W applied at the free end of the beam, calculate and draw the bending moment diagram, putting in the principal values.

Answer: $M_C = 0$, $M_B = Wl$, $M_A = -Wl/2$; linear distribution

P.15.16 The beam section of P.15.4 is subjected to a temperature rise of $4\,T_0$ in its upper flange 12, a temperature rise of $2T_0$ in both vertical webs, and a temperature rise of T_0 in its lower flange 34. Determine the changes in axial force and in the bending moments about the x and y axes. Young's modulus for the material of the beam is E and its coefficient of linear expansion is α.

Answer: $N_T = 9E\alpha dt T_0$, $M_{xT} = 3E\alpha\, d^2 t\, T_0/2$, $M_{yT} = 3E\alpha\, d^2 t\, T_0/4$

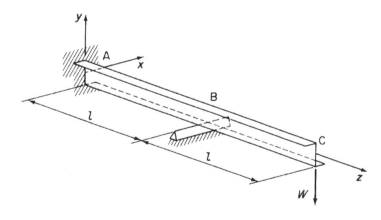

FIGURE P.15.15

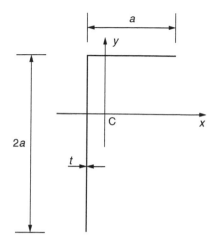

FIGURE P.15.17

P.15.17 The beam section shown in Fig. P.15.17 is subjected to a temperature change which varies with y such that $T = T_0 y/2a$. Determine the corresponding changes in the stress resultants. Young's modulus for the material of the beam is E, while its coefficient of linear expansion is α.

Answer: $N_T = 0$, $M_{xT} = 5E\alpha a^2 t\, T_0/3$, $M_{yT} = E\alpha\, a^2 t\, T_0/6$

Shear of beams

In Chapter 14, we developed the theory for the bending of beams by considering solid or thick beam sections and extended the theory to the thin-walled beam sections typical of aircraft structural components. In fact, it is only in the calculation of section properties that thin-walled sections subjected to bending are distinguished from solid and thick sections. However, for thin-walled beams subjected to shear, the theory is based on assumptions applicable only to thin-walled sections, so that we do not consider solid and thick sections; the relevant theory for such sections may be found in any text on structural and stress analysis[1]. The relationships between bending moments, shear forces, and load intensities derived in Section 16.2.5 still apply.

16.1 GENERAL STRESS, STRAIN, AND DISPLACEMENT RELATIONSHIPS FOR OPEN AND SINGLE-CELL CLOSED SECTION THIN-WALLED BEAMS

We establish in this section the equations of equilibrium and expressions for strain necessary for the analysis of open section beams supporting shear loads and closed section beams carrying shear and torsional loads. The analysis of open section beams subjected to torsion requires a different approach and is discussed separately in Chapter 17. The relationships are established from first principles for the particular case of thin-walled sections in preference to the adaption of Eqs. (1.6), (1.27), and (1.28), which refer to different coordinate axes; the form, however, will be seen to be the same. Generally, in the analysis, we assume that axial constraint effects are negligible, that the shear stresses normal to the beam surface may be neglected since they are zero at each surface and the wall is thin, that direct and shear stresses on planes normal to the beam surface are constant across the thickness, and finally that the beam is of uniform section so that the thickness may vary with distance around each section but is constant along the beam. In addition, we ignore squares and higher powers of the thickness t in the calculation of section properties (see Section 16.4.5).

The parameter s in the analysis is distance measured around the cross-section from some convenient origin.

An element $\delta s \times \delta z \times t$ of the beam wall is maintained in equilibrium by a system of direct and shear stresses, as shown in Fig. 16.1(a). The direct stress σ_z is produced by bending moments or by the bending action of shear loads, while the shear stresses are due to shear and/or torsion of a closed section beam or shear of an open section beam. The hoop stress σ_s is usually zero but may be caused, in closed section beams, by internal pressure. Although we specified that t may vary with s, this variation is small for most thin-walled structures, so that we may reasonably make the approximation that t is constant

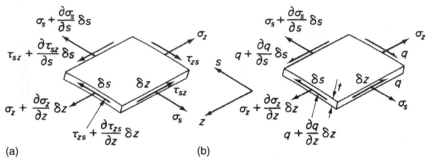

FIGURE 16.1 (a) General Stress System on an Element of a Closed or Open Section Beam; (b) Direct Stress and Shear Flow System on the Element

over the length δs. Also, from Eq. (1.4), we deduce that $\tau_{zs} = \tau_{sz} = \tau$, say. However, we find it convenient to work in terms of *shear flow* q, that is, shear force per unit length rather than in terms of shear stress. Hence, in Fig. 16.1(b),

$$q = \tau t \tag{16.1}$$

and is regarded as being positive in the direction of increasing s.

For equilibrium of the element in the z direction and neglecting body forces (see Section 1.2),

$$\left(\sigma_z + \frac{\partial \sigma_z}{\partial z}\delta z\right) t\delta s - \sigma_z t\delta s + \left(q + \frac{\partial q}{\partial s}\delta s\right)\delta z - q\delta z = 0$$

which reduces to

$$\frac{\partial q}{\partial s} + t\frac{\partial \sigma_z}{\partial z} = 0 \tag{16.2}$$

Similarly, for equilibrium in the s direction,

$$\frac{\partial q}{\partial z} + t\frac{\partial \sigma_s}{\partial s} = 0 \tag{16.3}$$

The direct stresses σ_z and σ_s produce direct strains ε_z and ε_s, while the shear stress τ induces a shear strain $\gamma(= \gamma_{zs} = \gamma_{sz})$. We now proceed to express these strains in terms of the three components of the displacement of a point in the section wall (see Fig. 16.2). Of these components v_t is a tangential displacement in the xy plane and is taken to be positive in the direction of increasing $s; v_n$ is a normal displacement in the xy plane and is positive outward; and w is an axial displacement defined in Section 16.2.1. Immediately, from the third of Eqs. (1.18), we have

$$\varepsilon_z = \frac{\partial w}{\partial z} \tag{16.4}$$

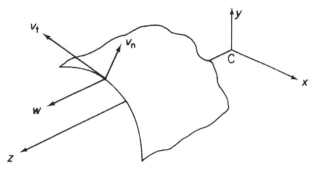

FIGURE 16.2 Axial, Tangential, and Normal Components of Displacement of a Point in the Beam Wall

It is possible to derive a simple expression for the direct strain ε_s in terms of v_t, v_n, s, and the curvature $1/r$ in the xy plane of the beam wall. However, as we do not require ε_s in the subsequent analysis, for brevity, we merely quote the expression

$$\varepsilon_s = \frac{\partial v_t}{\partial s} + \frac{v_n}{r} \tag{16.5}$$

The shear strain γ is found in terms of the displacements w and v_t by considering the shear distortion of an element $\delta s \times \delta z$ of the beam wall. From Fig. 16.3, we see that the shear strain is given by

$$\gamma = \phi_1 + \phi_2$$

or, in the limit as both δs and δz tend to zero,

$$\gamma = \frac{\partial w}{\partial s} + \frac{\partial v_t}{\partial z} \tag{16.6}$$

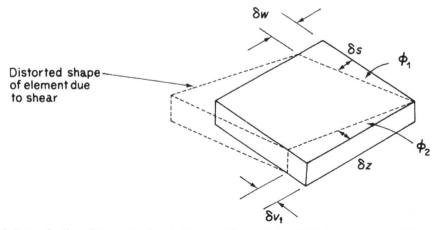

FIGURE 16.3 Determination of Shear Strain γ in Terms of Tangential and Axial Components of Displacement

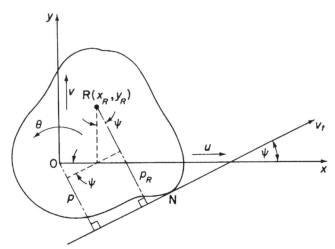

FIGURE 16.4 Establishment of Displacement Relationships and Position of Center of Twist of Beam (Open or Closed)

In addition to the assumptions specified in the earlier part of this section, we further assume that, during any displacement, the shape of the beam cross-section is maintained by a system of closely spaced diaphragms, which are rigid on their own plane but are perfectly flexible normal to their own plane (CSRD assumption). There is, therefore, no resistance to axial displacement w and the cross-section moves as a rigid body in its own plane, the displacement of any point being completely specified by translations u and v and a rotation θ (see Fig. 16.4).

At first sight this appears to be a rather sweeping assumption, but for aircraft structures of the thin-shell type described in Chapter 11 whose cross-sections are stiffened by ribs or frames positioned at frequent intervals along their lengths, it is a reasonable approximation of the actual behavior of such sections. The tangential displacement v_t of any point N in the wall of either an open or closed section beam is seen from Fig. 16.4 to be

$$v_t = p\theta + u \cos \psi + v \sin \psi \tag{16.7}$$

where clearly u, v, and θ are functions of z only (w may be a function of z and s).

The origin O of the axes in Fig. 16.4 has been chosen arbitrarily and the axes suffer displacements u, v, and θ. These displacements, in a loading case such as pure torsion, are equivalent to a pure rotation about some point $R(x_R, y_R)$ in the cross-section, where R is the *center of twist*. Therefore, in Fig. 16.4,

$$v_t = p_R\theta \tag{16.8}$$

and

$$p_R = p - x_R \sin \psi + y_R \cos \psi$$

which gives

$$v_t = p\theta - x_R\theta \sin \psi + y_R \, \theta \cos \psi$$

and

$$\frac{\partial v_t}{\partial z} = p\frac{d\theta}{dz} - x_R \sin\psi \frac{d\theta}{dz} + y_R \cos\psi \frac{d\theta}{dz} \tag{16.9}$$

Also, from Eq. (16.7),

$$\frac{\partial v_t}{\partial z} = p\frac{d\theta}{dz} + \frac{du}{dz}\cos\psi + \frac{dv}{dz}\sin\psi \tag{16.10}$$

Comparing the coefficients of Eqs. (16.9) and (16.10), we see that

$$x_R = -\frac{dv/dz}{d\theta/dz}, \quad y_R = \frac{du/dz}{d\theta/dz} \tag{16.11}$$

16.2 SHEAR OF OPEN SECTION BEAMS

The open section beam of arbitrary section shown in Fig. 16.5 supports shear loads S_x and S_y such that there is no twisting of the beam cross-section. For this condition to be valid, both the shear loads must pass through a particular point in the cross-section known as the *shear center*.

Since there are no hoop stresses in the beam, the shear flows and direct stresses acting on an element of the beam wall are related by Eq. (16.2), that is,

$$\frac{\partial q}{\partial s} + t\frac{\partial\sigma_z}{\partial z} = 0$$

We assume that the direct stresses are obtained with sufficient accuracy from basic bending theory so that, from Eq. (16.17),

$$\frac{\partial\sigma_z}{\partial z} = \frac{[(\partial M_y/\partial z)I_{xx} - (\partial M_x/\partial z)I_{xy}]}{I_{xx}I_{yy} - I_{xy}^2}x + \frac{[(\partial M_x/\partial z)I_{yy} - (\partial M_y/\partial z)I_{xy}]}{I_{xx}I_{yy} - I_{xy}^2}y$$

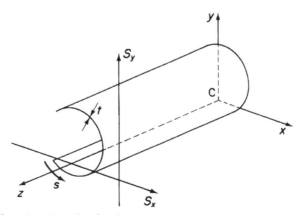

FIGURE 16.5 Shear Loading of an Open Section Beam

Using the relationships of Eqs. (16.22) and (16.23), that is, $\partial M_y/\partial z = S_x$, and so forth, this expression becomes

$$\frac{\partial \sigma_z}{\partial z} = \frac{(S_x I_{xx} - S_y I_{xy})}{I_{xx}I_{yy} - I_{xy}^2} x + \frac{(S_y I_{yy} - S_x I_{xy})}{I_{xx}I_{yy} - I_{xy}^2} y$$

Substituting for $\partial \sigma_z/\partial z$ in Eq. (16.2) gives

$$\frac{\partial q}{\partial s} = -\frac{(S_x I_{xx} - S_y I_{xy})}{I_{xx}I_{yy} - I_{xy}^2} tx - \frac{(S_y I_{yy} - S_x I_{xy})}{I_{xx}I_{yy} - I_{xy}^2} ty \tag{16.12}$$

Integrating Eq. (16.12) with respect to s from some origin for s to any point around the cross-section, we obtain

$$\int_0^s \frac{\partial q}{\partial s}\, ds = -\left(\frac{S_x I_{xx} - S_y I_{xy}}{I_{xx}I_{yy} - I_{xy}^2}\right)\int_0^s tx\, ds - \left(\frac{S_y I_{yy} - S_x I_{xy}}{I_{xx}I_{yy} - I_{xy}^2}\right)\int_0^s ty\, ds \tag{16.13}$$

If the origin for s is taken at the open edge of the cross-section, then $q = 0$ when $s = 0$, and Eq. (16.13) becomes

$$q_s = -\left(\frac{S_x I_{xx} - S_y I_{xy}}{I_{xx}I_{yy} - I_{xy}^2}\right)\int_0^s tx\, ds - \left(\frac{S_y I_{yy} - S_x I_{xy}}{I_{xx}I_{yy} - I_{xy}^2}\right)\int_0^s ty\, ds \tag{16.14}$$

For a section having either Cx or Cy as an axis of symmetry, $I_{xy} = 0$ and Eq. (16.14) reduces to

$$q_s = -\frac{S_x}{I_{yy}}\int_0^s tx\, ds - \frac{S_y}{I_{xx}}\int_0^s ty\, ds$$

Example 16.1

Determine the shear flow distribution in the thin-walled Z section shown in Fig. 16.6 due to a shear load S_y applied through the shear center of the section.

The origin for our system of reference axes coincides with the centroid of the section at the midpoint of the web. From antisymmetry, we also deduce by inspection that the shear center occupies the same position. Since S_y is applied through the shear center, no torsion exists and the shear flow distribution is given by Eq. (16.14), in which $S_x = 0$, that is,

$$q_s = \frac{S_y I_{xy}}{I_{xx}I_{yy} - I_{xy}^2}\int_0^s tx\, ds - \frac{S_y I_{yy}}{I_{xx}I_{yy} - I_{xy}^2}\int_0^s ty\, ds$$

or

$$q_s = \frac{S_y}{I_{xx}I_{yy} - I_{xy}^2}\left(I_{xy}\int_0^s tx\, ds - I_{yy}\int_0^s ty\, ds\right) \tag{i}$$

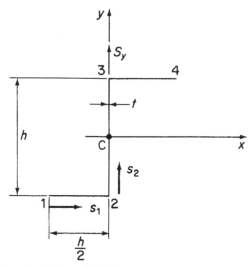

FIGURE 16.6 Shear Loaded *Z* Section of Example 16.1

The second moments of area of the section were determined in Example 16.14 and are

$$I_{xx} = \frac{h^3 t}{3}, \quad I_{yy} = \frac{h^3 t}{12}, \quad I_{xy} = \frac{h^3 t}{8}$$

Substituting these values in Eq. (i) we obtain

$$q_s = \frac{S_y}{h^3} \int_0^s (10.32x - 6.84y)\,\mathrm{d}s \tag{ii}$$

On the bottom flange 12, $y = -h/2$ and $x = -h/2 + s_1$, where $0 \le s_1 \le h/2$. Therefore,

$$q_{12} = \frac{S_y}{h^3} \int_0^{s_1} (10.32 s_1 - 1.74h)\,\mathrm{d}s_1$$

giving

$$q_{12} = \frac{S_y}{h^3}(5.16 s_1^2 - 1.74 h s_1) \tag{iii}$$

Hence, at 1, $(s_1 = 0)$, $q_1 = 0$ and, at 2, $(s_1 = h/2)$, $q_2 = 0.42 S_y/h$. Further examination of Eq. (iii) shows that the shear flow distribution on the bottom flange is parabolic with a change of sign (i.e., direction) at $s_1 = 0.336h$. For values of $s_1 < 0.336h$, q_{12} is negative and therefore in the opposite direction to s_1.

In the web 23, $y = -h/2 + s_2$, where $0 \le s_2 \le h$ and $x = 0$. Then,

$$q_{23} = \frac{S_y}{h^3} \int_0^{s_2} (3.42h - 6.84 s_2)\,\mathrm{d}s_2 + q_2 \tag{iv}$$

We note, in Eq. (iv), that the shear flow is not zero when $s_2 = 0$ but equal to the value obtained by inserting $s_1 = h/2$ in Eq. (iii), that is, $q_2 = 0.42 S_y/h$. Integration of Eq. (iv) yields

$$q_{23} = \frac{S_y}{h^3}(0.42 h^2 + 3.42 h s_2 - 3.42 s_2^2) \tag{v}$$

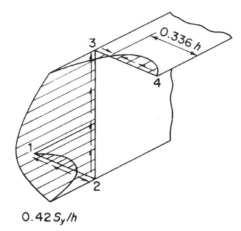

FIGURE 16.7 Shear Flow Distribution in Z Section of Example 16.1

This distribution is symmetrical about Cx with a maximum value at $s_2 = h/2(y = 0)$ and the shear flow is positive at all points in the web.

The shear flow distribution in the upper flange may be deduced from antisymmetry, so that the complete distribution is of the form shown in Fig. 16.7.

Example 16.2

Calculate the shear flow distribution in the thin-walled open section shown in Fig. 16.8 produced by a vertical shear load, S_y, acting through its shear centre.

The centroid of the section coincides with the centre of the circle. Also the Cx axis is an axis of symmetry so that $I_{xy} = 0$ and since $S_x = 0$ Eq. (16.14) reduces to

$$q_s = -\left(S_y/I_{xx}\right) \int_0^s ty \, ds \qquad (i)$$

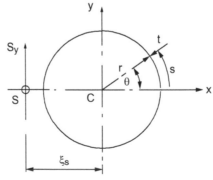

FIGURE 16.8 Beam Section of Example 16.2

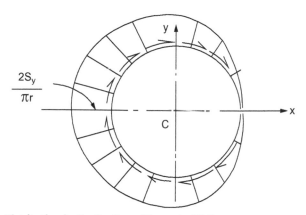

FIGURE 16.9 Shear Flow Distribution in the Section of Example 16.2

The second moment of area, I_{xx}, of the section about Cx may be deduced from the second moment of area of the semi-circular section shown in Fig. 16.33 and is $\pi r^3 t$. Then, at any point a distance s from one edge of the narrow slit

$$q_s = -(S_y/\pi r^3 t)\int_0^s ty \, ds \tag{ii}$$

Working with angular coordinates for convenience Eq. (ii) becomes

$$q_\theta = -(S_y/\pi r^3 t)\int_0^\theta tr\sin\theta r \, d\theta$$

i.e.,

$$q_\theta = -(S_y/\pi r)\int_0^\theta \sin\theta \, d\theta$$

which gives

$$q_\theta = -(S_y/\pi r)[\cos\theta]_0^\theta$$

Then

$$q_\theta = (S_y/\pi r)(\cos\theta - 1) \tag{iii}$$

From Eq. (iii), when $\theta = 0$, $q_\theta = 0$ (as expected at an open edge) and when $\theta = \pi$, $q_\theta = -2S_y/\pi r$. Further analysis of Eq. (iii) shows that q_θ is also zero at $\theta = 2\pi$ and that q_θ is a maximum when $\theta = \pi$. Also q_θ is negative, i.e., in the opposite sense to increasing values of θ, for all values of θ. The complete shear flow distribution is shown in Fig. 16.9.

16.2.1 Shear center

We defined the position of the shear center as that point in the cross-section through which shear loads produce no twisting. It may be shown by use of the reciprocal theorem that this point is also the center of twist of sections subjected to torsion. There are, however, some important exceptions to this general

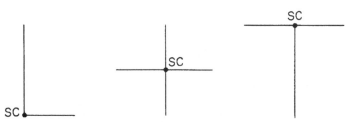

FIGURE 16.10 Shear Center Position for the Type of Open Section Beam Shown

rule, as we observe in Section 26.1. Clearly, in the majority of practical cases, it is impossible to guarantee that a shear load will act through the shear center of a section. Equally apparent is the fact that any shear load may be represented by the combination of the shear load applied through the shear center and a torque. The stresses produced by the separate actions of torsion and shear may then be added by superposition. It is therefore necessary to know the location of the shear center in all types of section or to calculate its position. Where a cross-section has an axis of symmetry, the shear center must, of course, lie on this axis. For cruciform or angle sections of the type shown in Fig. 16.10, the shear center is located at the intersection of the sides, since the resultant internal shear loads all pass through these points.

Example 16.3

Calculate the position of the shear center of the thin-walled channel section shown in Fig. 16.11. The thickness t of the walls is constant.

The shear center S lies on the horizontal axis of symmetry at some distance ξ_S, say, from the web. If we apply an arbitrary shear load S_y through the shear center, then the shear flow distribution is given by Eq. (16.14) and the moment about any point in the cross-section produced by these shear flows is *equivalent* to the moment of the applied shear load. S_y appears on both sides of the resulting equation and may therefore be eliminated to leave ξ_S.

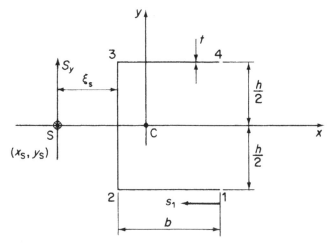

FIGURE 16.11 Determination of Shear Center Position of Channel Section of Example 16.3

For the channel section, Cx is an axis of symmetry so that $I_{xy} = 0$. Also $S_x = 0$ and therefore Eq. (16.14) simplifies to

$$q_s = -\frac{S_y}{I_{xx}} \int_0^s ty \, ds \tag{i}$$

where

$$I_{xx} = 2bt \left(\frac{h}{2}\right)^2 + \frac{th^3}{12} = \frac{h^3 t}{12} \left(1 + \frac{6b}{h}\right)$$

Substituting for I_{xx} in Eq. (i), we have

$$q_s = \frac{-12 S_y}{h^3 (1 + 6b/h)} \int_0^s y \, ds \tag{ii}$$

The amount of computation involved may be reduced by giving some thought to the requirements of the problem. In this case, we are asked to find the position of the shear center only, not a complete shear flow distribution. From symmetry, it is clear that the moments of the resultant shears on the top and bottom flanges about the midpoint of the web are numerically equal and act in the same rotational sense. Furthermore, the moment of the web shear about the same point is zero. We deduce that it is necessary to obtain the shear flow distribution only on either the top or bottom flange for a solution. Alternatively, choosing a web–flange junction as a moment center leads to the same conclusion.

On the bottom flange, $y = -h/2$, so that, from Eq. (ii), we have

$$q_{12} = \frac{6 S_y}{h^2 (1 + 6b/h)} s_1 \tag{iii}$$

Equating the clockwise moments of the internal shears about the midpoint of the web to the clockwise moment of the applied shear load about the same point gives

$$S_y \xi_s = 2 \int_0^b q_{12} \frac{h}{2} \, ds_1$$

or, by substitution from Eq. (iii),

$$S_y \xi_s = 2 \int_0^b \frac{6 S_y}{h^2 (1 + 6b/h)} \frac{h}{2} s_1 \, ds_1$$

from which

$$\xi_s = \frac{3b^2}{h(1 + 6b/h)} \tag{iv}$$

In the case of an unsymmetrical section, the coordinates (ξ_s, η_s) of the shear center referred to some convenient point in the cross-section are obtained by first determining ξ_s in a similar manner to that of Example 16.3 then finding η_s by applying a shear load S_x through the shear center. In both cases, the choice of a web–flange junction as a moment center reduces the amount of computation.

Example 16.4

Determine the position of the shear center of the open section beam of Example 16.2.

The shear centre S lies on the horizontal axis of symmetry a distance, ξ_S, say, from the centroid C. The shear flow distribution is given by Eq. (iii) of Ex. 16.2 , i.e., $q_\theta = S_y(\cos\theta - 1)/\pi r$. Therefore, taking moments about C

$$S_y\xi_S = -\int_0^{2\pi} q_\theta r\,\mathrm{d}s = -\int_0^{2\pi} q_\theta r^2\,\mathrm{d}\theta$$

Then

$$S_y\xi_S = -(S_y r/\pi)\int_0^{2\pi}(\cos\theta - 1)\,\mathrm{d}\theta \tag{i}$$

Note that the assumed direction of q_θ in Ex. 16.2 is in the direction of increasing θ so that we introduce a negative sign to allow for this when equating the moment of the external shear force to that of the internal shear flow. Integrating Eq. (i) we obtain

$$\xi_S = -(r/\pi)[\sin\theta - \theta]_0^{2\pi}$$

which gives

$$\xi_S = 2r$$

Example 16.5

Calculate the position of the shear center of the thin-walled section shown in Fig. 16.12; the thickness of the section is 2mm and is constant throughout.

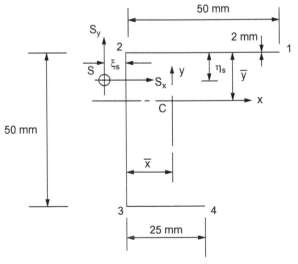

FIGURE 16.12 Beam Section of Example 16.5

Since we are only asked to find the position of the shear center of the section and not a complete shear flow distribution the argument of Ex. 16.3 applies in that if we refer the position of the shear centre to the web/flange junction 2 (or 3) it is only necessary to obtain the shear flow distribution in the flange 34 (or 12). First, we must calculate the section's properties.

Taking moments of area about the top flange

$$2(50 + 50 + 25)\bar{y} = 2 \times 50 \times 25 + 2 \times 25 \times 50$$

which gives

$$\bar{y} = 20 \text{ mm}$$

Now taking moments of area about the vertical web

$$2(50 + 50 + 25)\bar{x} = 2 \times 50 \times 25 + 2 \times 25 \times 12.5$$

from which

$$\bar{x} = 12.5 \text{ mm}$$

The second moments of area are then calculated using the methods of Section 16.4.5.

$$I_{xx} = 2 \times 50 \times 20^2 + (2 \times 50^3/12) + 2 \times 50 \times 5^2 + 2 \times 25 \times 30^2$$

i.e.,

$$I_{xx} = 108{,}333 \text{ mm}^4$$

$$I_{yy} = (2 \times 50^3/12) + 2 \times 50 \times 12.5^2 + 2 \times 50 \times 12.5^2 + (2 \times 25^3/12)$$

i.e.,

$$I_{yy} = 54{,}689 \text{ mm}^4$$

$$I_{xy} = 2 \times 50 \,(+12.5)(+20) + (-12.5)(-5)$$

i.e.,

$$I_{xy} = 31{,}250 \text{ mm}^4$$

Note that the flange 34 makes no contribution to I_{xy} since its centroid coincides with the y axis of the section.

Considering the horizontal position of the shear center we apply a vertical shear load, S_y, through the shear centre and determine the shear flow distribution in the flange 34. Since $S_x = 0$, Eq. (16.14) reduces to

$$q_s = \left[(S_y I_{xy})/(I_{xx}I_{yy} - I_{xy}^2)\right] \int_0^s tx \, ds - \left[(S_y I_{yy})/(I_{xx}I_{yy} - I_{xy}^2)\right] \int_0^s ty \, ds \qquad \text{(i)}$$

Substituting the values of I_{xx} etc. gives

$$q_s = 1.26 \times 10^{-5} S_y \int_0^s x \, ds - 2.21 \times 10^{-5} S_y \int_0^s ty \, ds \qquad \text{(ii)}$$

On the flange 34, $x = 12.5 - s$ and $y = -30$ mm. Eq. (ii) then becomes

$$q_{43} = S_y \times 10^{-5} \int_0^s (82.05 - 1.26s) \, ds$$

Integrating gives

$$q_{43} = S_y \times 10^{-5} \left(82.05s - 0.63s^2\right) \qquad \text{(iii)}$$

Now taking moments about the flange/web junction 2

$$S_y \xi_S = \int_0^{25} q_{43} \times 50 \, ds$$

Substituting for q_{43} from Eq. (iii) and integrating gives

$$\xi_S = 11.2 \text{ mm}$$

We now apply S_x only through the shear center and carry out the same procedure as above. Then

$$q_{43} = S_x \times 10^{-5} (2.19s^2 - 92.55s) \tag{iv}$$

Taking moments about the web/flange junction 2

$$S_x \eta_S = -\int_0^{25} q_{43} \times 50 \, ds$$

Note that in this case the moment of the resultant of the internal shear flows is in the opposite sense to that of the applied force. Substituting for q_{43} from Eq. (iv) and carrying out the integration gives

$$\eta_S = 7.2 \text{ mm}$$

■

16.3 SHEAR OF CLOSED SECTION BEAMS

The solution for a shear loaded closed section beam follows a similar pattern to that described in Section 16.2 for an open section beam but with two important differences. First, the shear loads may be applied through points in the cross-section other than the shear center, so that torsional as well as shear effects are included. This is possible, since, as we shall see, shear stresses produced by torsion in closed section beams have exactly the same form as shear stresses produced by shear, unlike shear stresses due to shear and torsion in open section beams. Second, it is generally not possible to choose an origin for s at which the value of shear flow is known. Consider the closed section beam of arbitrary section shown in Fig. 16.13. The shear loads S_x and S_y are applied through any point in the cross-section and, in general, cause direct bending stresses and shear flows related by the equilibrium equation (16.2). We assume that hoop stresses and body forces are absent. Therefore,

$$\frac{\partial q}{\partial s} + t \frac{\partial \sigma_z}{\partial z} = 0$$

From this point, the analysis is identical to that for a shear loaded open section beam, until we reach the stage of integrating Eq. (16.13), namely,

$$\int_0^s \frac{\partial q}{\partial s} \, ds = -\left(\frac{S_x I_{xx} - S_y I_{xy}}{I_{xx} I_{yy} - I_{xy}^2} \right) \int_0^s tx \, ds - \left(\frac{S_y I_{yy} - S_x I_{xy}}{I_{xx} I_{yy} - I_{xy}^2} \right) \int_0^s ty \, ds$$

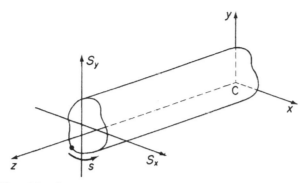

FIGURE 16.13 Shear of Closed Section Beams

Let us suppose that we choose an origin for s where the shear flow has the unknown value $q_{s,0}$. Integration of Eq. (16.13) then gives

$$q_s - q_{s,0} = -\left(\frac{S_x I_{xx} - S_y I_{xy}}{I_{xx} I_{yy} - I_{xy}^2}\right) \int_0^s tx \, ds - \left(\frac{S_y I_{yy} - S_x I_{xy}}{I_{xx} I_{yy} - I_{xy}^2}\right) \int_0^s ty \, ds$$

or

$$q_s = -\left(\frac{S_x I_{xx} - S_y I_{xy}}{I_{xx} I_{yy} - I_{xy}^2}\right) \int_0^s tx \, ds - \left(\frac{S_y I_{yy} - S_x I_{xy}}{I_{xx} I_{yy} - I_{xy}^2}\right) \int_0^s ty \, ds + q_{s,0} \qquad (16.15)$$

We observe by comparison of Eqs. (16.15) and (16.14) that the first two terms on the right-hand side of Eq. (16.15) represent the shear flow distribution in an open section beam loaded through its shear center. This fact indicates a method of solution for a shear loaded closed section beam. Representing this "open" section or "basic" shear flow by q_b, we write Eq. (16.15) in the form

$$q_s = q_b + q_{s,0} \qquad (16.16)$$

We obtain q_b by supposing that the closed beam section is "cut" at some convenient point, thereby producing an "open" section (see Fig. 16.14(b)). The shear flow distribution (q_b) around this open section is given by

$$q_b = -\left(\frac{S_x I_{xx} - S_y I_{xy}}{I_{xx} I_{yy} - I_{xy}^2}\right) \int_0^s tx \, ds - \left(\frac{S_y I_{yy} - S_x I_{xy}}{I_{xx} I_{yy} - I_{xy}^2}\right) \int_0^s ty \, ds$$

as in Section 16.2. The value of shear flow at the cut ($s = 0$) is found by equating applied and internal moments taken about some convenient moment center. Then, from Fig. 16.14(a),

$$S_x \eta_0 - S_y \xi_0 = \oint pq \, ds = \oint pq_b \, ds + q_{s,0} \oint p \, ds$$

where $\oint$ denotes integration completely around the cross-section. In Fig. 16.14(a),

$$\delta A = \frac{1}{2} \delta s p$$

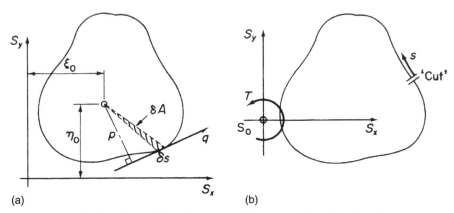

FIGURE 16.14 (a) Determination of $q_{s,0}$; (b) Equivalent Loading on an "Open" Section Beam

so that

$$\oint dA = \frac{1}{2}\oint p \, ds$$

Hence,

$$\oint p \, ds = 2A$$

where A is the area enclosed by the midline of the beam section wall. Hence,

$$S_x\eta_0 - S_y\xi_0 = \oint pq_b \, ds + 2Aq_{s,0} \tag{16.17}$$

If the moment center is chosen to coincide with the lines of action of S_x and S_y, then Eq. (16.17) reduces to

$$0 = \oint pq_b \, ds + 2Aq_{s,0} \tag{16.18}$$

The unknown shear flow $q_{s,0}$ follows from either Eq. (16.17) or (16.18).

It is worthwhile to consider some of the implications of this process. Equation (16.14) represents the shear flow distribution in an open section beam for the condition of zero twist. Therefore, by "cutting" the closed section beam of Fig. 16.14(a) to determine q_b, we are, in effect, replacing the shear loads of Fig. 16.14(a) by shear loads S_x and S_y acting through the shear center of the resulting "open" section beam together with a torque T, as shown in Fig. 16.14(b). We show in Section 18.1 that the application of a torque to a closed section beam results in a constant shear flow. In this case, the constant shear flow $q_{s,0}$ corresponds to the torque but has different values for different positions of the cut, since the corresponding various open'section beams have different locations for their shear centers. An additional effect of cutting the beam is to produce a statically determinate structure, since the q_b shear flows are obtained from statical equilibrium considerations. It follows that a single-cell closed section beam supporting shear loads is singly redundant.

Example 16.6

Determine the shear flow distribution in the walls of the thin-walled closed section beam shown in Fig. 16.15; the wall thickness, t, is constant throughout.

Since the x axis is an axis of symmetry, $I_{xy} = 0$, and since $S_x = 0$, Eq. (16.15) reduces to

$$q_s = -\left(S_y/I_{xx}\right) \int_0^s ty \, ds + q_{s,0} \tag{i}$$

where

$$I_{xx} = \left(\pi tr^3/2\right) + 2 \times 2rt \times r^2 + \left[t(2r)^3/12\right] = 6.24tr^3$$

(see Section 16.4.5)

We now "cut" the beam section at 1. Any point may be chosen for the "cut" but the amount of computation will be reduced if a point is chosen which coincides with the axis of symmetry. Then

$$q_{b,12} = -\left(S_y/I_{xx}\right) \int_0^\theta tr \sin\theta r \, d\theta$$

which gives

$$q_{b,12} = 0.16\left(S_y/r\right) \left[\cos\theta\right]_0^\theta$$

so that

$$q_{b,12} = 0.16\left(S_y/r\right)(\cos\theta - 1) \tag{ii}$$

When $\theta = \pi/2$, $q_{b,2} = -0.16(S_y/r)$. The q_b shear flow in the wall 23 is then

$$q_{b,23} = -\left(S_y/I_{xx}\right) \int_0^{s_1} tr \, ds_1 - 0.16\left(S_y/r\right)$$

i.e.,

$$q_{b,23} = -0.16\left(S_y/r^2\right)(s_1 + r) \tag{iii}$$

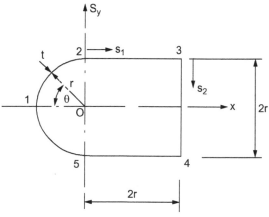

FIGURE 16.15 Beam Section of Example 16.6

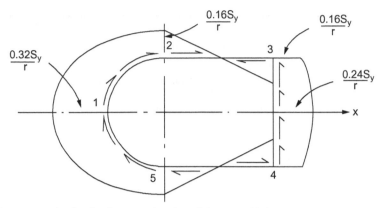

FIGURE 16.16 Shear Flow Distribution in Beam Section of Example 16.6

and when $s_1 = 2r$, $q_{b,3} = -0.48(S_y/r)$. Then, in the wall 34

$$q_{b,34} = -0.16(S_y/tr^3) \int_0^{s_2} t(r - s_2)\, ds_2 - 0.48(S_y/r)$$

i.e.,

$$q_{b,34} = -0.16(S_y/r^3)(rs_2 - 0.5s_2^2 + 3r^2) \tag{iv}$$

The remaining distribution follows from symmetry. Now taking moments about O and using Eq. (16.18)

$$0 = 2\left[\int_0^{\pi/2} q_{b,12}r^2\, d\theta + \int_0^{2r} q_{b,23}r ds_1 + \int_0^r q_{b,34}2r ds_2\right] + 2[4r^2 + (\pi r^2/2)]q_{s,0} \tag{v}$$

Substituting in Eq. (v) for $q_{b,12}$ etc from Eqs. (ii), (iii) and (iv) gives

$$q_{s,0} = 0.32S_y/r$$

Adding $q_{s,0}$ to the q_b distributions of Eqs. (ii), (iii) and (iv) gives

$$q_{12} = 0.16(S_y/r^3)(r^2\cos\theta + r^2)$$
$$q_{23} = 0.16(S_y/r^3)(r^2 - rs_1)$$
$$q_{34} = 0.16(S_y/r^3)(0.5s_2^2 - rs_2 - r^2)$$

Note that q_{23} changes sign at $s_1 = r$. The shear flow distribution in the lower half of the section follows from symmetry and the complete distribution is shown in Fig. 16.16.

16.3.1 Twist and warping of shear loaded closed section beams

Shear loads that are not applied through the shear center of a closed section beam cause cross-sections to twist and warp; that is, in addition to rotation, they suffer out-of-plane axial displacements. Expressions for these quantities may be derived in terms of the shear flow distribution q_s as follows. Since

$q = \tau t$ and $\tau = G\gamma$ (see Chapter 1), we can express q_s in terms of the warping and tangential displacements w and v_t of a point in the beam wall by using Eq. (16.6). Thus

$$q_s = Gt\left(\frac{\partial w}{\partial s} + \frac{\partial v_t}{\partial z}\right) \tag{16.19}$$

Substituting for $\partial v_t/\partial z$ from Eq. (16.10), we have

$$\frac{q_s}{Gt} = \frac{\partial w}{\partial s} + p\frac{d\theta}{dz} + \frac{du}{dz}\cos\psi + \frac{dv}{dz}\sin\psi \tag{16.20}$$

Integrating Eq. (16.20) with respect to s from the chosen origin for s and noting that G may also be a function of s, we obtain

$$\int_0^s \frac{q_s}{Gt}ds = \int_0^s \frac{\partial w}{\partial s}ds + \frac{d\theta}{dz}\int_0^s p\,ds + \frac{du}{dz}\int_0^s \cos\psi\,ds + \frac{dv}{dz}\int_0^s \sin\psi\,ds$$

or

$$\int_0^s \frac{q_s}{Gt}ds = \int_0^s \frac{\partial w}{\partial s}ds + \frac{d\theta}{dz}\int_0^s p\,ds + \frac{du}{dz}\int_0^s dx + \frac{dv}{dz}\int_0^s dy$$

which gives

$$\int_0^s \frac{q_s}{Gt}ds = (w_s - w_0) + 2A_{Os}\frac{d\theta}{dz} + \frac{du}{dz}(x_s - x_0) + \frac{dv}{dz}(y_s - y_0) \tag{16.21}$$

where A_{Os} is the area swept out by a generator, center at the origin of axes, O, from the origin for s to any point s around the cross-section. Continuing the integration completely around the cross-section yields, from Eq. (16.21),

$$\oint \frac{q_s}{Gt}ds = 2A\frac{d\theta}{dz}$$

from which

$$\frac{d\theta}{dz} = \frac{1}{2A}\oint \frac{q_s}{Gt}ds \tag{16.22}$$

Substituting for the rate of twist in Eq. (16.21) from Eq. (16.22) and rearranging, we obtain the warping distribution around the cross-section:

$$w_s - w_0 = \int_0^s \frac{q_s}{Gt}ds - \frac{A_{Os}}{A}\oint \frac{q_s}{Gt}ds - \frac{du}{dz}(x_s - x_0) - \frac{dv}{dz}(y_s - y_0) \tag{16.23}$$

Using Eqs. (16.11) to replace du/dz and dv/dz in Eq. (16.23), we have

$$w_s - w_0 = \int_0^s \frac{q_s}{Gt}ds - \frac{A_{Os}}{A}\oint \frac{q_s}{Gt}ds - y_R\frac{d\theta}{dz}(x_s - x_0) + x_R\frac{d\theta}{dz}(y_s - y_0) \tag{16.24}$$

The last two terms in Eq. (16.24) represent the effect of relating the warping displacement to an arbitrary origin that itself suffers axial displacement due to warping. In the case where the origin coincides with the center of twist R of the section, Eq. (16.24) simplifies to

$$w_s - w_0 = \int_0^s \frac{q_s}{Gt} \, ds - \frac{A_{Os}}{A} \oint \frac{q_s}{Gt} \, ds \qquad (16.25)$$

In problems involving singly or doubly symmetrical sections, the origin for s may be taken to coincide with a point of zero warping, which occurs where an axis of symmetry and the wall of the section intersect. For unsymmetrical sections, the origin for s may be chosen arbitrarily. The resulting warping distribution has exactly the same form as the actual distribution but is displaced axially by the unknown warping displacement at the origin for s. This value may be found by referring to the torsion of closed section beams subject to axial constraint (see Section 26.3). In the analysis of such beams, it is assumed that the direct stress distribution set up by the constraint is directly proportional to the free warping of the section; that is,

$$\sigma = \text{constant} \times w$$

Also, since a pure torque is applied, the result of any internal direct stress system must be zero; in other words, it is self-equilibrating. Thus,

$$\text{Resultant axial load} = \oint \sigma t \, ds$$

where σ is the direct stress at any point in the cross-section. Then, from the preceding assumption,

$$0 = \oint wt \, ds$$

or

$$0 = \oint (w_s - w_0)t \, ds$$

so that

$$w_0 = \frac{\oint w_s t \, ds}{\oint t \, ds} \qquad (16.26)$$

16.3.2 Shear center

The shear center of a closed section beam is located in a similar manner to that described in Section 16.2.1 for open section beams. Therefore, to determine the coordinate ξ_S (referred to any convenient point in the cross-section) of the shear center S of the closed section beam shown in Fig. 16.17, we apply an arbitrary shear load S_y through S, calculate the distribution of shear flow q_s due to S_y and equate internal and external moments. However, a difficulty arises in obtaining $q_{s,0}$ since, at this stage, it is impossible to equate internal and external moments to produce an equation similar to Eq. (16.17) as the position of S_y is unknown. We therefore use the condition that a

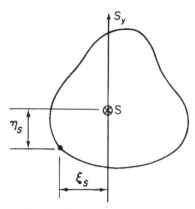

FIGURE 16.17 Shear Center of a Closed Section Beam

shear load acting through the shear center of a section produces zero twist. It follows that $d\theta/dz$ in Eq. (16.22) is zero, so that

$$0 = \oint \frac{q_s}{Gt}\,ds$$

or

$$0 = \oint \frac{1}{Gt}(q_b + q_{s,0})\,ds$$

which gives

$$q_{s,0} = -\frac{\oint (q_b/Gt)\,ds}{\oint ds/Gt} \tag{16.27}$$

If $Gt = $ constant, then Eq. (16.27) simplifies to

$$q_{s,0} = -\frac{\oint q_b\,ds}{\oint ds} \tag{16.28}$$

The coordinate η_S is found in a similar manner by applying S_x through S.

Example 16.7

A thin-walled closed section beam has the singly symmetrical cross-section shown in Fig. 16.18. Each wall of the section is flat and has the same thickness t and shear modulus G. Calculate the distance of the shear center from point 4.

The shear center clearly lies on the horizontal axis of symmetry, so that it is necessary only to apply a shear load S_y through S and to determine ξ_S. If we take the x reference axis to coincide with the axis of symmetry, then $I_{xy} = 0$, and since $S_x = 0$, Eq. (16.15) simplifies to

$$q_s = -\frac{S_y}{I_{xx}} \int_0^s ty\,ds + q_{s,0} \tag{i}$$

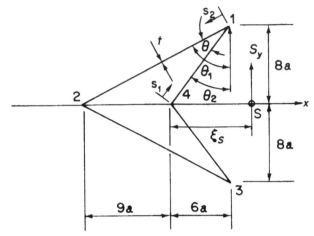

FIGURE 16.18 Closed Section Beam of Example 16.7

in which

$$I_{xx} = 2 \left[\int_0^{10a} t \left(\frac{8}{10} s_1 \right)^2 ds_1 + \int_0^{17a} t \left(\frac{8}{17} s_2 \right)^2 ds_2 \right]$$

Evaluating this expression gives $I_{xx} = 1{,}152a^3 t$.

The basic shear flow distribution q_b is obtained from the first term in Eq. (i). Then, for the wall 41,

$$q_{b,41} = \frac{-S_y}{1{,}152a^3 t} \int_0^{s_1} t \left(\frac{8}{10} s_1 \right) ds_1 = \frac{-S_y}{1{,}152a^3} \left(\frac{2}{5} s_1^2 \right) \tag{ii}$$

In the wall 12,

$$q_{b,12} = \frac{-S_y}{1{,}152a^3} \left[\int_0^{s_2} (17a - s_2) \frac{8}{17} ds_2 + 40a^2 \right] \tag{iii}$$

which gives

$$q_{b,12} = \frac{-S_y}{1{,}152a^3} \left(-\frac{4}{17} s_2^2 + 8as_2 + 40a^2 \right) \tag{iv}$$

The q_b distributions on the walls 23 and 34 follow from symmetry. Hence, from Eq. (16.28),

$$q_{s,0} = \frac{2S_y}{54a \times 1{,}152a^3} \left[\int_0^{10a} \frac{2}{5} s_1^2 ds_1 + \int_0^{17a} \left(-\frac{4}{17} s_2^2 + 8as_2 + 40a^2 \right) ds_2 \right]$$

giving

$$q_{s,0} = \frac{S_y}{1{,}152a^3} (58.7a^2) \tag{v}$$

Taking moments about the point 2, we have

$$S_y(\xi_S + 9a) = 2\int_0^{10a} q_{41} 17a \sin\theta \, ds_1$$

or

$$S_y(\xi_S + 9a) = \frac{S_y 34a \sin\theta}{1{,}152a^3} \int_0^{10a} \left(-\frac{2}{5}s_1^2 + 58.7a^2\right) ds_1 \tag{vi}$$

We may replace $\sin\theta$ by $\sin(\theta_1 - \theta_2) = \sin\theta_1 \cos\theta_2 - \cos\theta_1 \sin\theta_2$, where $\sin\theta_1 = 15/17$, $\cos\theta_2 = 8/10$, $\cos\theta_1 = 8/17$, and $\sin\theta_2 = 6/10$. Substituting these values and integrating Eq. (v) gives

$$\xi_s = -3.35a$$

which means that the shear center is inside the beam section.

■

Reference

[1] Megson THG. Structural and stress analysis. 2nd ed. Oxford: Elsevier; 2005.

PROBLEMS

P.16.1. A beam has the singly symmetrical, thin-walled cross-section shown in Fig. P.16.1. The thickness t of the walls is constant throughout. Show that the distance of the shear center from the web is given by

$$\xi_S = -d\frac{\rho^2 \sin\alpha \cos\alpha}{1 + 6\rho + 2\rho^3 \, \sin^2\alpha}$$

where

$$\rho = d/h$$

P.16.2. A beam has the singly symmetrical, thin-walled cross-section shown in Fig. P.16.2. Each wall of the section is flat and has the same length a and thickness t. Calculate the distance of the shear center from the point 3.

Answer: $5a \cos\alpha/8$

P.16.3. Determine the position of the shear center S for the thin-walled, open cross-section shown in Fig. P.16.3. The thickness t is constant.

Answer: $\pi r/3$

P.16.4. Figure P.16.4 shows the cross-section of a thin, singly symmetrical I section. Show that the distance ξ_S of the shear center from the vertical web is given by

$$\frac{\xi_S}{d} = \frac{3\rho(1-\beta)}{(1+12\rho)}$$

where $\rho = d/h$. The thickness t is taken to be negligibly small in comparison with the other dimensions.

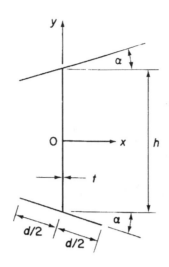

FIGURE P.16.1

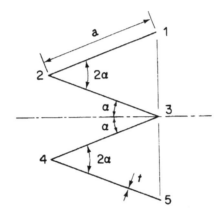

FIGURE P.16.2

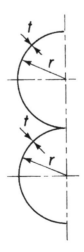

FIGURE P.16.3

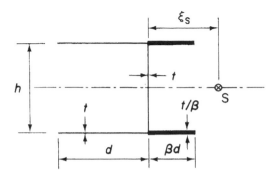

P.16.5. A thin-walled beam has the cross-section shown in Fig. P.16.5. The thickness of each flange varies linearly from t_1 at the tip to t_2 at the junction with the web. The web itself has a constant thickness t_3. Calculate the distance ξ_S from the web to the shear center S.

    Answer: $d^2 (2 t_1 + t_2)/[3d (t_1 + t_2) + ht_3]$

P.16.6. Figure P.16.6 shows the singly symmetrical cross-section of a thin-walled open section beam of constant wall thickness t, which has a narrow longitudinal slit at the corner 15. Calculate and sketch the distribution of shear flow due to a vertical shear force S_y acting through the shear center S and note the principal values. Show also that the distance ξ_S of the shear center from the nose of the section is $\xi_S = l/2(1 + a/b)$.

    Answer: $q_2 = q_4 = 3bS_y/2h(b + a)$, $q_3 = 3S_y/2h$, parabolic distributions

P.16.7. Show that the position of the shear center S with respect to the intersection of the web and lower flange of the thin-walled section shown in Fig. P.16.7, is given by

$$\xi_s = -45a/97, \quad \eta_s = 46a/97$$

P.16.8. Determine the position of the shear centre of the beam section shown in Fig. P.16.8.

    Answer: 13.5 mm to the left of the web 24 and 32.7 mm below the flange 123.

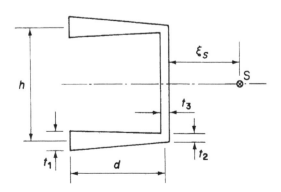

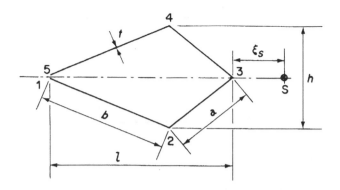

FIGURE P.16.6

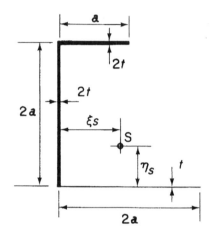

FIGURE P.16.7

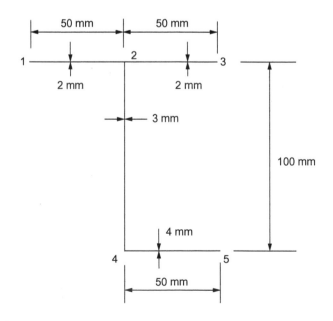

FIGURE P.16.8

P.16.9. Define the term *shear center* of a thin-walled open section and determine the position of the shear center of the thin-walled open section shown in Fig. P.16.9.

Answer: 2.66r from center of semi-circular wall

P.16.10. Determine the position of the shear center of the cold-formed, thin-walled section shown in Fig. P.16.10. The thickness of the section is constant throughout.

Answer: 87.5 mm above center of semi-circular wall

P.16.11. Find the position of the shear center of the thin-walled beam section shown in Fig. P.16.11.

Answer: 1.2r on axis of symmetry to the left of the section

P.16.12. Calculate the position of the shear center of the thin-walled section shown in Fig. P.16.12.

Answer: 20.2 mm to the left of the vertical web on axis of symmetry

P.16.12. MATLAB If the total length of the vertical web of the thin-walled section shown in Fig. P.16.12 is labeled L_{web}, use MATLAB to repeat Problem P.16.12 for values of L_{web} ranging from 40 mm to 80 mm in increments of 5 mm (i.e., 40, 45, 50, ..., 80).

Answer: (i) L_{web} = 40 mm: 16.2 mm to the left of the vertical web on axis of symmetry
(ii) L_{web} = 45 mm: 16.3 mm to the left of the vertical web on axis of symmetry

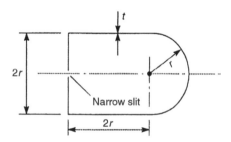

FIGURE P.16.9

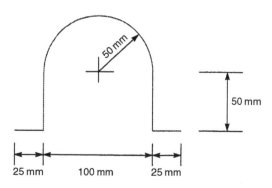

FIGURE P.16.10

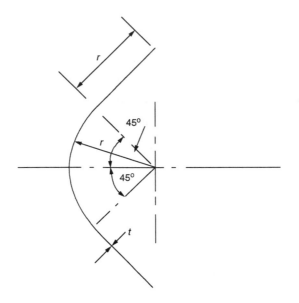

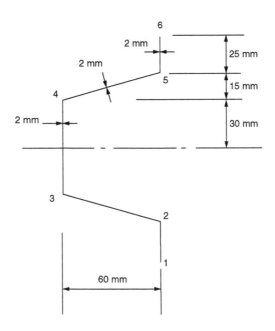

(iii) L_{web} = 50 mm: 18.3 mm to the left of the vertical web on axis of symmetry
(iv) L_{web} = 55 mm: 19.3 mm to the left of the vertical web on axis of symmetry
(v) L_{web} = 60 mm: 20.2 mm to the left of the vertical web on axis of symmetry
(vi) L_{web} = 65 mm: 21.0 mm to the left of the vertical web on axis of symmetry
(vii) L_{web} = 70 mm: 21.8 mm to the left of the vertical web on axis of symmetry
(viii) L_{web} = 75 mm: 22.5 mm to the left of the vertical web on axis of symmetry
(ix) L_{web} = 80 mm: 23.1 mm to the left of the vertical web on axis of symmetry

P.16.13. Determine the horizontal distance from O of the shear center of the beam section shown in Fig. P.16.13. Note that the position of the centroid of area, C, is given.

Answer: 2.0r to the left of O.

P.16.14. Calculate the horizontal distance from the vertical flange 12 of the shear center of the beam section shown in Fig. P.16.14. What would be the horizontal movement of the shear center if the circular arc 34 was removed? Note that the section properties are given.

Answer: 0.36r. 0.36r to the left.

P.16.15. A thin-walled closed section beam of constant wall thickness t has the cross-section shown in Fig. P.16.15. Assuming that the direct stresses are distributed according to the basic theory of bending, calculate and sketch the shear flow distribution for a vertical shear force S_y applied tangentially to the curved part of the beam.

Answer: $q_{01} = S_y(1.61 \cos \theta - 0.80)/r$

$$q_{12} = \frac{S_y}{r^3}\left(0.57s^2 - 1.14rs + 0.33r^2\right)$$

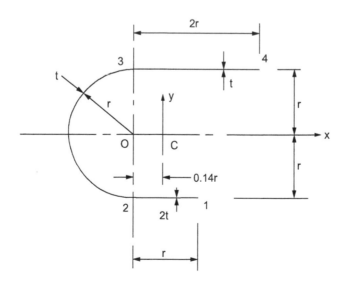

FIGURE P.16.13

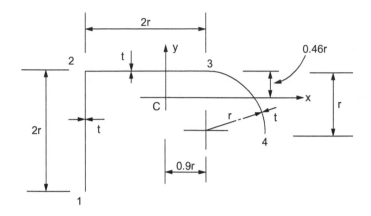

FIGURE P.16.14

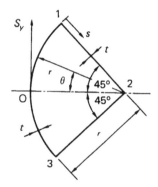

FIGURE P.16.15

P.16.16. A uniform thin-walled beam of constant wall thickness t has a cross-section in the shape of an isosceles triangle and is loaded with a vertical shear force S_y applied at the apex. Assuming that the distribution of shear stress is according to the basic theory of bending, calculate the distribution of shear flow over the cross-section. Illustrate your answer with a suitable sketch, marking in carefully with arrows the direction of the shear flows and noting the principal values.

Answer: $q_{12} = S_y(3s_1^2/d - h - 3d)/h(h + 2d)$

$q_{23} = S_y(-6s_2^2 + 6hs_2 - h^2)/h^2(h + 2d)$

P.16.17. Figure P.16.17 shows the regular hexagonal cross-section of a thin-walled beam of sides a and constant wall thickness t. The beam is subjected to a transverse shear force S, its line of action being along a side of the hexagon, as shown. Plot the shear flow distribution around the section, with values in terms of S and a.

Answer: $q_1 = -0.52S/a$, $q_2 = q_8 = -0.47S/a$, $q_3 = q_7 = -0.17S/a$,

$q_4 = q_6 = 0.13S/a$, $q_5 = 0.18S/a$

Parabolic distributions, q positive clockwise

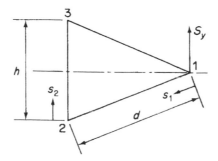

FIGURE P.16.16

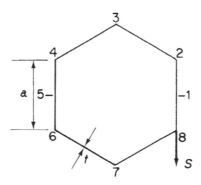

FIGURE P.16.17

P.16.18. A box girder has the singly symmetrical trapezoidal cross-section shown in Fig. P.16.18. It supports a vertical shear load of 500 kN applied through its shear center and in a direction perpendicular to its parallel sides. Calculate the shear flow distribution and the maximum shear stress in the section.

Answer: $q_{OA} = 0.25s_A$
$q_{AB} = 0.21s_B - 2.14 \times 10^{-4}s_B^2 + 250$
$q_{BC} = -0.17s_C + 246$
$\tau_{max} = 30.2\ \text{N/mm}^2$

P.16.19. Calculate the position of the shear center of the beam section shown in Fig. P.16.15.

Answer: $0.61r$ from the corner 2.

P.16.20. Determine the radius of the inscribed circle of the beam section shown in Fig. P.16.16 and show that the shear center of the section coincides with its center.

Answer: $h[(2d - h)(2d + h)]^{1/2}/2$

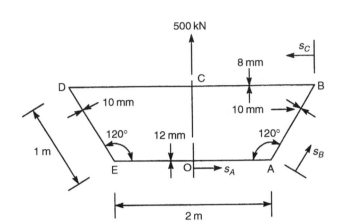

FIGURE P.16.18

Torsion of beams

In Chapter 3, we developed the theory for the torsion of solid sections using both the Prandtl stress function approach and the St. Venant warping function solution. From that point, we looked, via the membrane analogy, at the torsion of a narrow rectangular strip. We use the results of this analysis to investigate the torsion of thin-walled open section beams, but first we shall examine the torsion of thin-walled closed section beams, since the theory for this relies on the general stress, strain, and displacement relationships, which we established in Chapter 16.

17.1 TORSION OF CLOSED SECTION BEAMS

A closed section beam subjected to a pure torque T, as shown in Fig. 17.1, does not, in the absence of an axial constraint, develop a direct stress system. It follows that the equilibrium conditions of Eqs. (17.2) and (17.3) reduce to $\partial q/\partial s = 0$ and $\partial q/\partial z = 0$, respectively. These relationships may be satisfied simultaneously only by a constant value of q. We deduce, therefore, that the application of a pure torque to a closed section beam results in the development of a constant shear flow in the beam wall. However, the shear stress τ may vary around the cross-section, since we allow the wall thickness t to be a function of s. The relationship between the applied torque and this constant shear flow is simply derived by considering the torsional equilibrium of the section shown in Fig. 17.2. The torque produced by the shear flow acting on an element δs of the beam wall is $pq\delta s$. Hence,

$$T = \oint pq \, ds$$

or, since q is constant and $\oint p \, ds = 2A$ (see Section 17.3),

$$T = 2Aq \tag{17.1}$$

Note that the origin O of the axes in Fig. 17.2 may be positioned in or outside the cross-section of the beam, since the moment of the internal shear flows (whose resultant is a pure torque) is the same about any point in their plane. For an origin outside the cross-section, the term $\oint p \, ds$ involves the summation of positive and negative areas. The sign of an area is determined by the sign of p, which itself is associated with the sign convention for torque as follows. If the movement of the foot of p along the tangent at any point in the positive direction of s leads to a counterclockwise rotation of p about the origin of axes, p is positive. The positive direction of s is in the positive direction of q, which is counterclockwise (corresponding to a positive torque). Thus, in Fig. 17.3, a generator OA, rotating about O, initially sweeps out a negative area, since p_A is negative. At B, however, p_B is positive, so that the area swept out by the generator has changed sign (at the point where the tangent passes through O and $p = 0$).

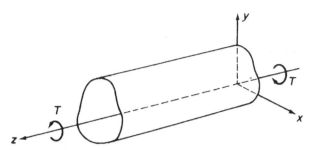

FIGURE 17.1 Torsion of a Closed Section Beam

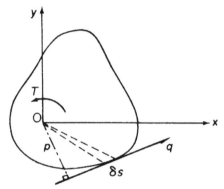

FIGURE 17.2 Determination of the Shear Flow Distribution in a Closed Section Beam Subjected to Torsion

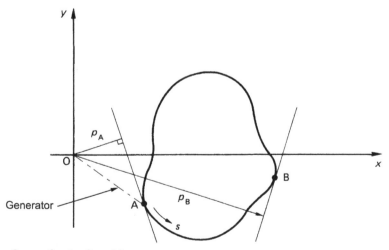

FIGURE 17.3 Sign Convention for Swept Areas

Positive and negative areas cancel each other out, as they overlap, so that, as the generator moves completely around the section, starting and returning to A, say, the resultant area is that enclosed by the profile of the beam.

The theory of the torsion of closed section beams is known as the *Bredt–Batho theory* and Eq. (17.1) is often referred to as the *Bredt–Batho formula*.

17.1.1 Displacements associated with the Bredt–Batho shear flow

The relationship between q and shear strain γ, established in Eq. (17.19) namely,

$$q = Gt \left(\frac{\partial w}{\partial s} + \frac{\partial v_t}{\partial z} \right)$$

is valid for the pure torsion case where q is constant. Differentiating this expression with respect to z, we have

$$\frac{\partial q}{\partial z} = Gt \left(\frac{\partial^2 w}{\partial z \partial s} + \frac{\partial^2 v_t}{\partial z^2} \right) = 0$$

or

$$\frac{\partial}{\partial s} \left(\frac{\partial w}{\partial z} \right) + \frac{\partial^2 v_t}{\partial z^2} = 0 \tag{17.2}$$

In the absence of direct stresses, the longitudinal strain $\partial w / \partial z \ (= \varepsilon_z)$ is zero, so that

$$\frac{\partial^2 v_t}{\partial z^2} = 0$$

Hence, from Eq. (17.7),

$$p \frac{d^2\theta}{dz^2} + \frac{d^2 u}{dz^2} \cos \psi + \frac{d^2 v}{dz^2} \sin \Psi = 0 \tag{17.3}$$

For Eq. (17.3) to hold for all points around the section wall, in other words for all values of Ψ,

$$\frac{d^2\theta}{dz^2} = 0, \quad \frac{d^2 u}{dz^2} = 0, \quad \frac{d^2 v}{dz^2} = 0$$

It follows that $\theta = Az + B$, $u = Cz + D$, $v = Ez + F$, where $A, B, C, D, E,$ and F are unknown constants. Thus, θ, u, and v are all linear functions of z.

Equation (17.22), relating the rate of twist to the variable shear flow q_s developed in a shear loaded closed section beam, is also valid for the case $q_s = q = $ constant. Hence,

$$\frac{d\theta}{dz} = \frac{q}{2A} \oint \frac{ds}{Gt}$$

which becomes, on substituting for q from Eq. (17.1),

$$\frac{d\theta}{dz} = \frac{T}{4A^2} \oint \frac{ds}{Gt} \tag{17.4}$$

The warping distribution produced by a varying shear flow, as defined by Eq. (17.25) for axes having their origin at the center of twist, is also applicable to the case of a constant shear flow. Thus,

$$w_s - w_0 = q \int_0^s \frac{ds}{Gt} - \frac{A_{Os}}{A} q \oint \frac{ds}{Gt}$$

Replacing q from Eq. (17.1), we have

$$w_s - w_0 = \frac{T\delta}{2A} \left(\frac{\delta_{Os}}{\delta} - \frac{A_{Os}}{A} \right) \tag{17.5}$$

where

$$\delta = \oint \frac{ds}{Gt} \quad \text{and} \quad \delta_{Os} = \int_0^s \frac{ds}{Gt}$$

The sign of the warping displacement in Eq. (17.5) is governed by the sign of the applied torque T and the signs of the parameters δ_{Os} and A_{Os}. Having specified initially that a positive torque is counterclockwise, the signs of δ_{Os} and A_{Os} are fixed, in that δ_{Os} is positive when s is positive, that is, s is taken as positive in a counterclockwise sense, and A_{Os} is positive when, as before, p (see Fig. 17.3) is positive.

We noted that the longitudinal strain ε_z is zero in a closed section beam subjected to a pure torque. This means that all sections of the beam must possess identical warping distributions. In other words, longitudinal generators of the beam surface remain unchanged in length although subjected to axial displacement.

Example 17.1

A thin-walled circular section beam has a diameter of 200 mm and is 2 m long; it is firmly restrained against rotation at each end. A concentrated torque of 30 kNm is applied to the beam at its mid-span point. If the maximum shear stress in the beam is limited to 200 N/mm^2 and the maximum angle of twist to 2°, calculate the minimum thickness of the beam walls. Take $G = 25{,}000$ N/mm^2. See Ex. 1.1.

The minimum thickness of the beam corresponding to the maximum allowable shear stress of 200 N/mm^2 is obtained directly using Eq. (17.1), in which $T_{max} = 15$ kNm.

Then,

$$t_{min} = \frac{15 \times 10^6 \times 4}{2 \times \pi \times 200^2 \times 200} = 1.2 \text{ mm}$$

The rate of twist along the beam is given by Eq. (17.4), in which

$$\oint \frac{ds}{t} = \frac{\pi \times 200}{t_{min}}$$

Hence,

$$\frac{d\theta}{dz} = \frac{T}{4A^2G} \times \frac{\pi \times 200}{t_{min}} \tag{i}$$

Taking the origin for z at one of the fixed ends and integrating Eq. (i) for half the length of the beam, we obtain

$$\theta = \frac{T}{4A^2G} \times \frac{200\pi}{t_{min}} z + C_1$$

where C_1 is a constant of integration. At the fixed end, where $z = 0$, $\theta = 0$, so that $C_1 = 0$. Hence,

$$\theta = \frac{T}{4A^2 G} \times \frac{200\pi}{t_{min}} \, z$$

The maximum angle of twist occurs at the mid-span of the beam, where $z = 1$ m. Hence,

$$t_{min} = \frac{15 \times 10^6 \times 200 \times \pi \times 1 \times 10^3 \times 180}{4 \times (\pi \times 200^2 / 4)^2 \times 25,000 \times 2 \times \pi} = 2.7 \text{ mm}$$

The minimum allowable thickness that satisfies both conditions is therefore 2.7 mm.

Example 17.2

Determine the warping distribution in the doubly symmetrical rectangular, closed section beam, shown in Fig. 17.4, when subjected to a counterclockwise torque T. See Ex. 1.1.

From symmetry, the center of twist R coincides with the mid-point of the cross-section and points of zero warping lie on the axes of symmetry at the mid-points of the sides. We therefore take the origin for s at the mid-point of side 14 and measure s in the positive, counterclockwise, sense around the section. Assuming the shear modulus G to be constant, we rewrite Eq. (17.5) in the form

$$w_s - w_0 = \frac{T\delta}{2AG} \left(\frac{\delta_{Os}}{\delta} - \frac{A_{Os}}{A} \right) \tag{i}$$

where

$$\delta = \oint \frac{ds}{t} \quad \text{and} \quad \delta_{Os} = \int_0^s \frac{ds}{t}$$

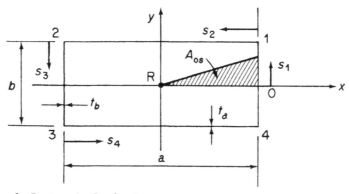

FIGURE 17.4 Torsion of a Rectangular Section Beam

In Eq. (i),

$$w_0 = 0, \quad \delta = 2\left(\frac{b}{t_b} + \frac{a}{t_a}\right), \quad \text{and} \quad A = ab$$

From 0 to 1, $0 \leq s_1 \leq b/2$ and

$$\delta_{Os} = \int_0^s \frac{ds_1}{t_b} = \frac{s_1}{t_b}, \quad A_{Os} = \frac{as_1}{4} \tag{ii}$$

Note that both δ_{Os} and A_{Os} are positive.

Substitution for δ_{Os} and A_{Os} from Eq. (ii) in (i) shows that the warping distribution in the wall 01, w_{01}, is linear. Also,

$$w_1 = \frac{T}{2abG} 2\left(\frac{b}{t_b} + \frac{a}{t_a}\right)\left[\frac{b/2t_b}{2(b/t_b + a/t_a)} - \frac{ab/8}{ab}\right]$$

which gives

$$w_1 = \frac{T}{8abG}\left(\frac{b}{t_b} - \frac{a}{t_a}\right) \tag{iii}$$

The remainder of the warping distribution may be deduced from symmetry and the fact that the warping must be zero at points where the axes of symmetry and the walls of the cross-section intersect. It follows that

$$w_2 = -w_1 = -w_3 = w_4$$

giving the distribution shown in Fig. 17.5. Note that the warping distribution takes the form shown in Fig. 17.5 as long as T is positive and $b/t_b > a/t_a$. If *either* of these conditions is reversed, w_1 and w_3 become negative and w_2 and w_4 positive. In the case when $b/t_b = a/t_a$, the warping is zero at all points in the cross-section.

Suppose now that the origin for s is chosen arbitrarily at, say, point 1. Then, from Fig. 17.6, δ_{Os} in the wall $12 = s_1/t_a$ and $A_{Os} = \frac{1}{2}s_1b/2 = s_1b/4$ and both are positive.

Substituting in Eq. (i) and setting $w_0 = 0$,

$$w'_{12} = \frac{T\delta}{2abG}\left(\frac{s_1}{\delta t_a} - \frac{s_1}{4a}\right) \tag{iv}$$

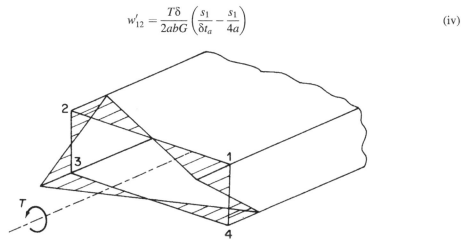

FIGURE 17.5 Warping Distribution in the Rectangular Section Beam of Example 17.2

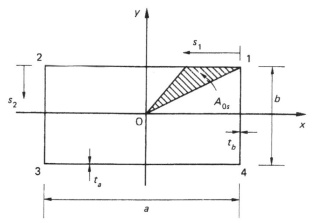

FIGURE 17.6 Arbitrary Origin for s

so that w'_{12} varies linearly from zero at 1 to

$$w'_2 = \frac{T}{2abG} 2 \left(\frac{b}{t_b} + \frac{a}{t_a} \right) \left[\frac{a}{2(b/t_b + a/t_a)t_a} - \frac{1}{4} \right]$$

at 2. Thus,

$$w'_2 = \frac{T}{4abG} \left(\frac{a}{t_a} - \frac{b}{t_b} \right)$$

or

$$w'_2 = -\frac{T}{4abG} \left(\frac{b}{t_b} - \frac{a}{t_a} \right) \tag{v}$$

Similarly,

$$w'_{23} = \frac{T\delta}{2abG} \left[\frac{1}{\delta} \left(\frac{a}{t_a} + \frac{s_2}{t_b} \right) - \frac{1}{4b} (b + s_2) \right] \tag{vi}$$

The warping distribution therefore varies linearly from a value $-T\,(b/t_b - a/t_a)/4abG$ at 2 to zero at 3. The remaining distribution follows from symmetry, so that the complete distribution takes the form shown in Fig. 17.7.

Comparing Figs. 17.5 and 17.7, it can be seen that the form of the warping distribution is the same but that, in the latter case, the complete distribution has been displaced axially. The actual value of the warping at the origin for s is found using Eq. (17.26).

Thus,

$$w_0 = \frac{2}{2(at_a + bt_b)} \left(\int_0^a w'_{12} t_a \; ds_1 + \int_0^b w'_{23} t_b \; ds_2 \right) \tag{vii}$$

Substituting in Eq. (vii) for w'_{12} and w'_{23} from Eqs. (iv) and (vi), respectively, and evaluating gives

$$w_0 = -\frac{T}{8abG} \left(\frac{b}{t_b} - \frac{a}{t_a} \right) \tag{viii}$$

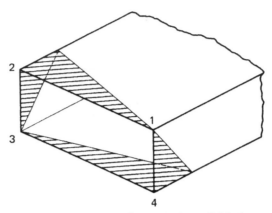

FIGURE 17.7 Warping Distribution Produced by Selecting an Arbitrary Origin for *s*

Subtracting this value from the values of $w_1'(= 0)$ and $w_2'(= -T(b/t_b - a/t_a)/4abG$, we have

$$w_1 = \frac{T}{8abG}\left(\frac{b}{t_b} - \frac{a}{t_a}\right), \quad w_2 = -\frac{T}{8abG}\left(\frac{b}{t_b} - \frac{a}{t_a}\right)$$

as before. Note that setting $w_0 = 0$ in Eq. (i) implies that w_0, the actual value of warping at the origin for s, has been added to all warping displacements. This value must therefore be *subtracted* from the calculated warping displacements (i.e., those based on an arbitrary choice of origin) to obtain true values.

It is instructive at this stage to examine the mechanics of warping to see how it arises. Suppose that each end of the rectangular section beam of Example 17.2 rotates through opposite angles θ giving a total angle of twist 2θ along its length L. The corner 1 at one end of the beam is displaced by amounts $a\theta/2$ vertically and $b\theta/2$ horizontally, as shown in Fig. 17.8. Consider now the displacements of the web and cover of the beam due to rotation. From Figs. 17.8 and 17.9(a) and (b), it can be seen that the angles of rotation of the web and the cover are, respectively,

$$\phi_b = (a\theta/2)/(L/2) = a\theta/L$$

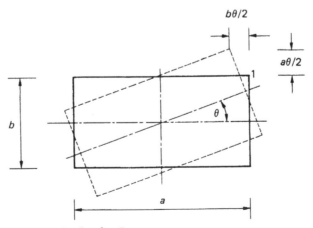

FIGURE 17.8 Twisting of a Rectangular Section Beam

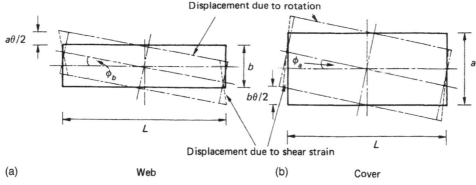

FIGURE 17.9 Displacements Due to Twist and Shear Strain

and

$$\phi_a = (b\theta/2)/(L/2) = b\theta/L$$

The axial displacements of the corner 1 in the web and cover are then

$$\frac{b}{2}\frac{a\theta}{L}, \quad \frac{a}{2}\frac{b\theta}{L}$$

respectively, as shown in Figs. 17.9(a) and (b). In addition to displacements produced by twisting, the webs and covers are subjected to shear strains γ_b and γ_a corresponding to the shear stress system given by Eq. (17.1). Due to γ_b, the axial displacement of corner 1 in the web is $\gamma_b b/2$ in the positive z direction, while in the cover the displacement is $\gamma_a a/2$ in the negative z direction. Note that the shear strains γ_b and γ_a correspond to the shear stress system produced by a positive counterclockwise torque. Clearly, the total axial displacement of the point 1 in the web and cover must be the same, so that

$$-\frac{b}{2}\frac{a\theta}{L} + \gamma_b\frac{b}{2} = \frac{a}{2}\frac{b\theta}{L} - \gamma_a\frac{a}{2}$$

from which

$$\theta = \frac{L}{2ab}(\gamma_a a + \gamma_b b)$$

The shear strains are obtained from Eq. (17.1) and are

$$\gamma_a = \frac{T}{2abGt_a}, \quad \gamma_b = \frac{T}{2abGt_b}$$

from which

$$\theta = \frac{TL}{4a^2b^2G}\left(\frac{a}{t_a} + \frac{b}{t_b}\right)$$

The total angle of twist from end to end of the beam is 2θ, therefore,

$$\frac{2\theta}{L} = \frac{TL}{4a^2b^2G}\left(\frac{2a}{t_a} + \frac{2b}{t_b}\right)$$

or

$$\frac{d\theta}{dz} = \frac{T}{4A^2G} \oint \frac{ds}{t}$$

as in Eq. (17.4).

Substituting for θ in either of the expressions for the axial displacement of the corner 1 gives the warping w_1 at 1. Thus,

$$w_1 = \frac{ab}{2L} \frac{TL}{4a^2b^2G} \left(\frac{a}{t_a} + \frac{b}{t_b}\right) - \frac{T}{2abGt_a} \frac{a}{2}$$

that is,

$$w_1 = \frac{T}{8abG} \left(\frac{b}{t_b} - \frac{a}{t_a}\right)$$

as before. It can be seen that the warping of the cross-section is produced by a combination of the displacements caused by twisting and the displacements due to the shear strains; these shear strains correspond to the shear stresses whose values are fixed by statics. The angle of twist must therefore be such as to ensure compatibility of displacement between the webs and covers.

17.1.2 Condition for zero warping at a section

The geometry of the cross-section of a closed section beam subjected to torsion may be such that no warping of the cross-section occurs. From Eq. (17.5), we see that this condition arises when

$$\frac{\delta_{Os}}{\delta} = \frac{A_{Os}}{A}$$

or

$$\frac{1}{\delta} \int_0^s \frac{ds}{Gt} = \frac{1}{2A} \int_0^s p_R \, ds \tag{17.6}$$

Differentiating Eq. (17.6) with respect to s gives

$$\frac{1}{\delta Gt} = \frac{p_R}{2A}$$

or

$$p_R Gt = \frac{2A}{\delta} = \text{constant} \tag{17.7}$$

A closed section beam for which $p_R Gt = \text{constant}$ does not warp and is known as a *Neuber beam*. For closed section beams having a constant shear modulus, the condition becomes

$$p_R t = \text{constant} \tag{17.8}$$

Examples of such beams are a circular section beam of constant thickness, a rectangular section beam for which $at_b = bt_a$ (see Example 17.2), and a triangular section beam of constant thickness. In the last case, the shear center and hence the center of twist may be shown to coincide with the center of the inscribed circle so that p_R for each side is the radius of the inscribed circle.

17.2 TORSION OF OPEN SECTION BEAMS

An approximate solution for the torsion of a thin-walled open section beam may be found by applying the results obtained in Section 3.4 for the torsion of a thin rectangular strip. If such a strip is bent to form an open section beam, as shown in Fig. 17.10(a), and if the distance s measured around the cross-section is large compared with its thickness t, then the contours of the membrane, that is, lines of shear stress, are still approximately parallel to the inner and outer boundaries. It follows that the shear lines in an element δs of the open section must be nearly the same as those in an element δy of a rectangular strip, as demonstrated in Fig. 17.10(b). Equations (3.27)–(3.29) may therefore be applied to the open beam but with reduced accuracy. Referring to Fig. 17.10(b), we observe that Eq. (3.27) becomes

$$\tau_{zs} = 2Gn\frac{d\theta}{dz}, \quad \tau_{zn} = 0 \tag{17.9}$$

Equation (3.28) becomes

$$\tau_{zs,\,max} = \pm Gt\frac{d\theta}{dz} \tag{17.10}$$

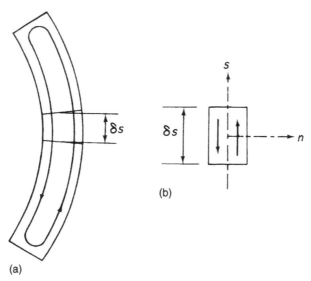

FIGURE 17.10 (a) Shear Lines in a Thin-Walled Open Section Beam Subjected to Torsion; (b) Approximation of Elemental Shear Lines to Those in a Thin Rectangular Strip

and Eq. (3.29) is

$$J = \sum \frac{st^3}{3} \quad \text{or} \quad J = \frac{1}{3} \int_{\text{sect}} t^3 \, ds \qquad (17.11)$$

In Eq. (17.11), the second expression for the torsion constant is used if the cross-section has a variable wall thickness. Finally, the rate of twist is expressed in terms of the applied torque by Eq. (3.12), namely,

$$T = GJ \frac{d\theta}{dz} \qquad (17.12)$$

The shear stress distribution and the maximum shear stress are sometimes more conveniently expressed in terms of the applied torque. Therefore, substituting for $d\theta/dz$ in Eqs. (17.9) and (17.10) gives

$$\tau_{zs} = \frac{2n}{J} T, \quad \tau_{zs,\text{max}} = \pm \frac{tT}{J} \qquad (17.13)$$

We assume, in open beam torsion analysis, that the cross-section is maintained by the system of closely spaced diaphragms described in Section 17.1 and that the beam is of uniform section. Clearly, in this problem, the shear stresses vary across the thickness of the beam wall, whereas other stresses, such as axial constraint stresses which we shall discuss in Chapter 27, are assumed constant across the thickness.

17.2.1 Warping of the cross-section

We saw, in Section 3.4, that a thin rectangular strip suffers warping across its thickness when subjected to torsion. In the same way, a thin-walled open section beam warps across its thickness. This warping, w_t, may be deduced by comparing Fig. 17.10(b) with Fig. 3.10 and using Eq. (3.32), thus,

$$w_t = ns \frac{d\theta}{dz} \qquad (17.14)$$

In addition to warping across the thickness, the cross-section of the beam warps in a manner similar to that of a closed section beam. From Fig. 17.3,

$$\gamma_{zs} = \frac{\partial w}{\partial s} + \frac{\partial v_t}{\partial z} \qquad (17.15)$$

Referring the tangential displacement v_t to the center of twist R of the cross-section, we have, from Eq. (17.8),

$$\frac{\partial v_t}{\partial z} = p_R \frac{d\theta}{dz} \qquad (17.16)$$

Substituting for $\partial v_t/\partial z$ in Eq. (17.15) gives

$$\gamma_{zs} = \frac{\partial w}{\partial s} + p_R \frac{d\theta}{dz}$$

from which

$$\tau_{zs} = G\left(\frac{\partial w}{\partial s} + p_R \frac{d\theta}{dz}\right) \tag{17.17}$$

On the midline of the section wall $\tau_{zs} = 0$ (see Eq. (17.9)), so that, from Eq. (17.17),

$$\frac{\partial w}{\partial s} = -p_R \frac{d\theta}{dz}$$

Integrating this expression with respect to s and taking the lower limit of integration to coincide with the point of zero warping, we obtain

$$w_s = -\frac{d\theta}{dz}\int_0^s p_R \, ds \tag{17.18}$$

From Eqs. (17.14) and (17.18), it can be seen that two types of warping exist in an open section beam. Equation (17.18) gives the warping of the midline of the beam; this is known as *primary warping* and is assumed to be constant across the wall thickness. Equation (17.14) gives the warping of the beam across its wall thickness. This is called *secondary warping*, is very much less than primary warping, and is usually ignored in the thin-walled sections common to aircraft structures.

Equation (17.18) may be rewritten in the form

$$w_s = -2A_R \frac{d\theta}{dz} \tag{17.19}$$

or, in terms of the applied torque,

$$w_s = -2A_R \frac{T}{GJ} \qquad \text{(see Eq. (17.12))} \tag{17.20}$$

in which $A_R = \frac{1}{2}\int_0^s p_R \, ds$ is the area swept out by a generator, rotating about the center of twist, from the point of zero warping, as shown in Fig. 17.11. The sign of w_s, for a given direction of torque, depends

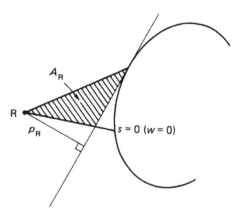

FIGURE 17.11 Warping of an Open Section Beam

upon the sign of A_R, which in turn depends upon the sign of p_R, the perpendicular distance from the center of twist to the tangent at any point. Again, as for closed section beams, the sign of p_R depends upon the assumed direction of a positive torque, in this case counterclockwise. Therefore, p_R (and therefore A_R) is positive if movement of the foot of p_R along the tangent in the assumed direction of s leads to a counterclockwise rotation of p_R about the center of twist. Note that, for open section beams, the positive direction of s may be chosen arbitrarily, since, for a given torque, the sign of the warping displacement depends only on the sign of the swept area A_R.

Example 17.3

Determine the maximum shear stress and the warping distribution in the channel section shown in Fig. 17.12 when it is subjected to a counterclockwise torque of 10 Nm. $G = 25,000$ N/mm². See Ex. 1.1.

From the second of Eqs. (17.13), it can be seen that the maximum shear stress occurs in the web of the section where the thickness is greatest. Also, from the first of Eqs. (17.11),

$$J = \frac{1}{3}(2 \times 25 \times 1.5^3 + 50 \times 2.5^3) = 316.7 \text{ mm}^4$$

so that

$$\tau_{max} = \pm\frac{2.5 \times 10 \times 10^3}{316.7} = \pm 78.9 \text{ N/mm}^2$$

The warping distribution is obtained using Eq. (17.20), in which the origin for s (and hence A_R) is taken at the intersection of the web and the axis of symmetry where the warping is zero. Further, the center of twist R of the section coincides with its shear center S, whose position is found using the method described in Section 17.2.1, this gives $\xi_S = 8.04$ mm. In the wall O2,

$$A_R = \frac{1}{2} \times 8.04s_1 \quad (p_R \text{ is positive})$$

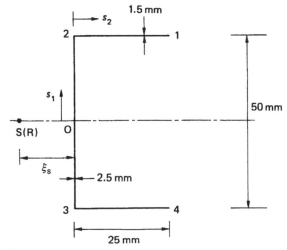

FIGURE 17.12 Channel Section of Example 17.3

so that

$$w_{O2} = -2 \times \frac{1}{2} \times 8.04s_1 \times \frac{10 \times 10^3}{25{,}000 \times 316.7} = -0.01s_1 \tag{i}$$

that is, the warping distribution is linear in O2 and

$$w_2 = -0.01 \times 25 = -0.25 \text{ mm}$$

In the wall 21,

$$A_R = \frac{1}{2} \times 8.04 \times 25 - \frac{1}{2} \times 25s_2$$

in which the area swept out by the generator in the wall 21 provides a negative contribution to the total swept area A_R. Thus

$$w_{21} = -25(8.04 - s_2)\frac{10 \times 10^3}{25{,}000 \times 316.7}$$

or

$$w_{21} = -0.03(8.04 - s_2) \tag{ii}$$

Again, the warping distribution is linear and varies from –0.25 mm at 2 to + 0.54 mm at 1. Examination of Eq. (ii) shows that w_{21} changes sign at $s_2 = 8.04$ mm. The remaining warping distribution follows from symmetry and the complete distribution is shown in Fig. 17.13. In unsymmetrical section beams, the position of the point of zero warping is not known but may be found using the method described in Section 27.2 for the restrained warping of an open section beam. From the derivation of Eq. (27.3), we see that

$$2A_R' = \frac{\int_{\text{sect}} 2A_{R,O}t \, ds}{\int_{\text{sect}} t \, ds} \tag{iii}$$

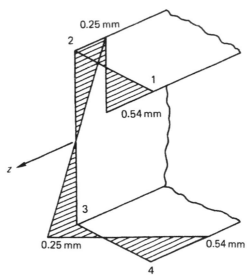

FIGURE 17.13 Warping Distribution in Channel Section of Example 17.3

in which $A_{R,O}$ is the area swept out by a generator rotating about the center of twist from some convenient origin and A'_R is the value of $A_{R,O}$ at the point of zero warping. As an illustration, we apply the method to the beam section of Example 17.3.

Suppose that the position of the center of twist (i.e., the shear center) has already been calculated and suppose also that we choose the origin for s to be at the point 1. Then, in Fig. 17.14,

$$\int_{sect} t \, ds = 2 \times 1.5 \times 25 + 2.5 \times 50 = 200 \text{ mm}^2$$

In the wall 12,

$$A_{12} = \frac{1}{2} \times 25 s_1 \quad (A_{R,O} \text{ for the wall } 12) \tag{iv}$$

from which

$$A_2 = \frac{1}{2} \times 25 \times 25 = 312.5 \text{ mm}^2$$

Also,

$$A_{23} = 312.5 - \frac{1}{2} \times 8.04 s_2 \tag{v}$$

and

$$A_3 = 312.5 - \frac{1}{2} \times 8.04 \times 50 = 111.5 \text{ mm}^2$$

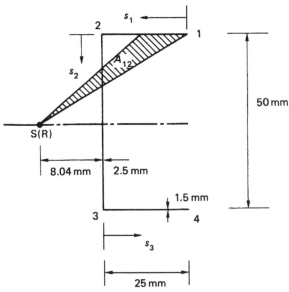

FIGURE 17.14 Determination of Points of Zero Warping

Finally,

$$A_{34} = 111.5 + \frac{1}{2} \times 25s_3 \tag{vi}$$

Substituting for A_{12}, A_{23}, and A_{34} from Eqs. (iv)–(vi) in Eq. (iii), we have

$$2A'_R = \frac{1}{200} \left[\int_0^{25} 25 \times 1.15s_1 \, ds_1 + \int_0^{50} 2(312.5 - 4.02s_2)2.5 \, ds_2 + \int_0^{25} 2(111.5 + 12.5s_3)1.5 \, ds_3 \right] \tag{vii}$$

Evaluation of Eq. (vii) gives

$$2A'_R = 424 \text{ mm}^2$$

We now examine each wall of the section in turn to determine points of zero warping. Suppose that, in the wall 12, a point of zero warping occurs at a value of s_1 equal to $s_{1,0}$. Then,

$$2 \times \frac{1}{2} \times 25s_{1,0} = 424$$

from which

$$s_{1,0} = 16.96 \text{ mm}$$

so that a point of zero warping occurs in the wall 12 at a distance of 8.04 mm from the point 2 as before. In the web 23, let the point of zero warping occur at $s_2 = s_{2,0}$. Then,

$$2 \times \frac{1}{2} \times 25 \times 25 - 2 \times \frac{1}{2} \times 8.04s_{2,0} = 424$$

which gives $s_{2,0} = 25$ mm (i.e., on the axis of symmetry). Clearly, from symmetry, a further point of zero warping occurs in the flange 34 at a distance of 8.04 mm from the point 3. The warping distribution is then obtained directly using Eq. (17.20), in which

$$A_R = A_{R,0} - A'_R$$

PROBLEMS

P.17.1 A uniform, thin-walled, cantilever beam of closed rectangular cross-section has the dimensions shown in Fig. P.17.1. The shear modulus G of the top and bottom covers of the beam is 18,000 N/mm^2 while that of the vertical webs is 26,000 N/mm^2. The beam is subjected to a uniformly distributed torque of 20 N m/mm along its length. Calculate the maximum shear stress according to the Bred–Batho theory of torsion. Calculate also, and sketch, the distribution of twist along the length of the cantilever, assuming that axial constraint effects are negligible.

Answer: $\tau_{max} = 83.3$ N/mm^2, $\quad \theta = 8.14 \times 10^{-9} \left(2,500z - \frac{z^2}{2} \right)$ rad

P.17.2 A single-cell, thin-walled beam with the double trapezoidal cross-section shown in Fig. P.17.2, is subjected to a constant torque $T = 90,500$ N m and is constrained to twist about an axis through the point R. Assuming that the shear stresses are distributed according to the Bredt–Batho theory of torsion,

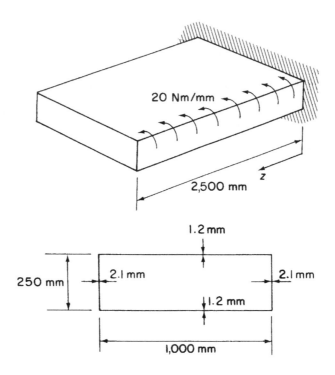

FIGURE P.17.1

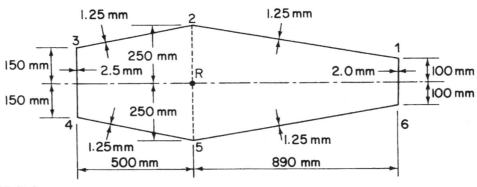

FIGURE P.17.2

calculate the distribution of warping around the cross-section. Illustrate your answer clearly by means of a sketch and insert the principal values of the warping displacements. The shear modulus $G = 27,500$ N/mm² and is constant throughout.

Answer: $w_1 = -w_6 = -0.53$ mm, $w_2 = -w_5 = 0.05$ mm, $w_3 = -w_4 = 0.38$ mm

Linear distribution.

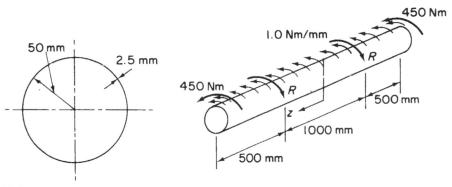

FIGURE P.17.3

P.17.3 A uniform thin-walled beam is circular in cross-section and has a constant thickness of 2.5 mm. The beam is 2,000 mm long, carrying end torques of 450 N m and, in the same sense, a distributed torque loading of 1.0 N m/mm. The loads are reacted by equal couples R at sections 500 mm distant from each end (Fig. P.17.3). Calculate the maximum shear stress in the beam and sketch the distribution of twist along its length. Take $G = 30,000$ N/mm^2 and neglect axial constraint effects.

Answer: $\tau_{max} = 24.2$ N/mm^2, $\theta = -0.85 \times 10^{-8} z^2$ rad, $0 \leq z \leq 500$ mm,
$\theta = 1.7 \times 10^{-8}(1,450z - z^2/2) - 12.33 \times 10^{-3}$ rad, $500 \leq z \leq 1,000$ mm.

P.17.4 The thin-walled box section beam ABCD shown in Fig. P.17.4 is attached at each end to supports that allow rotation of the ends of the beam in the longitudinal vertical plane of symmetry but prevent rotation of the ends in vertical planes perpendicular to the longitudinal axis of the beam. The beam is subjected to a uniform torque loading of 20 N m/mm over the portion BC of its span. Calculate the maximum shear stress in the cross-section of the beam and the distribution of angle of twist along its length, $G = 70,000$ N/mm^2.

Answer: 71.4 N/mm^2, $\theta_B = \theta_C = 0.36°$, θ at mid-span $= 0.72°$

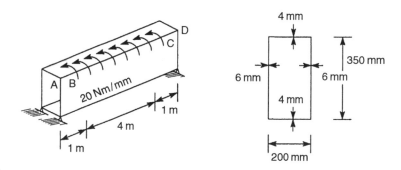

FIGURE P.17.4

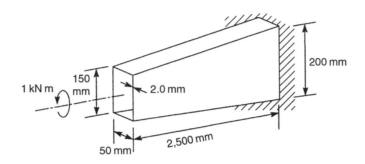

P.17.5 Figure P.17.5 shows a thin-walled cantilever box beam having a constant width of 50 mm and a depth which decreases linearly from 200 mm at the built-in end to 150 mm at the free end. If the beam is subjected to a torque of 1 kN m at its free end, plot the angle of twist of the beam at 500 mm intervals along its length and determine the maximum shear stress in the beam section. Take $G = 25,000$ N/mm^2.

Answer: $\tau_{max} = 33.3$ N/mm^2

P.17.6 A uniform closed section beam, of the thin-walled section shown in Fig. P.17.6, is subjected to a twisting couple of 4,500 N m. The beam is constrained to twist about a longitudinal axis through the center C of the semi-circular arc 12. For the curved wall 12, the thickness is 2 mm and the shear modulus is 22,000 N/mm^2. For the plane walls 23, 34, and 41, the corresponding figures are 1.6 mm and 27,500 N/mm^2. (Note: $Gt = $ constant.) Calculate the rate of twist in radians per millimetre. Give a sketch illustrating the distribution of warping displacement in the cross-section and quote values at points 1 and 4.

Answer: $d\theta/dz = 29.3 \times 10^{-6}$ rad/mm,

$w_3 = -w_4 = -0.19$ mm, $\quad w_2 = -w_1 = -0.056$ mm

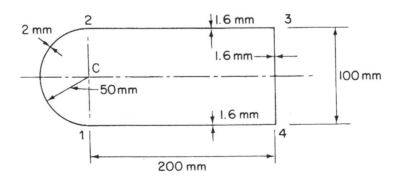

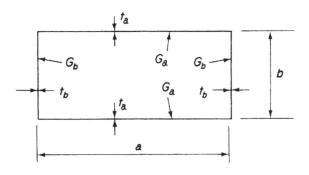

FIGURE P.17.7

P.17.7 A uniform beam with the doubly symmetrical cross-section shown in Fig. P.17.7, has horizontal and vertical walls made of different materials, which have shear moduli G_a and G_b, respectively. If, for any material, the ratio mass density/shear modulus is constant, find the ratio of the wall thicknesses t_a and t_b, so that, for a given torsional stiffness and given dimensions a,b, the beam has minimum weight per unit span. Assume the Bredt–Batho theory of torsion is valid. If this thickness requirement is satisfied find the a/b ratio (previously regarded as fixed), which gives minimum weight for given torsional stiffness.

Answer: $t_b/t_a = G_a/G_b, \quad b/a = 1$

P.17.8 The cold-formed section shown in Fig. P.17.8 is subjected to a torque of 50 Nm. Calculate the maximum shear stress in the section and its rate of twist. $G = 25,000 \text{ N/mm}^2$.

Answer: $\tau_{\max} = 220.6 \text{ N/mm}^2, \quad d\theta/dz = 0.0044 \text{ rad/mm}$

P.17.9 Determine the rate of twist per unit torque of the beam section shown in Fig. P.17.11 if the shear modulus G is 25,000 N/mm^2. (Note that the shear center position is calculated in P.17.11.)

Answer: $6.42 \times 10^{-8} \text{ rad/mm}$

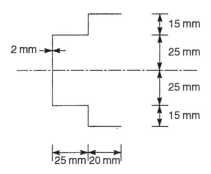

FIGURE P.17.8

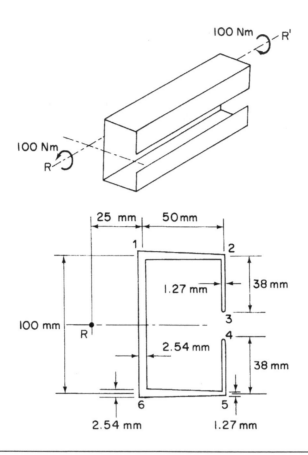

P.17.10 Figure P.17.10 shows the cross-section of a thin-walled beam in the form of a channel with lipped flanges. The lips are of constant thickness 1.27 mm, while the flanges increase linearly in thickness from 1.27 mm where they meet the lips to 2.54 mm at their junctions with the web. The web has a constant thickness of 2.54 mm. The shear modulus G is 26,700 N/mm^2 throughout. The beam has an enforced axis of twist RR$'$ and is supported in such a way that warping occurs freely but is zero at the mid-point of the web. If the beam carries a torque of 100 Nm, calculate the maximum shear stress according to the St. Venant theory of torsion for thin-walled sections. Ignore any effects of stress concentration at the corners. Find also the distribution of warping along the middle line of the section, illustrating your results by means of a sketch.

Answer: $\tau_{max} = \pm 297.4$ N/mm^2, $w_1 = -5.48$ mm $= -w_6$
$w_2 = 5.48$ mm $= -w_5$, $w_3 = 17.98$ mm $= -w_4$

P.17.11 The thin-walled section shown in Fig. P.17.11 is symmetrical about the x axis. The thickness t_0 of the center web 34 is constant, while the thickness of the other walls varies linearly from t_0 at points 3 and 4 to zero at the open ends 1, 6, 7, and 8. Determine the St. Venant torsion constant J for the section and also the maximum value of the shear stress due to a torque T. If the section is constrained to twist about an axis through the origin O, plot the relative warping displacements of the section per unit rate of twist.

Answer: $J = 4at_0^3/3$, $\tau_{max} = \pm 3T/4at_0^2$,
$$w_1 = +a^2(1 + 2\sqrt{2}),\quad w_2 = +\sqrt{2}a^2,\quad w_7 = -a^2,\quad w_3 = 0$$

P.17.11 MATLAB If the length of the vertical web of the section shown in Fig. P.17.11 is labeled L_{web}, use MATLAB to repeat the calculation of J and τ_{max} in Problem P.17.11 for values of L_{web} ranging from a to $4a$ in increments of $a/2$ (i.e., a, $3a/2$, $2a$, ... , $4a$).

Answer: (i) $L_{web} = a$: $J = at_0^3/3$, $\tau_{max} = \pm T/at_0^2$
(ii) $L_{web} = 3a/2$: $J = 7at_0^3/6$, $\tau_{max} = \pm 6T/7at_0^2$
(iii) $L_{web} = 2a$: $J = 4at_0^3/3$, $\tau_{max} = \pm 3T/4at_0^2$
(iv) $L_{web} = 5a/2$: $J = 3at_0^3/2$, $\tau_{max} = \pm 2T/3at_0^2$
(v) $L_{web} = 3a$: $J = 5at_0^3/3$, $\tau_{max} = \pm 3T/5at_0^2$
(vi) $L_{web} = 7a/2$: $J = 11at_0^3/6$, $\tau_{max} = \pm 6T/11at_0^2$
(vii) $L_{web} = 4a$: $J = 2at_0^3$, $\tau_{max} = \pm T/2at_0^2$

P.17.12 The thin-walled section shown in Fig. P.17.12 is constrained to twist about an axis through R, the center of the semi-circular wall 34. Calculate the maximum shear stress in the section per unit torque and the warping distribution per unit rate of twist. Also compare the value of warping displacement at the point 1 with that corresponding to the section being constrained to twist about an axis through the

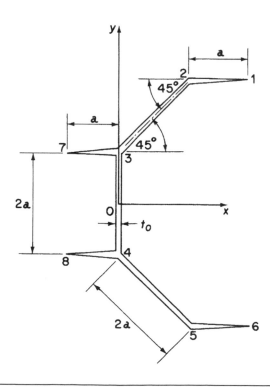

FIGURE P.17.11

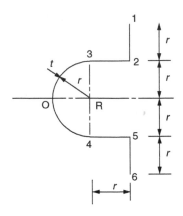

point O and state what effect this movement has on the maximum shear stress and the torsional stiffness of the section.

Answer: Maximum shear stress is $\pm 0.42/rt^2$ per unit torque

$$w_{03} = +r^2\theta, \quad w_{32} = +\frac{r}{2}(\pi r + 2s_1), \quad w_{21} = -\frac{r}{2}(2s_2 - 5.142r)$$

With center of twist at O, $w_1 = -0.43r^2$. Maximum shear stress is unchanged, torsional stiffness increased since warping reduced.

P.17.13 Determine the maximum shear stress in the beam section shown in Fig. P.17.13, stating clearly the point at which it occurs. Determine also the rate of twist of the beam section if the shear modulus G is 25,000 N/mm².

Answer: 70.2 N/mm² on the underside of 24 at 2 or on the upper surface of 32 at 2. 9.0×10^{-4} rad/mm

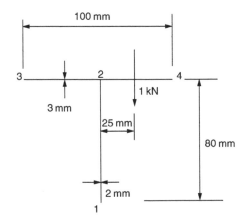

FIGURE P.17.13

Combined open and closed section beams

So far, in Chapters 15–17, we analyzed thin-walled beams that consist of either completely closed cross-sections or completely open cross-sections. Frequently, aircraft components comprise combinations of open and closed section beams. For example, the section of a wing in the region of an undercarriage bay could take the form shown in Fig. 18.1. Clearly, part of the section is an open channel section, while the nose portion is a single-cell closed section. We now examine the methods of analysis of such sections when subjected to bending, shear, and torsional loads.

18.1 BENDING

It is immaterial what form the cross-section of a beam takes; the direct stresses due to bending are given by either Eq. (16.17) or (16.18).

Example 18.1

Determine the direct stress distribution in the wing section shown in Fig. 18.2 when it is subjected to a bending moment of 5,000 Nm applied in a vertical plane.

The beam section has the x axis as an axis of symmetry so that $I_{xy} = 0$ and, also, since $M_y = 0$ Eq. (16.18) reduces to

$$\sigma_z = (M_x/I_{xx})y \tag{i}$$

From Section 16.4.5 the second moment of area of the semicircular nose portion is

$$I_{xx}(\text{nose portion}) = \pi \times 50^3 \times 2/2 = 0.4 \times 10^6 \text{ mm}^4$$

For the rectangular box

$$I_{xx}(\text{rect.box}) = \left(2 \times 4 \times 100^3\right)/12 + 2 \times 200 \times 2 \times 50^2 = 2.7 \times 10^6 \text{ mm}^4$$

For the arcs 23 and 56

$$I_{xx}(\text{arcs } 23, 56) = 2 \int_{\pi/4}^{\pi/2} 2(50 \sin \theta)^2 \, 50 \, d\theta$$

which gives

$$I_{xx}(\text{arcs}) = 0.3 \times 10^6 \text{ mm}^4$$

Therefore for the complete beam section

$$I_{xx} = 3.4 \times 10^6 \text{ mm}^4$$

Introduction to Aircraft Structural Analysis, Second Edition
Copyright © 2014, 2007 T.H.G. Megson. Published by Elsevier Ltd. All rights reserved.

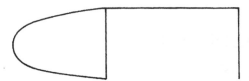

FIGURE 18.1 Wing Section Comprising Open and Closed Components

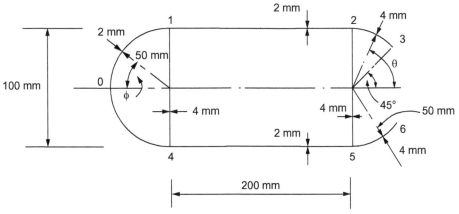

FIGURE 18.2 Beam Section of Example 18.1

and Eq. (i) becomes

$$\sigma_z = \left(5{,}000 \times 10^3/3.4 \times 10^6\right)y = 1.47y \tag{ii}$$

In 01

$$\sigma_{z,01} = 1.47 \times 50 \sin \phi = 73.5 \sin\phi \tag{iii}$$

so that when $\phi = 0$, $\sigma_{z,01} = 0$ and when $\phi = 90°$, $\sigma_{z,1} = +73.5$ N/mm^2.

In 71

$$\sigma_{z,71} = 1.47y \tag{iv}$$

When $y = 0$, $\sigma_{z,7} = 0$ and $y = 50$ mm, $\sigma_{z,1} = +73.5$ N/mm^2.

By inspection, $\sigma_{z,12} = +73.5$ N/mm^2.

In 32

$$\sigma_{z,32} = 1.47y = 1.47 \times 50 \sin \theta = 73.5 \sin\theta \tag{v}$$

When $\theta = \pi/4$,

$$\sigma_{z,3} = +52.0 \text{ N/mm}^2.$$

The direct distribution in the lower half of the wing section follows from symmetry; the complete distribution is shown in Fig. 18.3.

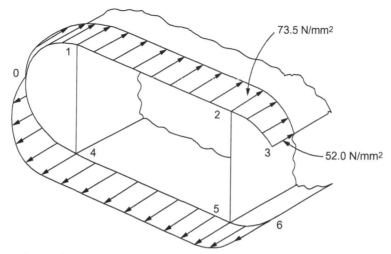

73.5 N/mm²

52.0 N/mm²

FIGURE 18.3 Direct Stress Distribution in Beam Section of Example 18.1

18.2 SHEAR

The methods described in Sections 17.2 and 17.3 are used to determine the shear stress distribution, although, unlike the completely closed section case, shear loads must be applied through the shear center of the combined section; otherwise, shear stresses of the type described in Section 18.2 due to torsion arise. Where shear loads do not act through the shear center, its position must be found and the loading system replaced by shear loads acting through the shear center together with a torque; the two loading cases are then analyzed separately. Again, we assume that the cross-section of the beam remains undistorted by the loading.

Example 18.2

Determine the shear flow distribution in the beam section shown in Fig. 18.2 when it is subjected to a shear load in its vertical plane of symmetry. The thickness of the walls of the section is 2 mm throughout.

The centroid of area C lies on the axis of symmetry at some distance $\bar{y}$ from the upper surface of the beam section. Taking moments of area about this upper surface,

$$(4 \times 100 \times 2 + 4 \times 200 \times 2)\bar{y} = 2 \times 100 \times 2 \times 50 + 2 \times 200 \times 2 \times 100 + 200 \times 2 \times 200$$

which gives $\bar{y} = 75$ mm.

The second moment of area of the section about Cx is given by

$$I_{xx} = 2\left(\frac{2 \times 100^3}{12} + 2 \times 100 \times 25^2\right) + 400 \times 2 \times 75^2 + 200 \times 2 \times 125^2$$

$$+ 2\left(\frac{2 \times 200^3}{12} + 2 \times 200 \times 25^2\right)$$

that is,

$$I_{xx} = 14.5 \times 10^6 \text{ mm}^4$$

The section is symmetrical about Cy so that $I_{xy} = 0$ and, since $S_x = 0$, the shear flow distribution in the closed section 3456 is, from Eq. (17.15),

$$q_s = -\frac{S_y}{I_{xx}} \int_0^s ty \ ds + q_{s,0} \tag{i}$$

Also, the shear load is applied through the shear center of the complete section, that is, along the axis of symmetry, so that, in the open portions 123 and 678, the shear flow distribution is, from Eq. (17.14),

$$q_s = \frac{S_y}{I_{xx}} \int_0^s ty \ ds \tag{ii}$$

We note that the shear flow is zero at the points 1 and 8 and therefore the analysis may conveniently, though not necessarily, begin at either of these points. Thus, referring to Fig. 18.4,

$$q_{12} = -\frac{100 \times 10^3}{14.5 \times 10^6} \int_0^{s_1} 2(-25 + s_1) \ ds_1$$

that is,

$$q_{12} = -69.0 \times 10^{-4}\left(-50s_1 + s_1^2\right) \tag{iii}$$

from which $q_2 = -34.5$ N/mm.

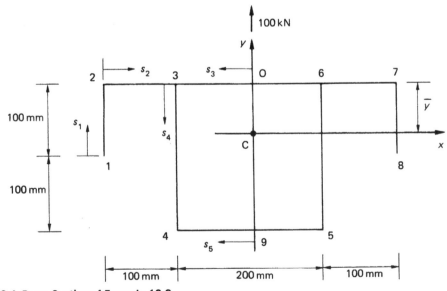

FIGURE 18.4 Beam Section of Example 18.2

Examination of Eq. (iii) shows that q_{12} is initially positive and changes sign when $s_1 = 50$ mm. Further, q_{12} has a turning value $(dq_{12}/ds_1 = 0)$ at $s_1 = 25$ mm of 4.3 N/mm. In the wall 23,

$$q_{23} = -69.0 \times 10^{-4} \int_0^{s_2} 2 \times 75 \ ds_2 - 34.5$$

that is,

$$q_{23} = -1.04 s_2 - 34.5 \tag{iv}$$

Hence, q_{23} varies linearly from a value of -34.5 N/mm at 2 to -138.5 N/mm at 3 in the wall 23.

The analysis of the open part of the beam section is now complete, since the shear flow distribution in the walls 67 and 78 follows from symmetry. To determine the shear flow distribution in the closed part of the section, we must use the method described in Section 17.3, in which the line of action of the shear load is known. Thus, we "cut" the closed part of the section at some convenient point, obtain the q_b or "open section" shear flows for the complete section then take moments as in Eq. (17.16) or (17.17). However, in this case, we may use the symmetry of the section and loading to deduce that the final value of shear flow must be zero at the midpoints of the walls 36 and 45, that is, $q_s = q_{s,0} = 0$ at these points. Hence,

$$q_{03} = -69.0 \times 10^{-4} \int_0^{s_3} 2 \times 75 \ ds_3$$

so that

$$q_{03} = -1.04 s_3 \tag{v}$$

and $q_3 = -104$ N/mm in the wall 03. It follows that, for equilibrium of shear flows at 3, q_3, in the wall 34, must be equal to $-138.5 - 104 = -242.5$ N/mm. Hence,

$$q_{34} = -69.0 \times 10^{-4} \int_0^{s_4} 2(75 - s_4) ds_4 - 242.5$$

which gives

$$q_{34} = -1.04 s_4 + 69.0 \times 10^{-4} s_4^2 - 242.5 \tag{vi}$$

Examination of Eq. (vi) shows that q_{34} has a maximum value of -281.7 N/mm at $s_4 = 75$ mm; also, $q_4 = -172.5$ N/mm. Finally, the distribution of shear flow in the wall 94 is given by

$$q_{94} = -69.0 \times 10^{-4} \int_0^{s_5} 2(-125) \ ds_5$$

that is,

$$q_{94} = 1.73 s_5 \tag{vii}$$

The complete distribution is shown in Fig. 18.5.

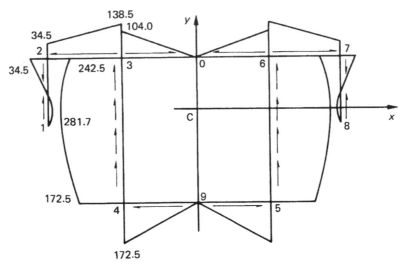

FIGURE 18.5 Shear Flow Distribution in Beam of Example 18.2 (All Shear Flows in N/mm)

18.3 TORSION

Generally, in the torsion of composite sections, the closed portion is dominant, since its torsional stiffness is far greater than that of the attached open section portion, which frequently may be ignored in the calculation of torsional stiffness; shear stresses should, however, be checked in this part of the section.

Example 18.3

Find the angle of twist per unit length in the wing whose cross-section is shown in Fig. 18.6 when it is subjected to a torque of 10 kN m. Find also the maximum shear stress in the section. $G = 25,000$ N/mm^2.

Wall 12 (outer) = 900 mm; nose cell area = 20,000 mm^2

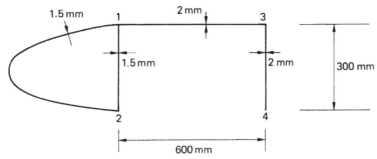

FIGURE 18.6 Wing Section of Example 18.3

It may be assumed, in a simplified approach, that the torsional rigidity GJ of the complete section is the sum of the torsional rigidities of the open and closed portions. For the closed portion, the torsional rigidity is, from Eq. (18.4),

$$(GJ)_{cl} = \frac{4A^2 G}{\oint ds/t} = \frac{4 \times 20{,}000^2 \times 25{,}000}{(900 + 300)/1.5}$$

which gives

$$(GJ)_{cl} = 5{,}000 \times 10^7 \text{ N mm}^2$$

The torsional rigidity of the open portion is found using Eq. (18.11), thus,

$$(GJ)_{op} = G \sum \frac{st^3}{3} = \frac{25{,}000 \times 900 \times 2^3}{3}$$

that is,

$$(GJ)_{op} = 6 \times 10^7 \text{ N mm}^2$$

The torsional rigidity of the complete section is then

$$GJ = 5{,}000 \times 10^7 + 6 \times 10^7 = 5{,}006 \times 10^7 \text{ N mm}^2$$

In all unrestrained torsion problems, the torque is related to the rate of twist by the expression

$$T = GJ \frac{d\theta}{dz}$$

The angle of twist per unit length is therefore given by

$$\frac{d\theta}{dz} = \frac{T}{GJ} = \frac{10 \times 10^6}{5{,}006 \times 10^7} = 0.0002 \text{ rad/mm}$$

Substituting for T in Eq. (18.1) from Eq. (18.4), we obtain the shear flow in the closed section. Thus,

$$q_{cl} = \frac{(GJ)_{cl}}{2A} \frac{d\theta}{dz} = \frac{5{,}000 \times 10^7}{2 \times 20{,}000} \times 0.0002$$

from which

$$q_{cl} = 250 \text{ N/mm}$$

The maximum shear stress in the closed section is then $250/1.5 = 166.7 \text{ N/mm}^2$.

In the open portion of the section, the maximum shear stress is obtained directly from Eq. (18.10) and is

$$\tau_{max,op} = 25{,}000 \times 2 \times 0.0002 = 10 \text{ N/mm}^2$$

It can be seen from this that, in terms of strength and stiffness, the closed portion of the wing section dominates. This dominance may be used to determine the warping distribution. Having first found the position of the center of twist (the shear center), the warping of the closed portion is calculated using the method described in Section 18.1. The warping in the walls 13 and 34 is then determined using Eq. (18.19), in which the origin of the swept area A_R is taken at the point 1 and the value of warping is that previously calculated for the closed portion at 1.

Example 18.4

The wing section shown in Fig. 18.7 is subjected to a vertical shear load of 50 kN in the plane of the web 43. Determine the rate of twist of the section and the maximum shear stress. Take $G=25{,}000 \text{ N/mm}^2$.

The 50 kN load acting in the plane of the web 43 must be replaced by a 50 kN load acting through the shear center of the section together with a torque. Initially, therefore, we must find the position of the shear center which will, however, lie on the horizontal axis of symmetry.

From Section 16.4.5

$$I_{xx} = (\pi \times 50^3 \times 2)/2 + (4 \times 100^3/12) + 2\left[2 \times 100 \times 50^2 + (2 \times 25^3/12) + 2 \times 25 \times 37.5^2\right]$$

i.e.,

$$I_{xx} = 1.87 \times 10^6 \text{ mm}^4$$

To find the position of the shear center we apply an arbitrary shear load, S_y, through the shear center S as shown in Fig. 18.7. Then, since $S_x=0$ and $I_{xy}=0$, Eq. (17.15) reduces to

$$q_s = -(S_y/I_{xx}) \int_0^s ty \, ds + q_{s,0} \tag{i}$$

For the open part of the wing section $q_{s,0}=0$ at the open edges 1 and 6 while for the closed part of the section $q_{s,0}$ may be found using Eq. (17.27) since the rate of twist will be zero for a shear load applied through the shear center. Then

$$q_{12} = -(S_y/I_{xx}) \int_0^{s_1} ty \, ds_1$$

i.e.,

$$q_{12} = -(S_y/I_{xx}) \int_0^{s_1} 2(25 + s_1) \, ds_1$$

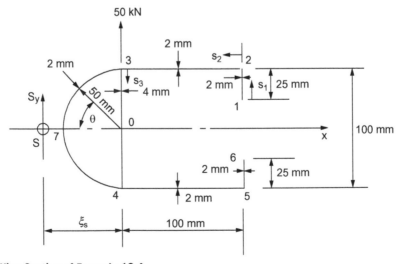

FIGURE 18.7 Wing Section of Example 18.4

which gives

$$q_{12} = -(S_y/I_{xx})(50s_1 + s_1{}^2) \tag{ii}$$

At the point 2, when $s_1 = 25$ mm, $q_2 = -1{,}875(S_y/I_{xx})$.

In the wall 23

$$q_{23} = -(S_y/I_{xx})\Big(\int_0^{s_2} 2 \times 50 \, ds_2 + 1{,}875\Big)$$

so that

$$q_{23} = -(S_y/I_{xx})(100s_2 + 1{,}875) \tag{iii}$$

When $s_2 = 100$ mm, $q_3 = -11{,}875(S_y/I_{xx})$.

Now we "cut" the closed part of the section at 7. Then

$$q_{b,73} = -(S_y/I_{xx})\int_0^{\theta} 2 \times 50 \, \sin\theta \times 50 \, d\theta$$

which gives

$$q_{b,73} = 5{,}000(S_y/I_{xx})(\cos\theta - 1) \tag{iv}$$

When $\theta = \pi/2$ at 3, $q_{b,3} = -5{,}000(S_y/I_{xx})$.

In the wall 34

$$q_{b,34} = -(S_y/I_{xx})\Big[\int_0^{s_3} 4(50 - s_3)ds_3 + 5{,}000 + 11{,}875\Big]$$

which gives

$$q_{b,34} = (S_y/I_{xx})(2s_3{}^2 - 200s_3 - 16{,}875) \tag{v}$$

From Eq. (17.27)

$$q_{s,0} = -[\oint(q_b/t)ds]/(\oint(ds/t) \tag{vi}$$

where $\oint ds/t = (\pi \times 50/2) + (100/4) = 103.5$ and

$$\oint(q_b/t)ds = 2(S_y/I_{xx})\Big[\int_0^{\pi/2}(5{,}000/2)(\cos\theta - 1)50 \, d\theta + \int_0^{50}(1/4)(2s_3{}^2 - 200s_3 - 16{,}875)ds_3\Big]$$

Integrating and substituting the limits gives

$$\oint(q_b/t)ds = -647{,}908(S_y/I_{xx})$$

so that from Eq.(vi)

$$q_{s,0} = 6{,}260(S_y/I_{xx})$$

Then

$$q_{73} = (S_y/I_{xx})(5{,}000 \, \cos\theta + 1{,}260) \tag{vii}$$

and

$$q_{34} = (S_y/I_{xx})(2s_3{}^2 - 200s_3 - 10{,}615) \tag{viii}$$

Now taking moments about the center of the web 34

$$S_y \xi_S = 2[\int_0^{\pi/2} q_{73} \times 50^2 \, d\theta - \int_0^{25} q_{12} \times 100 \, ds_1 - \int_0^{100} q_{23} \times 50 \, ds_2]$$

Substituting for q_{73} etc. from Eqs. (vii), (ii), and (iii) and carrying out the integration gives

$$\xi_S = 57.7 \text{ mm}$$

The actual loading on the wing section of 50 kN in the plane of the web 43 is therefore equivalent to a 50 kN load acting through the shear center together with an anticlockwise torque of $50 \times 10^3 \times 57.7 = 2885$ Nm.

The shear flow distribution due to the 50 kN load acting through the shear center is given by Eqs. (ii), (iii), (vii), and (viii) in which $S_y = 50$ kN. Then

$$q_{12} = -(50 \times 10^3/1.87 \times 10^6)(50s_1 + s_1{}^2) = -0.027(50s_1 + s_1{}^2) \tag{ix}$$

$$q_{23} = -0.027(100s_2 + 1{,}875) \tag{x}$$

$$q_{73} = 0.027(5{,}000 \cos \theta + 1{,}260) \tag{xi}$$

$$q_{34} = 0.027(2s_3{}^2 - 200s_3 - 10{,}615) \tag{xii}$$

The maximum shear flow in the wall 12 occurs, from Eq. (ix), at $s_1 = 25$ mm and is equal to -50.1 N/mm. From Eq. (x) the maximum shear flow in the wall 23 occurs when $s_2 = 100$ mm and is equal to -320.6 N/mm. From Eq. (xi) the maximum shear flow in the wall 73 occurs when $\theta = 0$ and is $+169.0$ N/mm. At 3 in the wall 73 the shear flow is 34.0 N/mm. Also, in the wall 34 at 3, the shear flow, from Eq. (xii), is 286.6 N/mm and the maximum value of the shear flow in the wall 34 occurs when $s_3 = 50$ mm and is equal to -421.6 N/mm.

We shall assume, as in Ex. 18.3, that the torsional rigidity , GJ, of the complete section is the sum of the torsional rigidities of the closed and open sections. For the closed portion, from Eq. (18.4)

$$(GJ)_{cl} = (4A^2/G)/(\oint ds/t) = [4(\pi \times 50^2/2)^2 \times 25{,}000]/103.5 = 1{,}487 \times 10^7 \text{ Nmm}^2$$

From Eq. (18.11)

$$(GJ)_{op} = 25{,}000 \times 2(100 \times 2^3 + 25 \times 2^3)/3 = 1.7 \times 10^7 \text{ Nmm}^2$$

The torsional rigidity of the complete section is then

$$GJ = 1{,}487 \times 10^7 + 1.7 \times 10^7 = 1{,}489 \times 10^7 \text{ Nmm}^2$$

The rate of twist in the combined section is given by

$$(d\theta/dz) = (T/GJ) = 2{,}885 \times 10^3/1{,}489 \times 10^7 = 1.94 \times 10^{-4} \text{ rad/mm}$$

Substituting for the torque, T, in Eq. (18.1) from Eq.(18.4) we obtain the shear flow in the closed part of the section, i.e.,

$$q_{cl} = [(GJ)_{cl}/2A](d\theta/dz) = [1{,}487 \times 10^7/2(\pi \times 50^2/2)] \times 1.94 \times 10^{-4} = 367.3 \text{ N/mm}$$

which is due to torsion.

In the open part of the section the maximum shear stress is obtained directly from Eq. (18.10) and is

$$\tau_{\text{max,op}} = 25{,}000 \times 2 \times 1.94 \times 10^{-4} = 9.7 \text{ N/mm}^2$$

We shall now examine the complete shear stress distribution to determine its maximum value. The maximum shear flow due to shear in the open part of the section occurs at 3 in the wall 23 and is 320.6 N/mm so that the corresponding shear stress is $320.6/2 = 160.3$ N/mm^2 giving a maximum shear stress of $160.3 + 9.7 = 170.0$ N/mm^2. The maximum shear flow due to shear in the closed part of the section is 421.6 N/mm in the direction 43 while the shear flow due to torsion is 363.5 N/mm in the direction 34 giving a resultant shear flow of $421.6 - 363.5 = 58.1$ N/mm in the direction 43; the corresponding shear stress is $58.1/4 = 14.5$ N/mm^2. At 3 in the web 34 the combined shear flow due to shear and torsion is $363.5 - 286.6 = 76.9$ N/mm in the direction 34; the corresponding shear stress is $76.9/4 = 19.2$ N/mm^2. In the wall 73 the maximum shear flow due to shear and torsion is $169.0 + 363.5 = 532.5$ N/mm giving a shear stress of $532.5/2 = 266.3$ N/mm^2.

We conclude, therefore, that the maximum shear stress in the combined section is 266.3 N/mm^2 and occurs at the point 7 in the wall 73.

PROBLEMS

P.18.1. The beam section of Example 18.2 (see Fig. 18.4) is subjected to a bending moment in a vertical plane of 20 kN m. Calculate the maximum direct stress in the cross-section of the beam.

Answer: 172.5 N/mm^2

P.18.2. A wing box has the cross-section shown diagrammatically in Fig. P.18.2 and supports a shear load of 100 kN in its vertical plane of symmetry. Calculate the shear stress at the midpoint of the web 36 if the thickness of all walls is 2 mm.

Answer: 89.7 N/mm^2

P.18.3. If the wing box of P.18.2 is subjected to a torque of 100 kN m, calculate the rate of twist of the section and the maximum shear stress. The shear modulus G is 25,000 N/mm^2.

Answer: 18.5×10^{-6} rad/mm, 170 N/mm^2

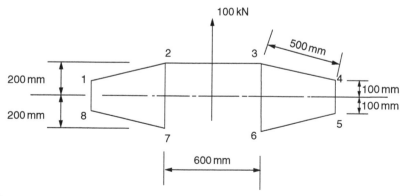

FIGURE P.18.2

P.18.3. MATLAB Use MATLAB to repeat Problem P.18.3 for values of t ranging from 1 to 3 mm in increments of 0.25 mm.

Answer:
(i) $t = 1$ mm: 37×10^{-6} rad/mm, 340 N/mm^2
(ii) $t = 1.25$ mm: 29.6×10^{-6} rad/mm, 272 N/mm^2
(iii) $t = 1.5$ mm: 24.7×10^{-6} rad/mm, 227 N/mm^2
(iv) $t = 1.75$ mm: 21.2×10^{-6} rad/mm, 194 N/mm^2
(v) $t = 2$ mm: 18.5×10^{-6} rad/mm, 170 N/mm^2
(vi) $t = 2.25$ mm: 16.5×10^{-6} rad/mm, 151 N/mm^2
(vii) $t = 2.5$ mm: 14.8×10^{-6} rad/mm, 136 N/mm^2
(viii) $t = 2.75$ mm: 13.5×10^{-6} rad/mm, 124 N/mm^2
(ix) $t = 3$ mm: 12.3×10^{-6} rad/mm, 113 N/mm^2

P.18.4. If the shear load of 100 kN acting in the vertical plane of symmetry of the wing box shown in Fig. P.18.2 now acts in the plane of the web 72, calculate the rate of twist in the section and the maximum shear stress.

Answer: 5.6×10^{-6} rad/mm, 140.7 N/mm^2.

P.18.5. The singly symmetrical wing section shown in Fig. P.18.5 carries a vertical shear load of 50 kN in the plane of the web 63. Calculate the rate of twist of the section and the maximum shear stress in its walls. Take $G = 25,000$ N/mm^2.

Answer: 3.23×10^{-4} rad/mm, 441.3 N/mm^2.

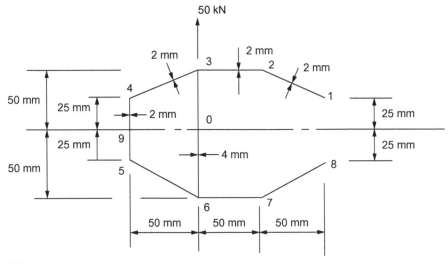

FIGURE P.18.5

Structural idealization

So far, we have been concerned with relatively uncomplicated structural sections which, in practice, are formed from thin plate or by the extrusion process. While these sections exist as structural members in their own right, they are frequently used, as we saw in Chapter 11, to stiffen more complex structural shapes, such as fuselages, wings, and tail surfaces. Thus, a two spar wing section could take the form shown in Fig. 19.1, in which Z-section stringers are used to stiffen the thin skin, while angle sections form the spar flanges. Clearly, the analysis of a section of this type is complicated and tedious unless some simplifying assumptions are made. Generally, the number and nature of these simplifying assumptions determine the accuracy and the degree of complexity of the analysis; the more complex the analysis, the greater the accuracy obtained. The degree of simplification introduced is governed by the particular situation surrounding the problem. For a preliminary investigation, speed and simplicity are often of greater importance than extreme accuracy; on the other hand, a final solution must be as exact as circumstances allow.

Complex structural sections may be idealized into simpler "mechanical model" forms, which behave, under given loading conditions, in the same, or very nearly the same, way as the actual structure. We shall see, however, that different models of the same structure are required to simulate actual behavior under different systems of loading.

19.1 PRINCIPLE

In the wing section of Fig. 19.1, the stringers and spar flanges have small cross-sectional dimensions compared with the complete section. Therefore, the variation in stress over the cross-section of a stringer due to, say, bending of the wing is small. Furthermore, the difference between the distances of the stringer centroids and the adjacent skin from the wing section axis is small. It is reasonable to assume therefore that the direct stress is constant over the stringer cross-sections. We therefore replace the stringers and spar flanges by concentrations of area, known as *booms*, over which the direct stress is constant and which are located along the mid-line of the skin, as shown in Fig. 19.2. In wing and fuselage sections of the type shown in Fig. 19.1, the stringers and spar flanges carry most of the direct stresses while the skin is effective mainly in resisting shear stresses, although it also carries some of the direct stresses. The idealization shown in Fig. 19.2 may therefore be taken a stage further by assuming that all direct stresses are carried by the booms while the skin is effective only in shear. The direct stress carrying capacity of the skin may be allowed for by increasing each boom area by an area equivalent to the direct stress carrying capacity of the adjacent skin panels. The calculation of these equivalent areas generally depends upon an initial assumption as to the form of the distribution of direct stress in a boom–skin panel.

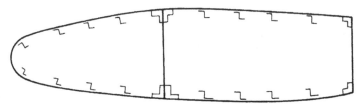

FIGURE 19.1 Typical Wing Section

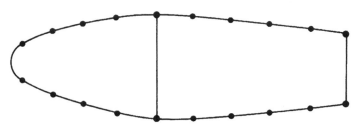

FIGURE 19.2 Idealization of a Wing Section

19.2 IDEALIZATION OF A PANEL

Suppose we wish to idealize the panel of Fig. 19.3(a) into a combination of direct stress carrying booms and shear stress only carrying skin, as shown in Fig. 19.3(b). In Fig. 19.3(a), the direct stress carrying thickness t_D of the skin is equal to its actual thickness t; while, in Fig. 19.3(b), $t_D = 0$. Suppose also that the direct stress distribution in the actual panel varies linearly from an unknown value σ_1 to an unknown value σ_2. Clearly, the analysis should predict the extremes of stress σ_1 and σ_2, although the distribution of direct stress is obviously lost. Since the loading producing the direct stresses in the actual and idealized panels must be the same, we can equate moments to obtain expressions for the boom areas B_1 and B_2. Thus, taking moments about the right-hand edge of each panel,

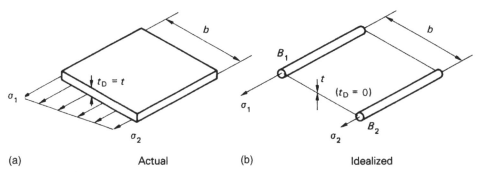

(a)   Actual   (b)   Idealized

FIGURE 19.3 Idealization of a Panel

$$\sigma_2 t_{\mathrm{D}} \frac{b^2}{2} + \frac{1}{2}(\sigma_1 - \sigma_2) t_{\mathrm{D}} b \frac{2}{3} b = \sigma_1 B_1 b$$

from which

$$B_1 = \frac{t_{\mathrm{D}} b}{6}\left(2 + \frac{\sigma_2}{\sigma_1}\right) \tag{19.1}$$

Similarly,

$$B_2 = \frac{t_{\mathrm{D}} b}{6}\left(2 + \frac{\sigma_1}{\sigma_2}\right) \tag{19.2}$$

In Eqs. (19.1) and (19.2), the ratio of σ_1 to σ_2, if not known, may frequently be assumed.

The direct stress distribution in Fig. 19.3(a) is caused by a combination of axial load and bending moment. For axial load only, $\sigma_1/\sigma_2 = 1$ and $B_1 = B_2 = t_{\mathrm{D}} b/2$; for a pure bending moment, $\sigma_1/\sigma_2 = -1$ and $B_1 = B_2 = t_{\mathrm{D}} b/6$. Thus, different idealizations of the same structure are required for different loading conditions.

Example 19.1

Part of a wing section is in the form of the two-cell box shown in Fig. 19.4(a), in which the vertical spars are connected to the wing skin through angle sections all having a cross-sectional area of 300 mm². Idealize the section into an arrangement of direct stress carrying booms and shear stress only carrying panels suitable for resisting bending moments in a vertical plane. Position the booms at the spar/skin junctions. See Ex. 1.1.

The idealized section is shown in Fig. 19.4(b), in which, from symmetry, $B_1 = B_6$, $B_2 = B_5$, $B_3 = B_4$. Since the section is required to resist bending moments in a vertical plane, the direct stress at any point in the actual wing section is directly proportional to its distance from the horizontal axis of symmetry. Further, the distribution of direct stress in all the panels is linear, so that either Eq. (19.1) or (19.2) may be used. We note that, in addition to contributions from adjacent panels, the boom areas include the existing spar flanges. Hence.

$$B_1 = 300 + \frac{3.0 \times 400}{6}\left(2 + \frac{\sigma_6}{\sigma_1}\right) + \frac{2.0 \times 600}{6}\left(2 + \frac{\sigma_2}{\sigma_1}\right)$$

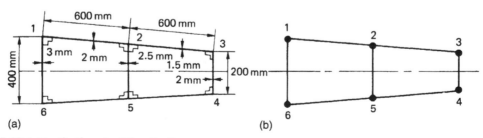

FIGURE 19.4 Idealization of a Wing Section

or

$$B_1 = 300 + \frac{3.0 \times 400}{6}(2-1) + \frac{2.0 \times 600}{6}\left(2 + \frac{150}{200}\right)$$

which gives

$$B_1(= B_6) = 1,050 \text{ mm}^2$$

Also,

$$B_2 = 2 \times 300 + \frac{2.0 \times 600}{6}\left(2 + \frac{\sigma_1}{\sigma_2}\right) + \frac{2.5 \times 300}{6}\left(2 + \frac{\sigma_5}{\sigma_2}\right) + \frac{1.5 \times 600}{6}\left(2 + \frac{\sigma_3}{\sigma_2}\right)$$

that is,

$$B_2 = 2 \times 300 + \frac{2.0 \times 600}{6}\left(2 + \frac{200}{150}\right) + \frac{2.5 \times 300}{6}(2-1) + \frac{1.5 \times 600}{6}\left(2 + \frac{100}{150}\right)$$

from which

$$B_2 (= B_5) = 1,791.7 \text{ mm}^2$$

Finally,

$$B_3 = 300 + \frac{1.5 \times 600}{6}\left(2 + \frac{\sigma_2}{\sigma_3}\right) + \frac{2.0 \times 200}{6}\left(2 + \frac{\sigma_4}{\sigma_3}\right)$$

that is,

$$B_3 = 300 + \frac{1.5 \times 600}{6}\left(2 + \frac{150}{100}\right) + \frac{2.0 \times 200}{6}(2-1)$$

so that

$$B_3 (= B_4) = 891.7 \text{ mm}^2$$

19.3 EFFECT OF IDEALIZATION ON THE ANALYSIS OF OPEN AND CLOSED SECTION BEAMS

The addition of direct stress carrying booms to open and closed section beams clearly modifies the analyses presented in Chapters 15–17. Before considering individual cases, we shall discuss the implications of structural idealization. Generally, in any idealization, different loading conditions require different idealizations of the same structure. In Example 19.1, the loading is applied in a vertical plane. If, however, the loading had been applied in a horizontal plane, the assumed stress distribution in the panels of the section would have been different, resulting in different values of boom area.

Suppose that an open or closed section beam is subjected to given bending or shear loads and that the required idealization has been completed. The analysis of such sections usually involves the determination of the neutral axis position and the calculation of sectional properties. The position of the neutral axis is derived from the condition that the resultant load on the beam cross-section is zero, that is,

$$\int_A \sigma_z \; dA = 0 \qquad (\text{see Eq.(16.3)})$$

The area A in this expression is clearly the direct stress carrying area. It follows that the centroid of the cross-section is the centroid of the direct stress carrying area of the section, depending on the degree and method of idealization. The sectional properties, I_{xx}, and so forth, must also refer to the direct stress carrying area.

19.3.1 Bending of open and closed section beams

The analysis presented in Sections 16.1 and 16.2 applies and the direct stress distribution is given by any of Eqs. (16.9), (16.17), or (16.18), depending on the beam section being investigated. In these equations the coordinates (x, y) of points in the cross-section are referred to axes having their origin at the centroid of the direct stress carrying area. Furthermore, the section properties I_{xx}, I_{yy}, and I_{xy} are calculated for the direct stress carrying area only.

In the case where the beam cross-section has been completely idealized into direct stress carrying booms and shear stress only carrying panels, the direct stress distribution consists of a series of direct stresses concentrated at the centroids of the booms.

Example 19.2

The fuselage section shown in Fig. 19.5 is subjected to a bending moment of 100 kNm applied in the vertical plane of symmetry. If the section has been completely idealized into a combination of direct stress carrying booms and shear stress only carrying panels, determine the direct stress in each boom. See Ex. 1.1.

The section has Cy as an axis of symmetry and resists a bending moment $M_x = 100$ kNm. Equation (16.17) therefore reduces to

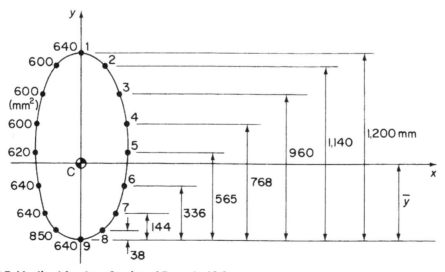

FIGURE 19.5 Idealized fuselage Section of Example 19.2

Table 19.1 Example 19.1

① Boom	② y (mm)	③ B (mm²)	④ $\Delta I_{xx} = By^2$ (mm²)	⑤ σ_z (N/mm²)
1	+660	640	278 × 10⁶	35.6
2	+600	600	216 × 10⁶	32.3
3	+420	600	106 × 10⁶	22.6
4	+228	600	31 × 10⁶	12.3
5	+25	620	0.4 × 10⁶	1.3
6	−204	640	27 × 10⁶	−11.0
7	−396	640	100 × 10⁶	−21.4
8	−502	850	214 × 10⁶	−27.0
9	−540	640	187 × 10⁶	−29.0

$$\sigma_z = \frac{M_x}{I_{xx}} y \tag{i}$$

The origin of axes Cxy coincides with the position of the centroid of the direct stress carrying area, which, in this case, is the centroid of the boom areas. Thus, taking moments of area about boom 9,

$$(6 \times 640 + 6 \times 600 + 2 \times 620 + 2 \times 850)\bar{y}$$
$$= 640 \times 1,200 + 2 \times 600 \times 1140 + 2 \times 600 \times 960 + 2 \times 600 \times 768$$
$$+ 2 \times 620 \times 565 + 2 \times 640 \times 336 + 2 \times 640 \times 144 + 2 \times 850 \times 38$$

which gives

$$\bar{y} = 540 \text{ mm}$$

The solution is now completed in Table 19.1.
From column ④.

$$I_{xx} = 1,854 \times 10^6 \text{ mm}^4$$

and column ⑤ is completed using Eq. (i).

19.3.2 Shear of open section beams

The derivation of Eq. (17.14) for the shear flow distribution in the cross-section of an open section beam is based on the equilibrium equation (17.2). The thickness t in this equation refers to the direct stress carrying thickness t_D of the skin. Equation (17.14) may therefore be rewritten

$$q_s = -\left(\frac{S_x I_{xx} - S_y I_{xy}}{I_{xx} I_{yy} - I_{xy}^2}\right) \int_0^s t_D x \, ds - \left(\frac{S_y I_{yy} - S_x I_{xy}}{I_{xx} I_{yy} - I_{xy}^2}\right) \int_0^s t_D y \, ds \tag{19.3}$$

in which $t_D = t$ if the skin is fully effective in carrying direct stress or $t_D = 0$ if the skin is assumed to carry only shear stresses. Again, the section properties in Eq. (19.3) refer to the direct stress carrying area of the section, since they are those which feature in Eqs. (16.17) and (16.18).

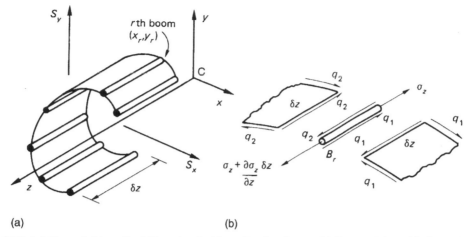

FIGURE 19.6 (a) Elemental Length of Shear Loaded Open Section Beam with Booms; (b) Equilibrium of Boom Element

Equation (19.3) makes no provision for the effects of booms which cause discontinuities in the skin and therefore interrupt the shear flow. Consider the equilibrium of the rth boom in the elemental length of beam shown in Fig. 19.6(a), which carries shear loads S_x and S_y acting through its shear center S. These shear loads produce direct stresses due to bending in the booms and skin and shear stresses in the skin. Suppose that the shear flows in the skin adjacent to the rth boom of cross-sectional area B_r are q_1 and q_2. Then, from Fig. 19.6(b),

$$\left(\sigma_z + \frac{\partial \sigma_s}{\partial z}\delta z\right)B_r - \sigma_z B_r + q_2\delta z - q_1\delta z = 0$$

which simplifies to

$$q_2 - q_1 = -\frac{\partial \sigma_z}{\partial z}B_r \tag{19.4}$$

Substituting for σ_z in Eq. (19.4) from (16.17), we have

$$q_2 - q_1 = -\left[\frac{(\partial M_y/\partial z)I_{xx} - (\partial M_x/\partial z)I_{xy}}{I_{xx}I_{yy} - I_{xy}^2}\right]B_r x_r$$
$$- \left[\frac{(\partial M_x/\partial z)I_{yy} - (\partial M_y/\partial z)I_{xy}}{I_{xx}I_{yy} - I_{xy}^2}\right]B_r y_r$$

or, using the relationships of Eqs. (16.22) and (16.23),

$$q_2 - q_1 = -\left(\frac{S_x I_{xx} - S_y I_{xy}}{I_{xx}I_{yy} - I_{xy}^2}\right)B_r x_r - \left(\frac{S_y I_{yy} - S_x I_{xy}}{I_{xx}I_{yy} - I_{xy}^2}\right)B_r y_r \tag{19.5}$$

Equation (19.5) gives the change in shear flow induced by a boom which itself is subjected to a direct load ($\sigma_z B_r$). Each time a boom is encountered, the shear flow is incremented by this amount, so that if, at any distance s around the profile of the section, n booms have been passed, the shear flow at the point is given by

$$q_s = -\left(\frac{S_x I_{xx} - S_y I_{xy}}{I_{xx} I_{yy} - I_{xy}^2}\right)\left(\int_0^s t_D x \, ds + \sum_{r=1}^n B_r x_r\right)$$
$$-\left(\frac{S_y I_{yy} - S_x I_{xy}}{I_{xx} I_{yy} - I_{xy}^2}\right)\left(\int_0^s t_D y \, ds + \sum_{r=1}^n B_r y_r\right) \tag{19.6}$$

Example 19.3

Calculate the shear flow distribution in the channel section shown in Fig. 19.7 produced by a vertical shear load of 4.8 kN acting through its shear center. Assume that the walls of the section are effective in resisting only shear stresses, while the booms, each of area 300 mm², carry all the direct stresses. See Ex. 1.1.

The effective direct stress carrying thickness t_D of the walls of the section is zero so that the centroid of area and the section properties refer to the boom areas only. Since Cx (and Cy as far as the boom areas are concerned) is an axis of symmetry, $I_{xy} = 0$; also $S_x = 0$ and Eq. (19.6) thereby reduces to

$$q_s = -\frac{S_y}{I_{xx}}\sum_{r=1}^n B_r y_r \tag{i}$$

in which $I_{xx} = 4 \times 300 \times 200^2 = 48 \times 10^6$ mm⁴. Substituting the values of S_y and I_{xx} in Eq. (i) gives

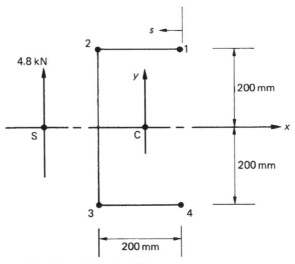

FIGURE 19.7 Idealized Channel Section of Example 19.3

$$q_s = -\frac{4.8 \times 10^3}{48 \times 10^6} \sum_{r=1}^{n} B_r y_r = -10^{-4} \sum_{r=1}^{n} B_r y_r \tag{ii}$$

At the outside of boom 1, $q_s = 0$. As boom 1 is crossed, the shear flow changes by an amount given by

$$\Delta q_1 = -10^{-4} \times 300 \times 200 = -6 \, \text{N/mm}$$

Hence, $q_{12} = -6$ N/mm, since, from Eq. (i), it can be seen that no further changes in shear flow occur until the next boom (2) is crossed. Hence,

$$q_{23} = -6 - 10^{-4} \times 300 \times 200 = -12 \, \text{N/mm}$$

Similarly,

$$q_{34} = -12 - 10^{-4} \times 300 \times (-200) = -6 \, \text{N/mm}$$

while, finally, at the outside of boom 4, the shear flow is

$$-6 - 10^{-4} \times 300 \times (-200) = 0$$

as expected. The complete shear flow distribution is shown in Fig. 19.8.

It can be seen, from Eq. (i) in Example 19.3, that the analysis of a beam section which is idealized into a combination of direct stress carrying booms and shear stress only carrying skin gives constant values of the shear flow in the skin between the booms; the actual distribution of shear flows is therefore lost. What remains is in fact the average of the shear flow, as can be seen by referring to Example 19.3. Analysis of the unidealized channel section results in a parabolic distribution of shear flow in the web 23, whose resultant is statically equivalent to the externally applied shear load of 4.8 kN. In Fig. 19.8, the resultant of the constant shear flow in the web 23 is $12 \times 400 = 4800 \, \text{N} = 4.8 \, \text{kN}$. It follows that this constant value of shear flow is the average of the parabolically distributed shear flows in the unidealized section.

The result, from the idealization of a beam section, of a constant shear flow between booms may be used to advantage in parts of the analysis. Suppose that the curved web 12 in Fig. 19.9 has booms at its extremities and that the shear flow q_{12} in the web is constant. The shear force on an element δs of the

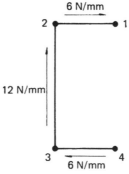

FIGURE 19.8 Shear Flow in Channel Section of Example 19.3

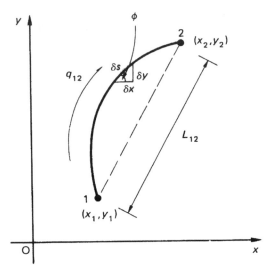

FIGURE 19.9 Curved Web with Constant Shear Flow

web is $q_{12}\delta s$, whose components horizontally and vertically are $q_{12}\delta s \cos \phi$ and $q_{12}\delta s \sin \phi$. The resultant, parallel to the x axis, S_x, of q_{12} is therefore given by

$$S_x = \int_1^2 q_{12} \cos \phi \, ds$$

or

$$S_x = q_{12} \int_1^2 \cos \phi \, ds$$

which, from Fig. 19.9, may be written

$$S_x = q_{12} \int_1^2 dx = q_{12}(x_2 - x_1) \tag{19.7}$$

Similarly, the resultant of q_{12} parallel to the y axis is

$$S_y = q_{12}(y_2 - y_1) \tag{19.8}$$

Thus, the resultant, in a given direction, of a constant shear flow acting on a web is the value of the shear flow multiplied by the projection on that direction of the web.

The resultant shear force S on the web of Fig. 19.9 is

$$S = \sqrt{S_x^2 + S_y^2} = q_{12}\sqrt{(x_2 - x_1)^2 + (y_2 - y_1)^2}$$

that is,

$$S = q_{12}L_{12} \tag{19.9}$$

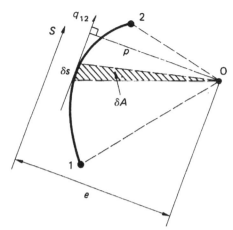

FIGURE 19.10 Moment Produced by a Constant Shear Flow

Therefore, the resultant shear force acting on the web is the product of the shear flow and the length of the straight line joining the ends of the web; clearly, the direction of the resultant is parallel to this line.

The moment M_q produced by the shear flow q_{12} about any point O in the plane of the web is, from Fig. 19.10,

$$M_q = \int_1^2 q_{12}p \, ds = q_{12} \int_1^2 2 \, dA$$

or

$$M_q = 2Aq_{12} \tag{19.10}$$

in which A is the area enclosed by the web and the lines joining the ends of the web to the point O. This result may be used to determine the distance of the line of action of the resultant shear force from any point. From Fig. 19.10,

$$Se = 2Aq_{12}$$

from which

$$e = \frac{2A}{S}q_{12}$$

Substituting for q_{12} from Eq. (19.9) gives

$$e = \frac{2A}{L_{12}}$$

Example 19.3 MATLAB

Use MATLAB to repeat Example 19.3. See Ex. 1.1.

The shear flow distribution is obtained through the following MATLAB file:

```
% Declare any needed variables
S_y = 4.8e3;
B = 300*ones(1,4);

% Define the lengths of all open section walls
L = [200 400 200];

% Calculate the section properties
y_r = [L(2)/2 L(2)/2 -L(2)/2 -L(2)/2];
I_xx = sum(B.*y_r.^2); % I_yy and I_xy not needed due to symmetry and S_x=0

% Due to symmetry and since S_x,t_D=0, Eq. (19.6) simplifies
% Calculate the shear flow in wall 1-2 using the reduced Eq. (19.6)
A = B.*y_r;
q_12 = -S_y/I_xx*sum(A(1));

% Calculate the subsequent shear flow in walls 2-3 and 3-4
q_23 = -S_y/I_xx*sum(A(1:2));
q_34 = -S_y/I_xx*sum(A(1:3));

% Output the calculated shear flows to the Command Window
disp(['q_12 =' num2str(q_12) 'N/mm'])
disp(['q_23 =' num2str(q_23) 'N/mm'])
disp(['q_34 =' num2str(q_34) 'N/mm'])
```

The Command Window outputs resulting from this MATLAB file are as follows:

```
q_12 = -6 N/mm
q_23 = -12 N/mm
q_34 = -6 N/mm
```

19.3.3 Shear loading of closed section beams

Arguments identical to those in the shear of open section beams apply in this case. Thus, the shear flow at any point around the cross-section of a closed section beam comprising booms and skin of direct stress carrying thickness t_D is, by a comparison of Eqs. (19.6) and (17.15),

$$
q_s = -\left(\frac{S_x I_{xx} - S_y I_{xy}}{I_{xx} I_{yy} - I_{xy}^2}\right)\left(\int_0^s t_D x \, ds + \sum_{r=1}^n B_r x_r\right)
$$
$$
-\left(\frac{S_y I_{yy} - S_x I_{xy}}{I_{xx} I_{yy} - I_{xy}^2}\right)\left(\int_0^s t_D y \, ds + \sum_{r=1}^n B_r y_r\right) + q_{s,0}
$$

(19.11)

Note that the zero value of the "basic" or "open section" shear flow at the "cut" in a skin for which $t_D = 0$ extends from the cut to the adjacent booms.

Example 19.4

The thin-walled single cell beam shown in Fig. 19.11 has been idealized into a combination of direct stress carrying booms and shear stress only carrying walls. If the section supports a vertical shear load of 10 kN acting in a vertical plane through booms 3 and 6, calculate the distribution of shear flow around the section.

Boom areas: $B_1 = B_8 = 200 \text{ mm}^2$, $B_2 = B_1 = 250 \text{ mm}^2$, $B_3 = B_6 = 400 \text{ mm}^2$, $B_4 = B_5 = 100 \text{ mm}^2$

The centroid of the direct stress carrying area lies on the horizontal axis of symmetry, so that $I_{xy} = 0$. Also, since $t_D = 0$ and only a vertical shear load is applied, Eq. (19.11) reduces to

$$q_s = -\frac{S_y}{I_{xx}} \sum_{r=1}^{n} B_r y_r + q_{s,0} \tag{i}$$

in which

$$I_{xx} = 2\left(200 \times 30^2 + 250 \times 100^2 + 400 \times 100^2 + 100 \times 50^2\right) = 13.86 \times 10^6 \text{ mm}^4$$

Equation (i) then becomes

$$q_s = -\frac{10 \times 10^3}{13.86 \times 10^6} \sum_{r=1}^{n} B_r y_r + q_{s,0}$$

that is,

$$q_s = -7.22 \times 10^{-4} \sum_{r=1}^{n} B_r y_r + q_{s,0} \tag{ii}$$

"Cutting" the beam section in the wall 23 (any wall may be chosen) and calculating the basic shear flow distribution q_b from the first term on the right-hand side of Eq. (ii), we have

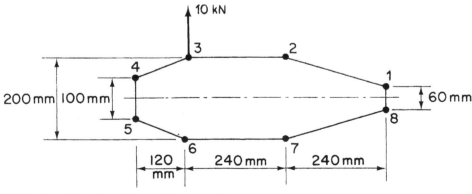

FIGURE 19.11 Closed Section of Beam of Example 19.4

$$q_{b,23} = 0$$
$$q_{b,34} = -7.22 \times 10^{-4}(400 \times 100) = -28.9 \text{ N/mm}$$
$$q_{b,45} = -28.9 - 7.22 \times 10^{-4}(100 \times 50) = -32.5 \text{ N/mm}$$
$$q_{b,56} = q_{b,34} = -28.9 \text{ N/mm (by symmetry)}$$
$$q_{b,67} = q_{b,23} = 0 \text{ (by symmetry)}$$
$$q_{b,21} = -7.22 \times 10^{-4}(250 \times 100) = -18.1 \text{ N/mm}$$
$$q_{b,18} = -18.1 - 7.22 \times 10^{-4}(200 \times 30) = -22.4 \text{ N/mm}$$
$$q_{b,87} = q_{b,21} = -18.1 \text{ N/mm (by symmetry)}$$

Taking moments about the intersection of the line of action of the shear load and the horizontal axis of symmetry and referring to the results of Eqs. (19.7) and (19.8), we have, from Eq. (17.18),

$$0 = [q_{b,81} \times 60 \times 480 + 2q_{b,12}(240 \times 100 + 70 \times 240) + 2q_{b,23} \times 240 \times 100$$
$$- 2q_{b,43} \times 120 \times 100 - q_{b,54} \times 100 \times 120] + 2 \times 97,200q_{s,0}$$

Substituting these values of q_b in this equation gives

$$q_{s,0} = -5.4 \text{ N/mm}$$

the negative sign indicating that $q_{s,0}$ acts in a clockwise sense.

In any wall, the final shear flow is given by $q_s = q_b + q_{s,0}$, so that

$$q_{21} = -18.1 + 5.4 = -12.7 \text{ N/mm} = q_{87}$$
$$q_{23} = -5.4 \text{ N/mm} = q_{67}$$
$$q_{34} = -34.3 \text{ N/mm} = q_{56}$$
$$q_{45} = -37.9 \text{ N/mm}$$

and

$$q_{81} = 17.0 \text{ N/mm}$$

giving the shear flow distribution shown in Fig. 19.12.

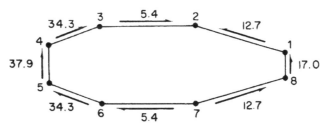

FIGURE 19.12 Shear Flow Distribution N/mm in Walls of the Beam Section of Example 19.4

19.3.4 Alternative method for the calculation of shear flow distribution

Equation (19.4) may be rewritten in the form

$$q_2 - q_1 = \frac{\partial P_r}{\partial z} \qquad (19.12)$$

in which P_r is the direct load in the rth boom. This form of the equation suggests an alternative approach to the determination of the effect of booms on the calculation of shear flow distributions in open and closed section beams.

Let us suppose that the boom load varies linearly with z. This is the case for a length of beam over which the shear force is constant. Equation (19.12) then becomes

$$q_2 - q_1 = -\Delta P_r \qquad (19.13)$$

in which ΔP_r is the *change* in boom load over unit length of the rth boom. ΔP_r may be calculated by first determining the *change* in bending moment between two sections of a beam a unit distance apart then calculating the corresponding change in boom stress using either Eq. (16.17) or (16.18); the change in boom load follows by multiplying the change in boom stress by the boom area B_r. Note that the section properties contained in Eqs. (16.17) and (16.18) refer to the direct stress carrying area of the beam section. In cases where the shear force is not constant over the unit length of beam, the method is approximate.

We illustrate the method by applying it to Example 19.3. In Fig. 19.7, the shear load of 4.8 kN is applied to the face of the section, which is seen when a view is taken along the z axis toward the origin. Thus, when considering unit length of the beam, we must ensure that this situation is unchanged. Figure 19.13 shows a unit (1 mm, say) length of beam. The change in bending moment between the front and rear faces of the length of beam is 4.8×1 kN mm, which produces a change in boom load given by (see Eq. (16.17))

$$\Delta P_r = \frac{4.8 \times 10^3 \times 200}{48 \times 10^6} \times 300 = 6 \text{ N}$$

The change in boom load is compressive in booms 1 and 2 and tensile in booms 3 and 4.

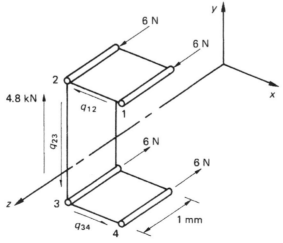

FIGURE 19.13 Alternative Solution to Example 19.3

Equation (19.12), and hence Eq. (19.13), is based on the tensile load in a boom increasing with increasing z. If the tensile load increases with decreasing z, the right-hand side of these equations is positive. It follows that in the case where a compressive load increases with decreasing z, as for booms 1 and 2 in Fig. 19.13, the right-hand side is negative; similarly for booms 3 and 4, the right-hand side is positive. Thus,

$$q_{12} = -6 \text{ N/mm}$$
$$q_{23} = -6 + q_{12} = -12 \text{ N/mm}$$

and

$$q_{34} = +6 + q_{23} = -6 \text{ N/mm}$$

giving the same solution as before. Note that, if the unit length of beam had been taken to be 1 m, the solution would have been $q_{12} = -6000$ N/m, $q_{23} = -12000$ N/m, $q_{34} = -6000$ N/m.

19.3.5 Torsion of open and closed section beams

No direct stresses are developed in either open or closed section beams subjected to a pure torque unless axial constraints are present. The shear stress distribution is therefore unaffected by the presence of booms and the analyses presented in Chapter 17 apply.

19.4 DEFLECTION OF OPEN AND CLOSED SECTION BEAMS

Bending, shear, and torsional deflections of thin-walled beams are readily obtained by application of the unit load method described in Section 5.5.

The displacement in a given direction due to torsion is given directly by the last of Eqs. (5.21), thus,

$$\Delta_T = \int_L \frac{T_0 T_1}{GJ} \, \mathrm{d}z \tag{19.14}$$

where J, the torsion constant, depends on the type of beam under consideration. For an open section beam, J is given by either of Eqs. (18.11), whereas in the case of a closed section beam, $J = 4A^2/(\oint \mathrm{d}s/t)$ (Eq. (18.4)) for a constant shear modulus.

Expressions for the bending and shear displacements of unsymmetrical thin-walled beams may also be determined by the unit load method. They are complex for the general case and are most easily derived from first principles by considering the complementary energy of the elastic body in terms of stresses and strains rather than loads and displacements. In Chapter 5, we observed that the theorem of the principle of the stationary value of the total complementary energy of an elastic system is equivalent to the application of the principle of virtual work where virtual forces act through real displacements. We may therefore specify that, in our expression for total complementary energy, the displacements are the actual displacements produced by the applied loads, while the virtual force system is the unit load.

Considering deflections due to bending, we see, from Eq. (5.6), that the increment in total complementary energy due to the application of a virtual unit load is

$$-\int_L \left(\int_A \sigma_{z,1} \varepsilon_{z,0} \, dA \right) dz + 1\Delta_M$$

where $\sigma_{z,1}$ is the direct bending stress at any point in the beam cross-section corresponding to the unit load and $\varepsilon_{z,0}$ is the strain at the point produced by the actual loading system. Further, Δ_M is the actual displacement due to bending at the point of application and in the direction of the unit load. Since the system is in equilibrium under the action of the unit load this expression must equal zero (see Eq. (5.6)). Hence,

$$\Delta_M = \int_L \left(\int_A \sigma_{z,1} \varepsilon_{z,0} \, dA \right) dz \tag{19.15}$$

From Eq. (16.17) and the third of Eqs. (1.42),

$$\sigma_{z,1} = \left(\frac{M_{y,1}I_{xx} - M_{x,1}I_{xy}}{I_{xx}I_{yy} - I_{xy}^2} \right) x + \left(\frac{M_{x,1}I_{yy} - M_{y,1}I_{xy}}{I_{xx}I_{yy} - I_{xy}^2} \right) y$$

$$\varepsilon_{z,0} = \frac{1}{E} \left[\left(\frac{M_{y,0}I_{xx} - M_{x,0}I_{xy}}{I_{xx}I_{yy} - I_{xy}^2} \right) x + \left(\frac{M_{x,0}I_{yy} - M_{y,0}I_{xy}}{I_{xx}I_{yy} - I_{xy}^2} \right) y \right]$$

where the suffixes 1 and 0 refer to the unit and actual loading systems and x, y are the coordinates of any point in the cross-section referred to a centroidal system of axes. Substituting for $\sigma_{z,1}$ and $\varepsilon_{z,0}$ in Eq. (19.15) and remembering that $\int_A x^2 dA = I_{yy}$, $\int_A y^2 dA = I_{xx}$, and $\int_A xy dA = I_{xy}$, we have

$$\Delta_M = \frac{1}{E\left(I_{xx}I_{yy} - I_{xy}^2\right)^2} \int_L \left\{ \left(M_{y,1}I_{xx} - M_{x,1}I_{xy}\right)\left(M_{y,0}I_{xx} - M_{x,0}I_{xy}\right)I_{yy} \right.$$
$$+ \left(M_{x,1}I_{yy} - M_{y,1}I_{xy}\right)\left(M_{x,0}I_{yy} - M_{y,0}I_{xy}\right)I_{xx} \tag{19.16}$$
$$+ \left[\left(M_{y,1}I_{xx} - M_{x,1}I_{xy}\right)\left(M_{x,0}I_{yy} - M_{y,0}I_{xy}\right)\right.$$
$$\left. + \left(M_{x,1}I_{yy} - M_{y,1}I_{xy}\right)\left(M_{y,0}I_{xx} - M_{x,0}I_{xy}\right)\right]I_{xy}\right\} dz$$

For a section having either the x or y axis as an axis of symmetry, $I_{xy} = 0$ and Eq. (19.16) reduces to

$$\Delta_M = \frac{1}{E} \int_L \left(\frac{M_{y,1}M_{y,0}}{I_{yy}} + \frac{M_{x,1}M_{x,0}}{I_{xx}} \right) dz \tag{19.17}$$

The derivation of an expression for the shear deflection of thin-walled sections by the unit load method is achieved in a similar manner. By comparison with Eq. (19.15), we deduce that the deflection Δ_S, due to shear of a thin-walled open or closed section beam of thickness t, is given by

$$\Delta_S = \int_L \left(\int_{sect} \tau_1 \gamma_0 t \, ds \right) dz \tag{19.18}$$

where τ_1 is the shear stress at an arbitrary point s around the section produced by a unit load applied at the point and in the direction Δ_S, and γ_0 is the shear strain at the arbitrary point corresponding to the actual loading system. The integral in parentheses is taken over all the walls of the beam. In fact, both the applied and unit shear loads must act through the shear center of the cross-section, otherwise additional torsional displacements occur. Where shear loads act at other points, these must be replaced by shear loads at the shear center plus a torque. The thickness t is the actual skin thickness and may vary around the cross-section but is assumed to be constant along the length of the beam. Rewriting Eq. (19.18) in terms of shear flows q_1 and q_0, we obtain

$$\Delta_S = \int_L \left(\int_{\text{sect}} \frac{q_0 q_1}{Gt} \, ds \right) dz \tag{19.19}$$

where again the suffixes refer to the actual and unit loading systems. In the cases of both open and closed section beams, the general expressions for shear flow are long and are best evaluated before substituting in Eq. (19.19). For an open section beam comprising booms and walls of direct stress carrying thickness t_D, we have, from Eq. (19.6),

$$
\begin{aligned}
q_0 = - &\left(\frac{S_{x,0} I_{xx} - S_{y,0} I_{xy}}{I_{xx} I_{yy} - I_{xy}^2} \right) \left(\int_0^s t_D x \, ds + \sum_{r=1}^n B_r x_r \right) \\
- &\left(\frac{S_{y,0} I_{yy} - S_{x,0} I_{xy}}{I_{xx} I_{yy} - I_{xy}^2} \right) \left(\int_0^s t_D y \, ds + \sum_{r=1}^n B_r y_r \right)
\end{aligned}
\tag{19.20}
$$

and

$$
\begin{aligned}
q_1 = - &\left(\frac{S_{x,1} I_{xx} - S_{y,1} I_{xy}}{I_{xx} I_{yy} - I_{xy}^2} \right) \left(\int_0^s t_D x \, ds + \sum_{r=1}^n B_r x_r \right) \\
- &\left(\frac{S_{y,1} I_{yy} - S_{x,1} I_{xy}}{I_{xx} I_{yy} - I_{xy}^2} \right) \left(\int_0^s t_D y \, ds + \sum_{r=1}^n B_r y_r \right)
\end{aligned}
\tag{19.21}
$$

Similar expressions are obtained for a closed section beam from Eq. (19.11).

Example 19.5

Calculate the deflection of the free end of a cantilever 2,000 mm long having a channel section identical to that in Example 19.3 and supporting a vertical, upward load of 4.8 kN acting through the shear center of the section. The effective direct stress carrying thickness of the skin is zero while its actual thickness is 1 mm. Young's modulus E and the shear modulus G are 70,000 and 30,000 N/mm², respectively. See Ex. 1.1.

The section is doubly symmetrical (i.e., the direct stress carrying area) and supports a vertical load producing a vertical deflection. Thus, we apply a unit load through the shear center of the section at the tip of the cantilever and in the same direction as the applied load. Since the load is applied through the shear center, there is no twisting of the section and the total deflection is given, from Eqs. (19.17), (19.19), (19.20), and (19.21), by

$$\Delta = \int_0^L \frac{M_{x,0} M_{x,1}}{EI_{xx}} \, dz + \int_0^L \left(\int_{\text{sect}} \frac{q_0 q_1}{Gt} \, ds \right) dz \tag{i}$$

where $M_{x,0} = -4.8 \times 10^3 \, (2000 - z)$, $M_{x,1} = -1(2000 - z)$,

$$q_0 = -\frac{4.8 \times 10^3}{I_{xx}} \sum_{r=1}^{n} B_r y_r, \quad q_1 = -\frac{1}{I_{xx}} \sum_{r=1}^{n} B_r y_r$$

and z is measured from the built-in end of the cantilever. The actual shear flow distribution was calculated in Example 19.3. In this case, the q_1 shear flows may be deduced from the actual distribution shown in Fig. 19.8, that is,

$$q_1 = q_0/4.8 \times 10^3$$

Evaluating the bending deflection, we have

$$\Delta_M = \int_0^{2000} \frac{4.8 \times 10^3 (2,000 - z)^2 \, dz}{70,000 \times 48 \times 10^6} = 3.81 \text{ mm}$$

The shear deflection Δ_S is given by

$$\Delta_S = \int_0^{2000} \frac{1}{30,000 \times 1} \left[\frac{1}{4.8 \times 10^3} (6^2 \times 200 + 12^2 \times 400 + 6^2 \times 200) \right] dz$$
$$= 1.0 \text{ mm}$$

The total deflection Δ is then $\Delta_M + \Delta_S = 4.81$ mm in a vertical upward direction.

PROBLEMS

P.19.1. Idealize the box section shown in Fig. P.19.1 into an arrangement of direct stress carrying booms positioned at the four corners and panels that are assumed to carry only shear stresses. Hence, determine the distance of the shear center from the left-hand web.

Answer: 225 mm

P.19.1. MATLAB Assuming the width of the section shown in Fig. P.19.1 is labeled L_1, use MATLAB to repeat Problem P.19.1 for values of L_1 from 300 to 700 mm in increments of 50 mm.

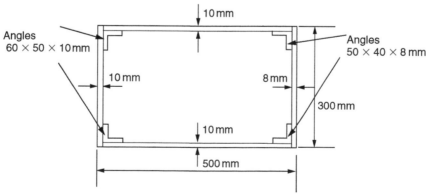

FIGURE P.19.1

Answer: (i) $L_1 = 300$ mm: 133 mm from left-hand web
(ii) $L_1 = 350$ mm: 156 mm from left-hand web
(iii) $L_1 = 400$ mm: 178 mm from left-hand web
(iv) $L_1 = 450$ mm: 201 mm from left-hand web
(v) $L_1 = 500$ mm: 225 mm from left-hand web
(vi) $L_1 = 550$ mm: 248 mm from left-hand web
(vii) $L_1 = 600$ mm: 271 mm from left-hand web
(viii) $L_1 = 650$ mm: 295 mm from left-hand web
(ix) $L_1 = 700$ mm: 319 mm from left-hand web

P.19.2. The beam section shown in Fig. P.19.2 has been idealized into an arrangement of direct stress carrying booms and shear stress only carrying panels. If the beam section is subjected to a vertical shear load of 1,495 N through its shear center, each of booms 1, 4, 5, and 8 has an area of 200 mm², and each of booms 2, 3, 6, and 7 has an area of 250 mm², determine the shear flow distribution and the position of the shear center.

Answer: Wall 12, 1.86 N/mm; 43, 1.49 N/mm; 32, 5.21 N/mm; 27, 10.79 N/mm; remaining distribution follows from symmetry. 122 mm to the left of the web 27

P.19.3 Figure P.19.3 shows the cross-section of a single-cell, thin-walled beam with a horizontal axis of symmetry. The direct stresses are carried by the booms B_1 to B_4, while the walls are effective only in carrying shear stresses. Assuming that the basic theory of bending is applicable, calculate the position of the shear center S. The shear modulus G is the same for all walls:

Cell area $= 135,000$ mm²; Boom areas: $B_1 = B_4 = 450$ mm², $B_2 = B_3 = 550$ mm²

Answer: 197.2 mm from vertical through booms 2 and 3.

Wall	Length (mm)	Thickness (mm)
12, 34	500	0.8
23	580	1.0
41	200	1.2

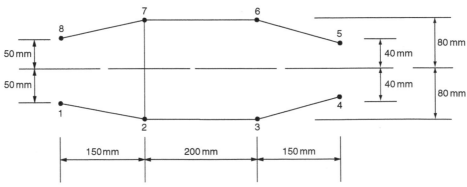

FIGURE P.19.2

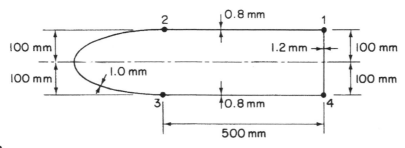

FIGURE P.19.3

P.19.3 MATLAB

Use MATLAB to repeat Problem P.19.3 for the following boom area combinations:

	(i)	(ii)	(iii)	(iv)	(v)	(vi)
B_1 (mm^2)	550	450	550	450	550	450
B_2 (mm^2)	550	450	450	550	450	550
B_3 (mm^2)	450	550	550	450	450	550
B_4 (mm^2)	450	550	450	550	550	450

Answer:
    (i)   197.2 mm from vertical through booms 2 and 3
    (ii) 230.9 mm from vertical through booms 2 and 3
   (iii) 215.4 mm from vertical through booms 2 and 3
   (iv) 215.4 mm from vertical through booms 2 and 3
    (v) 209.9 mm from vertical through booms 2 and 3
   (vi) 218.2 mm from vertical through booms 2 and 3

P.19.4 Find the position of the shear center of the rectangular four boom beam section shown in Fig. P.19.4. The booms carry only direct stresses, but the skin is fully effective in carrying both shear and direct stress. The area of each boom is 100 mm^2.

Answer: 142.5 mm from side 23

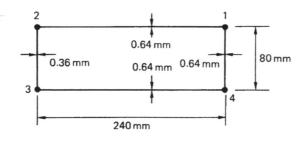

FIGURE P.19.4

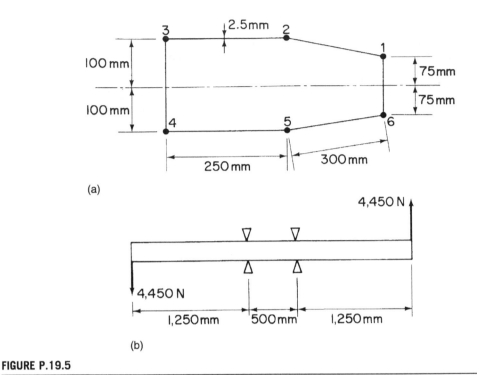

FIGURE P.19.5

P.19.5 A uniform beam with the cross-section shown in Fig. P.19.5(a) is supported and loaded as shown in Fig. P.19.5(b). If the direct and shear stresses are given by the basic theory of bending, the direct stresses being carried by the booms and the shear stresses by the walls, calculate the vertical deflection at the ends of the beam when the loads act through the shear centers of the end cross-sections, allowing for the effect of shear strains. Take $E = 69,000$ N/mm^2 and $G = 26,700$ N/mm^2. Boom areas: 1, 3, 4, 6 = 650 mm^2; 2, 5 = 1,300 mm^2.

Answer: 3.4 mm

P.19.6 A cantilever, length L, has a hollow cross-section in the form of a doubly symmetric wedge, as shown in Fig. P.19.6. The chord line is of length c, wedge thickness is t, the length of a sloping side is $a/2$, and the wall thickness is constant and equal to t_0. Uniform pressure distributions of the magnitudes shown act on the faces of the wedge. Find the vertical deflection of point A due to this given loading. If $G = 0.4E$, $t/c = 0.05$, and $L = 2c$, show that this deflection is approximately $5,600 p_0 c^2/Et_0$.

P.19.7 A rectangular section thin-walled beam of length L and breadth $3b$, depth b, and wall thickness t is built-in at one end (Fig. P.19.7). The upper surface of the beam is subjected to a pressure which varies linearly across the breadth from a value p_0 at edge AB to zero at edge CD. Thus, at any given value of x, the pressure is constant in the z direction. Find the vertical deflection of point A.

Answer: $p_0 L^2(9L^2/80Eb^2 + 1,609/2,000G)/t$

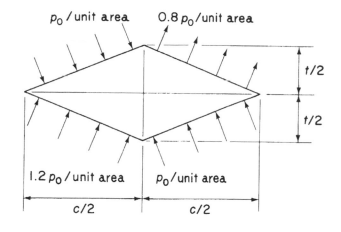

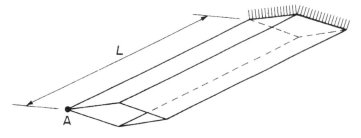

FIGURE P.19.6

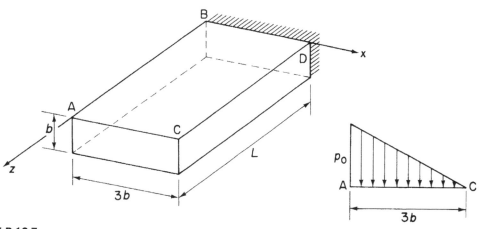

FIGURE P.19.7

Stress analysis of aircraft components

B4

CHAPTER

Wing spars and box beams

20

In Chapters 15–17, we established the basic theory for the analysis of open and closed section thin-walled beams subjected to bending, shear, and torsional loads. In addition, in Chapter 19, we saw how complex stringer stiffened sections could be idealized into sections more amenable to analysis. We now extend this analysis to actual aircraft components, including, in this chapter, wing spars and box beams. In subsequent chapters, we investigate the analysis of fuselages, wings, frames, and ribs and consider the effects of cut-outs in wings and fuselages. Finally, in Chapter 24, an introduction is given to the analysis of components fabricated from composite materials.

Aircraft structural components are, as we saw in Chapter 11, complex, consisting usually of thin sheets of metal stiffened by arrangements of stringers. These structures are highly redundant and require some degree of simplification or idealization before they can be analyzed. The analysis presented here is therefore approximate and the degree of accuracy obtained depends on the number of simplifying assumptions made. A further complication arises in that factors such as warping restraint, structural and loading discontinuities, and shear lag significantly affect the analysis; we investigate these effects in some simple structural components in Chapters 26 and 27. Generally, a high degree of accuracy can be obtained only by using computer-based techniques, such as the finite element method (see Chapter 6). However, the simpler, quicker, and cheaper approximate methods can be used to advantage in the preliminary stages of design, when several possible structural alternatives are being investigated; they also provide an insight into the physical behavior of structures which computer-based techniques do not.

Major aircraft structural components such as wings and fuselages are usually tapered along their lengths for greater structural efficiency. Thus, wing sections are reduced both chordwise and in depth along the wing span toward the tip and fuselage sections aft of the passenger cabin taper to provide a more efficient aerodynamic and structural shape.

The analysis of open and closed section beams, presented in Chapters 15–17, assumes that the beam sections are uniform. The effect of taper on the prediction of direct stresses produced by bending is minimal if the taper is small and the section properties are calculated at the particular section being considered; Eqs. (16.17)–(16.21) may therefore be used with reasonable accuracy. On the other hand, the calculation of shear stresses in beam webs can be significantly affected by taper.

20.1 TAPERED WING SPAR

Consider first the simple case of a beam, for example, a wing spar, positioned in the yz plane and comprising two flanges and a web: an elemental length δz of the beam is shown in Fig. 20.1. At the section z, the beam is subjected to a positive bending moment M_x and a positive shear force S_y. The bending moment resultants $P_{z,1}$ and $P_{z,2}$ are parallel to the z axis of the beam. For a beam in which the flanges

Introduction to Aircraft Structural Analysis, Second Edition
Copyright © 2014, 2007 T.H.G. Megson. Published by Elsevier Ltd. All rights reserved.

605

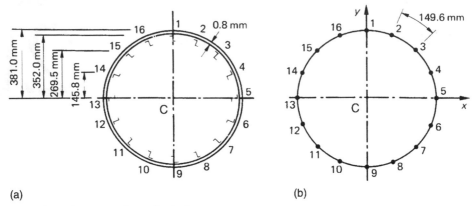

(a)   (b)

FIGURE 20.1 Effect of Taper on Beam Analysis

are assumed to resist all the direct stresses, $P_{z,1} = M_x/h$ and $P_{z,2} = -M_x/h$. In the case where the web is assumed to be fully effective in resisting direct stress, $P_{z,1}$ and $P_{z,2}$ are determined by multiplying the direct stresses $\sigma_{z,1}$ and $\sigma_{z,2}$, found using Eq. (16.17) or (16.18), by the flange areas B_1 and B_2. $P_{z,1}$ and $P_{z,2}$ are the components in the z direction of the axial loads P_1 and P_2 in the flanges. These have components $P_{y,1}$ and $P_{y,2}$ parallel to the y axis given by

$$P_{y,1} = P_{z,1}\frac{\delta y_1}{\delta z}, \quad P_{y,2} = -P_{z,2}\frac{\delta y_2}{\delta z} \tag{20.1}$$

in which, for the direction of taper shown, δy_2 is negative. The axial load in flange ① is given by

$$P_1 = (P_{z,1}^2 + P_{y,1}^2)^{1/2}$$

Substituting for $P_{y,1}$ from Eq. (20.1), we have

$$P_1 = P_{z,1}\frac{(\delta z^2 + \delta y_1^2)^{1/2}}{\delta z} = \frac{P_{z,1}}{\cos\alpha_1} \tag{20.2}$$

Similarly,

$$P_2 = \frac{P_{z,2}}{\cos\alpha_2} \tag{20.3}$$

The internal shear force S_y comprises the resultant $S_{y,w}$ of the web shear flows together with the vertical components of P_1 and P_2. Thus,

$$S_y = S_{y,w} + P_{y,1} - P_{y,2}$$

or

$$S_y = S_{y,w} + P_{z,1}\frac{\delta y_1}{\delta z} + P_{z,2}\frac{\delta y_2}{\delta z} \tag{20.4}$$

so that

$$S_{y,w} = S_y - P_{z,1}\frac{\delta y_1}{\delta z} - P_{z,2}\frac{\delta y_2}{\delta z} \tag{20.5}$$

Again we note that δy_2 in Eqs. (20.4) and (20.5) is negative. Equation (20.5) may be used to determine the shear flow distribution in the web. For a completely idealized beam, the web shear flow is constant through the depth and is given by $S_{y,w}/h$. For a beam in which the web is fully effective in resisting direct stresses, the web shear flow distribution is found using Eq. (20.6) in which S_y is replaced by $S_{y,w}$ and which, for the beam of Fig. 20.1, simplifies to

$$q_s = -\frac{S_{y,w}}{I_{xx}}\left(\int_0^s t_D y \, ds + B_1 y_1\right) \tag{20.6}$$

or

$$q_s = -\frac{S_{y,w}}{I_{xx}}\left(\int_0^s t_D y \, ds + B_2 y_2\right) \tag{20.7}$$

Example 20.1

Determine the shear flow distribution in the web of the tapered beam shown in Fig. 20.2, at a section midway along its length. The web of the beam has a thickness of 2 mm and is fully effective in resisting direct stress. The beam tapers symmetrically about its horizontal centroidal axis and the cross-sectional area of each flange is 400 mm². See Ex. 1.1.

The internal bending moment and shear load at the section AA produced by the externally applied load are, respectively,

$$M_x = 20 \times 1 = 20 \text{ kNm}, \quad S_y = -20 \text{ kN}$$

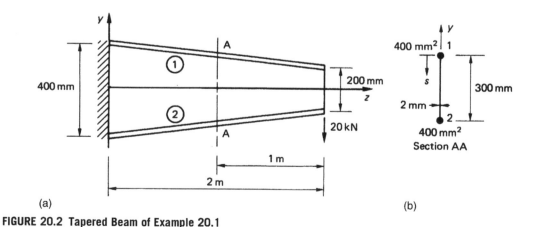

FIGURE 20.2 Tapered Beam of Example 20.1

The direct stresses parallel to the z axis in the flanges at this section are obtained from either Eq. (16.17) or (16.18) in which $M_y = 0$ and $I_{xy} = 0$. Thus, from Eq. (16.17),

$$\sigma_z = \frac{M_x y}{I_{xx}} \tag{i}$$

in which

$$I_{xx} = 2 \times 400 \times 150^2 + 2 \times 300^3 / 12$$

that is,

$$I_{xx} = 22.5 \times 10^6 \text{ mm}^4$$

Hence,

$$\sigma_{z,1} = -\sigma_{z,2} = \frac{20 \times 10^6 \times 150}{22.5 \times 10^6} = 133.3 \text{ N/mm}^2$$

The components parallel to the z axis of the axial loads in the flanges are therefore

$$P_{z,1} = -P_{z,2} = 133.3 \times 400 = 53,320 \text{ N}$$

The shear load resisted by the beam web is then, from Eq. (20.5),

$$S_{y,w} = -20 \times 10^3 - 53,320 \frac{\delta y_1}{\delta z} + 53,320 \frac{\delta y_2}{\delta z}$$

in which, from Figs. 20.1 and 20.2, we see that

$$\frac{\delta y_1}{\delta z} = \frac{-100}{2 \times 10^3} = -0.05, \quad \frac{\delta y_2}{\delta z} = \frac{100}{2 \times 10^3} = 0.05$$

Hence,

$$S_{y,w} = -20 \times 10^3 + 53,320 \times 0.05 + 53,320 \times 0.05 = -14,668 \text{ N}$$

The shear flow distribution in the web follows from either Eq. (20.6) or (20.7) and is (see Fig. 20.2(b))

$$q_{12} = \frac{14,668}{22.5 \times 10^6} \left(\int_0^s 2(150 - s)ds + 400 \times 150 \right)$$

that is,

$$q_{12} = 6.52 \times 10^{-4} \left(-s^2 + 300s + 60,000 \right) \tag{ii}$$

The maximum value of q_{12} occurs when $s = 150$ mm and q_{12} (max) $= 53.8$ N/mm. The values of shear flow at points 1 ($s = 0$) and 2 ($s = 300$ mm) are $q_1 = 39.1$ N/mm and $q_2 = 39.1$ N/mm; the complete distribution is shown in Fig. 20.3.

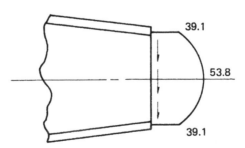

FIGURE 20.3 Shear Flow (N/mm) Distribution at Section AA in Example 20.1

20.2 OPEN AND CLOSED SECTION BEAMS

We now consider the more general case of a beam tapered in two directions along its length and comprising an arrangement of booms and skin. Practical examples of such a beam are complete wings and fuselages. The beam may be of open or closed section; the effects of taper are determined in an identical manner in either case.

Figure 20.4(a) shows a short length δz of a beam carrying shear loads S_x and S_y at the section z; S_x, and S_y are positive when acting in the directions shown. Note that, if the beam is of open cross-section, the shear loads are applied through its shear center, so that no twisting of the beam occurs. In addition to shear loads, the beam is subjected to bending moments M_x and M_y, which produce direct stresses σ_z in the booms and skin. Suppose that, in the rth boom, the direct stress in a direction parallel to the z axis is $\sigma_{z,r}$, which may be found using either Eq. (16.17) or (16.18). The component $P_{z,r}$ of the axial load P_r in the rth boom is then given by

$$P_{z,r} = \sigma_{z,r}B_r \tag{20.8}$$

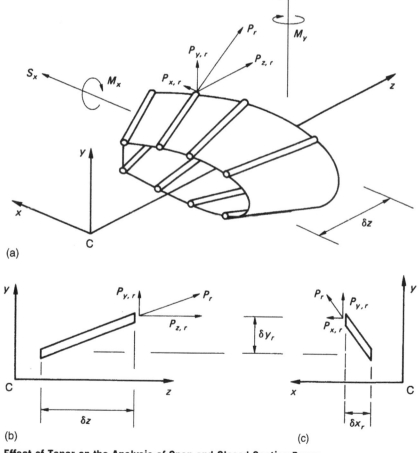

(a)

(b)

(c)

FIGURE 20.4 Effect of Taper on the Analysis of Open and Closed Section Beams

where B_r is the cross-sectional area of the rth boom.

From Fig. 20.4(b),

$$P_{y,r} = P_{z,r} \frac{\delta y_r}{\delta z} \tag{20.9}$$

Further, from Fig. 20.4(c),

$$P_{x,r} = P_{y,r} \frac{\delta x_r}{\delta y_r}$$

or, substituting for $P_{y,r}$ from Eq. (20.9),

$$P_{x,r} = P_{z,r} \frac{\delta x_r}{\delta z} \tag{20.10}$$

The axial load P_r is then given by

$$P_r = \left(P_{x,r}^2 + P_{y,r}^2 + P_{z,r}^2 \right)^{1/2} \tag{20.11}$$

or, alternatively,

$$P_r = P_{z,r} \frac{(\delta x_r^2 + \delta y_r^2 + \delta z^2)^{1/2}}{\delta z} \tag{20.12}$$

The applied shear loads S_x and S_y are reacted by the resultants of the shear flows in the skin panels and webs, together with the components $P_{x,r}$ and $P_{y,r}$ of the axial loads in the booms. Therefore, if $S_{x,w}$ and $S_{y,w}$ are the resultants of the skin and web shear flows and there is a total of m booms in the section,

$$S_x = S_{x,w} + \sum_{r=1}^{m} P_{x,r}, \qquad S_y = S_{y,w} + \sum_{r=1}^{m} P_{y,r} \tag{20.13}$$

Substituting in Eq. (20.13) for $P_{x,r}$ and $P_{y,r}$ from Eqs. (20.10) and (20.9), we have

$$S_x = S_{x,w} + \sum_{r=1}^{m} P_{z,r} \frac{\delta x_r}{\delta z}, \qquad S_y = S_{y,w} + \sum_{r=1}^{m} P_{z,r} \frac{\delta y_r}{\delta z} \tag{20.14}$$

Hence,

$$S_{x,w} = S_x - \sum_{r=1}^{m} P_{z,r} \frac{\delta x_r}{\delta z}, \qquad S_{y,w} = S_y - \sum_{r=1}^{m} P_{z,r} \frac{\delta y_r}{\delta z} \tag{20.15}$$

The shear flow distribution in an open section beam is now obtained using Eq. (20.6), in which S_x is replaced by $S_{x,w}$ and S_y by $S_{y,w}$ from Eq. (20.15). Similarly, for a closed section beam, S_x and S_y in Eq. (20.11) are replaced by $S_{x,w}$ and $S_{y,w}$. In the latter case, the moment equation (Eq. (17.17)) requires modification, due to the presence of the boom load components $P_{x,r}$ and $P_{y,r}$. Thus, from Fig. 20.5, we see that Eq. (17.17) becomes

$$S_x \eta_0 - S_y \xi_0 = \oint q_b p \, ds + 2A q_{s,0} - \sum_{r=1}^{m} P_{x,r} \eta_r + \sum_{r=1}^{m} P_{y,r} \xi_r \tag{20.16}$$

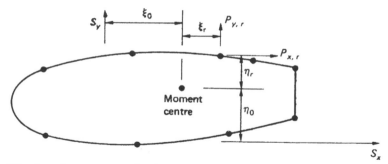

FIGURE 20.5 Modification of Moment Equation in Shear of Closed Section Beams Due to Boom Load

Equation (20.16) is directly applicable to a tapered beam subjected to forces positioned in relation to the moment center, as shown. Care must be taken in a particular problem to ensure that the moments of the forces are given the correct sign.

Example 20.2

The cantilever beam shown in Fig. 20.6 is uniformly tapered along its length in both x and y directions and carries a load of 100 kN at its free end. Calculate the forces in the booms and the shear flow distribution in the walls at a section 2 m from the built-in end if the booms resist all the direct stresses while the walls are effective only in shear. Each corner boom has a cross-sectional area of 900 mm², while both central booms have cross-sectional areas of 1,200 mm². See Ex. 1.1.

The internal force system at a section 2 m from the built-in end of the beam is

$$S_y = 100 \text{ kN}, \ S_x = 0, \ M_x = -100 \times 2 = -200 \text{ kNm}, \ M_y = 0$$

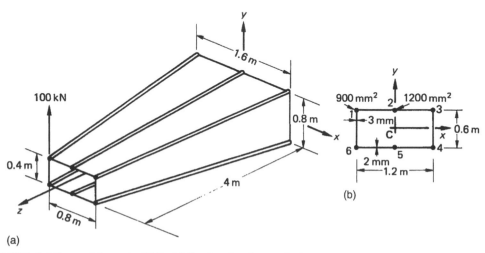

(a)

FIGURE 20.6 (a) Beam of Example 20.2; (b) Section 2 m from Built-in End

The beam has a doubly symmetrical cross-section, so that $I_{xy} = 0$ and Eq. (16.17) reduces to

$$\sigma_z = \frac{M_x y}{I_{xx}} \tag{i}$$

in which, for the beam section shown in Fig. 20.6(b),

$$I_{xx} = 4 \times 900 \times 300^2 + 2 \times 1,200 \times 300^2 = 5.4 \times 10^8 \text{ mm}^4$$

Then,

$$\sigma_{z,r} = \frac{-200 \times 10^6}{5.4 \times 10^8} y_r$$

or

$$\sigma_{z,r} = -0.37 y_r \tag{ii}$$

Hence,

$$P_{z,r} = -0.37 y_r B_r \tag{iii}$$

The value of $P_{z,r}$ is calculated from Eq. (iii) in column ② of Table 20.1; $P_{x,r}$ and $P_{y,r}$ follow from Eqs. (20.10) and (20.9), respectively, in columns ⑤ and ⑥. The axial load P_r, column ⑦, is given by $[②^2 + ⑤^2 + ⑥^2]^{1/2}$ and has the same sign as $P_{z,r}$ (see Eq. (20.12)). The moments of $P_{x,r}$ and $P_{y,r}$ are calculated for a moment center at the center of symmetry with counterclockwise moments taken as positive. Note that, in Table 20.1, $P_{x,r}$ and $P_{y,r}$ are positive when they act in the positive directions of the section x and y axes, respectively; the distances η_r and ξ_r of the lines of action of $P_{x,r}$ and $P_{y,r}$ from the moment center are not given signs, since it is simpler to determine the sign of each moment, $P_{x,r}\eta_r$ and $P_{y,r}\xi_r$, by referring to the directions of $P_{x,r}$ and $P_{y,r}$ individually.
    From column ⑥,

$$\sum_{r=1}^{6} P_{y,r} = 33.4 \text{ kN}$$

From column ⑩,

$$\sum_{r=1}^{6} P_{x,r}\eta_r = 0$$

Table 20.1 Example 20.2

① Boom	② $P_{z,r}$ (kN)	③ $\delta x_r/\delta z$	④ $\delta y_r/\delta z$	⑤ $P_{x,r}$ (kN)	⑥ $P_{y,r}$ (kN)	⑦ P_r (kN)	⑧ ξ_r (m)	⑨ η_r (m)	$P_{x,r}\eta_r$ (kNm)	$P_{y,r}\xi_r$ (kNm)
1	−100	0.1	−0.05	−10	5	−101.3	0.6	0.3	3	−3
2	−133	0	−0.05	0	6.7	−177.3	0	0.3	0	0
3	−100	−0.1	−0.05	10	5	−101.3	0.6	0.3	−3	3
4	100	−0.1	0.05	−10	5	101.3	0.6	0.3	−3	3
5	133	0	0.05	0	6.7	177.3	0	0.3	0	0
6	100	0.1	0.05	10	5	101.3	0.6	0.3	3	−3

From column ,

$$\sum_{r=1}^{6} P_{y,r}\xi_r = 0$$

From Eq. (20.15),

$$S_{x,w} = 0, \quad S_{y,w} = 100 - 33.4 = 66.6 \text{ kN}$$

The shear flow distribution in the walls of the beam is now found using the method described in Section 20.3. Since, for this beam, $I_{xy} = 0$ and $S_x = S_{x,w} = 0$, Eq. (20.11) reduces to

$$q_s = \frac{-S_{y,w}}{I_{xx}}\sum_{r=1}^{n} B_r y_r + q_{s,0} \tag{iv}$$

We now "cut" one of the walls, say 16. The resulting "open section" shear flow is given by

$$q_b = -\frac{66.6 \times 10^3}{5.4 \times 10^8}\sum_{r=1}^{n} B_r y_r$$

or

$$q_b = -1.23 \times 10^{-4}\sum_{r=1}^{n} B_r y_r \tag{v}$$

Thus,

$$q_{b,16} = 0$$
$$q_{b,12} = 0 - 1.23 \times 10^{-4} \times 900 \times 300 = -33.2 \text{ N/mm}$$
$$q_{b,23} = -33.2 - 1.23 \times 10^{-4} \times 1,200 \times 300 = -77.5 \text{ N/mm}$$
$$q_{b,34} = -77.5 - 1.23 \times 10^{-4} \times 900 \times 300 = -110.7 \text{ N/mm}$$
$$q_{b,45} = -77.5 \text{ N/mm (from symmetry)}$$
$$q_{b,56} = -33.2 \text{ N/mm (from symmetry)}$$

giving the distribution shown in Fig. 20.7. Taking moments about the center of symmetry, we have, from Eq. (20.16),

$$-100 \times 10^3 \times 600 = 2 \times 33.2 \times 600 \times 300 + 2 \times 77.5 \times 600 \times 300 + 110.7 \times 600 \times 600 + 2 \times 1,200 \times 600 q_{s,0}$$

from which $q_{s,0} = -97.0$ N/mm (i.e., clockwise). The complete shear flow distribution is found by adding the value of $q_{s,0}$ to the q_b shear flow distribution of Fig. 20.7 and is shown in Fig. 20.8.

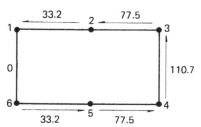

FIGURE 20.7 "Open Section" Shear Flow (N/mm) Distribution in Beam Section of Example 20.2

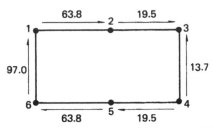

FIGURE 20.8 Shear Flow (N/mm) Distribution in Beam Section of Example 20.2

20.3 BEAMS HAVING VARIABLE STRINGER AREAS

In many aircraft, structural beams, such as wings, have stringers whose cross-sectional areas vary in the spanwise direction. The effects of this variation on the determination of shear flow distribution cannot therefore be found by the methods described in Section 20.3, which assume constant boom areas. In fact, as we noted in Section 20.3, if the stringer stress is made constant by varying the area of cross-section, there is no change in shear flow, as the stringer–boom is crossed.

The calculation of shear flow distributions in beams having variable stringer areas is based on the alternative method for the calculation of shear flow distributions described in Section 20.3 and illustrated in the alternative solution to Example 20.3. The stringer loads $P_{z,1}$ and $P_{z,2}$ are calculated at two sections z_1 and z_2 of the beam a convenient distance apart. We assume that the stringer load varies linearly along its length so that the change in stringer load per unit length of beam is given by

$$\Delta P = \frac{P_{z,1} - P_{z,2}}{z_1 - z_2}$$

The shear flow distribution follows as previously described.

Example 20.3

Solve Example 20.2 by considering the differences in boom load at sections of the beam either side of the specified section. See Ex. 1.1.

In this example, the stringer areas do not vary along the length of the beam but the method of solution is identical.

We are required to find the shear flow distribution at a section 2 m from the built-in end of the beam. We therefore calculate the boom loads at sections, say, 0.1 m either side of this section. Thus, at a distance 2.1 m from the built-in end,

$$M_x = -100 \times 1.9 = -190 \text{ kNm}$$

The dimensions of this section are easily found by proportion and are width $= 1.18$ m, depth $= 0.59$ m. Therefore, the second moment of area is

$$I_{xx} = 4 \times 900 \times 295^2 + 2 \times 1,200 \times 295^2 = 5.22 \times 10^8 \text{ mm}^4$$

and

$$\sigma_{z,r} = \frac{-190 \times 10^6}{5.22 \times 10^8} y_r = -0.364 y_r$$

Hence,

$$P_1 = P_3 = -P_4 = -P_6 = -0.364 \times 295 \times 900 = -96{,}642 \text{ N}$$

and

$$P_2 = -P_5 = -0.364 \times 295 \times 1{,}200 = -128{,}856 \text{ N}$$

At a section 1.9 m from the built-in end,

$$M_x = -100 \times 2.1 = -210 \text{ kNm}$$

and the section dimensions are width $= 1.22$ m, depth $= 0.61$ m, so that

$$I_{xx} = 4 \times 900 \times 305^2 + 2 \times 1{,}200 \times 305^2 = 5.58 \times 10^8 \text{ mm}^4$$

and

$$\sigma_{z,r} = \frac{-210 \times 10^6}{5.58 \times 10^8} y_r = -0.376 y_r$$

Hence,

$$P_1 = P_3 = -P_4 = -P_6 = -0.376 \times 305 \times 900 = -103{,}212 \text{ N}$$

and

$$P_2 = -P_5 = -0.376 \times 305 \times 1{,}200 = -137{,}616 \text{ N}$$

Thus, there is an increase in compressive load of $103{,}212 - 96{,}642 = 6{,}570$ N in booms 1 and 3 and an increase in tensile load of 6,570 N in booms 4 and 6 between the two sections. Also, the compressive load in boom 2 increases by $137{,}616 - 128{,}856 = 8{,}760$ N, while the tensile load in boom 5 increases by 8,760 N. Therefore, the change in boom load per unit length is given by

$$\Delta P_1 = \Delta P_3 = -\Delta P_4 = -\Delta P_6 = \frac{6{,}570}{200} = 32.85 \text{ N}$$

and

$$\Delta P_2 = -\Delta P_5 = \frac{8{,}760}{200} = 43.8 \text{ N}$$

The situation is illustrated in Fig. 20.9. Suppose now that the shear flows in the panels 12, 23, 34, and so forth are q_{12}, q_{23}, q_{34}, and so on and consider the equilibrium of boom 2, as shown in Fig. 20.10, with adjacent portions of the panels 12 and 23. Thus,

$$q_{23} + 43.8 - q_{12} = 0$$

or

$$q_{23} = q_{12} - 43.8$$

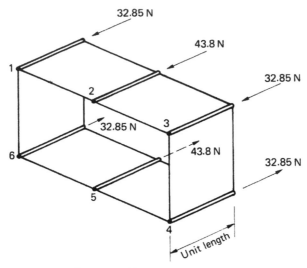

FIGURE 20.9 Change in Boom Loads/Unit Length of Beam

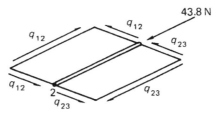

FIGURE 20.10 Equilibrium of Boom

Similarly,

$$q_{34} = q_{23} - 32.85 = q_{12} - 76.65$$
$$q_{45} = q_{34} + 32.85 = q_{12} - 43.8$$
$$q_{56} = q_{45} + 43.8 = q_{12}$$
$$q_{61} = q_{45} + 32.85 = q_{12} + 32.85$$

The moment resultant of the internal shear flows, together with the moments of the components $P_{y,r}$ of the boom loads about any point in the cross-section, is equivalent to the moment of the externally applied load about the same point. We note, from Example 20.2, that for moments about the center of symmetry,

$$\sum_{r=1}^{6} P_{x,r}\eta_r = 0, \quad \sum_{r=1}^{6} P_{y,r}\xi_r = 0$$

Therefore, taking moments about the center of symmetry,

$$100 \times 10^3 \times 600 = 2q_{12} \times 600 \times 300 + 2(q_{12} - 43.8)600 \times 300 + (q_{12} - 76.65)600 \times 600$$
$$+ (q_{12} + 32.85)600 \times 600$$

from which

$$q_{12} = 62.5 \text{ N/mm}$$

whence

$$q_{23} = 19.7 \text{ N/mm}, \qquad q_{34} = -13.2 \text{ N/mm}, \qquad q_{45} = 19.7 \text{ N/mm},$$
$$q_{56} = 63.5 \text{ N/mm}, \qquad q_{61} = 96.4 \text{ N/mm}$$

so that the solution is almost identical to the longer exact solution of Example 20.2.

The shear flows q_{12}, q_{23}, and so forth induce complementary shear flows q_{12}, q_{23}, and so on in the panels in the longitudinal direction of the beam; these are, in fact, the average shear flows between the two sections considered. For a complete beam analysis, this procedure is applied to a series of sections along the span. The distance between adjacent sections may be taken to be any convenient value; for actual wings, distances of the order of 350–700 mm are usually chosen. However, for very small values, small percentage errors in $P_{z,1}$ and $P_{z,2}$ result in large percentage errors in ΔP. On the other hand, if the distance is too large, the average shear flow between two adjacent sections may not be quite equal to the shear flow midway between the sections.

PROBLEMS

P.20.1. A wing spar has the dimensions shown in Fig. P.20.1 and carries a uniformly distributed load of 15 kN/m along its complete length. Each flange has a cross-sectional area of 500 mm^2 with the top flange being horizontal. If the flanges are assumed to resist all direct loads while the spar web is effective only in shear, determine the flange loads and the shear flows in the web at sections 1 and 2 m from the free end.

Answer: 1 m from free end: $P_U = 25$ kN (tension), $P_L = 25.1$ kN (compression),
$q = 41.7 \text{ N/mm}$
2 m from free end: $P_U = 75$ kN (tension), $P_L = 75.4$ kN (copmpression),
$q = 56.3 \text{ N/mm}$

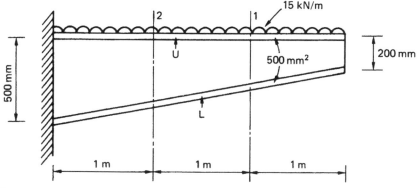

FIGURE P.20.1

P.20.2. If the web in the wing spar of Problem 20.1 has a thickness of 2 mm and is fully effective in resisting direct stresses, calculate the maximum value of shear flow in the web at a section 1 m from the free end of the beam.

Answer: 46.8 N/mm

P.20.2 MATLAB Use MATLAB to repeat Problem P.20.2 for web thickness values (t) from 1 to 2.6 mm in increments of 0.2 mm.

Answer:
(i) $t = 1$ mm : 44 N/mm^2
(ii) $t = 1.2$ mm : 45.1 N/mm^2
(iii) $t = 1.4$ mm : 45.4 N/mm^2
(iv) $t = 1.6$ mm : 45.6 N/mm^2
(v) $t = 1.8$ mm : 46.7 N/mm^2
(vi) $t = 2$ mm : 46.8 N/mm^2
(vii) $t = 2.2$ mm : 46.9 N/mm^2
(viii) $t = 2.4$ mm : 47.9 N/mm^2
(ix) $t = 2.6$ mm : 48 N/mm^2

P.20.3. Calculate the shear flow distribution and the stringer and flange loads in the beam shown in Fig. P.20.3 at a section 1.5 m from the built-in end. Assume that the skin and web panels are effective in resisting shear stress only; the beam tapers symmetrically in a vertical direction about its longitudinal axis.

Answer: $q_{13} = q_{42} = 36.9$ N/mm, $q_{35} = q_{64} = 7.3$ N/mm, $q_{21} = 96.2$ N/mm,
$q_{65} = 22.3$ N/mm
$P_2 = -P_1 = 133.3$ kN, $P_4 = P_6 = -P_3 = -P_5 = 66.7$ kN

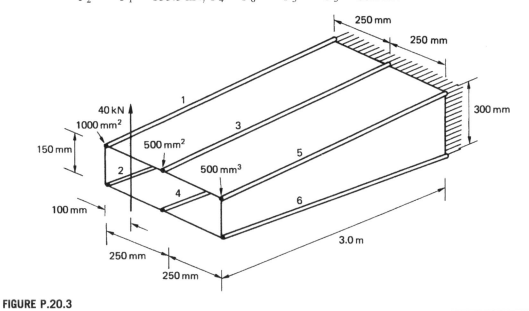

FIGURE P.20.3

Fuselages

21

Aircraft fuselages consist, as we saw in Chapter 11, of thin sheets of material stiffened by large numbers of longitudinal stringers together with transverse frames. Generally, they carry bending moments, shear forces, and torsional loads, which induce axial stresses in the stringers and skin, together with shear stresses in the skin; the resistance of the stringers to shear forces is generally ignored. Also, the distance between adjacent stringers is usually small, so that the variation in shear flow in the connecting panel is small. It is therefore reasonable to assume that the shear flow is constant between adjacent stringers, so that the analysis simplifies to the analysis of an idealized section in which the stringers/booms carry all the direct stresses while the skin is effective only in shear. The direct stress carrying capacity of the skin may be allowed for by increasing the stringer/boom areas as described in Section 20.3. The analysis of fuselages therefore involves the calculation of direct stresses in the stringers and the shear stress distributions in the skin; the latter are also required in the analysis of transverse frames, as we shall see in Chapter 23.

21.1 BENDING

The skin–stringer arrangement is idealized into one comprising booms and skin, as described in Section 20.3. The direct stress in each boom is then calculated using either Eq. (16.17) or (16.18), in which the reference axes and the section properties refer to the direct stress carrying areas of the cross-section.

Example 21.1

The fuselage of a light passenger carrying aircraft has the circular cross-section shown in Fig. 21.1(a). The cross-sectional area of each stringer is 100 mm^2 and the vertical distances given in Fig. 21.1(a) are to the mid-line of the section wall at the corresponding stringer position. If the fuselage is subjected to a bending moment of 200 kN m applied in the vertical plane of symmetry, at this section, calculate the direct stress distribution. See Ex. 1.1.

The section is first idealized using the method described in Section 20.3. As an approximation, we assume that the skin between adjacent stringers is flat, so that we may use either Eq. (20.1) or (20.2) to determine the boom areas. From symmetry, $B_1 = B_9$, $B_2 = B_8 = B_{10} = B_{16}$, $B_3 = B_7 = B_{11} = B_{15}$, $B_4 = B_6 = B_{12} = B_{14}$, and $B_5 = B_{13}$. From Eq. (20.1),

$$B_1 = 100 + \frac{0.8 \times 149.6}{6}\left(2 + \frac{\sigma_2}{\sigma_1}\right) + \frac{0.8 \times 149.6}{6}\left(2 + \frac{\sigma_{16}}{\sigma_1}\right)$$

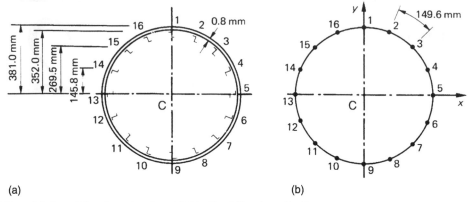

FIGURE 21.1 (a) Actual Fuselage Section; (b) Idealized Fuselage Section

that is,

$$B_1 = 100 + \frac{0.8 \times 149.6}{6}\left(2 + \frac{352.0}{381.0}\right) \times 2 = 216.6 \text{ mm}^2$$

Similarly $B_2 = 216.6$ mm^2, $B_3 = 216.6$ mm^2, $B_4 = 216.7$ mm^2. We note that stringers 5 and 13 lie on the neutral axis of the section and are therefore unstressed; the calculation of boom areas B_5 and B_{13} does not then arise. For this particular section, $I_{xy} = 0$, since Cx (and Cy) is an axis of symmetry. Further, $M_y = 0$, so that Eq. (16.17) reduces to

$$\sigma_z = \frac{M_x y}{I_{xx}}$$

in which

$$I_{xx} = 2 \times 216.6 \times 381.0^2 + 4 \times 216.6 \times 352.0^2 + 4 \times 216.6 \times 2,695^2 + 4 \times 216.7 \times 145.8^2$$
$$= 2.52 \times 10^8 \text{ mm}^4$$

The solution is completed in Table 21.1.

Table 21.1 Example 21.1		
Stringer/boom	**y (mm)**	**σ_z (N/mm^2)**
1	381.0	302.4
2,16	352.0	279.4
3,15	269.5	213.9
4,14	145.8	115.7
5,13	0	0
6,12	−145.8	−115.7
7,11	−269.5	−213.9
8,10	−352.0	−279.4
9	−381.0	−302.4

21.2 SHEAR

For a fuselage having a cross-section of the type shown in Fig. 21.1(a), the determination of the shear flow distribution in the skin produced by shear is basically the analysis of an idealized single-cell closed section beam. The shear flow distribution is therefore given by Eq. (20.11), in which the direct stress carrying capacity of the skin is assumed to be zero, that is, $t_D = 0$, thus,

$$q_s = -\left(\frac{S_x I_{xx} - S_y I_{xy}}{I_{xx} I_{yy} - I_{xy}^2}\right) \sum_{r=1}^{n} B_r y_r - \left(\frac{S_y I_{yy} - S_x I_{xy}}{I_{xx} I_{yy} - I_{xy}^2}\right) \sum_{r=1}^{n} B_r x_r + q_{s,0} \qquad (21.1)$$

Equation (21.1) is applicable to loading cases in which the shear loads are not applied through the section shear center, so that the effects of shear and torsion are included simultaneously. Alternatively, if the position of the shear center is known, the loading system may be replaced by shear loads acting through the shear center together with a pure torque, and the corresponding shear flow distributions may be calculated separately and superimposed to obtain the final distribution.

Example 21.2

The fuselage of Example 21.1 is subjected to a vertical shear load of 100 kN applied at a distance of 150 mm from the vertical axis of symmetry, as shown, for the idealized section, in Fig. 21.2. Calculate the distribution of shear flow in the section. See Ex. 1.1.

As in Example 21.1, $I_{xy} = 0$ and, since $S_x = 0$, Eq. (21.1) reduces to

$$q_s = -\frac{S_y}{I_{xx}} \sum_{r=1}^{n} B_r y_r + q_{s,0} \qquad (i)$$

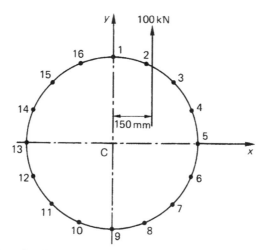

FIGURE 21.2 Idealized Fuselage Section of Example 21.2

in which $I_{xx} = 2.52 \times 10^8$ mm^4 as before. Then,

$$q_s = \frac{-100 \times 10^3}{2.52 \times 10^8} \sum_{r=1}^{n} B_r y_r + q_{s,0}$$

or

$$q_s = -3.97 \times 10^{-4} \sum_{r=1}^{n} B_r y_r + q_{s,0} \tag{ii}$$

The first term on the right-hand side of Eq. (ii) is the "open section" shear flow q_b. We therefore "cut" one of the skin panels, say, 12, and calculate q_b. The results are presented in Table 21.2.

Note that in Table 21.2, the column headed Boom indicates the boom that is crossed when the analysis moves from one panel to the next. Note also that, as would be expected, the q_b shear flow distribution is symmetrical about the Cx axis. The shear flow $q_{s,0}$ in panel 12 is now found by taking moments about a convenient moment center, say C. Therefore, from Eq. (17.17),

$$100 \times 10^3 \times 150 = \oint q_b p\, ds + 2A q_{s,0} \tag{iii}$$

in which $A = \pi \times 381.0^2 = 4.56 \times 10^5$ mm^2. Since the q_b shear flows are constant between the booms, Eq. (iii) may be rewritten in the form (see Eq. (20.10))

$$100 \times 10^3 \times 150 = -2A_{12}q_{b,12} - 2A_{23}q_{b,23} - \cdots - 2A_{161}q_{b,161} + 2A q_{s,0} \tag{iv}$$

in which $A_{12}, A_{23}, \cdots, A_{161}$ are the areas subtended by the skin panels 12, 23, $\cdots$, 161 at the center C of the circular cross-section and counterclockwise moments are taken as positive. Clearly, $A_{12} = A_{23} = \ldots = A_{161} = 4.56 \times 10^5/16 = 28{,}500$ mm^2. Equation (iv) then becomes

$$100 \times 10^3 \times 150 = 2 \times 28{,}500(-q_{b12} - q_{b23} - \cdots - q_{b161}) + 2 \times 4.56 \times 10^5 q_{s,0} \tag{v}$$

Table 21.2 Example 21.2

Skin panel		Boom	B_r (mm^2)	y_r (mm)	q_b (N/mm)
1	2	—	—	—	0
2	3	2	216.6	352.0	−30.3
3	4	3	216.6	269.5	−53.5
4	5	4	216.7	145.8	−66.0
5	6	5	—	0	−66.0
6	7	6	216.7	−145.8	−53.5
7	8	7	216.6	−269.5	−30.3
8	9	8	216.6	−352.0	0
1	16	1	216.6	381.0	−32.8
16	15	16	216.6	352.0	−63.1
15	14	15	216.6	269.5	−86.3
14	13	14	216.6	145.8	−98.8
13	12	13	—	0	−98.8
12	11	12	216.7	−145.8	−86.3
11	10	11	216.6	−269.5	−63.1
10	9	10	216.6	−352.0	−32.8

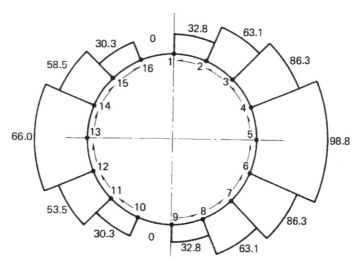

FIGURE 21.3 Shear Flow (N/mm) Distribution in the Fuselage Section of Example 21.2

Substituting the values of q_b from Table 21.2 in Eq. (v), we obtain

$$100 \times 10^3 \times 150 = 2 \times 28,500(-262.4) + 2 \times 4.56 \times 10^5 q_{s,0}$$

from which

$$q_{s,0} = 32.8 \,\text{N/mm (acting in a counterclockwise sense)}$$

The complete shear flow distribution follows by adding the value of $q_{s,0}$ to the q_b shear flow distribution, giving the final distribution shown in Fig. 21.3. The solution may be checked by calculating the resultant of the shear flow distribution parallel to the Cy axis. Thus,

$$2[(98.8 + 66.0)145.8 + (86.3 + 53.5)123.7 + (63.1 + 30.3)82.5 + (32.8 - 0)29.0] \times 10^{-3} = 99.96 \,\text{kN}$$

which agrees with the applied shear load of 100 kN. The analysis of a fuselage tapered along its length is carried out using the method described in Section 21.2 and illustrated in Example 21.2.

21.3 TORSION

A fuselage section is basically a single-cell closed section beam. The shear flow distribution produced by a pure torque is therefore given by Eq. (18.1) and is

$$q = \frac{T}{2A} \tag{21.2}$$

It is immaterial whether or not the section has been idealized, since, in both cases, the booms are assumed not to carry shear stresses.

Equation (21.2) provides an alternative approach to that illustrated in Example 21.2 for the solution of shear loaded sections in which the position of the shear center is known. In Fig. 21.1, the shear center coincides with the center of symmetry, so that the loading system may be replaced

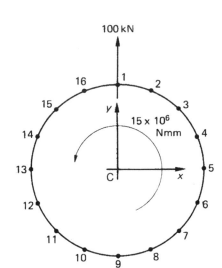

FIGURE 21.4 Alternative Solution of Example 21.2

by the shear load of 100kN acting through the shear center together with a pure torque equal to $100 \times 10^3 \times 150 = 15 \times 10^6$ Nmm, as shown in Fig. 21.4. The shear flow distribution due to the shear load may be found using the method of Example 21.2 but with the left-hand side of the moment equation (iii) equal to zero for moments about the center of symmetry. Alternatively, use may be made of the symmetry of the section and the fact that the shear flow is constant between adjacent booms. Suppose that the shear flow in the panel 21 is q_{21}. Then, from symmetry and using the results of Table 21.2,

$$q_{9\,8} = q_{9\,10} = q_{16\,1} = q_{2\,1}$$
$$q_{3\,2} = q_{8\,7} = q_{10\,11} = q_{15\,16} = 30.3 + q_{2\,1}$$
$$q_{4\,3} = q_{7\,6} = q_{11\,12} = q_{14\,15} = 53.5 + q_{2\,1}$$
$$q_{5\,4} = q_{6\,5} = q_{12\,13} = q_{13\,14} = 66.0 + q_{2\,1}$$

The resultant of these shear flows is statically equivalent to the applied shear load, so that

$$4(29.0q_{2\,1} + 82.5q_{3\,2} + 123.7q_{4\,3} + 145.8q_{5\,4}) = 100 \times 10^3$$

Substituting for $q_{3\,2}$, $q_{4\,3}$ and $q_{5\,4}$ from the preceding, we obtain

$$4(381q_{2\,1} + 18,740.5) = 100 \times 10^3$$

from which

$$q_{2\,1} = 16.4 \text{N/mm}$$

and

$$q_{3\,2} = 46.7 \text{N/mm}, \quad q_{4\,3} = 69.9 \text{N/mm}, \quad q_{5\,4} = 83.4 \text{ N/mm, and so forth}$$

The shear flow distribution due to the applied torque is, from Eq. (21.2),

$$q = \frac{15 \times 10^6}{2 \times 4.56 \times 10^5} = 16.4 \text{N/mm}$$

acting in a counterclockwise sense completely around the section. This value of shear flow is now superimposed on the shear flows produced by the shear load; this gives the solution shown in Fig. 21.3; that is,

$$q_{2\,1} = 16.4 + 16.4 = 32.8 \text{ N/mm}$$
$$q_{16\,1} = 16.4 - 16.4 = 0, \text{ and so on}$$

21.4 CUT-OUTS IN FUSELAGES

So far, we have considered fuselages to be closed sections stiffened by transverse frames and longitudinal stringers. In practice, it is necessary to provide openings in these closed stiffened shells for, for example, doors, cockpits, bomb bays, and windows in passenger cabins. These openings or "cut-outs" produce discontinuities in the otherwise continuous shell structure, so that loads are redistributed in the vicinity of the cut-out, thereby affecting loads in the skin, stringers, and frames. Frequently, these regions must be heavily reinforced, resulting in unavoidable weight increases. In some cases, for example, door openings in passenger aircraft, it is not possible to provide rigid fuselage frames on each side of the opening, because the cabin space must not be restricted. In such situations, a rigid frame is placed around the opening to resist shear loads and to transmit loads from one side of the opening to the other.

The effects of smaller cut-outs, such as those required for rows of windows in passenger aircraft, may be found approximately as follows. Figure 21.5 shows a fuselage panel provided with cut-outs for

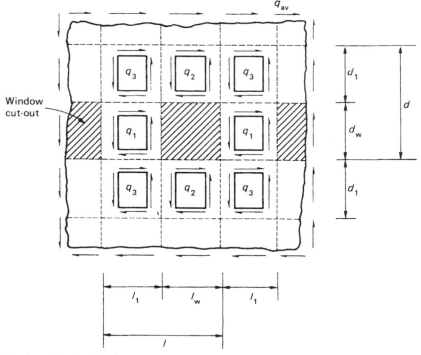

FIGURE 21.5 Fuselage Panel with Windows

windows, which are spaced a distance l apart. The panel is subjected to an average shear flow q_{av}, which would be the value of the shear flow in the panel without cut-outs. Considering a horizontal length of the panel through the cut-outs, we see that

$$q_1 l_1 = q_{av} l$$

or

$$q_1 = \frac{l}{l_1} q_{av} \tag{21.3}$$

Now, considering a vertical length of the panel through the cut-outs,

$$q_2 d_1 = q_{av} d$$

or

$$q_2 = \frac{d}{d_1} q_{av} \tag{21.4}$$

The shear flows q_3 may be obtained by considering either vertical or horizontal sections not containing the cut-out. Thus,

$$q_3 l_1 + q_2 l_w = q_{av} l$$

Substituting for q_2 from Eq. (21.3) and noting that $l = l_1 + l_w$ and $d = d_1 + d_w$, we obtain

$$q_3 = \left(1 - \frac{d_w}{d_1} \frac{l_w}{l_1} \right) q_{av} \tag{21.5}$$

PROBLEMS

P.21.1 The doubly symmetrical fuselage section shown in Fig. P.21.1 has been idealized into an arrangement of direct stress carrying booms and shear stress carrying skin panels; all the boom areas are 150 mm^2. Calculate the direct stresses in the booms and the shear flows in the panels when the section is subjected to a shear load of 50 kN and a bending moment of 100 kNm.

Answer: $\sigma_{z,1} = -\sigma_{z,6} = 180 \text{ N/mm}^2$, $\sigma_{z,2} = \sigma_{z,10} = -\sigma_{z,5} = -\sigma_{z,7} = 144.9 \text{ N/mm}^2$,
$\sigma_{z,3} = \sigma_{z,9} = -\sigma_{z,4} = -\sigma_{z,8} = 60 \text{ N/mm}^2$

$q_{21} = q_{65} = 1.9\text{N/mm}$, $q_{32} = q_{54} = 12.8 \text{ N/mm}$, $q_{43} = 17.3 \text{ N/mm}$,
$q_{67} = q_{101} = 11.6 \text{ N/mm}$, $q_{78} = q_{910} = 22.5 \text{ N/mm}$, $q_{89} = 27.0 \text{ N/mm}$

P.21.1 MATLAB Use MATLAB to repeat Problem P.21.1 for the following shear load (S_y) and bending moment (M_x) combinations.

	(i)	(ii)	(iii)	(iv)	(v)
S_y (kN)	40	40	50	60	50
M_x (kNm)	150	100	75	200	250

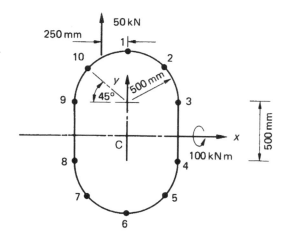

FIGURE P.21.1

Answer: (i) $[\sigma_{z,1},\ \sigma_{z,2},\ldots,\ \sigma_{z,10}] = [262.5,\ 211.2,\ 87.5,\ -87.5,\ -211.2,\ -262.5,\ -211.2,\\ -87.5,\ 87.5,\ 211.2]\ \mathrm{N/mm^2}$
$[q_{1\ 2},q_{2\ 3},\ldots,q_{10\ 1}] = [-1.4,\ -9.8,\ -13.3,\ -9.8,\ -1.4,\ 9.1,\ 17.5,\ 21,\\ 17.5,\ 9.1]\ \mathrm{N/mm}$

(ii) $[\sigma_{z,1},\ \sigma_{z,2},\ldots,\ \sigma_{z,10}] = [180,\ 144.9,\ 60,\ -60,\ -144.9,\ -180,\ -144.9,\ -60,\\ 60,\ 144.9]\ \mathrm{N/mm^2};$
$[q_{1\ 2},q_{2\ 3},\ldots,q_{10\ 1}] = [-1.9,\ -12.8,\ -17.3,\ -12.8,\ -1.9,\ 11.6,\ 22.5,\ 27,\\ 22.5,\ 11.6]\ \mathrm{N/mm}$

(iii) $[\sigma_{z,1},\ \sigma_{z,2},\ldots,\ \sigma_{z,10}] = [135,\ 108.6,\ 45,\ -45,\ -108.6,\ -135,\ -108.6,\ -45,\\ 45,\ 108.6]\ \mathrm{N/mm^2};$
$[q_{1\ 2},q_{2\ 3},\ldots,q_{10\ 1}] = [-2,\ -14.7,\ -20,\ -14.8,\ -2.1,\ 13.7,\ 26.4,\ 31.7,\ 26.4,\\ 13.8]\ \mathrm{N/mm}$

(iv) $[\sigma_{z,1},\ \sigma_{z,2},\ldots,\ \sigma_{z,10}] = [352.5,\ 283.7,\ 117.5,\ -117.5,\ -283.7,\ -352.5,\ -283.7,\\ -117.5,\ 117.5,\ 283.7]\ \mathrm{N/mm^2};$
$[q_{1\ 2},q_{2\ 3},\ldots,q_{10\ 1}] = [-1.9,\ -12.8,\ -17.3,\ -12.8,\ -1.9,\ 11.6,\ 22.5,\ 27,\\ 22.5,\ 11.6]\ \mathrm{N/mm}$

(v) $[\sigma_{z,1},\ \sigma_{z,2},\ldots,\ \sigma_{z,10}] = [442.5,\ 356.1,\ 147.5,\ -147.5,\ -356.1,\ -442.5,\\ -356.1,\ -147.5,\ 147.5,\ 356.1]\ \mathrm{N/mm^2};$
$[q_{1\ 2},q_{2\ 3},\ldots,q_{10\ 1}] = [-1.4,\ -9.8,\ -13.3,\ -9.8,\ -1.4,\ 9.1,\ 17.5,\ 21,\\ 17.5,\ 9.1]\ \mathrm{N/mm}$

P.21.2 Determine the shear flow distribution in the fuselage section of P.21.1 by replacing the applied load by a shear load through the shear center together with a pure torque.

Wings

We have seen, in Chapters 11 and 19, that wing sections consist of thin skins stiffened by combinations of stringers, spar webs, and caps and ribs. The resulting structure frequently comprises one, two, or more cells and is highly redundant. However, as in the case of fuselage sections, the large number of closely spaced stringers allows the assumption of a constant shear flow in the skin between adjacent stringers, so that a wing section may be analyzed as though it were completely idealized, as long as the direct stress carrying capacity of the skin is allowed for by additions to the existing stringer/boom areas. We shall investigate the analysis of multicellular wing sections subjected to bending, torsional, and shear loads, although, initially, it is instructive to examine the special case of an idealized three-boom shell.

22.1 THREE-BOOM SHELL

The wing section shown in Fig. 22.1 has been idealized into an arrangement of direct stress carrying booms and shear stress only carrying skin panels. The part of the wing section aft of the vertical spar 31 performs only an aerodynamic role and is therefore unstressed. Lift and drag loads, S_y and S_x, induce shear flows in the skin panels, which are constant between adjacent booms, since the section has been completely idealized. Therefore, resolving horizontally and noting that the resultant of the internal shear flows is equivalent to the applied load, we have

$$S_x = -q_{12}l_{12} + q_{23}l_{23} \tag{22.1}$$

Now resolving vertically,

$$S_y = q_{31}(h_{12} + h_{23}) - q_{12}h_{12} - q_{23}h_{23} \tag{22.2}$$

Finally, taking moments about, say, boom 3,

$$S_x\eta_0 + S_y\xi_0 = -2A_{12}q_{12} - 2A_{23}q_{23} \tag{22.3}$$

(see Eqs. (20.9) and (20.10)). In the preceding, there are three unknown values of shear flow, q_{12}, q_{23}, q_{31} and three equations of statical equilibrium. We conclude therefore that a three-boom idealized shell is statically determinate.

We shall return to the simple case of a three-boom wing section when we examine the distributions of direct load and shear flows in wing ribs. Meanwhile, we consider the bending, torsion, and shear of multicellular wing sections.

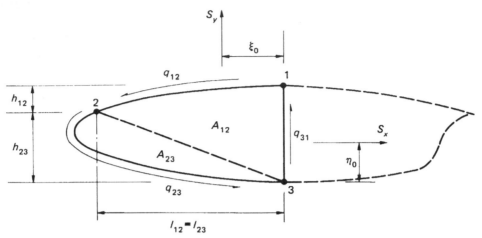

FIGURE 22.1 Three-Boom Wing Section

22.2 BENDING

Bending moments at any section of a wing are usually produced by shear loads at other sections of the wing. The direct stress system for such a wing section (Fig. 22.2) is given by either Eq. (16.17) or (16.18), in which the coordinates (x,y) of any point in the cross-section and the sectional properties are referred to axes Cxy, in which the origin C coincides with the centroid of the direct stress carrying area.

Example 22.1

The wing section shown in Fig. 22.3 has been idealized such that the booms carry all the direct stresses. If the wing section is subjected to a bending moment of 300 kNm applied in a vertical plane, calculate the direct stresses in the booms:

Boom areas: $B_1 = B_6 = 2,580 \text{ mm}^2$, $B_2 = B_5 = 3,880 \text{ mm}^2$, $B_3 = B_4 = 3,230 \text{ mm}^2$

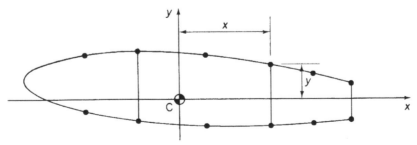

FIGURE 22.2 Idealized Section of a Multicell Wing

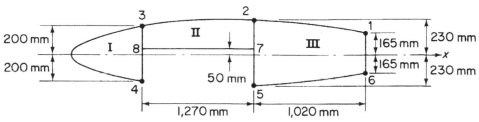

FIGURE 22.3 Wing Section of Example 22.1

Table 22.1	Example 22.1	
Boom	y **(mm)**	σ_z **(N/mm²)**
1	165	61.2
2	230	85.3
3	200	74.2
4	−200	−74.2
5	−230	−85.3
6	−165	−61.2

We note that the distribution of the boom areas is symmetrical about the horizontal x axis. Hence, in Eq. (16.17), $I_{xy} = 0$. Further, $M_x = 300$ kNm and $M_y = 0$, so that Eq. (16.17) reduces to

$$\sigma_z = \frac{M_x y}{I_{xx}} \tag{i}$$

in which

$$I_{xx} = 2\left(2,580 \times 165^2 + 3,880 \times 230^2 + 3,230 \times 200^2\right) = 809 \times 10^6 \ \text{mm}^4$$

Hence,

$$\sigma_z = \frac{300 \times 10^6}{809 \times 10^6} y = 0.371 y \tag{ii}$$

The solution is now completed in Table 22.1, in which positive direct stresses are tensile and negative direct stresses compressive.

22.3 TORSION

The chordwise pressure distribution on an aerodynamic surface may be represented by shear loads (lift and drag loads) acting through the aerodynamic center together with a pitching moment M_0 (see Section 12.1). This system of shear loads may be transferred to the shear center of the section in the form of shear loads S_x and S_y together with a torque T. The pure torsion case is considered here.

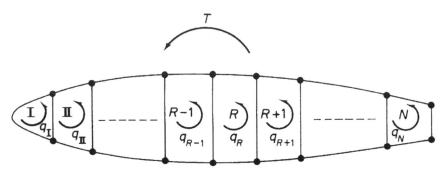

FIGURE 22.4 Multicell Wing Section Subjected to Torsion

In the analysis, we assume that no axial constraint effects are present and that the shape of the wing section remains unchanged by the load application. In the absence of axial constraint, there is no development of direct stress in the wing section, so that only shear stresses are present. It follows that the presence of booms does not affect the analysis in the pure torsion case.

The wing section shown in Fig. 22.4 comprises N cells and carries a torque T, which generates individual but unknown torques in each of the N cells. Each cell therefore develops a constant shear flow $q_I, q_{II}, \ldots, q_R, \ldots, q_N$ given by Eq. (18.1).

The total is therefore

$$T = \sum_{R=1}^{N} 2A_R q_R \qquad (22.4)$$

Although Eq. (22.4) is sufficient for the solution of the special case of a single-cell section, which is therefore statically determinate, additional equations are required for an N-cell section. These are obtained by considering the rate of twist in each cell and the compatibility of displacement condition that all N cells possess the same rate of twist $d\theta/dz$; this arises directly from the assumption of an undistorted cross-section.

Consider the Rth cell of the wing section, shown in Fig. 22.5. The rate of twist in the cell is, from Eq. (17.22),

$$\frac{d\theta}{dz} = \frac{1}{2A_R G} \oint_R q \frac{ds}{t} \qquad (22.5)$$

The shear flow in Eq. (22.5) is constant along each wall of the cell and has the values shown in Fig. 22.5. Writing $\int ds/t$ for each wall as δ, Eq. (22.5) becomes

$$\frac{d\theta}{dz} = \frac{1}{2A_R G} [q_R \delta_{12} + (q_R - q_{R-1})\delta_{23} + q_R \delta_{34} + (q_R - q_{R+1})\delta_{41}]$$

or, rearranging the terms in square brackets,

$$\frac{d\theta}{dz} = \frac{1}{2A_R G} [-q_{R-1} \delta_{23} + q_R (\delta_{12} + \delta_{23} + \delta_{34} + \delta_{41}) - q_{R+1} \delta_{41}]$$

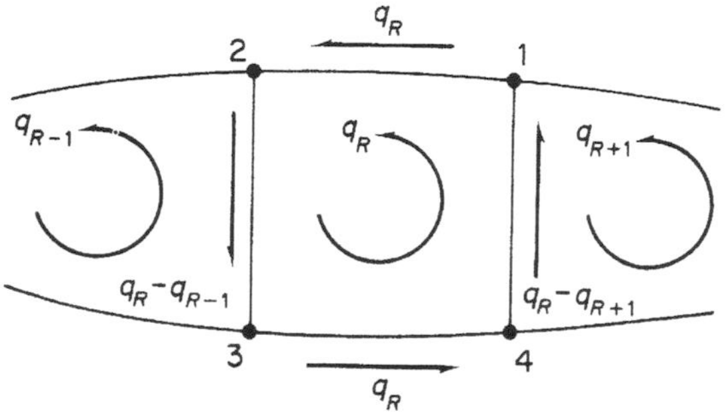

FIGURE 22.5 Shear Flow Distribution in the *R*th Cell of an *N*-Cell Wing Section

In general terms, this equation may be rewritten in the form

$$\frac{d\theta}{dz} = \frac{1}{2A_R G}\left(-q_{R-1}\delta_{R-1,R} + q_R\delta_R - q_{R+1}\delta_{R+1,R}\right) \tag{22.6}$$

in which $\delta_{R-1,R}$ is $\int ds/t$ for the wall common to the Rth and $(R-1)$th cells, δ_R is $\int ds/t$ for all the walls enclosing the Rth cell, and $\delta_{R+1,R}$ is $\int ds/t$ for the wall common to the Rth and $(R+1)$th cells.

The general form of Eq. (22.6) is applicable to multicell sections in which the cells are connected consecutively, that is, cell I is connected to cell II, cell II to cells I and III, and so on. In some cases, cell I may be connected to cells II and III, and so forth (see Problem P.22.4), so that Eq. (22.6) cannot be used in its general form. For this type of section, the term $\int q(ds/t)$ should be computed by considering $\int q(ds/t)$ for each wall of a particular cell in turn.

There are N equations of the type (22.6), which, with Eq. (22.4), comprise the $N+1$ equations required to solve for the N unknown values of shear flow and the one unknown value of $d\theta/dz$.

Frequently, in practice, the skin panels and spar webs are fabricated from materials possessing different properties such that the shear modulus G is not constant. The analysis of such sections is simplified if the actual thickness t of a wall is converted to a modulus-weighted thickness t^* as follows. For the Rth cell of an N-cell wing section in which G varies from wall to wall, Eq. (22.5) takes the form

$$\frac{d\theta}{dz} = \frac{1}{2A_R}\oint_R q\frac{ds}{Gt}$$

This equation may be rewritten as

$$\frac{d\theta}{dz} = \frac{1}{2A_R G_{\text{REF}}}\oint_R q\frac{ds}{(G/G_{\text{REF}})t} \tag{22.7}$$

in which G_{REF} is a convenient reference value of the shear modulus. Equation (22.7) is now rewritten as

$$\frac{d\theta}{dz} = \frac{1}{2A_R G_{\text{REF}}}\oint_R q\frac{ds}{t^*} \tag{22.8}$$

in which the modulus-weighted thickness t^* is given by

$$t^* = \frac{G}{G_{REF}} t \qquad (22.9)$$

Then, in Eq. (22.6), δ becomes $\int ds/t^*$.

Example 22.2

Calculate the shear stress distribution in the walls of the three-cell wing section shown in Fig. 22.6 when it is subjected to a counterclockwise torque of 11.3 kN m. The data are in Table 22.2. See Ex. 1.1.

Since the wing section is loaded by a pure torque, the presence of the booms has no effect on the analysis. Choosing $G_{REF} = 27,600$ N/mm^2, then, from Eq. (22.9),

$$t^*_{12^\circ} = \frac{24,200}{27,600} \times 1.22 = 1.07 \text{ mm}$$

Similarly,

$$t^*_{13} = t^*_{24} = 1.07 \text{ mm}, \quad t^*_{35} = t^*_{46} = t^*_{56} = 0.69 \text{ mm}$$

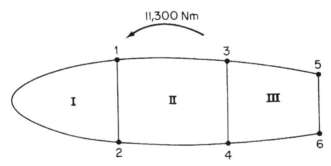

FIGURE 22.6 Wing Section of Example 22.2

Table 22.2 Example 22.2

Wall	Length (mm)	Thickness (mm)	G (N/mm²)	Cell area (mm²)
12°	1,650	1.22	24,200	$A_I = 258,000$
12^i	508	2.03	27,600	$A_{II} = 355,000$
13, 24	775	1.22	24,200	$A_{III} = 161,000$
34	380	1.63	27,600	
35, 46	508	0.92	20,700	
56	254	0.92	20,700	

Note: The superscript symbols o and i are used to distinguish between outer and inner walls connecting the same two booms.

Hence,

$$\delta_{12^\circ} = \int_{12^\circ} \frac{ds}{t^*} = \frac{1,650}{1.07} = 1542$$

Similarly,

$$\delta_{12^i} = 250, \quad \delta_{13} = \delta_{24} = 725, \quad \delta_{34} = 233, \quad \delta_{35} = \delta_{46} = 736, \quad \delta_{56} = 368$$

Substituting the appropriate values of δ in Eq. (22.6) for each cell in turn gives the following:

- For cell I,

$$\frac{d\theta}{dz} = \frac{1}{2 \times 258{,}000 G_{REF}} [q_I(1{,}542 + 250) - 250q_{II}] \tag{i}$$

- For cell II,

$$\frac{d\theta}{dz} = \frac{1}{2 \times 355{,}000 G_{REF}} [-250q_I + q_{II}(250 + 725 + 233 + 725) - 233q_{III}] \tag{ii}$$

- For cell III,

$$\frac{d\theta}{dz} = \frac{1}{2 \times 161{,}000 G_{REF}} [-233q_{II} + q_{III}(736 + 233 + 736 + 368)] \tag{iii}$$

In addition, from Eq. (22.4),

$$11.3 \times 10^6 = 2(258{,}000q_I + 355{,}000q_{II} + 161{,}000q_{III}) \tag{iv}$$

Solving Eqs. (i)–(iv) simultaneously gives

$$q_I = 7.1 \, \text{N/mm}, \quad q_{II} = 8.9 \, \text{N/mm}, \quad q_{III} = 4.2 \, \text{N/mm}$$

The shear stress in any wall is obtained by dividing the shear flow by the *actual* wall thickness. Hence, the shear stress distribution is as shown in Fig. 22.7.

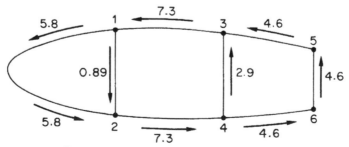

FIGURE 22.7 Shear Stress (N/mm^2) Distribution in the Wing Section of Example 22.2

22.4 SHEAR

Initially, we consider the general case of an N-cell wing section comprising booms and skin panels, the latter being capable of resisting both direct and shear stresses. The wing section is subjected to shear loads S_x and S_y, whose lines of action do not necessarily pass through the shear center S (see Fig. 22.8); the resulting shear flow distribution is therefore due to the combined effects of shear and torsion.

The method for determining the shear flow distribution and the rate of twist is based on a simple extension of the analysis of a single-cell beam subjected to shear loads (Sections 17.3 and 20.3). Such a beam is statically indeterminate, the single redundancy being selected as the value of shear flow at an arbitrarily positioned "cut." Thus, the N-cell wing section of Fig. 22.8 may be made statically determinate by cutting a skin panel in each cell, as shown. While the actual position of these cuts is theoretically immaterial, there are advantages to be gained from a numerical point of view if the cuts are made near the center of the top or bottom skin panel in each cell. Generally, at these points, the redundant shear flows ($q_{s,0}$) are small, so that the final shear flows differ only slightly from those of the determinate structure. The system of simultaneous equations from which the final shear flows are found will then be "well conditioned" and produce reliable results. The solution of an "ill-conditioned" system of equations would probably involve the subtraction of large numbers of a similar size, which therefore need to be expressed to a large number of significant figures for reasonable accuracy. Although this reasoning does not apply to a completely idealized wing section, since the calculated values of shear flow are constant between the booms, it is again advantageous to cut either the top or bottom skin panels for, in the special case of a wing section having a horizontal axis of symmetry, a cut in, say, the top skin panels results in the "'open section" shear flows (q_b) being zero in the bottom skin panels. This decreases the arithmetical labor and simplifies the derivation of the moment equation, as will become obvious in Example 22.4.

The open section shear flow q_b in the wing section of Fig. 22.8 is given by Eq. (20.6), that is,

$$q_b = -\left(\frac{S_x I_{xx} - S_y I_{xy}}{I_{xx}I_{yy} - I_{xy}^2}\right)\left(\int_0^s t_D x\,ds + \sum_{r=1}^n B_r x_r\right)$$
$$- \left(\frac{S_y I_{yy} - S_x I_{xy}}{I_{xx}I_{yy} - I_{xy}^2}\right)\left(\int_0^s t_D y\,ds + \sum_{r=1}^n B_r y_r\right)$$

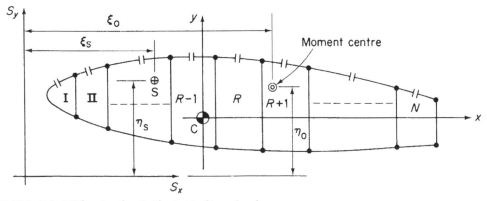

FIGURE 22.8 *N*-Cell Wing Section Subjected to Shear Loads

We are left with an unknown value of shear flow at each of the cuts, that is, $q_{s,0,\text{I}}$, $q_{s,0,\text{II}}$, $\ldots$, $q_{s,0,N}$, plus the unknown rate of twist $d\theta/dz$, which, from the assumption of an undistorted cross-section, is the same for each cell. Therefore, as in the torsion case, there are $N+1$ unknowns requiring $N+1$ equations for a solution.

Consider the Rth cell shown in Fig. 22.9. The complete distribution of shear flow around the cell is given by the summation of the open section shear flow q_b and the value of shear flow at the cut, $q_{s,0,R}$. We may therefore regard $q_{s,0,R}$ as a constant shear flow acting around the cell. The rate of twist is again given by Eq. (17.22); thus,

$$\frac{d\theta}{dz} = \frac{1}{2A_R G}\oint_R q\,\frac{ds}{t} = \frac{1}{2A_R G}\oint_R (q_b + q_{s,0,R})\frac{ds}{t}$$

By comparison with the pure torsion case, we deduce that

$$\frac{d\theta}{dz} = \frac{1}{2A_R G}\left(-q_{s,0,R-1}\delta_{R-1,R} + q_{s,0,R}\delta_R - q_{s,0,R+1}\delta_{R+1,R} + \oint_R q_b\frac{ds}{t}\right) \tag{22.10}$$

in which q_b has previously been determined. There are N equations of the type (22.10), so that a further equation is required to solve for the $N+1$ unknowns. This is obtained by considering the moment equilibrium of the Rth cell in Fig. 22.10.

The moment $M_{q,R}$ produced by the total shear flow about any convenient moment center O is given by

$$M_{q,R} = \oint q_R p_0\,ds \qquad \text{(see Section 18.1)}$$

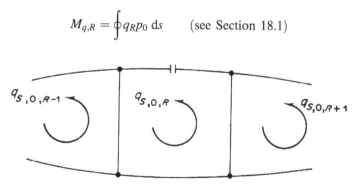

FIGURE 22.9 Redundant Shear Flow in the Rth Cell of an N-Cell Wing Section Subjected to Shear

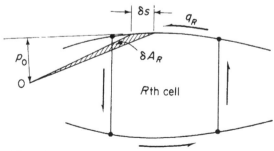

FIGURE 22.10 Moment Equilibrium of the Rth Cell

Substituting for q_R in terms of the open section shear flow q_b and the redundant shear flow $q_{s,0,R}$, we have

$$M_{q,R} = \oint_R q_b p_0 \, ds + q_{s,0,R} \oint_R p_0 \, ds$$

or

$$M_{q,R} = \oint_R q_b p_0 \, ds + 2A_R q_{s,0,R}$$

The sum of the moments from the individual cells is equivalent to the moment of the externally applied loads about the same point. Thus, for the wing section of Fig. 22.8,

$$S_x \eta_0 - S_y \xi_0 = \sum_{R=1}^{N} M_{q,R} = \sum_{R=1}^{N} \oint_R q_b p_0 \, ds + \sum_{R=1}^{N} 2A_R q_{s,0,R} \tag{22.11}$$

If the moment center is chosen to coincide with the point of intersection of the lines of action of S_x and S_y, Eq. (22.11) becomes

$$0 = \sum_{R=1}^{N} \oint_R q_b p_0 \, ds + \sum_{R=1}^{N} 2A_R q_{s,0,R} \tag{22.12}$$

Example 22.3

The wing section of Example 22.1 (Fig. 22.3) carries a vertically upward shear load of 86.8 kN in the plane of the web 572. The section has been idealized such that the booms resist all the direct stresses while the walls are effective only in shear. If the shear modulus of all walls is 27,600 N/mm^2 except for wall 78, for which it is three times this value, calculate the shear flow distribution in the section and the rate of twist. Additional data are given in Table 22.3. See Ex. 1.1.

Choosing G_{REF} as 27,600 N/mm^2, then, from Eq. (22.9),

$$t_{78}^* = \frac{3 \times 27,600}{27,600} \times 1.22 = 3.66 \text{ mm}$$

Table 22.3 Example 22.3

Wall	Length (mm)	Thickness (mm)	Cell area (mm^2)
12, 56	1,023	1.22	$A_I = 265,000$
23	1,274	1.63	$A_{II} = 213,000$
34	2,200	2.03	$A_{III} = 413,000$
483	400	2.64	
572	460	2.64	
61	330	1.63	
78	1,270	1.22	

Hence,

$$\delta_{78} = \frac{1,270}{3.66} = 347$$

Also,

$$\delta_{12} = \delta_{56} = 839, \; \delta_{23} - 782, \; \delta_{34} = 1,084, \; \delta_{38} = 57, \; \delta_{84} = 95, \; \delta_{87} = 347, \; \delta_{27} = 68, \; \delta_{75} = 106, \; \delta_{16} = 202$$

We now cut the top skin panels in each cell and calculate the open section shear flows using Eq. (20.6), which, since the wing section is idealized, singly symmetrical (as far as the direct stress carrying area is concerned), and subjected to a vertical shear load only, reduces to

$$q_b = \frac{-S_y}{I_{xx}} \sum_{r=1}^{n} B_r y_r \tag{i}$$

where, from Example 22.1, $I_{xx} = 809 \times 10^6$ mm^4. Thus, from Eq. (i),

$$q_b = -\frac{86.8 \times 10^3}{809 \times 10^6} \sum_{r=1}^{n} B_r y_r = -1.07 \times 10^{-4} \sum_{r=1}^{n} B_r y_r \tag{ii}$$

Since $q_b = 0$ at each "cut," $q_b = 0$ for the skin panels 12, 23, and 34. The remaining q_b shear flows are now calculated using Eq. (ii). Note that the order of the numerals in the subscript of q_b indicates the direction of movement from boom to boom.

$$q_{b,27} = -1.07 \times 10^{-4} \times 3,880 \times 230 = -95.5 \text{ N/mm}$$
$$q_{b,16} = -1.07 \times 10^{-4} \times 2,580 \times 165 = -45.5 \text{ N/mm}$$
$$q_{b,65} = -45.5 - 1.07 \times 10^{-4} \times 2,580 \times (-165) = 0$$
$$q_{b,57} = -1.07 \times 10^{-4} \times 3,880 \times (-230) = 95.5 \text{ N/mm}$$
$$q_{b,38} = -1.07 \times 10^{-4} \times 3,230 \times 200 = -69.1 \text{ N/mm}$$
$$q_{b,48} = -1.07 \times 10^{-4} \times 3,230 \times (-200) = 69.1 \text{ N/mm}$$

Therefore, as $q_{b,83} = q_{b,48}$ (or $q_{b,72} = q_{b,57}$), $q_{b,78} = 0$. The distribution of the q_b shear flows is shown in Fig. 22.11. The values of δ and q_b are now substituted in Eq. (22.10) for each cell in turn.

- For cell I,

$$d\theta/dz = [q_{s,0,I}(1,084 + 95 + 57) - 57q_{s,0,II} + 69.1 \times 95 + 69.1 \times 57]/(2 \times 265,000G_{REF}) \tag{iii}$$

- For cell II,

$$d\theta/dz = [-57q_{s,0,I} + q_{s,0,II}(782 + 57 + 347 + 68) - 68q_{s,0,III} + 95.5 \times 68 \\ -69.1 \times 57]/(2 \times 213,000G_{REF}) \tag{iv}$$

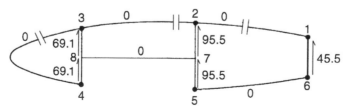

FIGURE 22.11 The q_b Distribution (N/mm)

- For cell III,

$$d\theta/dz = [-68q_{s,0,\text{II}} + q_{s,0,\text{III}}(839 + 68 + 106 + 839 + 202) + 45.5 \times 202 - 95.5 \times 68 \\ -95.5 \times 106]/(2 \times 413{,}000G_{\text{REF}}) \tag{v}$$

The solely numerical terms in Eqs. (iii)–(v) represent $\oint_R q_b(ds/t)$ for each cell. Care must be taken to ensure that the contribution of each q_b value to this term is interpreted correctly. The path of the integration follows the positive direction of $q_{s,0}$ in each cell, that is, counterclockwise. Thus, the positive contribution of $q_{b,83}$ to $\oint_{\text{II}} q_b(ds/t)$ becomes a negative contribution to $\oint_{\text{II}} q_b(ds/t)$ and so on.

The fourth equation required for a solution is obtained from Eq. (22.12) by taking moments about the intersection of the x axis and the web 572. Thus,

$$0 = -69.1 \times 250 \times 1{,}270 - 69.1 \times 150 \times 1{,}270 + 45.5 \times 330 \times 1{,}020 + 2 \times 265{,}000q_{s,0,\text{I}} \\ +2 \times 213{,}000q_{s,0,\text{II}} + 2 \times 413{,}000q_{s,0,\text{III}} \tag{vi}$$

Simultaneous solution of Eqs. (iii)–(vi) gives

$$q_{s,0,\text{I}} = 4.9\,\text{N/mm}, \quad q_{s,0,\text{II}} = 9.3\,\text{N/mm}, \quad q_{s,0,\text{III}} = 16.0\,\text{N/mm}$$

Superimposing these shear flows on the q_b distribution of Fig. 22.11, we obtain the final shear flow distribution:

$$q_{34} = 4.9\,\text{N/mm}, \quad q_{23} = q_{87} = 9.3\,\text{N/mm}, \quad q_{12} = q_{56} = 16.0\,\text{N/mm} \\ q_{61} = 61.5\,\text{N/mm}, \quad q_{57} = 79.5\,\text{N/mm}, \quad q_{72} = 88.7\,\text{N/mm} \\ q_{48} = 74.0\,\text{N/mm}, \quad q_{83} = 64.7\,\text{N/mm}$$

Finally, from any of Eqs. (iii)–(v),

$$d\theta/dz = 1.09 \times 10^{-6}\,\text{rad/mm}$$

Example 22.3 MATLAB

Use MATLAB to repeat Example 22.3. See Ex. 1.1.

The shear flow distribution and rate of twist in the section are obtained through the following MATLAB file:

```
% Declare any needed variables
syms q_sOI q_sOII q_sOIII
B = [2580 3880 3230 3230 3880 2580];
y = [165 230 200 -200 -230 -165];
e = 50;
w = 1020;
S_y = 86.8*10^3;
G = [27600 27600 27600 3*27600 27600 27600 27600 27600 27600 27600];
G_ref = 27600;

% Define the length, thickness, and cell areas provided
L = [1023 1274 2200 1270 1023 330 (y(2)-e) -(y(5)-e) (y(3)-e) -(y(4)-e)];
t = [1.22 1.63 2.03 1.22 1.22 1.63 2.64 2.64 2.64 2.64];
A = [265000 213000 413000];
```

```
% Calculate the modulus weighted thickness using Eq. (22.9)
t_s = t.*G/G_ref;

% Calculate the delta for each wall/cell
del = round(L./t_s);
delta_I = del(3)+sum(del(9:10));
delta_II = del(2)+del(9)+del(4)+del(7);
delta_III = del(1)+sum(del(5:8));


delta = [delta_I delta_II delta_III];
delta_I = [del(3) del(9) del(10)];
delta_II = [del(2) del(9) del(4) del(7)];
delta_III = [del(1) del(5) del(6) del(7) del(8)];

% Calculate the section properties
I_xx = sum(B.*y.^2);
 % I_yy and I_xy not needed due to symmetry and M_y=0
% Due to symmetry and since S_x=0, Eq. (20.6) simplifies
% Calculate the base shear flows in the section using the reduced Eq. (20.6)
B_i = B.*y;
C = round(-S_y/I_xx*10^6)/(10^6);
q_12b = 0; % Cut cell in wall 1-2
q_23b = 0; % Cut cell in wall 2-3
q_34b = 0; % Cut cell in wall 3-4
q_38b = round(C*B_i(3)*10)/10;
q_48b = round(C*B_i(4)*10)/10;
q_78b = q_48b + q_38b;
q_27b = round(C*B_i(2)*10)/10;
q_57b = round(C*B_i(5)*10)/10;
q_16b = round(C*B_i(1)*10)/10;
q_65b = q_16b + round(C*B_i(6)*10)/10;
q_b = [q_12b q_23b q_34b -q_78b -q_65b -q_16b q_27b q_57b q_38b -q_48b];
q_bI = [q_34b -q_38b q_48b];
q_bII = [q_23b q_38b q_78b -q_27b];
q_bIII = [q_12b -q_65b -q_16b q_27b -q_57b];

% Substitute the values of delta and q_b into Eq. (22.10) for each cell
eqI = (q_sOI*delta(1)-q_sOII*del(9)+sum(q_bI.*delta_I))/(2*A(1)*G_ref);
eqII = (-q_sOI*del(9)+q_sOII*delta(2)-q_sOIII*del(7)+sum(q_bII.*delta_II))/
(2*A(2)*G_ref);
eqIII = (-q_sOII*del(7)+q_sOIII*delta(3)+sum(q_bIII.*delta_III))/
(2*A(3)*G_ref);

% Take moments about the intersection of the x-axis and web 572 using Eq. (22.12)
L_M = [y(1) y(2) y(3) e y(5) w 0 0 L(4) L(4)];
eqIV = sum(q_b.*L_M.*L)+sum(2*A.*[q_sOI q_sOII q_sOIII]);
```

```
% Simultaneously solve eqI-eqIV
[q_s0i q_s0ii q_s0iii] = solve(eqI-eqII,eqII-eqIII,eqIV,q_s0I,q_s0II,q_s0III);
q_s0i = vpa(q_s0i,2);
q_s0ii = vpa(q_s0ii,2);
q_s0iii = vpa(q_s0iii,2);

% Calculate the shear flows using q_s0i, q_s0ii, and q_s0iii
q_12 = q_12b+q_s0iii;
q_23 = q_23b+q_s0ii;
q_34 = q_34b+q_s0i;
q_78 = q_78b-q_s0ii;
q_65 = q_65b-q_s0iii;
q_16 = q_16b-q_s0iii;
q_27 = q_27b+q_s0iii-q_s0ii;
q_57 = q_57b-q_s0iii;
q_38 = q_38b-q_s0i+q_s0ii;
q_48 = q_48b+q_s0i;
q = double([q_12 q_23 q_34 q_78 q_65 q_16 q_27 q_57 q_38 q_48]);
q = round(q*10)/10;

% Substitute q_s0i, q_s0ii, and q_s0iii back into any of eqI-eqIV
dtheta_dz = subs(subs(eqI,q_s0I,q_s0i),q_s0II,q_s0ii);
dtheta_dz = round(double(dtheta_dz)*10^8)/(10^8);

% Output the shear flows and angle of twist to the Command Window
disp('The shear flows (N/mm) are:')
text = '[q_12, q_23, q_34, q_78, q_65, q_16, q_27, q_57, q_38, q_48] = [';
c = length(text);
for j=1:1:10
  num = num2str(q(j));
  if j<10
    text(c+1:c+length(num)+2) = [num ','];
    c = length(text);
  else
    text(c+1:c+length(num)+1) = [num ']'];
    disp(text)
  end
end
disp(['The rate of twist in the section =' num2str(dtheta_dz) 'rad/mm'])
```

The Command Window outputs resulting from this MATLAB file are as follows:

```
The shear flows (N/mm) are:
[q_12, q_23, q_34, q_78, q_65, q_16, q_27, q_57, q_38, q_48] = [16, 9.3, 4.9, -9.3, -16,
-61.5, -88.7, 79.5, -64.7, 74]
The rate of twist in the section = 1.09e-006 rad/mm
```

22.5 SHEAR CENTER

The position of the shear center of a wing section is found in an identical manner to that described in Section 17.3. Arbitrary shear loads S_x and S_y are applied in turn through the shear center S, the corresponding shear flow distributions determined, and moments taken about some convenient point. The shear flow distributions are obtained as described previously in the shear of multicell wing sections, except that the N equations of the type (22.10) are sufficient for a solution, since the rate of twist $d\theta/dz$ is zero for shear loads applied through the shear center.

22.6 TAPERED WINGS

Wings are generally tapered in both spanwise and chordwise directions. The effects on the analysis of taper in a single-cell beam have been discussed in Section 21.2. In a multicell wing section, the effects are dealt with in an identical manner, except that the moment equation (21.16) becomes, for an N-cell wing section (see Figs 21.5 and 22.8),

$$S_x\eta_0 - S_y\xi_0 = \sum_{R=1}^{N}\oint_R q_b p_0 \, ds + \sum_{R=1}^{N}2A_R q_{s,0,R} - \sum_{r=1}^{m}P_{x,r}\eta_r + \sum_{r=1}^{m}P_{y,r}\xi_r \qquad (22.13)$$

Example 22.4

A two-cell beam has singly symmetrical cross-sections 1.2 m apart and tapers symmetrically in the y direction about a longitudinal axis (Fig. 22.12). The beam supports loads which produce a shear force $S_y = 10$ kN and a bending moment $M_x = 1.65$ kN m at the larger cross-section; the shear load is applied in the plane of the internal spar web. If booms 1 and 6 lie on a plane which is parallel to the yz plane, calculate the forces in the booms and the shear flow distribution in the walls at the larger cross-section. The booms are assumed to resist all the direct stresses while the walls are effective only in shear. The shear modulus is constant throughout, the vertical webs are all 1.0 mm thick, while the remaining walls are all 0.8 mm thick:

$$\text{Boom areas: } B_1 = B_3 = B_4 = B_6 = 600 \text{ mm}^2, \ B_2 = B_5 = 900 \text{ mm}^2$$

At the larger cross-section,

$$I_{xx} = 4 \times 600 \times 90^2 + 2 \times 900 \times 90^2 = 34.02 \times 10^6 \text{ mm}^4$$

The direct stress in a boom is given by Eq. (16.17) in which $I_{xy} = 0$ and $M_y = 0$; that is,

$$\sigma_{z,r} = \frac{M_x y_r}{I_{xx}}$$

from which

$$P_{z,r} = \frac{M_x y_r}{I_{xx}}B_r$$

or

$$P_{z,r} = \frac{1.65 \times 10^6 y_r B_r}{34.02 \times 10^6} = 0.08 y_r B_r \qquad (i)$$

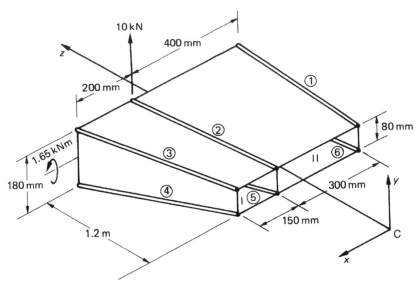

FIGURE 22.12 Tapered Beam of Example 22.4

The value of $P_{z,r}$ is calculated from Eq. (i) in column ② of Table 22.4; $P_{x,r}$ and $P_{y,r}$ follow from Eqs. (21.10) and (21.9), respectively, in columns ⑤ and ⑥. The axial load P_r is given by $[②^2 + ⑤^2 + ⑥^2]^{1/2}$ in column ⑦ and has the same sign as $P_{z,r}$ (see Eq. (21.12)). The moments of $P_{x,r}$ and $P_{y,r}$, columns ⑩ and , are calculated for a moment center at the mid-point of the internal web taking counterclockwise moments as positive.

From column ⑤,

$$\sum_{r=1}^{6} P_{x,r} = 0$$

(as would be expected from symmetry). From column ⑥,

$$\sum_{r=1}^{6} P_{y,r} = 764.4 \text{ N}$$

Table 22.4 Example 22.4

① Boom	② $P_{z,r}$ (N)	③ $\dfrac{\delta x_r}{\delta z}$	④ $\dfrac{\delta y_r}{\delta z}$	⑤ $P_{x,r}$ (N)	⑥ $P_{y,r}$ (N)	⑦ P_r (N)	⑧ ξ_r (mm)	⑨ η_r (mm)	$P_{x,r}\eta_r$ (N mm)	$P_{y,r}\xi_r$ (N mm)
1	2,619.0	0	0.0417	0	109.2	2,621.3	400	90	0	43,680
2	3,928.6	0.0833	0.0417	327.3	163.8	3,945.6	0	90	−29,457	0
3	2,619.0	0.1250	0.0417	327.4	109.2	2,641.6	200	90	−29,466	21,840
4	−2,619.0	0.1250	−0.0417	−327.4	109.2	−2,641.6	200	90	−29,466	21,840
5	−3,928.6	0.0833	−0.0417	−327.3	163.8	−3,945.6	0	90	−29,457	0
6	−2,619.0	0	−0.0417	0	109.2	−2,621.3	400	90	0	−43,680

From column ⑩,

$$\sum_{r=1}^{6} P_{x,r}\eta_r = -117,846 \text{ Nmm}$$

From column ,

$$\sum_{r=1}^{6} P_{y,r}\xi_r = -43,680 \text{ Nmm}$$

From Eq. (21.15),

$$S_{x,w} = 0, \quad S_{y,w} = 10 \times 10^3 - 764.4 = 9,235.6 \text{ N}$$

Also, since Cx is an axis of symmetry, $I_{xy} = 0$ and Eq. (20.6) for the "open section" shear flow reduces to

$$q_b = -\frac{S_{y,w}}{I_{xx}} \sum_{r=1}^{n} B_r y_r$$

or

$$q_b = \frac{9,235.6}{34.02 \times 10^6} \sum_{r=1}^{n} B_r y_r = -2.715 \times 10^{-4} \sum_{r=1}^{n} B_r y_r \tag{ii}$$

Cutting the top walls of each cell and using Eq. (ii), we obtain the q_b distribution shown in Fig. 22.13. Evaluating δ for each wall and substituting in Eq. (22.10) gives, for cell I,

$$\frac{d\theta}{dz} = \frac{1}{2 \times 36,000G}(760q_{s,0,I} - 180q_{s,0,II} - 1314) \tag{iii}$$

and, for cell II,

$$\frac{d\theta}{dz} = \frac{1}{2 \times 72,000G}(-180q_{s,0,I} + 1160q_{s,0,II} + 1314) \tag{iv}$$

Taking moments about the mid-point of web 25, we have, using Eq. (22.13),

$$0 = -14.7 \times 180 \times 400 + 14.7 \times 180 \times 200 + 2 \times 36,000q_{s,0,I} + 2 \times 72,000q_{s,0,II}$$
$$-117,846 - 43,680$$

or

$$0 = -690,726 + 72,000q_{s,0,I} + 144,000q_{s,0,II} \tag{v}$$

Solving Eqs (iii)–(v) gives

$$q_{s,0,I} = 4.6 \text{ N/mm}, \quad q_{s,0,II} = 2.5 \text{ N/mm}$$

and the resulting shear flow distribution is shown in Fig. 22.14.

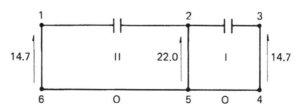

FIGURE 22.13 The q_b **(N/mm) Distribution in the Beam Section of Example 22.4 (viewed along z axis toward C)**

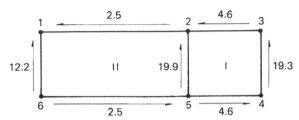

FIGURE 22.14 Shear Flow (N/mm) Distribution in the Tapered Beam of Example 22.4

22.7 DEFLECTIONS

Deflections of multicell wings may be calculated by the unit load method in an identical manner to that described in Section 20.4 for open and single-cell beams.

Example 22.5

Calculate the deflection at the free end of the two-cell beam shown in Fig. 22.15, allowing for both bending and shear effects. The booms carry all the direct stresses while the skin panels, of constant thickness throughout, are effective only in shear.

Take $E = 69,000 \text{ N/mm}^2$ and $G = 25,900 \text{ N/mm}^2$
Boom areas: $B_1 = B_3 = B_4 = B_6 = 650 \text{ mm}^2$, $B_2 = B_5 = 1,300 \text{ mm}^2$

The beam cross-section is symmetrical about a horizontal axis and carries a vertical load at its free end through the shear center. The deflection Δ at the free end is then, from Eqs. (20.17) and (20.19),

$$\Delta = \int_0^{2,000} \frac{M_{x,0} M_{x,1}}{EI_{xx}} \, dz + \int_0^{2,000} \left(\int_{\text{section}} \frac{q_0 q_1}{Gt} \, ds \right) dz \qquad (i)$$

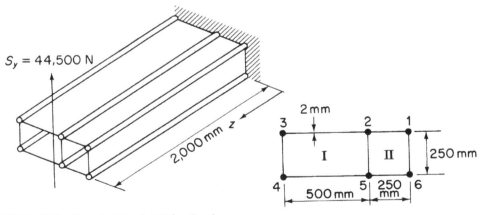

FIGURE 22.15 Deflection of a Two-Cell Wing Section

where

$$M_{x,0} = -44.5 \times 10^3(2,000 - z), \quad M_{x,1} = -(2,000 - z)$$

and

$$I_{xx} = 4 \times 650 \times 125^2 + 2 \times 13,00 \times 125^2 = 81.3 \times 10^6 \text{mm}^4$$

also

$$S_{y,0} = 44.5 \times 10^3 \text{ N}, \quad S_{y,1} = 1$$

The q_0 and q_1 shear flow distributions are obtained as previously described (note $d\theta/dz = 0$ for a shear load through the shear center) and are

$$q_{0,12} = 9.6 \text{ N/mm}, \quad q_{0,23} = -5.8 \text{ N/mm}, \quad q_{0,43} = 50.3 \text{ N/mm}$$
$$q_{0,45} = -5.8 \text{ N/mm}, \quad q_{0,56} = 9.6 \text{ N/mm}, \quad q_{0,61} = 54.1 \text{ N/mm}$$
$$q_{0,52} = 73.6 \text{ N/mm at all section of the beam}$$

The q_1 shear flows in this case are given by $q_0/44.5 \times 10^3$. Thus,

$$\int_{\text{section}} \frac{q_0 q_1}{Gt} ds = \frac{1}{25,900 \times 2 \times 44.5 \times 10^3} (9.6^2 \times 250 \times 2 + 5.8^2 \times 500 \times 2$$
$$+ 50.3^2 \times 250 + 54.1^2 \times 250 + 73.6^2 \times 250)$$
$$= 1.22 \times 10^{-3}$$

Hence, from Eq. (i),

$$\Delta = \int_0^{2,000} \frac{44.5 \times 10^3(2,000 - z)^2}{69,000 \times 81.3 \times 10^6} dz + \int_0^{2,000} 1.22 \times 10^{-3} dz$$

giving

$$\Delta = 23.5 \text{ mm}$$

22.8 CUT-OUTS IN WINGS

Wings, as well as fuselages, have openings in their surfaces to accommodate undercarriages, engine nacelles and weapons installations, and the like. In addition, inspection panels are required at specific positions, so that, as for fuselages, the loads in adjacent portions of the wing structure are modified.

Initially, we consider the case of a wing subjected to a pure torque in which one bay of the wing has the skin on its undersurface removed. The method is best illustrated by a numerical example.

Example 22.6

The structural portion of a wing consists of a three-bay rectangular section box which may be assumed to be firmly attached at all points around its periphery to the aircraft fuselage at its inboard end. The skin on the undersurface of the central bay has been removed and the wing is subjected to a torque of 10 kN m at its tip (Fig. 22.16). Calculate the shear flows in the skin panels and spar webs, the loads in the corner flanges, and the forces in the ribs on each

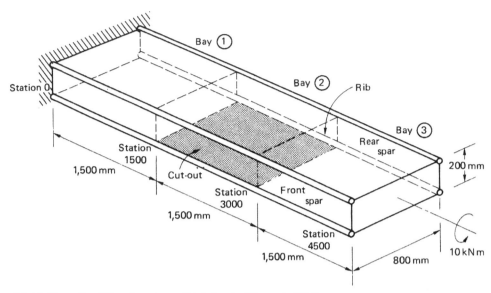

FIGURE 22.16 Three-Bay Wing Structure with Cut-out of Example 22.6

side of the cut-out, assuming that the spar flanges carry all the direct loads while the skin panels and spar webs are effective only in shear. See Ex. 1.1.

If the wing structure were continuous and the effects of restrained warping at the built-in end ignored, the shear flows in the skin panels would be given by Eq. (18.1), that is,

$$q = \frac{T}{2A} = \frac{10 \times 10^6}{2 \times 200 \times 800} = 31.3 \text{ N/mm}$$

and the flanges would be unloaded. However, the removal of the lower skin panel in bay ② results in a torsionally weak channel section for the length of bay ②, which must in any case still transmit the applied torque to bay ① and subsequently to the wing support points. Although open section beams are inherently weak in torsion (see Section 18.2), the channel section in this case is attached at its inboard and outboard ends to torsionally stiff closed boxes, so that, in effect, it is built-in at both ends. We shall examine the effect of axial constraint on open section beams subjected to torsion in Chapter 27. An alternative approach is to assume that the torque is transmitted across bay ② by the differential bending of the front and rear spars. The bending moment in each spar is resisted by the flange loads P as shown, for the front spar, in Fig. 22.17(a). The shear loads in the front and rear spars form a couple at any station in bay ② which is equivalent to the applied torque. Thus, from Fig. 22.17(b),

$$800S = 10 \times 10^6 \text{ Nmm}$$

that is,

$$S = 12,500 \text{ N}$$

The shear flow q_1 in Fig. 22.17(a) is given by

$$q_1 = \frac{12,500}{200} = 62.5 \text{ N/mm}$$

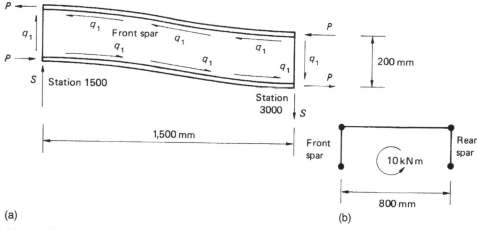

(a)   (b)

FIGURE 22.17 Differential Bending of Front Spar

Midway between stations 1500 and 3000 a point of contraflexure occurs in the front and rear spars, so that at this point the bending moment is zero. Hence,

$$200P = 12{,}500 \times 750 \text{ N mm}$$

so that

$$P = 46{,}875 \text{ N}$$

Alternatively, P may be found by considering the equilibrium of either of the spar flanges. Thus,

$$2P = 1{,}500q_1 = 1{,}500 \times 62.5 \text{ N}$$

from which

$$P = 46{,}875 \text{ N}$$

The flange loads P are reacted by loads in the flanges of bays ① and ③. These flange loads are transmitted to the adjacent spar webs and skin panels, as shown in Fig. 22.18 for bay ③, and modify the shear flow distribution given by Eq. (18.1). For equilibrium of flange 1,

$$1{,}500q_2 - 1{,}500q_3 = P = 46{,}875 \text{ N}$$

or

$$q_2 - q_3 = 31.3 \tag{i}$$

The result of the shear flows q_2 and q_3 must be equivalent to the applied torque. Hence, for moments about the center of symmetry at any section in bay ③ and using Eq. (20.10),

$$200 \times 800q_2 + 200 \times 800q_3 = 10 \times 10^6 \text{ Nmm}$$

or

$$q_2 + q_3 = 62.5 \tag{ii}$$

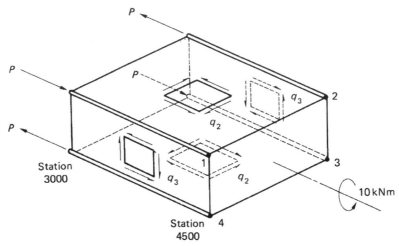

FIGURE 22.18 Loads on Bay ③ of the Wing of Example 22.6

Solving Eqs. (i) and (ii), we obtain

$$q_2 = 46.9 \text{ N/mm}, \quad q_3 = 15.6 \text{ N/mm}$$

Comparison with the results of Eq. (18.1) shows that the shear flows are increased by a factor of 1.5 in the upper and lower skin panels and decreased by a factor of 0.5 in the spar webs.

The flange loads are in equilibrium with the resultants of the shear flows in the adjacent skin panels and spar webs. Thus, for example, in the top flange of the front spar,

$$P(\text{st. }4500) = 0$$
$$P(\text{st. }3000) = 1{,}500q_2 - 1{,}500q_3 = 46{,}875 \text{ N (compression)}$$
$$P(\text{st. }2250) = 1{,}500q_2 - 1{,}500q_3 - 750q_1 = 0$$

The loads along the remainder of the flange follow from antisymmetry, giving the distribution shown in Fig. 22.19. The load distribution in the bottom flange of the rear spar is identical to that shown in Fig. 22.19, while the distributions in the bottom flange of the front spar and the top flange of the rear spar are reversed. We note that the flange loads are zero at the built-in end of the wing (station 0). Generally, however, additional stresses are induced

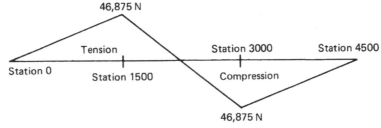

FIGURE 22.19 Distribution of Load in the Top Flange of the Front Spar of the Wing of Example 22.6

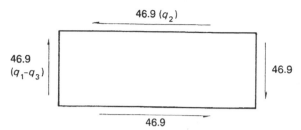

FIGURE 22.20 Shear Flows (N/mm) on Wing Rib at Station 3000 in the Wing of Example 22.6

by the warping restraint at the built-in end; these are investigated in Chapter 26. The loads on the wing ribs on either the inboard or outboard end of the cut-out are found by considering the shear flows in the skin panels and spar webs immediately inboard and outboard of the rib. Thus, for the rib at station 3000, we obtain the shear flow distribution shown in Fig. 22.20.

In Example 22.6, we implicitly assumed in the analysis that the local effects of the cut-out were completely dissipated within the length of the adjoining bays, which were equal in length to the cut-out bay. The validity of this assumption relies on St. Venant's principle (Section 2.4). It may generally be assumed therefore that the effects of a cut-out are restricted to spanwise lengths of the wing equal to the length of the cut-out on both the inboard and outboard ends of the cut-out bay.

We now consider the more complex case of a wing having a cut-out and subjected to shear loads which produce both bending and torsion. Again, the method is illustrated by a numerical example.

Example 22.7

A wing box has the skin panel on its undersurface removed between stations 2000 and 3000 and carries lift and drag loads which are constant between stations 1000 and 4000, as shown in Fig. 22.21(a). Determine the shear flows in the skin panels and spar webs and also the loads on the wing ribs at the inboard and outboard ends of the cut-out bay.

Assume that all bending moments are resisted by the spar flanges while the skin panels and spar webs are effective only in shear.

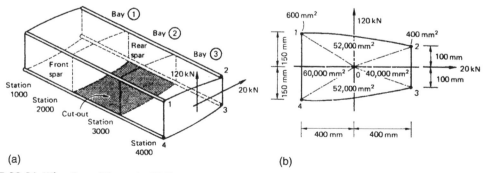

(a)   (b)

FIGURE 22.21 Wing Box of Example 22.7

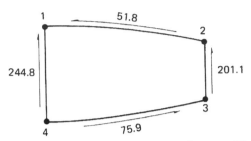

FIGURE 22.22 Shear Flow (N/mm) Distribution at Any Station in the Wing Box of Example 22.7 without the Cut-out

The simplest approach is first to determine the shear flows in the skin panels and spar webs as though the wing box were continuous and to apply an equal and opposite shear flow to that calculated around the edges of the cut-out. The shear flows in the wing box without the cut-out is the same in each bay and are calculated using the method described in Section 20.3 and illustrated in Example 20.4. This gives the shear flow distribution shown in Fig. 22.22.

We now consider bay ② and apply a shear flow of 75.9 N/mm in the wall 34 in the opposite sense to that shown in Fig. 22.22. This reduces the shear flow in the wall 34 to zero and, in effect, restores the cut-out to bay ②. The shear flows in the remaining walls of the cut-out bay no longer are equivalent to the externally applied shear loads, so that corrections are required. Consider the cut-out bay (Fig. 22.23) with the shear flow of 75.9 N/mm applied in the opposite sense to that shown in Fig. 22.22. The correction shear flows q'_{12}, q'_{32}, and q'_{14} may be found using statics. Thus, resolving forces horizontally, we have

$$800q'_{12} = 800 \times 75.9 \text{ N}$$

from which

$$q'_{12} = 75.9 \text{ N/mm}$$

Resolving forces vertically,

$$200q'_{32} = 50q'_{12} - 50 \times 75.9 - 300q'_{14} = 0 \tag{i}$$

and, taking moments about O in Fig. 22.21(b), we obtain

$$2 \times 52,000q'_{12} - 2 \times 40,000q'_{32} + 2 \times 52,000 \times 75.9 - 2 \times 60,000q'_{14} = 0 \tag{ii}$$

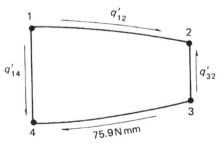

FIGURE 22.23 Correction Shear Flows in the Cut-out Bay of the Wing Box of Example 22.7

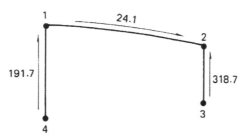

FIGURE 22.24 Final Shear Flows (N/mm) in the Cut-out Bay of the Wing Box of Example 22.7

Solving Eqs. (i) and (ii) gives

$$q'_{32} = 117.6 \, \text{N/mm}, \qquad q'_{14} = 53.1 \, \text{N/mm}$$

The final shear flows in bay ② are found by superimposing q'_{12}, q'_{32}, and q'_{14} on the shear flows in Fig. 22.22, giving the distribution shown in Fig. 22.24. Alternatively, these shear flows could be found directly by considering the equilibrium of the cut-out bay under the action of the applied shear loads.

The correction shear flows in bay ② (Fig. 22.23) also modify the shear flow distributions in bays ① and ③. The correction shear flows to be applied to those shown in Fig. 22.22 for bay ③ (those in bay ① are identical) may be found by determining the flange loads corresponding to the correction shear flows in bay ②.

It can be seen from the magnitudes and directions of these correction shear flows (Fig. 22.23) that, at any section in bay ②, the loads in the upper and lower flanges of the front spar are equal in magnitude but opposite in direction; similarly for the rear spar. Thus, the correction shear flows in bay ② produce an identical system of flange loads to that shown in Fig. 22.17 for the cut-out bays in the wing structure of Example 22.6.

It follows that these correction shear flows produce differential bending of the front and rear spars in bay ② and that the spar bending moments and hence the flange loads are zero at the mid-bay points. Therefore, at station 3000, the flange loads are

$$P_1 = (75.9 + 53.1) \times 500 = 64{,}500 \, \text{N (compression)}$$
$$P_4 = 64{,}500 \, \text{N (tension)}$$
$$P_2 = (75.9 + 117.6) \times 500 = 96{,}750 \, \text{N (tension)}$$
$$P_3 = 96{,}750 \, \text{N (tension)}$$

These flange loads produce correction shear flows q''_{21}, q''_{43}, q''_{23}, and q''_{41} in the skin panels and spar webs of bay ③, as shown in Fig. 22.25. Thus, for equilibrium of flange 1,

$$1{,}000q''_{41} + 1{,}000q''_{21} = 64{,}500 \, \text{N} \tag{iii}$$

and, for equilibrium of flange 2,

$$1{,}000q''_{21} + 1{,}000q''_{23} = 96{,}750 \, \text{N} \tag{iv}$$

For equilibrium in the chordwise direction at any section in bay ③,

$$800q''_{21} = 800q''_{43}$$

or

$$q''_{21} = q''_{43} \tag{v}$$

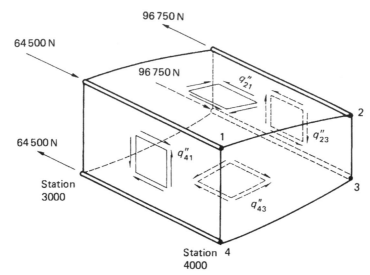

FIGURE 22.25 Correction Shear Flows in Bay ③ of the Wing Box of Example 22.7

Finally, for vertical equilibrium at any section in bay ③,

$$300q''_{41} + 50q''_{43} + 50q''_{23} - 200q''_{23} = 0 \qquad \text{(vi)}$$

Simultaneous solution of Eqs. (iii)–(vi) gives

$$q''_{21} = q''_{43} = 38.0 \, \text{N/mm}, \quad q''_{23} = 58.8 \, \text{N/mm}, \quad q''_{41} = 26.6 \, \text{N/mm}$$

Superimposing these correction shear flows on those shown in Fig. 22.22 gives the final shear flow distribution in bay ③, as shown in Fig. 22.26. The rib loads at stations 2000 and 3000 are found as before by adding algebraically the shear flows in the skin panels and spar webs on each side of the rib. Thus, at station 3000, we obtain the shear flows acting around the periphery of the rib, as shown in Fig. 22.27. The shear flows applied to the rib at the inboard end of the cut-out bay will be equal in magnitude but opposite in direction.

Note that, in this example, only the shear loads on the wing box between stations 1000 and 4000 are given. We cannot therefore determine the final values of the loads in the spar flanges, since we do not know the values of the bending moments at these positions caused by loads acting on other parts of the wing. ∎

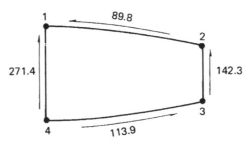

FIGURE 22.26 Final Shear Flows (N/mm) in Bay ③ (and Bay ①) of the Wing Box of Example 22.7

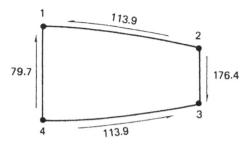

FIGURE 22.27 Shear Flows (N/mm) Applied to the Wing Rib at Station 3000 in the Wing Box of Example 22.7

PROBLEMS

P.22.1 The central cell of a wing has the idealized section shown in Fig. P.22.1. If the lift and drag loads on the wing produce bending moments of –120,000 Nm and –30,000 Nm, respectively, at the section shown, calculate the direct stresses in the booms. Neglect axial constraint effects and assume that the lift and drag vectors are in the vertical and horizontal planes.

$$\text{Boom areas: } B_1 = B_4 = B_5 = B_8 = 1000 \text{ mm}^2$$
$$B_2 = B_3 = B_6 = B_7 = 600 \text{ mm}^2$$

Answer: $\sigma_1 = -190.7 \text{ N/mm}^2$, $\sigma_2 = -181.7 \text{ N/mm}^2$, $\sigma_3 = -172.8 \text{ N/mm}^2$
$\sigma_4 = -163.8 \text{ N/mm}^2$, $\sigma_5 = 140 \text{ N/mm}^2$, $\sigma_6 = 164.8 \text{ N/mm}^2$
$\sigma_7 = 189.6 \text{ N/mm}^2$, $\sigma_8 = 214.4 \text{ N/mm}^2$.

P.22.1 MATLAB Assuming the left-hand side height of the section shown in Fig. P.22.1 is labeled h_1, use MATLAB to repeat Problem P.22.1 for values of h_1 ranging from 180 to 270 mm in increments of 10 mm. All other dimensions are as shown in Fig. P.22.1.

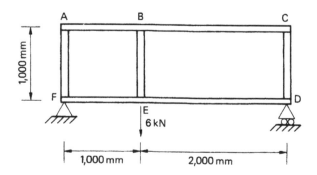

FIGURE P.22.1

Answer:

	h_1 (mm)	σ_1	σ_2	σ_3	σ_4	σ_5	σ_6	σ_7	σ_8
(i)	180	−196.4	−204.4	−212.4	−220.4	196.4	204.4	212.4	220.4
(ii)	190	−196.4	−200.7	−205	−209.3	185.5	197.1	208.7	220.3
(iii)	200	−196	−196.9	−197.9	−198.8	175.2	190	204.8	219.6
(iv)	210	−195.3	−193.2	−191.1	−189	165.8	183.4	201	218.6
(v)	220	−193.8	−189.3	−184.7	−180.2	156.4	176.8	197.2	217.6
(vi)	230	−192.2	−185.4	−178.6	−171.8	147.8	170.6	193.4	216.2
(vii)	240	−190.7	−181.7	−172.8	−163.8	140	164.8	189.6	214.4
(viii)	250	−189.3	−178.2	−167.1	−156	133	159.4	185.8	212.2
(ix)	260	−187.4	−174.6	−161.8	−149	126.2	154.2	182.2	210.2
(x)	270	−185.1	−171	−156.9	−142.7	119.5	149.1	178.7	208.3

The units for these direct stresses are N/mm^2.

P.22.2 Figure P.22.2 shows the cross-section of a two-cell torque box. If the shear stress in any wall must not exceed 140 N/mm^2, find the maximum torque which can be applied to the box. If this torque were applied at one end and resisted at the other end of such a box of span 2,500 mm, find the twist in degrees of one end relative to the other and the torsional rigidity of the box. The shear modulus $G = 26,600$ N/mm^2 for all walls. Data:

$$\text{Shaded areas:} \quad A_{34} = 6,450 \text{ mm}^2, \; A_{16} = 7,750 \text{ mm}^2$$
$$\text{Wall lengths:} \quad s_{34} = 250 \text{ mm}, \; s_{16} = 300 \text{ mm}$$
$$\text{Wall thickness:} \quad t_{12} = 1.63 \text{ mm}, \; t_{34} = 0.56 \text{ mm}$$
$$t_{23} = t_{45} = t_{56} = 0.92 \text{ mm}$$
$$t_{61} = 2.3 \text{ mm}$$
$$t_{25} = 2.54 \text{ mm}$$

Answer: $T = 102,417$ Nm, $\theta = 1.46°$, $GJ = 10 \times 10^{12}$ Nmm^2/rad.

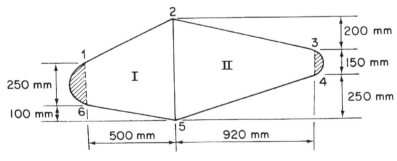

FIGURE P.22.2

P.22.3 Determine the torsional stiffness of the four-cell wing section shown in Fig. P.22.3. Data are in Table 22.5.

Answer: $522.5 \times 10^6 \, G \, \text{Nmm}^2/\text{rad}$

P.22.4 Determine the shear flow distribution for a torque of 56,500 Nm for the three-cell section shown in Fig. P.22.4 with the data shown in Table 22.6. The section has a constant shear modulus throughout.

Answer: $q_{12}^U = 25.4 \, \text{N/mm}, \quad q_{21}^L = 33.5 \, \text{N/mm}, \quad q_{14} = q_{32} = 8.1 \, \text{N/mm}$
$q_{43}^U = 13.4 \, \text{N/mm}, \quad q_{34}^L = 5.3 \, \text{N/mm}$

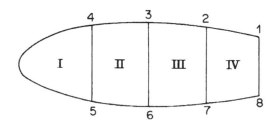

FIGURE P.22.3

Table 22.5 Problem 22.3

Wall	12	23	34					
	78	67	56	45°	45^i	36	27	18
Peripheral length (mm)	762	812	812	1,525	356	406	356	254
Thickness (mm)	0.915	0.915	0.915	0.711	1.220	1.625	1.220	0.915
Cell areas (mm^2)	$A_I = 161,500$		$A_{II} = 291,000$					
	$A_{III} = 291,000$		$A_{IV} = 226,000$					

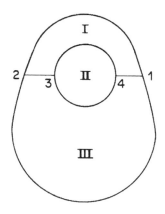

FIGURE P.22.4

Table 22.6 Problem 22.4

Wall	Length (mm)	Thickness (mm)	Cell	Area (mm^2)
12^U	1,084	1.220	I	108,400
12^L	2,160	1.625	II	202,500
14, 23	127	0.915	III	528,000
34^U	797	0.915		
34^L	797	0.915		

P.22.5 The idealized cross-section of a two-cell thin-walled wing box is shown in Fig. P.22.5 with the data in Table 22.7. If the wing box supports a load of 44,500 N acting along web 25, calculate the shear flow distribution. The shear modulus G is the same for all walls of the wing box. The cell areas are $A_I = 232,000$ mm^2, $A_{II} = 258,000$ mm^2.

Answer: $q_{16} = 33.9$ N/mm, $q_{65} = q_{21} = 1.1$ N/mm
$q_{45} = q_{23} = 7.2$ N/mm, $q_{34} = 20.8$ N/mm
$q_{25} = 73.4$ N/mm

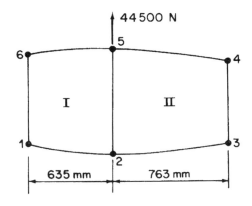

FIGURE P.22.5

Table 22.7 Problem 22.5

Wall	Length (mm)	Thickness (mm)	Boom	Area (mm^2)
16	254	1.625	1,6	1290
25	406	2.032	2,5	1936
34	202	1.220	3,4	645
12, 56	647	0.915		
23, 45	775	0.559		

P.22.6 Figure P.22.6 shows a singly symmetric, two-cell wing section in which all direct stresses are carried by the booms, shear stresses alone being carried by the walls. All walls are flat with the exception of the nose portion 45. Find the position of the shear center S and the shear flow distribution for a load of $S_y = 66,750$ N through S. Tabulated in Table 22.8 are lengths, thicknesses, and shear moduli of the shear carrying walls. Note that dotted line 45 is not a wall. Nose area $N_1 = 51,500$ mm^2.

Answer: $x_S = 164.1$ mm, $q_{12} = q_{78} = 17.8$ N/mm, $q_{32} = q_{76} = 18.5$ N/mm
$q_{63} = 88.4$ N/mm, $q_{43} = q_{65} = 2.7$ N/mm, $q_{54} = 39.0$ N/mm
$q_{81} = 90.4$ N/mm.

P.22.6 MATLAB Assuming the length of cell wall 34 (and consequently wall 56) of the section shown in Fig. P.22.6 is labeled L_I, use MATLAB to repeat Problem P.22.6 for values of L_I ranging from 350 to 410 mm in increments of 10 mm. The units for the shear flows (q) are N/mm; the units for L_I and x_S are mm.

Answer:

	L_I	x_S	q_{12}	q_{23}	q_{34}	q_{45}	q_{56}	q_{67}	q_{78}	q_{81}	q_{36}
(i)	350	166.4	17.8	−18.5	−2.8	−39.1	−2.8	−18.5	17.8	90.4	−88.3
(ii)	360	164.4	17.8	−18.5	−2.8	v39.1	−2.8	−18.5	17.8	90.4	−88.3
(iii)	370	163	17.8	−18.5	−2.7	−39	−2.7	−18.5	17.8	90.4	−88.3
(iv)	380	161.1	17.8	−18.5	−2.7	−39	−2.7	−18.5	17.8	90.4	−88.4
(v)	390	159.7	17.8	−18.5	−2.6	−38.9	−2.6	−18.5	17.8	90.4	−88.4
(vi)	400	157.8	17.8	−18.5	−2.6	−38.9	−2.6	−18.5	17.8	90.4	−88.5
(vii)	410	155.9	17.8	−18.5	−2.6	−38.9	−2.6	−18.5	17.8	90.4	−88.5

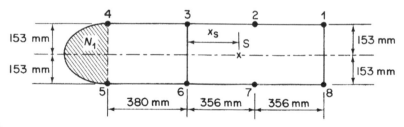

FIGURE P.22.6

Table 22.8 Problem 22.6					
Wall	**Length (mm)**	**Thickness (mm)**	**G(N/mm^2)**	**Boom**	**Area (mm^2)**
34, 56	380	0.915	20,700	1, 3, 6, 8	1,290
12, 23, 67, 78	356	0.915	24,200	2, 4, 5, 7	645
36, 81	306	1.220	24,800		
45	610	1.220	24,800		

P.22.7 A singly symmetric wing section consists of two closed cells and one open cell (see Fig. P.22.7 and Table 22.9). Webs 25, 34 and walls 12, 56 are straight, while all other walls are curved. All walls of the section are assumed to be effective in carrying shear stresses only, direct stresses being carried by booms 1–6. Calculate the distance x_S of the shear center S aft of web 34. The shear modulus G is the same for all walls.

    Answer: 241.4 mm

P.22.8 A portion of a tapered, three-cell wing has singly symmetrical idealized cross-sections 1,000 mm apart, as shown in Fig. P.22.8. A bending moment $M_x = 1,800$ Nm and a shear load $S_y = 12,000$ N in the plane of web 52 are applied at the larger cross-section. Calculate the forces in the booms and the shear

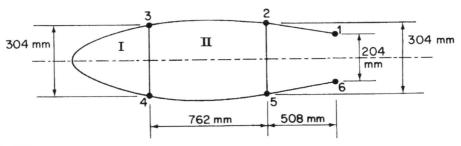

FIGURE P.22.7

Table 22.9 Problem 22.7

Wall	Length (mm)	Thickness (mm)	Boom	Area (mm²)	Cell	Area (mm²)
12, 56	510	0.559	1, 6	645	I	93,000
23, 45	765	0.915	2, 5	1,290	II	258,000
34°	1,015	0.559	3, 4	1,935		
34ⁱ	304	2.030				
25	304	1.625				

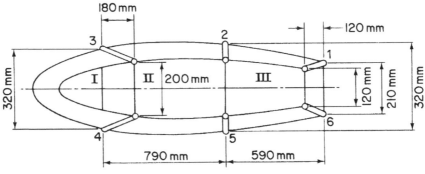

FIGURE P.22.8

flow distribution at this cross-section. The modulus G is constant throughout. Section dimensions at the larger cross-section are given in Table 22.10.

Answer: $P_1 = -P_6 = 1,200$ N, $P_2 = -P_5 = 2,424$ N, $P_3 = -P_4 = 2,462$ N
$q_{12} = q_{56} = 3.74$ N/mm, $q_{23} = q_{45} = 3.11$ N/mm, $q_{34°} = 0.06$ N/mm
$q_{43}{}^i = 12.16$ N/mm, $q_{52} = 14.58$ N/mm, $q_{61} = 11.22$ N/mm.

P.22.9 A portion of a wing box is built-in at one end and carries a shear load of 2,000 N through its shear center and a torque of 1,000 Nm, as shown in Fig. P.22.9. If the skin panel in the upper surface of the inboard bay is removed, calculate the shear flows in the spar webs and remaining skin panels, the distribution of load in the spar flanges, and the loading on the central rib. Assume that the spar webs and skin panels are effective in resisting shear stresses only.

Table 22.10 Problem 22.8

Wall	Length (mm)	Thickness (mm)	Boom	Area (mm²)	Cell	Area (mm²)
12, 56	600	1.0	1, 6	600	I	100,000
23, 45	800	1.0	2, 5	800	II	260,000
34°	1,200	0.6	3, 4	800	III	180,000
34^i	320	2.0				
25	320	2.0				
16	210	1.5				

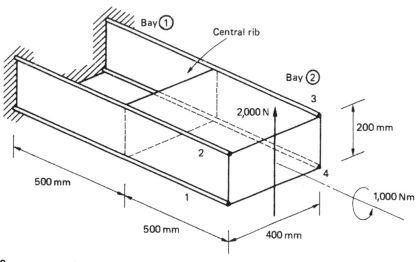

FIGURE P.22.9

Answer: Bay ①: q in spar webs $= 7.5$ N/mm
           Bay ②: q in spar webs $= 1.9$ N/mm, in skin panels $= 9.4$ N/mm
           Flange loads (2): at built-in end $= 1{,}875$ N (compression)
                              at central rib $= 5{,}625$ N (compression)
           Rib loads: q (horizontal edges) $= 9.4$ N/mm
                       q (vertical edges) $= 9.4$ N/mm.

Fuselage frames and wing ribs

23

Aircraft are constructed primarily from thin metal skins which are capable of resisting in-plane tension and shear loads but buckle under comparatively low values of in-plane compressive loads. The skins are therefore stiffened by longitudinal stringers which resist the in-plane compressive loads and, at the same time, resist small distributed loads normal to the plane of the skin. The effective length in compression of the stringers is reduced, in the case of fuselages, by transverse frames or bulkheads or, in the case of wings, by ribs. In addition, the frames and ribs resist concentrated loads in transverse planes and transmit them to the stringers and the plane of the skin. Therefore, cantilever wings may be bolted to fuselage frames at the spar caps while undercarriage loads are transmitted to the wing through spar and rib attachment points.

23.1 PRINCIPLES OF STIFFENER/WEB CONSTRUCTION

Generally, frames and ribs are themselves fabricated from thin sheets of metal and therefore require stiffening members to distribute the concentrated loads to the thin webs. If the load is applied in the plane of a web, the stiffeners must be aligned with the direction of the load. Alternatively, if this is not possible, the load should be applied at the intersection of two stiffeners, so that each stiffener resists the component of load in its direction. The basic principles of stiffener/web construction are illustrated in Example 23.1.

Example 23.1

A cantilever beam (Fig. 23.1) carries concentrated loads as shown. Calculate the distribution of stiffener loads and the shear flow distribution in the web panels, assuming that the latter are effective only in shear. See Ex. 1.1.

We note that stiffeners HKD and JK are required at the point of application of the 4,000 N load to resist its vertical and horizontal components. A further transverse stiffener GJC is positioned at the unloaded end J of the stiffener JK, since stress concentrations are produced if a stiffener ends in the center of a web panel. We note also that the web panels are effective only in shear, so that the shear flow is constant throughout a particular web panel; the assumed directions of the shear flows are shown in Fig. 23.1.

It is instructive at this stage to examine the physical role of the different structural components in supporting the applied loads. Generally, stiffeners are assumed to withstand axial forces only, so that the horizontal component of the load at K is equilibrated locally by the axial load in the stiffener JK and not by the bending of stiffener HKD. By the same argument, the vertical component of the load at K is resisted by the axial load in the stiffener HKD. These axial stiffener loads are equilibrated in turn by the resultants of the shear flows q_1 and q_2 in the web panels CDKJ and JKHG. Thus, we see that the web panels resist the shear component of the externally applied load and at the

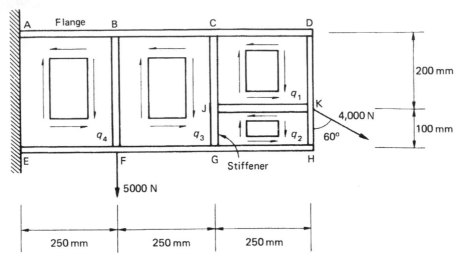

FIGURE 23.1 Cantilever Beam of Example 23.1

same time transmit the bending and axial load of the externally applied load to the beam flanges; subsequently, the flange loads are reacted at the support points A and E.

Consider the free body diagrams of the stiffeners JK and HKD shown in Fig. 23.2.

From the equilibrium of stiffener JK, we have

$$(q_1 - q_2) \times 250 = 4,000 \sin 60° = 3,464.1 \text{ N} \tag{i}$$

and, from the equilibrium of stiffener HKD,

$$200q_1 + 100q_2 = 4,000 \cos 60° = 2,000 \text{ N} \tag{ii}$$

Solving Eqs. (i) and (ii), we obtain

$$q_1 = 11.3 \text{ N/mm}, \quad q_2 = -2.6 \text{ N/mm}$$

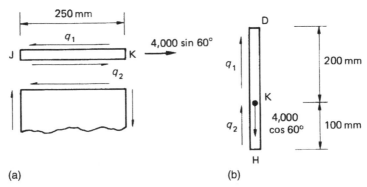

FIGURE 23.2 Free Body Diagrams of Stiffeners JK and HKD in the Beam of Example 23.1

The vertical shear force in the panel BCGF is equilibrated by the vertical resultant of the shear flow q_3. Thus,

$$300q_3 = 4,000 \cos 60° = 2,000 \text{ N}$$

from which

$$q_3 = 6.7 \text{ N/mm}$$

Alternatively, q_3 may be found by considering the equilibrium of the stiffener CJG. From Fig. 23.3,

$$300q_3 = 200q_1 + 100q_2$$

or

$$300q_3 = 200 \times 11.3 - 100 \times 2.6$$

from which

$$q_3 = 6.7 \text{ N/mm}$$

The shear flow q_4 in the panel ABFE may be found using either of the above methods. Thus, considering the vertical shear force in the panel,

$$300q_4 = 4,000 \cos 60° + 5,000 = 7,000 \text{ N}$$

from which

$$q_4 = 23.3 \text{ N/mm}$$

Alternatively, from the equilibrium of stiffener BF,

$$300q_4 - 300q_3 = 5,000 \text{ N}$$

from which

$$q_4 = 23.3 \text{ N/mm}$$

The flange and stiffener load distributions are calculated in the same way and are obtained from the algebraic summation of the shear flows along their lengths. For example, the axial load P_A at A in the flange ABCD is given by

$$P_A = 250q_1 + 250q_3 + 250q_4$$

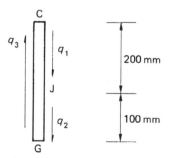

FIGURE 23.3 Equilibrium of Stiffener CJG in the Beam of Example 23.1

or

$$P_A = 250 \times 11.3 + 250 \times 6.7 + 250 \times 23.3 = 10{,}325 \text{ N (tension)}$$

Similarly,

$$P_E = -250q_2 - 250q_3 - 250q_4$$

that is,

$$P_E = 250 \times 2.6 - 250 \times 6.7 - 250 \times 23.3 = -6{,}850 \text{ N (compression)}$$

The complete load distribution in each flange is shown in Fig. 23.4. The stiffener load distributions are calculated in the same way and are shown in Fig. 23.5.

The distribution of flange load in the bays ABFE and BCGF could be obtained by considering the bending and axial loads on the beam at any section. For example, at the section AE, we can replace the actual loading system by a bending moment

$$M_{AE} = 5{,}000 \times 250 + 2{,}000 \times 750 - 3{,}464.1 \times 50 = 2{,}576{,}800 \text{ N mm}$$

and an axial load acting midway between the flanges (irrespective of whether or not the flange areas are symmetrical about this point) of

$$P = 3{,}464.1 \text{ N}$$

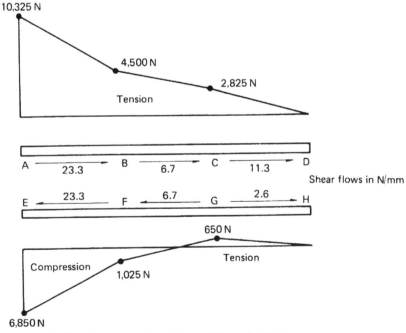

FIGURE 23.4 Load Distributions in Flanges of the Beam of Example 23.1

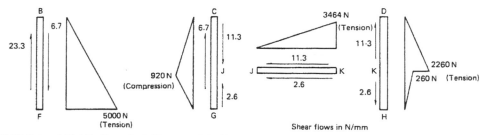

FIGURE 23.5 Load Distributions in Stiffeners of the Beam of Example 23.1

Thus,

$$P_A = \frac{2,576,800}{300} + \frac{3,464.1}{2} = 10,321 \text{ N (tension)}$$

and

$$P_E = \frac{-2,576,800}{300} + \frac{3,464.1}{2} = -6,857 \text{ N (compression)}$$

This approach cannot be used in the bay CDHG except at the section CJG, since the axial load in the stiffener JK introduces an additional unknown.

This analysis assumes that the web panels in beams of the type shown in Fig. 23.1 resist pure shear along their boundaries. In Chapter 9, we saw that thin webs may buckle under the action of such shear loads, producing tension field stresses which, in turn, induce additional loads in the stiffeners and flanges of beams. The tension field stresses may be calculated separately by the methods described in Chapter 9 and then superimposed on the stresses determined as just described.

So far, we have been concerned with web stiffener arrangements in which the loads have been applied on the plane of the web, so that two stiffeners are sufficient to resist the components of a concentrated load. Frequently, loads have an out-of-plane component; in which case, the structure should be arranged so that two webs meet at the point of load application with stiffeners aligned with

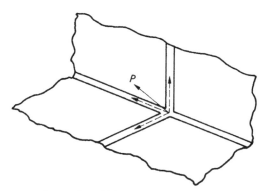

FIGURE 23.6 Structural Arrangement for an Out-of-Plane Load

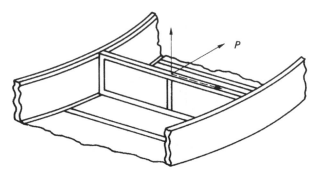

FIGURE 23.7 Support of Load Having a Component Normal to a Web

the three component directions (Fig. 23.6). In some situations, it is not practicable to have two webs meeting at the point of load application, so that a component normal to a web exists. If this component is small, it may be resisted in bending by an in-plane stiffener; otherwise, an additional member must be provided, spanning between adjacent frames or ribs, as shown in Fig. 23.7. In general, no normal loads should be applied to an unsupported web no matter how small their magnitude.

23.2 FUSELAGE FRAMES

We noted that fuselage frames transfer loads to the fuselage shell and provide column support for the longitudinal stringers. The frames generally take the form of open rings, so that the interior of the fuselage is not obstructed. They are connected continuously around their peripheries to the fuselage shell and are not necessarily circular in form but usually are symmetrical about a vertical axis.

A fuselage frame is in equilibrium under the action of any external loads and the reaction shear flows from the fuselage shell. Suppose that a fuselage frame has a vertical axis of symmetry and carries a vertical external load W, as shown in Fig. 23.8. The fuselage shell stringer section has been idealized

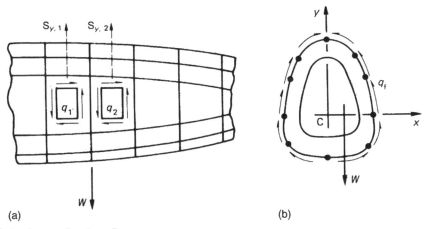

FIGURE 23.8 Loads on a Fuselage Frame

such that the fuselage skin is effective only in shear. Suppose also that the shear force in the fuselage immediately to the left of the frame is $S_{y,1}$ and that the shear force in the fuselage immediately to the right of the frame is $S_{y,2}$; clearly, $S_{y,2} = S_{y,1} - W$. $S_{y,1}$ and $S_{y,2}$ generate shear flow distributions q_1 and q_2, respectively, in the fuselage skin, each given by (Eq. 22.1), in which $S_{x,1} = S_{x,2} = 0$ and $I_{xy} = 0$ (Cy is an axis of symmetry). The shear flow q_f transmitted to the periphery of the frame is equal to the algebraic sum of q_1 and q_2; that is,

$$q_f = q_1 - q_2$$

Thus, substituting for q_1 and q_2 obtained from (Eq. 22.1) and noting that $S_{y,2} = S_{y,1} - W$, we have

$$q_f = \frac{-W}{I_{xx}} \sum_{r=1}^{n} B_r y_r + q_{s,0}$$

in which $q_{s,0}$ is calculated using (Eq. 17.17), where the shear load is W and

$$q_b = \frac{-W}{I_{xx}} \sum_{r=1}^{n} B_r y_r$$

The method of determining the shear flow distribution applied to the periphery of a fuselage frame is identical to the method of solution (or the alternative method) of Example 22.2.

Having determined the shear flow distribution around the periphery of the frame, the frame itself may be analyzed for distributions of bending moment, shear force, and normal force, as described in Section 5.4.

23.3 WING RIBS

Wing ribs perform similar functions to those performed by fuselage frames. They maintain the shape of the wing section, assist in transmitting external loads to the wing skin, and reduce the column length of the stringers. Their geometry, however, is usually different, in that they are frequently of unsymmetrical shape and possess webs which are continuous except for lightness holes and openings for control runs.

Wing ribs are subjected to loading systems similar to those applied to fuselage frames. External loads applied in the plane of the rib produce a change in shear force in the wing across the rib; this induces reaction shear flows around its periphery. These are calculated using the methods described in Chapters 16 and 22.

To illustrate the method of rib analysis, we use the example of a three-flange wing section in which, as we noted in Section 23.1, the shear flow distribution is statically determinate.

Example 23.2

Calculate the shear flows in the web panels and the axial loads in the flanges of the wing rib shown in Fig. 23.9. Assume that the web of the rib is effective only in shear while the resistance of the wing to bending moments is provided entirely by the three flanges 1, 2, and 3.

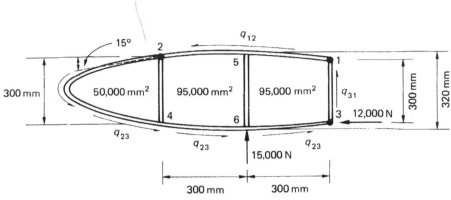

FIGURE 23.9 Wing Rib of Example 23.2

Since the wing bending moments are resisted entirely by the flanges 1, 2, and 3, the shear flows developed in the wing skin are constant between the flanges. Using the method described in Section 23.1 for a three-flange wing section, we have, resolving forces horizontally,

$$600q_{12} - 600q_{23} = 12,000 \text{ N} \qquad\qquad\text{(i)}$$

Resolving vertically,

$$300q_{31} - 300q_{23} = 15,000 \text{ N} \qquad\qquad\text{(ii)}$$

Taking moments about flange 3,

$$2(50,000 + 95,000)q_{23} + 2 \times 95,000q_{12} = -15,000 \times 300 \text{ N mm} \qquad\text{(iii)}$$

Solution of Eqs. (i)–(iii) gives

$$q_{12} = 13.0 \text{ N/mm}, \quad q_{23} = -7.0 \text{ N/mm}, \quad q_{31} = 43.0 \text{ N/mm}$$

Consider now the nose portion of the rib shown in Fig. 23.10 and suppose that the shear flow in the web immediately to the left of the stiffener 24 is q_1. The total vertical shear force $S_{y,1}$ at this section is given by

$$S_{y,1} = 7.0 \times 300 = 2,100 \text{ N}$$

The horizontal components of the rib flange loads resist the bending moment at this section. Thus,

$$P_{x,4} = P_{x,2} = \frac{2 \times 50,000 \times 7.0}{300} = 2,333.3 \text{ N}$$

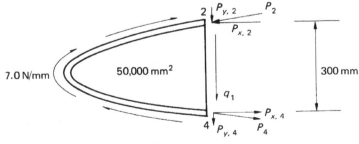

FIGURE 23.10 Equilibrium of Nose Portion of the Rib

The corresponding vertical components are then

$$P_{y,2} = P_{y,4} = 2,333.3 \tan 15° = 625.2 \text{ N}$$

Thus, the shear force carried by the web is $2,100 - 2 \times 625.2 = 849.6$ N. Hence,

$$q_1 = \frac{849.6}{300} = 2.8 \text{ N/mm}$$

The axial loads in the rib flanges at this section are given by

$$P_2 = P_4 = \left(2,333.3^2 + 625.2^2\right)^{1/2} = 2,415.6 \text{ N}$$

The rib flange loads and web panel shear flows, at a vertical section immediately to the left of the intermediate web stiffener 56, are found by considering the free body diagram shown in Fig. 23.11. At this section, the rib flanges have zero slope, so that the flange loads P_5 and P_6 are obtained directly from the value of bending moment at this section. Thus,

$$P_5 = P_6 = 2[(50,000 + 46,000) \times 7.0 - 49,000 \times 13.0]/320 = 218.8 \text{ N}$$

The shear force at this section is resisted solely by the web. Hence,

$$320q_2 = 7.0 \times 300 + 7.0 \times 10 - 13.0 \times 10 = 2,040 \text{ N}$$

so that

$$q_2 = 6.4 \text{ N/mm}$$

The shear flow in the rib immediately to the right of stiffener 56 is found most simply by considering the vertical equilibrium of stiffener 56, as shown in Fig. 23.12. Thus,

$$320q_3 = 6.4 \times 320 + 15,000$$

which gives

$$q_3 = 53.3 \text{ N/mm}$$

Finally, we consider the rib flange loads and the web shear flow at a section immediately forward of stiffener 31. From Fig. 23.13, in which we take moments about the point 3,

$$M_3 = 2[(50,000 + 95,000) \times 7.0 - 95,000 \times 13.0] + 15,000 \times 300 = 4.06 \times 10^6 \text{ N mm}$$

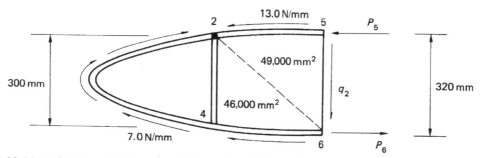

FIGURE 23.11 Equilibrium of Rib Forward of Intermediate Stiffener 56

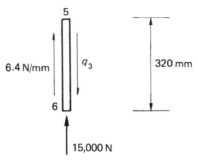

FIGURE 23.12 Equilibrium of Stiffener 56

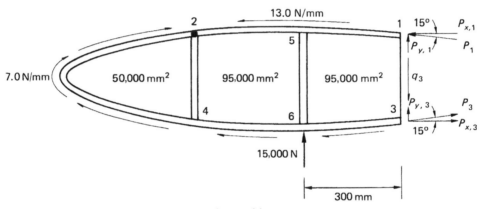

FIGURE 23.13 Equilibrium of the Rib Forward of Stiffener 31

The horizontal components of the flange loads at this section are then

$$P_{x,1} = P_{x,3} = \frac{4.06 \times 10^6}{300} = 13,533.3 \text{ N}$$

and the vertical components are

$$P_{y,1} = P_{y,3} = 3,626.2 \text{ N}$$

Hence,

$$P_1 = P_3 = \sqrt{13,533.3^2 + 3,626.2^2} = 14,010.7 \text{ N}$$

The total shear force at this section is $15,000 + 300 \times 7.0 = 17,100$ N. Therefore, the shear force resisted by the web is $17,100 - 2 \times 3,626.2 = 9,847.6$ N, so that the shear flow q_3 in the web at this section is

$$q_3 = \frac{9,847.6}{300} = 32.8 \text{ N/mm}$$

PROBLEMS

P.23.1 The beam shown in Fig. P.23.1 is simply supported at each end and carries a load of 6,000 N. If all direct stresses are resisted by the flanges and stiffeners and the web panels are effective only in shear, calculate the distribution of axial load in the flange ABC and the stiffener BE and the shear flows in the panels.

Answer: $q(\text{ABEF}) = 4\,\text{N/mm}, \quad q(\text{BCDE}) = 2\,\text{N/mm}$
P_{BE} increases linearly from zero at B to 6,000 N (tension) at E.
P_{AB} and P_{CB} increase linearly from zero at A and C to 4,000 N (compression) at B.

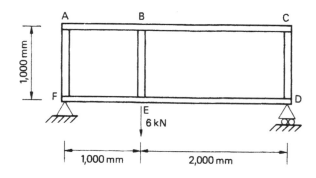

FIGURE P.23.1

P.23.2 Calculate the shear flows in the web panels and direct load in the flanges and stiffeners of the beam shown in Fig. P.23.2 if the web panels resist shear stresses only.

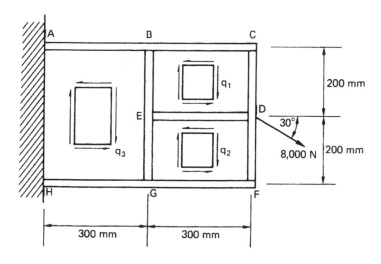

FIGURE P.23.2

Answer: $q_1 = 21.6\,\text{N/mm}$, $q_2 = -1.6\,\text{N/mm}$, $q_3 = 10\,\text{N/mm}$
$P_C = 0$, $P_B = 6,480\,\text{N}$ (tension), $P_A = 9,480\,\text{N}$ (tension)
$P_F = 0$, $P_G = 480\,\text{N}$ (tension), $P_H = 2,520\,\text{N}$ (compression)
P_E in BEG $= 2,320\,\text{N}$ (compression), P_D in ED $= 6,928\,\text{N}$ (tension)
P_D in CD $= 4,320\,\text{N}$ (tension), P_D in DF $= 320\,\text{N}$ (tension)

P.23.2 MATLAB Assuming the angle ($+ =$ clockwise) from horizontal of the 8,000 N load shown in Fig. P.23.2 is labeled α, use MATLAB to repeat the calculations of the web panel shear flows in Problem P.23.2 for values of α ranging from $-50°$ to $50°$ in increments of $10°$. All other dimensions and loads are as shown in Fig. P.23.2.

Answer:

	α	q_1	q_2	q_3
(i)	$-50°$	-6.8 N/mm	-23.9 N/mm	-15.3 N/mm
(ii)	$-40°$	-2.7 N/mm	-23.1 N/mm	-12.9 N/mm
(iii)	$-30°$	1.6 N/mm	-21.6 N/mm	-10 N/mm
(iv)	$-20°$	5.7 N/mm	-19.4 N/mm	-6.8 N/mm
(v)	$-10°$	9.7 N/mm	-16.6 N/mm	-3.5 N/mm
(vi)	$0°$	13.4 N/mm	-13.4 N/mm	0 N/mm
(vii)	$10°$	16.6 N/mm	-9.7 N/mm	3.5 N/mm
(viii)	$20°$	19.4 N/mm	-5.7 N/mm	6.8 N/mm
(ix)	$30°$	21.6 N/mm	1.6 N/mm	10 N/mm
(x)	$40°$	23.1 N/mm	2.7 N/mm	12.9 N/mm
(xi)	$50°$	23.9 N/mm	6.8 N/mm	15.3 N/mm

P.23.3 A three-flange wing section is stiffened by the wing rib shown in Fig. P.23.3. If the rib flanges and stiffeners carry all the direct loads while the rib panels are effective only in shear, calculate the shear flows in the panels and the direct loads in the rib flanges and stiffeners.

Answer: $q_1 = 4.0\,\text{N/mm}$, $q_2 = 26.0\,\text{N/mm}$, $q_3 = 6.0\,\text{N/mm}$
P_2 in $12 = -P_3$ in $43 = 1,200\,\text{N}$ (tension), P_5 in $154 = 2,000\,\text{N}$ (tension)
P_3 in $263 = 8,000\,\text{N}$ (compression), P_5 in $56 = 12,000\,\text{N}$ (tension)
P_6 in $263 = 6,000\,\text{N}$ (compression)

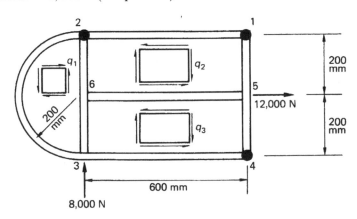

FIGURE P.23.3

Laminated composite structures

An increasingly large proportion of the structures of many modern aircraft are fabricated from composite materials. These, as we saw in Chapter 11, consist of laminas in which a stiff, high-strength filament, for example, carbon fiber, is embedded in a matrix such as epoxy or polyester. The use of composites can lead to considerable savings in weight over conventional metallic structures. They also have the advantage that the direction of the filaments in a multilamina structure may be aligned with the direction of the major loads at a particular point, resulting in a more efficient design.

There are two approaches to the analysis of composite materials. In the first, micromechanics, the constituent materials, that is, the fibers and resin (the matrix) are considered separately. The properties of the composite change from point to point in a particular direction, depending on whether the fiber or the resin is being examined. In the second approach, macromechanics, the composite material is regarded as a whole, so that the properties do not change from point to point in a particular direction. Generally, the design and analysis of composite materials are based on the macro rather than the micro approach.

Initially, but briefly, we consider the micro approach, in which the elastic constants of a lamina are determined in terms of the known properties of the constituent materials; we then determine the corresponding stresses.

24.1 ELASTIC CONSTANTS OF A SIMPLE LAMINA

A simple lamina of a composite structure can be considered as orthotropic with two principal material directions in its own plane: one parallel, the other perpendicular to the direction of the filaments; we designate the former the longitudinal direction (l), the latter the transverse direction (t).

In Fig. 24.1, a portion of a lamina containing a single filament is subjected to a stress, σ_l, in the longitudinal direction, which produces an extension Δl. If it is assumed that plane sections remain plane during deformation, then the strain ε_l corresponding to σ_l is given by

$$\varepsilon_l = \frac{\Delta l}{l} \tag{24.1}$$

and

$$\sigma_l = E_l \varepsilon_l \tag{24.2}$$

where E_l is the modulus of elasticity of the lamina in the direction of the filament. Also, using the suffixes f and m to designate filament and matrix parameters, we have

$$\sigma_f = E_f \varepsilon_l \quad \sigma_m = E_m \varepsilon_l \tag{24.3}$$

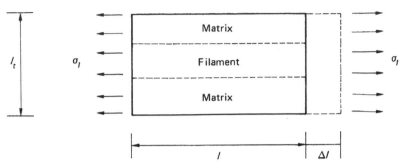

FIGURE 24.1 Determination of E_l

Further, if A is the total area of cross-section of the lamina in Fig. 24.1, A_f is the cross-sectional area of the filament and A_m the cross-sectional area of the matrix, then, for equilibrium in the direction of the filament,

$$\sigma_l A = \sigma_f A_f + \sigma_m A_m$$

or, substituting for σ_l, σ_f, and σ_m from Eqs. (24.2) and (24.3),

$$E_l \varepsilon_l A = E_f \varepsilon_f A_f + E_m \varepsilon_l A_m$$

so that

$$E_l = E_f \frac{A_f}{A} + E_m \frac{A_m}{A} \tag{24.4}$$

Writing $A_f/A = v_f$ and $A_m/A = v_m$, Eq. (24.4) becomes

$$E_l = v_f E_f + v_m E_m \tag{24.5}$$

Equation (24.5) is generally referred to as the *law of mixtures.*

A similar approach may be used to determine the modulus of elasticity in the transverse direction (E_t). In Fig. 24.2, the total extension in the transverse direction is produced by σ_t and is given by

$$\varepsilon_t l_t = \varepsilon_m l_m + \varepsilon_f l_f$$

or

$$\mu_{lt} = \frac{\varepsilon_t}{\varepsilon_l}$$

$$\frac{\sigma_t}{E_t} l_t = \frac{\sigma_t}{E_m} l_m + \frac{\sigma_t}{E_f} l_f$$

which gives

$$\frac{1}{E_t} = \frac{v_m}{E_m} + \frac{v_f}{E_f}$$

Rearranging this, we obtain

$$E_t = \frac{E_m E_f}{v_m E_f + v_f E_m} \tag{24.6}$$

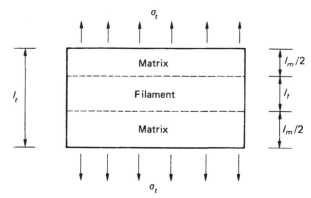

FIGURE 24.2 Determination of E_t

The major Poisson's ratio v_{lt} may be found by referring to the stress system of Fig. 24.1 and the dimensions given in Fig. 24.2. The total displacement in the transverse direction produced by σ_l is given by

$$\Delta_t = v_{lt}\varepsilon_l l_t$$

that is,

$$\Delta_t = v_{lt}\varepsilon_l l_t = v_m\varepsilon_l l_m + v_f\varepsilon_l l_f$$

from which

$$v_{lt} = v_m v_m + v_f v_f \qquad \mu_{lt} = v_m \mu_m + v_f \mu_f \qquad (24.7)$$

The minor Poisson's ratio v_{tl} is found by referring to Fig. 24.2. The strain in the longitudinal direction produced by the transverse stress σ_t is given by

$$v_{tl}\frac{\sigma_t}{E_t} = v_m\frac{\sigma_t}{E_m} = v_f\frac{\sigma_t}{E_f} \qquad (24.8)$$

From the last two of Eqs. (24.8),

$$v_f = \frac{E_f}{E_m}v_m$$

Substituting in Eq. (24.7),

$$v_{lt} = v_m\left(v_m + \frac{E_f}{E_m}v_f\right) = \frac{v_m}{E_m}(v_m E_m + v_f E_f)$$

or, from Eq. (24.5),

$$v_{lt} = v_m\frac{E_l}{E_m}$$

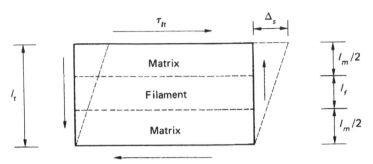

FIGURE 24.3 Determination of G_{lt}

Now, substituting for ν_m in the first two of Eqs. (24.8),

$$\frac{\nu_{tl}}{E_t} = \frac{\nu_{lt}}{E_l}$$

or, from Eq. (24.7),

$$\nu_{tl} = \frac{E_t}{E_l}\nu_{lt} = \frac{E_t}{E_l}(\nu_m\nu_m + \nu_f\nu_f) \tag{24.9}$$

Finally, the shear modulus $G_{lt}(= G_{tl})$ is determined by assuming that the constituent materials are subjected to the same shear stress τ_{lt} as shown in Fig. 24.3. The displacement Δ_s produced by shear is

$$\Delta_s = \frac{\tau_{lt}}{G_{lt}}l_t = \frac{\tau_{lt}}{G_m}l_m + \frac{\tau_{lt}}{G_f}l_f$$

in which G_m and G_f are the shear moduli of the matrix and filament, respectively. Then,

$$\frac{l_t}{G_{lt}} = \frac{l_m}{G_m} + \frac{l_f}{G_f}$$

from which

$$G_{lt} = \frac{G_m G_f}{\nu_m G_f + \nu_f G_m} \tag{24.10}$$

Example 24.1

A laminated bar, whose cross-section is shown in Fig. 24.4, is 500 mm long and is composed of an epoxy resin matrix reinforced by a carbon filament having moduli equal to 5,000 N/mm² and 200,000 N/mm², respectively; the corresponding values of Poisson's ratio are 0.2 and 0.3. If the bar is subjected to an axial tensile load of 100 kN, determine the lengthening of the bar and the reduction in its thickness. Calculate also the stresses in the epoxy resin and the carbon filament.

From Eq. (24.5), the modulus of the bar is given by

$$E_l = 200,000 \times \frac{80 \times 10}{80 \times 50} + 5,000 \times \frac{80 \times 40}{80 \times 50}$$

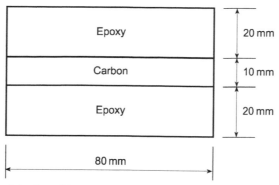

FIGURE 24.4 Cross-section of the Bar of Example 24.1

that is,

$$E_l = 44{,}000 \text{ N/mm}^2$$

The direct stress, σ_l, in the longitudinal direction is given by

$$\sigma_l = \frac{100 \times 10^3}{80 \times 50} = 25.0 \text{ N/mm}^2$$

Therefore, from Eq. (24.2), the longitudinal strain in the bar is

$$\varepsilon_l = \frac{25.0}{44{,}000} = 5.68 \times 10^{-4}$$

The lengthening, Δ_l, of the bar is then

$$\Delta_l = 5.68 \times 10^{-4} \times 500$$

that is,

$$\Delta_l = 0.284 \text{ mm}$$

The major Poisson's ratio for the bar is found from Eq. (24.7); that is,

$$v_{lt} = \frac{80 \times 40}{80 \times 50} \times 0.2 + \frac{80 \times 10}{80 \times 50} \times 0.3 = 0.22$$

The strain in the bar across its thickness is then

$$\varepsilon_t = -0.22 \times 5.68 \times 10^{-4} = -1.25 \times 10^{-4}$$

The reduction in thickness, Δ_t, of the bar is then

$$\Delta_t = 1.25 \times 10^{-4} \times 50$$

that is,

$$\Delta_t = 0.006 \text{ mm}$$

The stresses in the epoxy and the carbon are found using Eq. (24.3). Thus,

$$\sigma_m(\text{epoxy}) = 5{,}000 \times 5.68 \times 10^{-4} = 2.84 \text{ N/mm}^2$$
$$\sigma_f(\text{carbon}) = 200{,}000 \times 5.68 \times 10^{-4} = 113.6 \text{ N/mm}^2$$

Example 24.2

The wings of a replica biplane are connected by a system of composite cables, each of which is 1.5 m long and can only carry tensile loads. The cross-section of a cable is shown in Fig. 24.5 where the polyester matrix has a modulus of 2,800 N/mm² while that of the Kevlar filament is 145,000 N/mm² with corresponding Poisson's ratios of 0.15 and 0.30 respectively.

If the maximum tensile load in a cable is 10 kN determine the corresponding extension and the stresses in the polyester and Kevlar.

From Eq. (24.5) the modulus of the cable is

$$E_l = 145,000 \times \left(\pi \times 5^2/4\right)/\left(\pi \times 15^2/4\right) + 2,800 \times \left[\pi\left(15^2 - 5^2\right)/4\right]/\left(\pi \times 15^2/4\right)$$

which gives

$$E_l = 18,600 \text{ N/mm}^2.$$

The direct stress is given by

$$\sigma_l = 10 \times 10^3/\left(\pi \times 15^2/4\right) = 56.6 \text{ N/mm}^2$$

Therefore, from Eq. (24.2), the longitudinal strain in the cable is

$$\varepsilon_l = 56.6/18,600 = 3.0 \times 10^{-3}$$

so that the lengthening, Δ_l, of the cable is given by

$$\Delta_l = 3.0 \times 10^{-3} \times 1.5 \times 10^3 = 4.5 \text{ mm}$$

The stresses in the polyester and Kevlar are found using Eq. (24.3), i.e.,

$$\sigma_m \text{ (polyester)} = 2,800 \times 3.0 \times 10^{-3} = 8.4 \text{ N/mm}^2$$
$$\sigma_f \text{ (Kevlar)} = 145,000 \times 3.0 \times 10^{-3} = 435.0 \text{ N/mm}^2$$

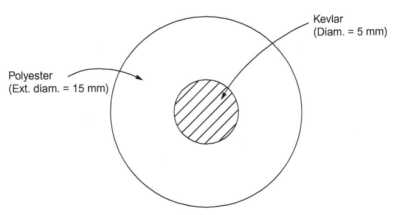

FIGURE 24.5 Cross-section of the Cable of Example 24.2

24.2 STRESS–STRAIN RELATIONSHIPS FOR AN ORTHOTROPIC PLY (MACRO APPROACH)

A single sheet of composite material in which the sheet has been preimpregnated with resin (a prepreg) and with the fibers aligned in one particular direction is called a *unidirectional ply* or *lamina* (Fig. 24.6(a)). On the other hand a woven ply has the fibers placed in two perpendicular directions (Fig. 24.6(b)); generally, the fiber reinforcement is the same in both directions. In plies where this is not the case, so that the material properties are different in the two mutually perpendicular directions, the ply is said to be *orthotropic* (see Section 11.7). Two cases of orthotropic plies arise. In the first, the directions of the applied loads coincide with directions of the plies; these are known as *specially orthotropic plies*. In the second, the applied loads are applied in any direction; these are termed *generally orthotropic plies*.

24.2.1 Specially orthotropic ply

Figure 24.7 shows an element of a specially orthotropic ply. The ply reference axes are the same as in Section 24.1, that is, longitudinal (subscript l) and transverse (subscript t). Of course, these axes do not have the same significance for a woven ply as they do for a unidirectional ply, but reference axes must

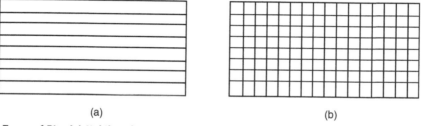

(a)   (b)

FIGURE 24.6 Types of Ply: (a) Unidirectional Ply; (b) Woven Ply

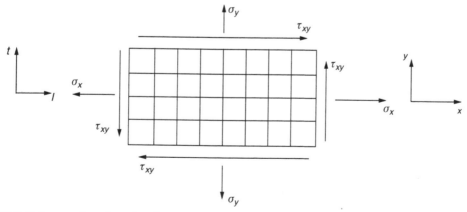

FIGURE 24.7 Reference Axes for a Specially Orthotropic Ply

be specified and these are as convenient as any. We also specify loading axes, x and y, which, for a specially orthotropic ply, coincide with the ply reference axes.

Suppose that the ply is subjected to direct stresses σ_x and σ_y, shear and complementary shear stresses τ_{xy} and that the elastic constants for the ply are E_l, E_t, G_{lt} ($= G_{tl}$), v_{lt}, and v_{tl} (see Eqs. (24.5)–(24.10)). Note that, unlike an isotropic material, the shear modulus G_{lt} is not related to the other elastic constants. From Section 1.15 (see Eqs. (1.52)), the strains in the longitudinal and transverse directions are given by

$$\left.\begin{aligned}
\varepsilon_l &= \frac{\sigma_x}{E_l} - \frac{v_{tl}\sigma_y}{E_t} \\[2mm]
\varepsilon_t &= \frac{\sigma_y}{E_t} - \frac{v_{lt}\sigma_x}{E_l}
\end{aligned}\right\} \tag{24.11}$$

Equations (24.11) may be written in matrix form; that is,

$$\left\{\begin{aligned} \varepsilon_l \\ \varepsilon_t \end{aligned}\right\} = \begin{bmatrix} \dfrac{1}{E_l} & -\dfrac{v_{tl}}{E_t} \\[3mm] -\dfrac{v_{lt}}{E_l} & \dfrac{1}{E_t} \end{bmatrix} \left\{\begin{aligned} \sigma_x \\ \sigma_y \end{aligned}\right\} \tag{24.12}$$

or, in general terms,

$$[\varepsilon] = [S][\sigma] \tag{24.13}$$

in which $[S]$ is frequently termed *the compliance matrix*. It may be shown, using an energy approach, that the compliance matrix $[S]$ must be symmetric about the leading diagonal. Therefore,

$$-\frac{v_{tl}}{E_t} = -\frac{v_{lt}}{E_l}$$

giving

$$\frac{v_{tl}}{E_t} = \frac{v_{lt}}{E_1} \tag{24.14}$$

so that, of the four elastic constants E_l, E_t, v_{lt}, and v_{tl}, only three are independent.

Equations (24.11) may be transposed (as in Section 1.15) to give stress–strain relationships. Then,

$$\left.\begin{aligned}
\sigma_x &= \frac{E_l}{1 - v_{lt}v_{tl}}\varepsilon_l + \frac{v_{tl}E_l}{1 - v_{lt}v_{tl}}\varepsilon_t \\[2mm]
\sigma_y &= \frac{E_t}{1 - v_{lt}v_{tl}}\varepsilon_t + \frac{v_{lt}E_t}{1 - v_{lt}v_{tl}}\varepsilon_l
\end{aligned}\right\} \tag{24.15}$$

From the last of Eqs. (1.52),

$$\gamma_{lt} = \frac{\tau_{xy}}{G_{lt}} \tag{24.16}$$

$$\tau_{xy} = \gamma_{lt}G_{lt}$$

Equations (24.15) and (24.16) may be written in matrix form, that is,

$$
\begin{Bmatrix} \sigma_x \\ \sigma_y \\ \tau_{xy} \end{Bmatrix} =
\begin{bmatrix}
\dfrac{E_l}{1 - \nu_{lt}\nu_{tl}} & \dfrac{\nu_{tl}E_l}{1 - \nu_{lt}\nu_{tl}} & 0 \\[2ex]
\dfrac{\nu_{lt}E_t}{1 - \nu_{lt}\nu_{tl}} & \dfrac{E_t}{1 - \nu_{lt}\nu_{tl}} & 0 \\[2ex]
0 & 0 & G_{lt}
\end{bmatrix}
\begin{Bmatrix} \varepsilon_l \\ \varepsilon_t \\ \gamma_{lt} \end{Bmatrix}
\tag{24.17}
$$

Example 24.3

A single sheet of woven ply is subjected to longitudinal and transverse direct stresses of 50 and 25 N/mm², respectively, together with a shear stress of 40 N/mm². The elastic constants for the ply are $E_l = 120{,}000$ N/mm², $E_t = 80{,}000$ N/mm², $G_{lt} = 5{,}000$ N/mm², and $\nu_{lt} = 0.3$. Calculate the direct strains in the longitudinal and transverse directions and the shear strain in the ply.

The value of the minor Poisson's ratio, ν_{tl}, is not given and must be calculated first. From Eq. (24.14),

$$
\nu_{tl} = \nu_{lt}\frac{E_t}{E_l} = 0.3 \times \frac{80{,}000}{120{,}000} = 0.2
$$

Therefore, from Eqs. (24.11),

$$
\varepsilon_l = \frac{50}{120{,}000} - \frac{0.2 \times 25}{80{,}000} = 3.54 \times 10^{-4}
$$

$$
\varepsilon_t = \frac{25}{80{,}000} - \frac{0.3 \times 50}{120{,}000} = 1.88 \times 10^{-4}
$$

and, from Eq. (24.16),

$$
\gamma_{lt} = \frac{40}{5{,}000} = 80.0 \times 10^{-4}
$$

24.2.2 Generally orthotropic ply

In Fig. 24.8, the direction of the fibers in the ply does not coincide with the loading axes x and y. We specify that the longitudinal fibers of the ply are inclined at an angle θ to the x axis; θ is positive when the fibers are rotated in a counterclockwise sense from the x axis.

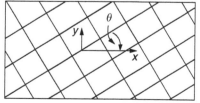

FIGURE 24.8 Generally Orthotropic Ply

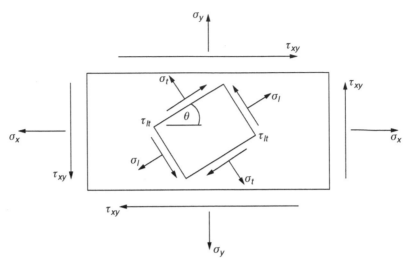

FIGURE 24.9 Stresses in a Generally Orthotropic Ply

Suppose that an element of the ply is subjected to stresses σ_x, σ_y, and τ_{xy}, as shown in Fig. 24.9. The stresses on an element of the ply in the directions of the fibers may be found in terms of the applied stresses using the method described in Section 1.6. Therefore, by comparison with Eq. (1.8),

$$\sigma_l = \sigma_x \cos^2 \theta + \sigma_y \sin^2 \theta + 2\tau_{xy} \cos \theta \sin \theta \tag{24.18}$$

Similarly,

$$\sigma_t = \sigma_x \sin^2 \theta + \sigma_y \cos^2 \theta - 2\tau_{xy} \cos \theta \sin \theta \tag{24.19}$$

and, by comparison with Eq. (1.9) but noting that τ_{lt} is in the opposite sense to τ,

$$\tau_{lt} = -\sigma_x \cos \theta \sin \theta + \sigma_y \cos \theta \sin \theta + \tau_{xy}(\cos^2 \theta - \sin^2 \theta) \tag{24.20}$$

If we write $m = \cos \theta$ and $n = \sin \theta$, Eqs. (24.18) through (24.20) become

$$\sigma_l = m^2 \sigma_x + n^2 \sigma_y + 2mn\tau_{xy} \tag{24.21}$$

$$\sigma_t = n^2 \sigma_x + m^2 \sigma_y - 2mn\tau_{xy} \tag{24.22}$$

$$\tau_{lt} = -mn\sigma_x + mn\sigma_y + (m^2 - n^2)\tau_{xy} \tag{24.23}$$

Writing Eqs. (24.21)–(24.23) in matrix form, we have

$$\begin{Bmatrix} \sigma_l \\ \sigma_t \\ \tau_{lt} \end{Bmatrix} = \begin{bmatrix} m^2 & n^2 & 2mn \\ n^2 & m^2 & -2mn \\ -mn & mn & m^2 - n^2 \end{bmatrix} \begin{Bmatrix} \sigma_x \\ \sigma_y \\ \tau_{xy} \end{Bmatrix} \tag{24.24}$$

Similarly, from Eqs. (1.31) and (1.34),

$$
\left\{ \begin{array}{c} \varepsilon_l \\ \varepsilon_t \\ \gamma_{lt} \end{array} \right\} = \left[\begin{array}{ccc} m^2 & n^2 & mn \\ n^2 & m^2 & -mn \\ -2mn & 2mn & m^2 - n^2 \end{array} \right] \left\{ \begin{array}{c} \varepsilon_x \\ \varepsilon_y \\ \gamma_{xy} \end{array} \right\}
\tag{24.25}
$$

Equations (24.24) may be transposed so that the applied stresses are expressed in terms of the ply stresses:

$$
\left\{ \begin{array}{c} \sigma_x \\ \sigma_y \\ \tau_{xy} \end{array} \right\} = \left[\begin{array}{ccc} m^2 & n^2 & -2mn \\ n^2 & m^2 & 2mn \\ mn & -mn & m^2 - n^2 \end{array} \right] \left\{ \begin{array}{c} \sigma_l \\ \sigma_t \\ \tau_{lt} \end{array} \right\}
\tag{24.26}
$$

In Eqs. (24.17), for a specially orthotropic ply, the ply stresses and loading stresses are identical, so that we may use this equation to relate the ply stresses in a generally orthotropic ply to the ply strains. Then,

$$
\left\{ \begin{array}{c} \sigma_l \\ \sigma_t \\ \tau_{lt} \end{array} \right\} = \left[\begin{array}{ccc} \dfrac{E_l}{1 - v_{lt}v_{tl}} & \dfrac{v_{tl}E_l}{1 - v_{lt}v_{tl}} & 0 \\[2ex] \dfrac{v_{lt}E_t}{1 - v_{lt}v_{tl}} & \dfrac{E_t}{1 - v_{lt}v_{tl}} & 0 \\[2ex] 0 & 0 & G_{lt} \end{array} \right] \left\{ \begin{array}{c} \varepsilon_l \\ \varepsilon_t \\ \gamma_{lt} \end{array} \right\}
\tag{24.27}
$$

Substituting for the ply stresses in Eqs. (24.26) from Eqs. (24.27), we express the applied stresses in terms of the ply strains; that is,

$$
\left\{ \begin{array}{c} \sigma_x \\ \sigma_y \\ \tau_{xy} \end{array} \right\} = \left[\begin{array}{ccc} m^2 & n^2 & -2mn \\ n^2 & m^2 & 2mn \\ mn & -mn & m^2 - n^2 \end{array} \right] \left[\begin{array}{ccc} \dfrac{E_l}{1 - v_{lt}v_{tl}} & \dfrac{v_{tl}E_l}{1 - v_{lt}v_{tl}} & 0 \\[2ex] \dfrac{v_{lt}E_t}{1 - v_{lt}v_{tl}} & \dfrac{E_t}{1 - v_{lt}v_{tl}} & 0 \\[2ex] 0 & 0 & G_{lt} \end{array} \right] \left\{ \begin{array}{c} \varepsilon_l \\ \varepsilon_t \\ \gamma_{lt} \end{array} \right\}
\tag{24.28}
$$

Finally, by substituting for the ply strains in Eqs. (24.28) from Eqs. (24.25), we obtain the applied stresses in terms of the strains referred to the xy axes; that is,

$$
\left\{ \begin{array}{c} \sigma_x \\ \sigma_y \\ \tau_{xy} \end{array} \right\} = \left[\begin{array}{ccc} m^2 & n^2 & -2mn \\ n^2 & m^2 & 2mn \\ mn & -mn & m^2 - n^2 \end{array} \right] \left[\begin{array}{ccc} \dfrac{E_l}{1 - v_{lt}v_{tl}} & \dfrac{v_{tl}E_l}{1 - v_{lt}v_{tl}} & 0 \\[2ex] \dfrac{v_{lt}E_t}{1 - v_{lt}v_{tl}} & \dfrac{E_t}{1 - v_{lt}v_{tl}} & 0 \\[2ex] 0 & 0 & G_{lt} \end{array} \right]
$$

$$
\times \left[\begin{array}{ccc} m^2 & n^2 & mn \\ n^2 & m^2 & -mn \\ -2mn & 2mn & m^2 - n^2 \end{array} \right] \left\{ \begin{array}{c} \varepsilon_x \\ \varepsilon_y \\ \gamma_{xy} \end{array} \right\}
\tag{24.29}
$$

Writing the individual terms of the central matrix as

$$k_{11} = \frac{E_l}{1 - v_{lt}v_{tl}}, \quad k_{12} = \frac{v_{tl}E_l}{1 - v_{lt}v_{tl}} = k_{21} \qquad \text{(from Eq.(25.14))}$$

$$k_{22} = \frac{E_t}{1 - v_{lt}v_{tl}}, \quad k_{33} = G_{lt}$$

k_{11}, k_{12}, etc., are sometimes referred to as *the reduced stiffnesses*. Eqs. (24.29) become

$$\left\{\begin{array}{c} \sigma_x \\ \sigma_y \\ \tau_{xy} \end{array}\right\} = \begin{bmatrix} m^2 & n^2 & -2mn \\ n^2 & m^2 & 2mn \\ mn & -mn & m^2 - n^2 \end{bmatrix} \begin{bmatrix} k_{11} & k_{12} & 0 \\ k_{12} & k_{22} & 0 \\ 0 & 0 & k_{33} \end{bmatrix}$$
$$\times \begin{bmatrix} m^2 & n^2 & mn \\ n^2 & m^2 & -mn \\ -2mn & 2mn & m^2 - n^2 \end{bmatrix} \left\{\begin{array}{c} \varepsilon_x \\ \varepsilon_y \\ \gamma_{xy} \end{array}\right\} \tag{24.30}$$

Carrying out the matrix multiplication in Eqs. (24.30), we obtain

$$\left\{\begin{array}{c} \sigma_x \\ \sigma_y \\ \tau_{xy} \end{array}\right\} = \begin{bmatrix} \begin{array}{c} m^4 k_{11} + m^2 n^2 (2k_{12} \\ +4k_{33}) + n^4 k_{22} \end{array} & \begin{array}{c} m^2 n^2 (k_{11} + k_{22} - 4k_{33}) \\ +(m^4 + n^4)k_{12} \end{array} & \begin{array}{c} m^3 n(k_{11} - k_{12} - 2k_{33}) \\ +mn^3(k_{12} - k_{22} + 2k_{33}) \end{array} \\ \begin{array}{c} m^2 n^2 (k_{11} + k_{22} - 4k_{33}) \\ +(m^4 + n^4)k_{12} \end{array} & \begin{array}{c} n^4 k_{11} + m^2 n^2 (2k_{12} \\ +4k_{33}) + m^4 k_{22} \end{array} & \begin{array}{c} mn^3(k_{11} - k_{12} - 2k_{33}) \\ +m^3 n(k_{12} - k_{22} + 2k_{33}) \end{array} \\ \begin{array}{c} m^3 n(k_{11} - k_{12} - 2k_{33}) \\ +mn^3(k_{12} + k_{22} + 2k_{33}) \end{array} & \begin{array}{c} mn^3(k_{11} - k_{12} - 2k_{33}) \\ +m^3 n(k_{12} - k_{22} + 2k_{33}) \end{array} & \begin{array}{c} m^2 n^2 (k_{11} - k_{22} - 2k_{12} \\ -2k_{33}) + (m^4 + n^4)k_{33} \end{array} \end{bmatrix} \left\{\begin{array}{c} \varepsilon_x \\ \varepsilon_y \\ \gamma_{xy} \end{array}\right\} \tag{24.31}$$

It can be seen that, for a specially orthotropic ply where $\theta = 0$, Eqs. (24.31) reduce to

$$\left\{\begin{array}{c} \sigma_x \\ \sigma_y \\ \tau_{xy} \end{array}\right\} = \begin{bmatrix} k_{11} & k_{12} & 0 \\ k_{12} & k_{22} & 0 \\ 0 & 0 & k_{33} \end{bmatrix} \left\{\begin{array}{c} \varepsilon_x \\ \varepsilon_y \\ \gamma_{xy} \end{array}\right\} \tag{24.32}$$

which are identical to Eqs. (24.17). Eqs (24.31) and (24.32) may be written in general terms as

$$[s] = [K][\varepsilon]$$

in which $[K]$ is *the reduced stiffness matrix*.

Having expressed the applied stresses in terms of the xy strains, Eqs. (24.31) may be transposed to obtain the xy strains in terms of the applied stresses. This may be shown to be

$$\left\{\begin{array}{c} \varepsilon_x \\ \varepsilon_y \\ \gamma_{xy} \end{array}\right\} = \begin{bmatrix} \begin{array}{c} m^4 s_{11} + n^4 s_{22} \\ +2m^2 n^2 s_{12} + m^2 n^2 s_{33} \end{array} & \begin{array}{c} m^2 n^2 s_{11} + m^2 n^2 s_{22} \\ +(m^4 + n^4)s_{12} \\ -m^2 n^2 s_{33} \end{array} & \begin{array}{c} 2m^3 n s_{11} - 2mn^3 s_{22} \\ +2(mn^3 - m^3 n)s_{12} \\ +(mn^3 - m^3 n)s_{33} \end{array} \\ \begin{array}{c} m^2 n^2 s_{11} + m^2 n^2 s_{22} \\ +(m^4 + n^4)s_{12} \\ -m^2 n^2 s_{33} \end{array} & \begin{array}{c} n^4 s_{11} + m^4 s_{22} \\ +2m^2 n^2 s_{12} + m^2 n^2 s_{33} \end{array} & \begin{array}{c} 2mn^3 s_{11} - 2m^3 n s_{22} \\ +2(m^3 n - mn^3)s_{12} \\ +(m^3 n - mn^3)s_{33} \end{array} \\ \begin{array}{c} 2m^3 n s_{11} - 2mn^3 s_{22} \\ +2(mn^3 - m^3 n)s_{12} \\ +(mn^3 - m^3 n)s_{33} \end{array} & \begin{array}{c} 2mn^3 s_{11} - 2m^3 n s_{22} \\ +2(m^3 n - mn^3)s_{12} \\ +(m^3 n - nm^3)s_{33} \end{array} & \begin{array}{c} 4m^2 n^2 s_{11} + 4m^2 n^2 s_{22} \\ -8m^2 n^2 s_{12} \\ +(m^2 - n^2)^2 s_{33} \end{array} \end{bmatrix} \left\{\begin{array}{c} \sigma_x \\ \sigma_y \\ \tau_{xy} \end{array}\right\} \tag{24.33}$$

in which

$$s_{11} = 1/E_1, \quad s_{12} = -v_{tl}/E_t, \quad s_{22} = 1/E_t, \quad s_{33} = 1/G_{lt}$$

For a specially orthotropic ply in which only direct stresses σ_x and σ_y are applied, Eqs. (24.33) reduce to

$$\begin{Bmatrix} \varepsilon_x \\ \varepsilon_y \end{Bmatrix} = \begin{bmatrix} s_{11} & s_{12} \\ s_{12} & s_{22} \end{bmatrix} \begin{Bmatrix} \sigma_x \\ \sigma_y \end{Bmatrix}$$

that is,

$$\begin{Bmatrix} \varepsilon_x \\ \varepsilon_y \end{Bmatrix} = \begin{bmatrix} \dfrac{1}{E_l} & -\dfrac{v_{tl}}{E_t} \\ -\dfrac{v_{lt}}{E_l} & \dfrac{1}{E_t} \end{bmatrix} \begin{Bmatrix} \sigma_x \\ \sigma_y \end{Bmatrix}$$

which are identical to Eqs. (24.12).

Example 24.4

A thin ply consists of an isotropic material having a Young's modulus of 80,000 N/mm^2 and a Poisson's ratio of 0.3. Determine the terms in the reduced stiffness and compliance matrices.

The reduced stiffnesses are

$$k_{11} = k_{22} = E/(1 - v^2) = 80,000/(1 - 0.3^2) = 87\ 912\ \text{N/mm}^2$$
$$k_{12} = vE/(1 - v^2) = 0.3 \times 80,000/(1 - 0.3^2) = 26\ 374\ \text{N/mm}^2$$

Also $k_{33} = G$ but since the ply is isotropic the shear modulus may be expressed in terms of Young's modulus and Poisson's ratio, i.e.,

$$G = E/2(1 + v) \quad \text{(see Eq. (1.50))}$$

so that

$$k_{33} = 80,000/2(1 + 0.3) = 30\ 769\ \text{N/mm}^2$$

The reduced compliances are

$$s_{11} = s_{22} = 1/E = 1/80,000 = 12.5 \times 10^{-6}$$
$$s_{12} = s_{21} = -v/E = -0.3/80,000 = -3.75 \times 10^{-6}$$
$$s_{33} = 1/G = 1/30,769 = 32.5 \times 10^{-6}$$

Example 24.5

If the reduced stiffnesses in a unidirectional ply are $k_{11} = 40,000$ N/mm^2, $k_{12} = k_{21} = 2,000$ N/mm^2, $k_{22} = 8,000$ N/mm^2 and $k_{33} = 5,000$ N/mm^2, determine the elastic constants of the ply.

From Eqs (24.27) and (24.32)

$$k_{11} = E_l/(1 - v_{lt}v_{tl}) \tag{i}$$

$$k_{12} = v_{tl}E_l/(1 - v_{lt}v_{tl}) \tag{ii}$$

Dividing Eq. (ii) by Eq. (i)

$$v_{tl} = k_{12}/k_{11} = 2{,}000/40{,}000 = 0.05$$

Also

$$k_{21} = v_{lt}E_t/(1 - v_{lt}v_{tl}) \tag{iii}$$

$$k_{22} = E_t/(1 - v_{lt}v_{tl}) \tag{iv}$$

Dividing Eq. (iii) by Eq. (iv)

$$v_{lt} = k_{21}/k_{22} = 2{,}000/8{,}000 = 0.25$$

Substituting in Eq. (i) for v_{tl} and v_{lt}

$$40{,}000 = E_l/(1 - 0.25 \times 0.05)$$

which gives

$$E_l = 39{,}500 \text{ N/mm}^2$$

From Eq. (iii)

$$E_t = 2{,}000(1 - 0.25 \times 0.05)/0.25 = 7{,}900 \text{ N/mm}^2$$

Also

$$G = k_{33} = 5{,}000 \text{ N/mm}^2.$$

Example 24.6

A generally orthotropic ply is subjected to direct stresses of 60 N/mm^2 parallel to the x reference axis and 40 N/mm^2 perpendicular to the x reference axis. If the longitudinal plies are inclined at an angle of 45° to the x axis and the elastic constants are $E_l = 150{,}000$ N/mm^2, $E_t = 90{,}000$ N/mm^2, $G_{lt} = 5{,}000$ N/mm^2, and $v_{lt} = 0.3$, calculate the direct strains parallel to the x and y directions and the shear strain referred to the xy axes.

We note that there is no applied shear stress, so it is unnecessary to calculate the terms in the third column of the matrix of Eqs. (24.33). Then,

$$s_{11} = \frac{1}{E_l} = \frac{1}{150{,}000} = 6.7 \times 10^{-6}$$

$$s_{22} = \frac{1}{E_t} = \frac{1}{90{,}000} = 11.1 \times 10^{-6}$$

$$s_{12} = -\frac{v_{lt}}{E_l} = -\frac{0.3}{150{,}000} = -2.0 \times 10^{-6}$$

$$s_{33} = \frac{1}{G_{lt}} = \frac{1}{5{,}000} = 200 \times 10^{-6}$$

Also,

$$\cos \theta = \sin \theta = \cos 45° = 1/\sqrt{2}$$

so that

$$m^2 = 0.5 = n^2, \quad m^4 = n^4 = 0.25, \quad m^2 n^2 = 0.25, \text{ and so forth}$$

Substituting these values in Eqs. (24.33), we have

$$\begin{Bmatrix} \varepsilon_x \\ \varepsilon_y \\ \gamma_{xy} \end{Bmatrix} = \begin{bmatrix} 53.45 & -46.55 & - \\ -46.55 & 53.45 & - \\ -2.2 & -2.2 & - \end{bmatrix} \begin{Bmatrix} 60 \\ 40 \\ 0 \end{Bmatrix}$$

which gives

$$\varepsilon_x = 1345 \times 10^{-6}$$
$$\varepsilon_y = -655 \times 10^{-6}$$
$$\gamma_{xy} = -220 \times 10^{-6}$$

It should be noted that the preceding is an introduction into the analysis and design of composite materials. Complete texts[1,2] are devoted to the subject in which multiply laminates, laminate failure, residual thermal stresses, and the like are considered.

Example 24.7
A generally orthotropic ply is subjected to stresses of $\sigma_x = 120 \text{ N/mm}^2$, $\sigma_y = 45 \text{ N/mm}^2$ and $\tau_{xy} = 70 \text{ N/mm}^2$ referred to an xy system of loading axes. If the ply angle is $45°$ determine the equivalent stresses referred to the material axes.

Since $\cos \theta = \sin \theta = \cos 45° = 1/\sqrt{2}$

$$m^2 = n^2 = mn = 0.5$$

Substituting in Eqs (24.24)

$$\begin{Bmatrix} \sigma_l \\ \sigma_t \\ \tau_{lt} \end{Bmatrix} = \begin{bmatrix} 0.5 & 0.5 & 1 \\ 0.5 & 0.5 & -1 \\ -0.5 & 0.5 & 0 \end{bmatrix} \begin{Bmatrix} 120 \\ 45 \\ 70 \end{Bmatrix}$$

which gives $\sigma_l = 152.5 \text{ N/mm}^2$, $\sigma_t = 12.5 \text{ N/mm}^2$, $\tau_{lt} = -37.5 \text{ N/mm}^2$.

Example 24.3 MATLAB
Use MATLAB to repeat Example 24.3.

The direct and shear strains are obtained through the following MATLAB file:

```
% Declare any needed variables
sig_x=60;
sig_y=40;
tau_xy=0;
theta=45*pi/180;
E_1=150e3;
```

```
E_t=90e3;
G_lt=5e3;
v_lt=0.3;

% Calculate variables needed to obtain the desired strains
m=cos(theta);
n=sin(theta);
s_11=round(1/E_l*10^7)/(10^7);
s_12=round(-v_lt/E_l*10^7)/(10^7);
s_22=round(1/E_t*10^7)/(10^7);
s_33=round(1/G_lt*10^7)/(10^7);
a_11=m^4*s_11+n^4*s_22+2*m^2*n^2*s_12+m^2*n^2*s_33;
a_12=m^2*n^2*(s_11+s_22-s_33)+(m^4+n^4)*s_12;
a_13=2*m^3*n*s_11-2*m*n^3*s_22+2*(m*n^3-m^3*n)*s_12+(m*n^3-m^3*n)*s_33;
a_21=a_12;
a_22=n^4*s_11+m^4*s_22+2*m^2*n^2*s_12+m^2*n^2*s_33;
a_23=2*m*n^3*s_11-2*m^3*n*s_22+2*(m^3*n-m*n^3)*s_12+(m^3*n-m*n^3)*s_33;
a_31=a_13;
a_32=a_23;
a_33=4*m^2*n^2*(s_11+s_22)-8*m^2*n^2*s_12+(m^2-n^2)*s_33;

% Substitute these variables intoEq. (24.33)
A=[a_11 a_12 a_13;
 a_21 a_22 a_23;
 a_31 a_32 a_33];
b=[sig_x sig_y tau_xy]';
strains=A*b;

% Output the direct and shear strains to the Command Window
disp(['e_x=' num2str(strains(1))])
disp(['e_y=' num2str(strains(2))])
disp(['gamma_xy=' num2str(strains(3))])
```

The Command Window outputs resulting from this MATLAB file are as follows:

```
e_x=0.001345
e_y=-0.000655
gamma_xy=-0.00022
```

24.3 THIN-WALLED COMPOSITE BEAMS

We noted, in Chapter 10, that some structural components in many modern aircraft are fabricated from composite materials. These components are generally in the form of laminates, which are stacks of plies bonded together. The orientation of each ply is different to that of its immediate neighbors so that the required strength and stiffness in a particular direction is obtained. The determination of the elastic

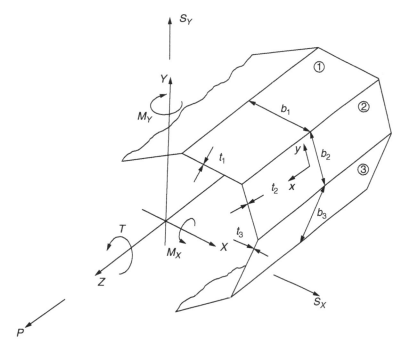

FIGURE 24.10 Composite Thin-Walled Section

properties of a laminate is discussed in Calcote[1] and Datoo[2] and is lengthy, so we assume that these are known and concentrate on the effects of the composite construction on the analysis.

In Chapters 15 through 17, we determined stresses and displacements in open and closed section thin-walled beams subjected to bending, shear, and torsional loads; the effect of axial load was considered in Chapter 1. We now re-examine these cases to determine the effect of composite construction.

Figure 24.10 shows a thin-walled beam that may be of either open or closed section and that is fabricated from laminates ①, ②, ③, ... The dimensions of each laminate are different, as are their elastic properties.

The beam is subjected to axial, bending, shear, and torsional loads that are positive in the directions shown (see also Fig 16.9). The beam axes XYZ are now in upper case letters to avoid confusion with the laminate axes xy.

24.3.1 Axial load

Suppose that the portion of the axial load P taken by the ith laminate is P_i. The longitudinal strain $\varepsilon_{x,i}$ in the laminate is equal to the longitudinal strain ε_z in the beam, since one of the basic assumptions of our analysis, except in the case of torsion, is that plane sections remain plane after the load is applied. Then, from Eq. (1.40),

$$\frac{P_i}{b_i t_i} = \varepsilon_{x,i} E_{x,i}$$

Therefore,

$$P_i = b_i t_i \varepsilon_{x,i} E_{x,i}$$

that is,

$$P_i = \varepsilon_z b_i t_i E_{x,i} \tag{24.34}$$

The total axial load on the beam is then given by

$$P = \varepsilon_z \sum_{i=1}^{n} b_i t_i E_{x,i} \tag{24.35}$$

Note that, in Eq. (24.35), ε_z is the longitudinal strain in the beam section and is therefore the same for every laminate; it may therefore be taken outside the summation. Further, the value of Young's modulus for a particular laminate is the same whether referred to the laminate x axis or the beam Z axis; we therefore refer it to the beam Z axis. Equation (24.35) may therefore be written

$$P = \varepsilon_z \sum_{i=1}^{n} b_i t_i E_{z,i} \tag{24.36}$$

from which

$$\varepsilon_z = \frac{P}{\sum\limits_{i=1}^{n} b_i t_i E_{z,i}} \tag{24.37}$$

Example 24.8
A beam has the singly symmetrical composite section shown in Fig. 24.11. The flange laminates are identical and have a Young's modulus, E_z, of 60,000 N/mm^2 while the vertical web has a Young's modulus, E_z, of 20,000 N/mm^2. If the beam is subjected to an axial load of 40 kN, determine the axial load in each laminate.

For each flange,

$$b_i t_i E_{z,i} = 100 \times 2.0 \times 60,000 = 12 \times 10^6$$

and, for the web,

$$b_i t_i E_{z,i} = 150 \times 1.0 \times 20,000 = 3 \times 10^6$$

Therefore,

$$\sum_{i=1}^{n} b_i t_i E_{z,i} = 2 \times 12 \times 10^6 + 3 \times 10^6 = 27 \times 10^6$$

Then, from Eq. (24.37),

$$\varepsilon_z = \frac{40 \times 10^3}{27 \times 10^6} = 1.48 \times 10^{-3}$$

Therefore, from Eq. (24.34),

$$P \text{ (flanges)} = 1.48 \times 10^{-3} \times 12 \times 10^6 = 17{,}760 \text{ N} = 17.76 \text{ kN}$$
$$P \text{ (web)} = 1.48 \times 10^{-3} \times 3 \times 10^6 = 4{,}440 \text{ N} = 4.44 \text{ kN}$$

Note that $2 \times 17.76 + 4.44 = 39.96$ kN, the discrepancy, 0.04 kN, is due to rounding off errors.

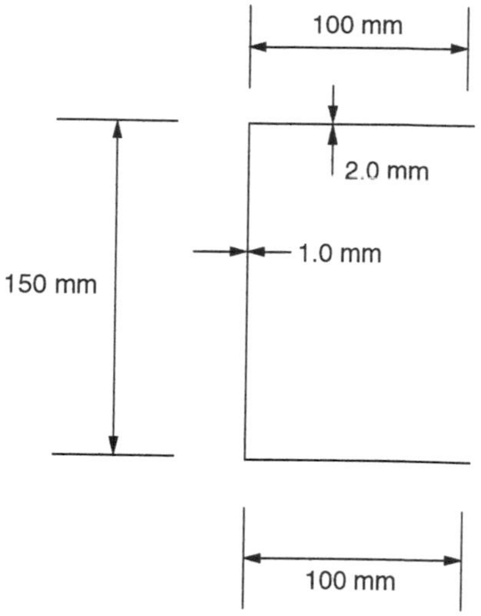

1) DRAW ALL SHEAR FLOWS
2) WRITE MOMENT
3) τ in EACH SECTION
$\tau = \frac{VQ}{Ib}$
-REMEMBER TO INCLUDE ENTIRE CUT SECTION FOR Q
4) CALCULATE V $\int \tau\, t\, dx$ usually
5) CALCULATE I
6) SOLVE FOR e

FIGURE 24.11 Beam Section of Example 24.8

24.3.2 Bending

In Section 16.2, we derived an expression for the direct stress distribution in a beam of unsymmetrical cross-section (Eq. (16.17) or (16.18)). In this derivation, the direct stress on an element of the beam cross-section was expressed in terms of Young's modulus, the radius of curvature of the beam, the coordinates of the element, and the inclination of the neutral axis to the section x axis (see Eq. (16.15)). The beam was assumed to be composed of homogenous material so that Young's modulus is a constant. This, as we have seen, is not necessarily the case for a composite beam where E can vary from laminate to laminate. We therefore rewrite Eqs. (16.16) in the form

$$M_x = \int_A \frac{E_{z,i}}{\rho}(x \sin\alpha + y\cos\alpha)y\, dA, \quad M_y = \int_A \frac{E_{z,i}}{\rho}(x \sin\alpha + y\cos\alpha)x\, dA,$$

or

$$M_x = \frac{\sin\alpha}{\rho}\int_A E_{z,i}xy\, dA + \frac{\cos\alpha}{\rho}\int_A E_{z,i}y^2\, dA,$$

$$M_y = \frac{\sin\alpha}{\rho}\int_A E_{z,i}x^2\, dA + \frac{\cos\alpha}{\rho}\int_A E_{z,i}xy\, dA.$$

We therefore define modified second moments of area that include the laminate value of Young's modulus, $E_{Z,i}$, and that are referred to the XYZ axes of Fig. 24.11. Then,

$$I'_{XX} = \int_A E_{Z,i}Y^2\, dA, \quad I'_{YY} = \int_A E_{Z,i}X^2\, dA, \quad I'_{XY} = \int_A E_{Z,i}XY\, dA \tag{24.38}$$

so that

$$M_X = \frac{\sin \alpha}{\rho} I'_{XY} + \frac{\cos \alpha}{\rho} I'_{XX}$$

$$M_Y = \frac{\sin \alpha}{\rho} I'_{YY} + \frac{\cos \alpha}{\rho} I'_{XY}$$

Solving, we obtain

$$\frac{\sin \alpha}{\rho} = \frac{M_Y I'_{XX} - M_X I'_{XY}}{I'_{XX} I'_{YY} - {I'_{XY}}^2}$$

$$\frac{\cos \alpha}{\rho} = \frac{M_X I'_{YY} - M_Y I'_{XY}}{I'_{XX} I'_{YY} - {I'_{XY}}^2}$$

Then, from Eq. (16.15),

$$\sigma_Z = E_{Z,i} \left[\left(\frac{M_Y I'_{XX} - M_X I'_{XY}}{I'_{XX} I'_{YY} - I'^2_{XY}} \right) x + \left(\frac{M_X I'_{YY} - M_Y I'_{XY}}{I'_{XX} I'_{YY} - I'^2_{XY}} \right) y \right] \tag{24.39}$$

Note that the preceding applies equally to open or closed section thin-walled beams.

Example 24.9

A thin-walled beam has the composite cross-section shown in Fig. 24.12 and is subjected to a bending moment of 1 kN m applied in a vertical plane. If the values of Young's modulus for the flange laminates are each 50,000 N/mm^2 and that of the web is 15,000 N/mm^2, determine the maximum value of direct stress in the cross-section of the beam.

From Section 16.4.5 and Eqs. (24.38),

$$I'_{XX} = 2 \times 50{,}000 \times 50 \times 2.0 \times 50^2 + 15{,}000 \times 1.0 \times \frac{100^3}{12} = 2.63 \times 10^{10} \text{ Nmm}^2$$

$$I'_{YY} = 50{,}000 \times 2.0 \times \frac{100^3}{12} = 0.83 \times 10^{10} \text{ Nmm}^2$$

$$I'_{XY} = 50{,}000 \times 50 \times 2.0(+50)(+50) + 50{,}000 \times 50 \times 2.0(-50)(-50)$$
$$= 2.50 \times 10^{10} \text{ Nmm}^2$$

Also, since $M_X = 1$ kNm and $M_Y = 0$, Eq. (24.39) becomes

$$\sigma_Z = E_{Z,i} \left[\frac{-1 \times 10^6 \times 2.5 \times 10^{10}}{10^{20}(2.63 \times 0.83 - 2.5^2)} X + \frac{1 \times 10^6 \times 0.83 \times 10^{10}}{10^{20}(2.63 \times 0.83 - 2.5^2)} Y \right]$$

that is,

$$\sigma_Z = E_{Z,i}(6.15 \times 10^{-5} X - 2.04 \times 10^{-5} Y) \tag{i}$$

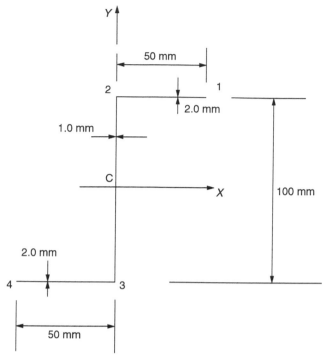

FIGURE 24.12 Beam Section of Example 24.9

On the top flange 12, $E_{Z,i}=50,000$ N/mm^2 and $Y=50$ mm, so that Eq. (i) becomes

$$\sigma_Z = 3.08X - 51.0$$

Then,

$$\sigma_{Z,1} = 3.08 \times 50 - 51.0 = 103.0 \text{ N/mm}^2$$

and

$$\sigma_{Z,2} = -51.0 \text{ N/mm}^2$$

In web 23, $E_{Z,i}=15,000$ N/mm^2 and $X=0$. Equation (i) then becomes

$$\sigma_Z = -0.31Y$$

and

$$\sigma_{Z,2} = -15.5 \text{ N/mm}^2$$

The remaining distribution follows from antisymmetry, so that the maximum direct stress in the beam cross-section is ± 103 N/mm^2.

24.3.3 Shear

Open section beams

In Section 17.2, we derived an expression for the shear flow distribution in an open section thin-walled beam subjected to shear loads (Eq. (17.14)). This is related to the direct stress distribution in the section (Eq. (17.2)), so that the arguments applied to composite section beams subjected to bending apply to the case of composite beams subjected to shear. Equation (17.14) then becomes

$$q_s = -E_{Z,i}\left[\left(\frac{S_X I'_{XX} - S_Y I'_{XY}}{I'_{XX}I'_{YY} - I'^2_{XY}}\right)\int_0^s t_i x\,ds + \left(\frac{S_Y I'_{YY} - S_X I'_{XX}}{I'_{XX}I'_{YY} - I'^2_{XY}}\right)\int_0^s t_i Y\,ds\right] \quad (24.40)$$

Note that, in Eq. (24.40), s is measured from an open edge in the beam section and the second moments of area are those defined in Eq. (24.38).

Example 24.10

Determine the position of the shear center of the channel section beam of Example 24.8.

As in Example 17.2 the shear center lies on the horizontal axis of symmetry so that we only need to apply an arbitrary shear load, S_y, through the shear center and calculate the shear flow distribution in, say, the bottom flange.

In this case, $S_x = 0$, $I'_{XX} = 0$ and Eq. (24.40) reduces to

$$q_s = -E_{Z,i}\left(S_y/I'_{XX}\right)\int_0^s t_i Y\,ds \quad (i)$$

Referring to Fig. 24.11 and Example 24.8

$$I'_{XX} = 2 \times 60,000 \times 100 \times 2 \times 75^2 + 20,000\left(1.0 \times 150^3/12\right) = 14.1 \times 10^{10}\,\text{N/mm}^2$$

On the bottom flange $Y = -75$ mm and $t_i = 2.0$ mm so that Eq.(i) becomes

$$q_s = 60,000\left(S_y/14.1 \times 10^{10}\right)\int_0^s 2 \times 75\,ds$$

where s is measured from the free edge of the flange. Then

$$q_s = 64 \times 10^{-5} S_y s \quad (ii)$$

Suppose that the shear center is a distance ξ_S to the left of the vertical web. Then, taking moments about the mid-point of the web

$$S_y \xi_S = 2S_y \int_0^{100} 6.4 \times 10^{-5} \times 75s\,ds$$

which gives

$$\xi_S = 48.0\,\text{mm}$$

Closed section beams

Again, the same arguments apply to the composite case as before and Eq. (17.15) becomes

$$q_s - -E_{Z,i}\left[\left(\frac{S_X I'_{XX} - S_Y I'_{XY}}{I'_{XX}I'_{YY} - I'^2_{XY}}\right)\int_0^s t_i x\,\mathrm{d}s + \left(\frac{S_Y I'_{YY} - S_X I'_{XX}}{I'_{XX}I'_{YY} - I'^2_{XY}}\right)\int_0^s t_i y\,\mathrm{d}s\right] + q_{s,0} \qquad (24.41)$$

In Eq. (24.41), the value of the shear flow, $q_{s,0}$, at the origin for s is found using either Eq. (17.17) or (17.18).

Example 24.11

The composite triangular section thin-walled beam shown in Fig. 24.13 carries a vertical shear load of 2 kN applied at the apex. If the walls 12 and 13 have a laminate Young's modulus of 45,000 N/mm^2 while that of the vertical web 23 is 20,000 N/mm^2, determine the shear flow distribution in the section.

The X axis is an axis of symmetry, so that $I'_{XY}=0$ and, since $S_X=0$, Eq. (24.41) reduces to

$$q_s = -E_{z,i}\frac{S_Y}{I'_{XX}}\int_0^s tY\,\mathrm{d}s + q_{s,0} \qquad (i)$$

From Section 16.4.5 and Eq. (24.38),

$$I'_{XX} = \frac{2\times45,000\times2.0\times250^3(150/250)^2}{12} + \frac{20,000\times1.5\times300^3}{12}$$

$$= 15.2\times10^{10}\text{ N mm}^2$$

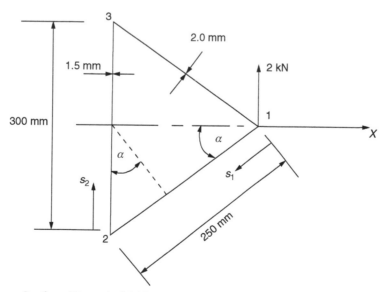

FIGURE 24.13 Beam Section of Example 24.11

"Cut" the section at 1. Then, from the first term on the right-hand side of Eq. (i),

$$q_{b,12} = -\frac{45{,}000 \times 2 \times 10^3}{15.2 \times 10^{10}} \int_0^{s_1} 2.0(-s_1 \sin \alpha)\, ds_1$$

in which $\sin \alpha = 150/250 = 0.6$. Therefore,

$$q_{b,12} = 3.6 \times 10^{-4} s_1^2 \tag{ii}$$

so that

$$q_{b,2} = 22.2 \text{ N/mm}$$

Also,

$$q_{b,23} = -\frac{20{,}000 \times 2 \times 10^3}{15.2 \times 10^{10}} \int_0^{s_2} 1.5(-150 + s_2)ds_2 + 22.2$$

from which

$$q_{b,23} = 0.06 s_2 - 1.95 \times 10^{-4} s_2^2 + 22.2 \tag{iii}$$

Taking moments about the mid-point of the wall 23 (or about point 1), we have

$$2 \times 10^3 \times 250 \cos \alpha = -2 \int_0^{250} q_{b,12} 150 \cos \alpha\, ds_2 + 2 \times \frac{300}{2} \times (250 \cos \alpha) q_{s,0}$$

which gives

$$q_{s,0} = 14.2 \text{ N/mm (in a counterclockwise sense)}$$

The shear flow distribution is then

$$q_{12} = 3.6 \times 10^{-4} s_1^2 - 14.2$$
$$q_{23} = -1.95 \times 10^{-4} s_2^2 + 0.06 s_2 + 8.0$$

24.3.4 Torsion

Closed section beams

We consider composite closed section beams first, since, as we saw in Chapters 16 and 17, the strain–displacement relationships derived for open and closed section beams subjected to shear loads apply to the torsion of closed section beams, so that the analysis follows logically on.

The shear flow distribution in a closed section thin-walled beam subjected to a torque in which the warping is unrestrained is given by Eq. (18.1); that is,

$$T = 2Aq$$

or

$$q = \frac{T}{2A} \tag{24.42}$$

The derivation of Eq. (24.42) is based purely on equilibrium considerations and does not, therefore, rely on the elastic properties of the beam section. Equation (24.42) therefore applies equally to composite and isotropic beam sections.

The rate of twist of a closed section beam subjected to a torque is given by Eq. (18.4); that is,

$$\frac{d\theta}{dZ} = \frac{T}{4A^2} \oint \frac{ds}{Gt}$$

This expression also applies to a composite closed section beam, provided that the shear modulus G remains within the integration and that the laminate shear modulus $G_{XY,i}$ is used as appropriate. Equation (18.4) then becomes

$$\frac{d\theta}{dZ} = \frac{T}{4A^2} \oint \frac{ds}{G_{XY,i}\, t_i} \tag{24.43}$$

Rearranging,

$$T = \frac{4A^2}{\oint \frac{ds}{G_{XY,i}\, t_i}} \frac{d\theta}{dZ} \tag{24.44}$$

We saw, in Chapter 3, Eq. (3.12), that the torque and rate of twist in a beam are related by the torsional stiffness GJ. Therefore, from Eq. (24.44), we see that the torsional stiffness of a composite closed section beam is given by

$$GJ = \frac{4A^2}{\oint \frac{ds}{G_{XY,i} t_i}} \tag{24.45}$$

These arguments apply to the determination of the warping distribution in a closed section composite beam. This is then given by (see the derivation of Eq. (18.5))

$$W_s - W_0 = q \int_0^s \frac{ds}{G_{XY,i}\, t_i} - \frac{A_{0s}}{A} q \oint \frac{ds}{G_{XY,i}\, t_i} \tag{24.46}$$

or, from Eq. (24.42) in terms of the applied torque,

$$W_s - W_0 = \frac{T}{2A} \left(\int_0^s \frac{ds}{G_{XY,i}\, t_i} - \frac{A_{0s}}{A} \oint \frac{ds}{G_{XY,i}\, t_i} \right) \tag{24.47}$$

Example 24.12

The rectangular section, thin-walled, composite beam shown in Fig. 24.14 is subjected to a torque of 10 kN m. If the laminate shear modulus of the covers is 20,000 N/mm^2 and that of the webs is 35,000 N/mm^2, determine the shear flow distribution in the section and the distribution of warping.

The shear flow distribution is obtained from Eq. (24.42) and is

$$q = \frac{10 \times 10^6}{2 \times 200 \times 100} = 250\ \text{N/mm}$$

The warping distribution is given by Eq. (24.47), in which

$$\oint \frac{ds}{G_{XY,i} t_i} = \frac{2 \times 200}{20,000 \times 2.0} + \frac{2 \times 100}{35,000 \times 1.0} = 0.0157$$

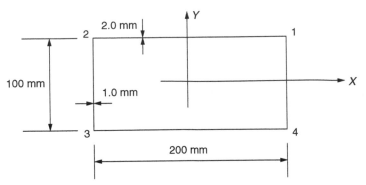

FIGURE 24.14 Beam Section of Example 24.12

Equation (24.47) then becomes

$$W_s - W_0 = 250 \left(\int_0^s \frac{ds}{G_{XY,i}\, t_i} - \frac{A_{0s}}{200 \times 100} \times 0.0157 \right)$$

or

$$W_2 - W_0 = 250 \left(\int_0^s \frac{ds}{G_{XY,i}t_i} - 0.785 \times 10^{-6} A_{0s} \right) \qquad (i)$$

We saw, in Example 18.2, that the warping distribution in a rectangular section thin-walled beam is linear with zero values at the mid-points of the webs and covers. The same situation applies in this example, so that it is necessary to calculate the value of warping only at, say, corner 1. Then, from Eq. (i),

$$W_1 = 250 \left(\frac{50}{35{,}000 \times 1.0} - 0.785 \times 10^{-6} \times 100 \times \frac{50}{2} \right)$$

which gives

$$W_1 = -0.13 \text{ mm}$$

The remaining distribution follows from symmetry.

Open section beams

The torsional stiffness of an open section thin-walled beam is, as for a closed section beam, GJ, but in which the torsion constant, J, is given by either of Eqs. (18.11). However, for a composite beam section, the shear modulus must be taken inside the summation or integral and is the laminate shear modulus $G_{XY,i}$. Then,

$$GJ = \sum_{i=1}^{n} G_{XY,i} \frac{s t_i^3}{3} \quad \text{or} \quad GJ = \frac{1}{3} \int_{\text{sect}} G_{XY,i}\, t_i^3 \, ds \qquad (24.48)$$

The rate of twist of a beam is related to the applied torque by Eq. (3.12). For a composite open section beam, the relationship holds but the torsional stiffness is given by either of Eqs. (24.48); that is,

$$T = \left(\sum_{i=1}^{n} G_{XY,i} \frac{s t_i^3}{3} \right) \frac{d\theta}{dZ} \quad \text{or} \quad T = \left(\frac{1}{3} \int_{\text{sect}} G_{XY,i} t_i^3 \, ds \right) \frac{d\theta}{dZ} \tag{24.49}$$

Having obtained the rate of twist, Eq. (18.9) gives the shear stress distribution across the thickness at any point round the beam section; that is,

$$\tau = 2 G_{XY,i} n \frac{d\theta}{dZ} \tag{24.50}$$

Again, the maximum shear stress occurs at the surface of the beam section, where $n = \pm t/2$.

The primary warping distribution follows from Eq. (18.19), in which the rate of twist is found from either of Eqs. (24.49).

Example 24.13

A composite channel section has the dimensions shown in Fig. 18.12 and is subjected to a torque of 10 Nm. If the flanges have a laminate shear modulus of 20,000 N/mm² and that of the web is 15,000 N/mm² determine the maximum shear stress in the beam section and the distribution of warping, assuming that the beam is constrained to twist about an axis through the mid-point of the web.

The torsional stiffness of the section is obtained from the first of Eqs, (24.48) and is

$$GJ = 2 \times 20,000 \times 25 \times \frac{1.5^3}{3} + 15,000 \times 50 \times \frac{2.5^3}{3} = 5.03 \times 10^6 \, \text{N mm}^2$$

Then, from Eq. (24.49),

$$\frac{d\theta}{dZ} = \frac{10 \times 10^3}{5.03 \times 10^6} = 1.99 \times 10^{-3}$$

and, from Eq. (24.50),

$$\tau_{\max}(12) = 2 \times 20,000 \times (1.5/2) \times 1.99 \times 10^{-3} = 59.7 \, \text{N/mm}^2$$
$$\tau_{\max}(23) = 2 \times 15,000 \times (2.5/2) \times 1.99 \times 10^{-3} = 74.6 \, \text{N/mm}^2$$

The maximum therefore occurs in the web and is 74.6 N/mm².

The section is constrained to twist about an axis through the mid-point of the web, so that W is zero everywhere in the web. Then, from Eq. (18.19),

$$W_1 = -2 \times \frac{1}{2} \times 25 \times 25 \times 1.99 \times 10^{-3} = -1.24 \, \text{mm}$$

The warping is linear along the flange 12, the warping along the flange 34 follows from symmetry.

Note that, if the axis of twist is not specified, the position of the shear center of the section has to be found using the method previously described.

References

[1] Calcote LR. The analysis of laminated composite structures. New York: Van Nostrand Reinhold; 1969.
[2] Datoo MH. Mechanics of fibrous composites. London: Elsevier Applied Science; 1991.

PROBLEMS

P.24.1. A bar, whose cross-section is shown in Fig. P.24.1, is composed of a polyester matrix and Kevlar filaments; the respective moduli are 3,000 and 140,000 N/mm^2 with corresponding Poisson's ratios of 0.16 and 0.28. If the bar is 1 m long and is subjected to a compressive axial load of 500 kN, determine the shortening of the bar, the increase in its thickness, and the stresses in the polyester and Kevlar.

Answer: 3.26 mm, 0.032 mm, 9.78 N/mm^2, 456.4 N/mm^2

P.24.1. MATLAB Use MATLAB to repeat Problem P.24.1 for compressive axial loads (P) ranging from 300 to 700 kN in increments of 50 kN.

Answer:		P	Δ_l	Δ_t	σ_m (polyester)	σ_f (Kevlar)
	(i)	300 kN	1.95 mm	0.019 mm	5.85 N/mm^2	273 N/mm^2
	(ii)	350 kN	2.28 mm	0.023 mm	6.84 N/mm^2	319.2 N/mm^2
	(iii)	400 kN	2.61 mm	0.026 mm	7.83 N/mm^2	365.4 N/mm^2
	(iv)	450 kN	2.93 mm	0.029 mm	8.79 N/mm^2	410.2 N/mm^2
	(v)	500 kN	3.26 mm	0.032 mm	9.78 N/mm^2	456.4 N/mm^2
	(vi)	550 kN	3.58 mm	0.035 mm	10.74 N/mm^2	501.2 N/mm^2
	(vii)	600 kN	3.91 mm	0.039 mm	11.73 N/mm^2	547.4 N/mm^2
	(viii)	650 kN	4.23 mm	0.042 mm	12.69 N/mm^2	592.2 N/mm^2
	(ix)	700 kN	4.56 mm	0.045 mm	13.68 N/mm^2	638.4 N/mm^2

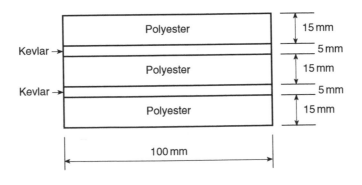

FIGURE P.24.1

P.24.2. If a thin isotropic ply has a Young's modulus of 60,000 N/mm^2 and a Poisson's ratio of 0.25, determine the terms in the reduced stiffness and compliance matrices.

Answer: $k_{11} = k_{22} = 64,000$ N/mm^2, $k_{12} = 16,000$ N/mm^2, $k_{33} = 24,000$ N/mm^2, $s_{11} = s_{22} = 1.67 \times 10^{-5}$, $s_{12} = s_{21} = -0.42 \times 10^{-5}$, $s_{33} = 4.17 \times 10^{-5}$.

P.24.3. The reduced stiffnesses in a unidirectional ply are $k_{11} = 50,000$ N/mm^2, $k_{12} = k_{21} = 4,000$ N/mm^2, $k_{22} = 15,000$ N/mm^2 and $k_{33} = 6,000$ N/mm^2. Calculate the elastic constants of the ply and also the reduced compliances.

Answer: $v_{tl} = 0.08$, $v_{lt} = 0.27$, $E_l = 48,920$ N/mm^2, $E_t = 14,495$ N/mm^2, $G = 6,000$ N/mm^2. $s_{11} = 2.04 \times 10^{-5}$, $s_{12} = -0.55 \times 10^{-5} = s_{21}$, $s_{22} = 6.90 \times 10^{-5}$, $s_{33} = 1.67 \times 10^{-4}$.

P.24.4. A generally orthotropic ply is subjected to direct stresses of 120 N/mm^2 and 60 N/mm^2 parallel to the x and y reference axes respectively together with a shear stress of 80 N/mm^2. If the ply angle is 45° determine the direct and shear stresses referred to the material axes.

Answer: $\sigma_l = 170$ N/mm^2, $\sigma_t = 10$ N/mm^2, $\tau_{lt} = -30$ N/mm^2.

P.24.5. The strains at a point in a generally orthotropic ply are $\varepsilon_x = 0.005$, $\varepsilon_y = 0.002$ and $\gamma_{xy} = 0.0002$ referred to the xy reference axis system. If the ply angle is 45°, calculate the strains in the directions of the material axes.

Answer: $\varepsilon_l = 3.6 \times 10^{-3}$, $\varepsilon_t = 3.4 \times 10^{-3}$, $\gamma_{lt} = -3.0 \times 10^{-3}$.

P.24.6. A generally orthotropic ply is subjected to direct stresses of 100 N/mm^2 and 50 N/mm^2 parallel to the x and y reference axis system respectively. If the ply also carries a shear stress of 75 N/mm^2 referred to the xy axis system and the ply angle is 45° calculate the direct and shear stresses referred to the material axes and the corresponding strains. The elastic constants are $E_l = 150,000$ N/mm^2, $E_t = 90,000$ N/mm^2, $G_{lt} = 5000$ N/mm^2, $v_{lt} = 0.3$ and $v_{tl} = 0.1$.

Answer: $\sigma_l = 150$ N/mm^2, $\sigma_t = 0$, $\tau_{lt} = -25$ N/mm^2. $\varepsilon_l = 10.0 \times 10^{-4}$, $\varepsilon_t = -3.03 \times 10^{-4}$, $\gamma_{lt} = -50.0 \times 10^{-4}$.

P.24.7. A box beam has the thin-walled composite cross-section shown in Fig. P.24.7. The cover laminates are identical and have a Young's modulus of 20,000 N/mm^2 while that of the vertical webs is 60,000 N/mm^2. If the beam is subjected to an axial load of 40 kN, determine the axial force in each laminate.

Answer: Covers, 4 kN; webs, 16 kN

P.24.8. If the thin-walled box beam of Fig. P.24.7 carries a bending moment of 1 kNm applied in a vertical plane, determine the maximum direct stress in the cross-section of the beam.

Answer: 85.8 N/mm^2

P.24.9. If the thin-walled composite beam of Example 24.9 is subjected to a bending moment of 0.5 kNm applied in a horizontal plane, calculate the maximum value of direct stress in the beam section.

Answer: 76.8 N/mm^2

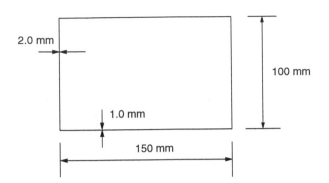

P.24.10. The thin-walled composite beam section of Example 24.9 carries a vertical shear load of 2kN applied on the plane of the web. Determine the shear flow distribution.

Answer: $q_{12} = 0.00575s_1{}^2 - 0.385s_1$
$q_{23} = 0.0287s_2 - 2.865 \times 10^{-4}s_2{}^2 - 4.875.$

P.24.11. The closed, composite section, thin-walled beam shown in Fig. P.24.11 is subjected to a vertical shear load of 20 kN applied through its center of symmetry. If the laminate elastic properties are, for the covers, $E_{z,i} = 54,100$ N/mm^2; for the webs, $E_{z,i} = 17,700$ N/mm^2, determine the distribution of shear flow around the cross-section.

Answer: $q_{01} = -1.98s_1$, $q_{12} = 6.5 \times 10^{-3}s_2{}^2 - 0.325s_2 - 198.$

P.24.12. The beam section shown in Fig. P.24.11 is subjected to a counterclockwise torque of 1 kNm. If the laminate shear modulus of the covers is 20,700 N/mm^2 and that of the webs is 36,400 N/mm^2, determine the maximum shear stress in the section, its rate of twist, and the distribution of warping.

Answer: 100 N/mm^2, 6.25×10^{-5} rad/mm, -0.086 mm (at 4, zero at 0)

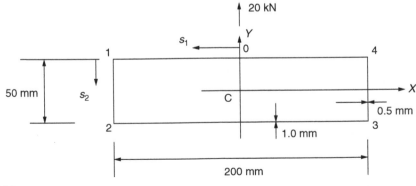

FIGURE P.24.11

P.24.12. MATLAB Assuming the dimension for the web height shown in Fig. P.24.11 is labeled h, use MATLAB to repeat Problem P.24.12 for values of h ranging from 20 to 80 mm in increments of 10 mm.

Answer:	h	τ_{max}	$d\theta/dz$	W
(x)	20 mm	250 N/mm²	33.33×10^{-5} rad/mm	0.268 mm (at 4, zcro at 0)
(xi)	30 mm	166.7 N/mm²	16.67×10^{-5} rad/mm	0.167 mm (at 4, zero at 0)
(xii)	40 mm	125 N/mm²	9.09×10^{-5} rad/mm	0.117 mm (at 4, zero at 0)
(xiii)	50 mm	100 N/mm²	6.25×10^{-5} rad/mm	0.086 mm (at 4, zero at 0)
(xiv)	60 mm	83.3 N/mm²	4.55×10^{-5} rad/mm	0.066 mm (at 4, zero at 0)
(xv)	70 mm	71.4 N/mm²	3.45×10^{-5} rad/mm	0.052 mm (at 4, zero at 0)
(xvi)	80 mm	62.5 N/mm²	2.78×10^{-5} rad/mm	0.041 mm (at 4, zero at 0)

P.24.13. The thin-walled, composite beam section shown in Fig. P.24.13 has laminate shear moduli of 16,300 N/mm² for the flanges and 20,900 N/mm² for the web. If the beam is subjected to a torque of 0.5 kNmm determine the rate of twist in the section, the maximum shear stress, and the value of warping at the point 1.

Answer: 0.8×10^{-3} rad/mm, ± 13 N/mm² (in flanges), 2.0 mm

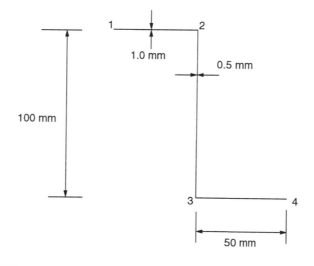

FIGURER P.24.13

Index

Note: Page numbers followed by *f* indicate figures and *b* indicate boxes.

Edwards Brothers Malloy
Ann Arbor MI. USA
January 25, 2015